SIGNAL TRANSDUCTION PATHWAYS, PART A
Apoptotic and Extracellular Signaling

ANNALS OF THE NEW YORK ACADEMY OF SCIENCES
Volume 1090

SIGNAL TRANSDUCTION PATHWAYS, PART A
Apoptotic and Extracellular Signaling

Edited by Marc Diederich

Published by Blackwell Publishing on behalf of the New York Academy of Sciences
Boston, Massachusetts
2006

Library of Congress Cataloging-in-Publication Data

Signal transduction pathways. Part A, Apoptotic and extracellular Signaling / edited by Marc Diederich.
  p. ; cm. – (Annals of the New York Academy of Sciences, ISSN 0077-8923 ; v. 1090)
  Includes bibliographical references and index.
  ISBN-13: 978-1-57331-645-3 (alk. paper)
  ISBN-10: 1-57331-645-8 (alk. paper)
  1. Pathology, Molecular–Congresses. 2. Cellular signal transduction–Congresses. 3. Chemoprevention–Congresses.
  I. Diederich, Marc. II. New York Academy of Sciences.
  III. Title: Apoptotic and extracellular Signaling. IV. Series.
  [DNLM: 1. Signal Transduction–physiology–Congresses.
  2. Carcinoma–physiopathology–Congresses. 3. Extracellular Matrix–physiology–Congresses. 4. Oxidative Stress–physiology
  –Congresses. W1 AN626YL v. 1090 2006 / QU 375 S5786 2006]

  RB113.S526 2006
  500 s–dc22
  [616.07]
                                                        2006033063

The *Annals of the New York Academy of Sciences* (ISSN: 0077-8923 [print]; ISSN: 1749-6632 [online]) is published 28 times a year on behalf of the New York Academy of Sciences by Blackwell Publishing, with offices located at 350 Main Street, Malden, Massachusetts 02148 USA, PO Box 1354, Garsington Road, Oxford OX4 2DQ UK, and PO Box 378 Carlton South, 3053 Victoria Australia.

Information for subscribers: Subscription prices for 2006 are: Premium Institutional: $3850.00 (US) and £2139.00 (Europe and Rest of World). Customers in the UK should add VAT at 5%. Customers in the EU should also add VAT at 5% or provide a VAT registration number or evidence of entitlement to exemption. Customers in Canada should add 7% GST or provide evidence of entitlement to exemption. The Premium Institutional price also includes online access to full-text articles from 1997 to present, where available. For other pricing options or more information about online access to Blackwell Publishing journals, including access information and terms and conditions, please visit www.blackwellpublishing.com/nyas.

Membership information: Members may order copies of the *Annals* volumes directly from the Academy by visiting www.nyas.org/annals, emailing membership@nyas.org, faxing 212-298-3650, or calling 800-843-6927 (US only), or +1 212-298-8640, ext. 345 (International). For more information on becoming a member of the New York Academy of Sciences, please visit www.nyas.org/membership.

Journal Customer Services: For ordering information, claims, and any inquiry concerning your institutional subscription, please contact your nearest office:
UK: Email: customerservices@blackwellpublishing.com; Tel: +44 (0) 1865 778315; Fax +44 (0) 1865 471775
US: Email: customerservices@blackwellpublishing.com; Tel: +1 781 388 8599 or 1 800 835 6770 (Toll free in the USA); Fax: +1 781 388 8232
Asia: Email: customerservices@blackwellpublishing.com; Tel: +65 6511 8000; Fax: +61 3 8359 1120
Members: Claims and inquiries on member orders should be directed to the Academy at email: membership@nyas.org or Tel: +1 212 838 0230 (International) or 800-843-6927 (US only).

Printed in the USA.
Printed on acid-free paper.

Mailing: The *Annals of the New York Academy of Sciences* are mailed Standard Rate. **Postmaster:** Send all address changes to *Annals of the New York Academy of Sciences*, Blackwell Publishing, Inc., Journals Subscription Department, 350 Main Street, Malden, MA 01248-5020. Mailing to rest of world by DHL Smart and Global Mail.

Copyright and Photocopying
© 2006 The New York Academy of Sciences. All rights reserved. No part of this publication may be reproduced, stored, or transmitted in any form or by any means without the prior permission in writing from the copyright holder. Authorization to photocopy items for internal and personal use is granted by the copyright holder for libraries and other users registered with their local Reproduction Rights Organization (RRO), e.g. Copyright Clearance Center (CCC), 222 Rosewood Drive, Danvers, MA 01923, USA (www.copyright.com), provided the appropriate fee is paid directly to the RRO. This consent does not extend to other kinds of copying such as copying for general distribution, for advertising or promotional purposes, for creating new collective works, or for resale. Special requests should be addressed to Blackwell Publishing at journalsrights@oxon.blackwellpublishing.com.

Disclaimer: The Publisher, the New York Academy of Sciences, and the Editors cannot be held responsible for errors or any consequences arising from the use of information contained in this publication; the views and opinions expressed do not necessarily reflect those of the Publisher, the New York Academy of Sciences, or the Editors.

Annals are available to subscribers online at the New York Academy of Sciences and also at Blackwell Synergy. Visit www.annalsnyas.org or www.blackwell-synergy.com to search the articles and register for table of contents e-mail alerts. Access to full text and PDF downloads of *Annals* articles are available to nonmembers and subscribers on a pay-per-view basis at www.annalsnyas.org.

The paper used in this publication meets the minimum requirements of the National Standard for Information Sciences Permanence of Paper for Printed Library Materials, ANSI Z39.48.1984.

ISSN: 0077-8923 (print); 1749-6632 (online)
ISBN-10: 1-57331-645-8 (paper); ISBN-13: 978-1-57331-645-3 (paper)

A catalogue record for this title is available from the British Library.

Digitization of the *Annals of the New York Academy of Sciences*

An agreement has recently been reached between Blackwell Publishing and the New York Academy of Sciences to digitize the entire run of the *Annals of the New York Academy of Sciences* back to volume one.

The back files, which have been defined as all of those issues published before 1997, will be sold to libraries as part of Blackwell Publishing's Legacy Sales Program and hosted on the Blackwell Synergy website.

Copyright of all material will remain with the rights holder. Contributors: Please contact Blackwell Publishing if you do not wish an article or picture from the *Annals of the New York Academy of Sciences* to be included in this digitization project.

ANNALS OF THE NEW YORK ACADEMY OF SCIENCES

Volume 1090
December 2006

SIGNAL TRANSDUCTION PATHWAYS, PART A
Apoptotic and Extracellular Signaling

Editor
MARC DIEDERICH

This volume is the result of a meeting entitled **Cell Signaling World: Signal Transduction Pathways as Therapeutic Targets**, held January 25–28, 2006, in Luxembourg. Parts B, C, and D can be found in Volumes 1091, 1095, and 1096, respectively.

CONTENTS

Preface. *By* MARC DIEDERICH xxvii

Part I. Apoptotic Cell Signaling Mechanisms

Pleiotropic Effects of PI-3′ Kinase/Akt Signaling in Human Hepatoma Cell Proliferation and Drug-Induced Apoptosis. *By* CATHERINE ALEXIA, MARLÈNE BRAS, GUILLAUME FALLOT, NATHALIE VADROT, FANNY DANIEL, MALIKA LASFER, HOUDA TAMOUZA, AND ANDRÉ GROYER 1

Expression of Bcl-xL, Bax, and p53 in Primary Tumors and Lymph Node Metastases in Oral Squamous Cell Carcinoma. *By* MAREK BALTAZIAK, EWA DURAJ, MARIUSZ KODA, ANDRZEJ WINCEWICZ, MARCIN MUSIATOWICZ, LUIZA KANCZUGA-KODA, MAGDALENA SZYMANSKA, TOMASZ LESNIEWICZ, AND BOGUSLAW MUSIATOWICZ 18

Reactive Structural Dynamics of Synaptic Mitochondria in Ischemic Delayed Neuronal Death. *By* CARLO BERTONI-FREDDARI, PATRIZIA FATTORETTI, TIZIANA CASOLI, GIUSEPPINA DI STEFANO, MORENO SOLAZZI, ELISA PERNA, AND CLARA DE ANGELIS 26

Selective Resistance of Tetraploid Cancer Cells against DNA Damage-Induced Apoptosis. *By* MARIA CASTEDO, ARNAUD COQUELLE, ILIO VITALE, SONIA VIVET, SHAHUL MOUHAMAD, SOPHIE VIAUD, LAURENCE ZITVOGEL, AND GUIDO KROEMER 35

Molecular Determinants Involved in the Increase of Damage-Induced
Apoptosis and Delay of Secondary Necrosis due to Inhibition of
Mono(ADP-Ribosyl)ation. *By* CLAUDIA CERELLA, CRISTINA MEARELLI,
SERGIO AMMENDOLA, MILENA DE NICOLA, MARIA D'ALESSIO,
ANDREA MAGRINI, ANTONIO BERGAMASCHI, AND LINA GHIBELLI 50

Magnetic Fields Protect from Apoptosis via Redox Alteration. *By*
M. DE NICOLA, S. CORDISCO, C. CERELLA, M.C. ALBERTINI,
M. D'ALESSIO, A. ACCORSI, A. BERGAMASCHI, A. MAGRINI, AND
L. GHIBELLI 59

The Cleavage Mode of Apoptotic Nuclear Vesiculation Is Related to Plasma
Membrane Blebbing and Depends on Actin Reorganization. *By*
M. DE NICOLA, C. CERELLA, M. D'ALESSIO, S. COPPOLA, A. MAGRINI,
A. BERGAMASCHI, AND L. GHIBELLI 69

Experimental Apoptosis Provides Clues about the Role of Mitochondrial
Changes in Neuronal Death. *By* PATRIZIA FATTORETTI,
CARLO BERTONI-FREDDARI, RINA RECCHIONI, BELINDA GIORGETTI,
MARTA BALIETTI, YESSICA GROSSI, MORENO SOLAZZI, TIZIANA CASOLI,
GIUSEPPINA DI STEFANO, AND FIORELLA MARCHESELLI 79

Alterations in mRNA Expression of Apoptosis-Related Genes *BCL2*, *BAX*,
FAS, *Caspase-3*, and the Novel Member *BCL2L12* after Treatment of
Human Leukemic Cell Line HL60 with the Antineoplastic Agent
Etoposide. *By* KOSTAS V. FLOROS, HELLINIDA THOMADAKI,
DIMITRA FLOROU, MAROULIO TALIERI, AND ANDREAS SCORILAS 89

Using Janus Green B to Study Paraquat Toxicity in Rat Liver Mitochondria:
Role of ACE Inhibitors (Thiol and Nonthiol ACEi). *By*
M. GHAZI-KHANSARI, A. MOHAMMADI-BARDBORI, AND
M-J. HOSSEINI 98

Membrane Fluidity Changes Are Associated with Benzo[*a*]Pyrene-Induced
Apoptosis in F258 Cells: Protection by Exogenous Cholesterol. *By*
MORGANE GORRIA, XAVIER TEKPLI, ODILE SERGENT, LAURENCE HUC,
FRANÇOIS GABORIAU, MARY RISSEL, MARTINE CHEVANNE,
MARIE-THÉRÈSE DIMANCHE-BOITREL, AND
DOMINIQUE LAGADIC-GOSSMANN 108

Metal-Containing Proteins in the Apoptosis and Redox Processes in the Rat
Prostate and Human Prostate Cells. *By* I. GRBAVAC, C. WOLF,
N. WENDA, D. ALBER, M. KÜHBACHER, D. BEHNE, AND
A. KYRIAKOPOULOS 113

Does Transduced p27 Induce Apoptosis in Human Tumor Cell Lines? *By*
MIRA GRDIŠA, ANA-MATEA MIKECIN, AND MIROSLAV POZNIC 120

Apoptotic Cell Signaling in Lymphocytes from HIV[+] Patients during
Successful Therapy. *By* SANDRO GRELLI, EMANUELA BALESTRIERI,
CLAUDIA MATTEUCCI, ANTONELLA MINUTOLO, GABRIELLA D'ETTORRE,
FILIPPO LAURIA, FRANCESCO MONTELLA, VINCENZO VULLO,
STEFANO VELLA, CARTESIO FAVALLI, ANTONIO MASTINO, AND
BEATRICE MACCHI 130

Capacitation and Acrosome Reaction in Nonapoptotic Human Spermatozoa. *By*
SONJA GRUNEWALD, THOMAS BAUMANN, UWE PAASCH, AND
HANS-JUERGEN GLANDER 138

Caspase Activation and Extracellular Signal-Regulated Kinase/Akt Inhibition Were Involved in Luteolin-Induced Apoptosis in Lewis Lung Carcinoma Cells. *By* Jin-Hyung Kim, Eun-Ok Lee, Hyo-Jung Lee, Jin-Sook Ku, Min-Ho Lee, Deok-Chun Yang, and Sung-Hoon Kim 147

Ceramide Modulation of Antigen-Triggered Ca^{2+} Signals and Cell Fate: Diversity in the Responses of Various Immunocytes. *By* Endre Kiss, Gabriella Sármay, and János Matkó 161

Caspase-3 Activation, Bcl-2 Contents, and Soluble FAS-Ligand Are Not Related to the Inflammatory Marker Profile in Patients with Sepsis and Septic Shock. *By* Fabian Kriebel, Silke Wittemann, Hsin-Yun Hsu, Thomas Joos, Manfred Weiss, and E. Marion Schneider 168

Two Forms of the Nuclear Matrix–Bound p53 Protein in HEK293 Cells. *By* Maria A. Lapshina, Igor I. Parkhomenko, and Alexei A. Terentiev 177

Jpk, a Novel Cell Death Inducer, Regulates the Expression of Hoxa7 in F9 Teratocarcinoma Cells, but not during Apoptosis. *By* Eun Young Lee and Myoung Hee Kim 182

HGF/SF Regulates Expression of Apoptotic Genes in MCF-10A Human Mammary Epithelial Cells. *By* Catherine Leroy, Julien Deheuninck, Sylvie Reveneau, Bénédicte Foveau, Zongling Ji, Céline Villenet, Sabine Quief, David Tulasne, Jean-Pierre Kerckaert, and Véronique Fafeur 188

Arsenic Trioxide Represses NF-κB Activation and Increases Apoptosis in ATRA-Treated APL Cells. *By* Julie Mathieu and Françoise Besançon 203

Cytotoxicity of TRAIL/Anticancer Drug Combinations in Human Normal Cells. *By* Olivier Meurette, Anne Fontaine, Amelie Rebillard, Gwenaelle Le Moigne, Thierry Lamy, Dominique Lagadic-Gossmann, and Marie-Therese Dimanche-Boitrel 209

Hyperpolarization of Plasma Membrane of Tumor Cells Sensitive to Antiapoptotic Effects of Magnetic Fields. *By* S. Nuccitelli, C. Cerella, S. Cordisco, M.C. Albertini, A. Accorsi, M. De Nicola, M. D'Alessio, F. Radogna, A. Magrini, A. Bergamaschi, and L. Ghibelli 217

Melatonin as an Apoptosis Antagonist. *By* Flavia Radogna, Laura Paternoster, Maria Cristina Albertini, Augusto Accorsi, Claudia Cerella, Maria D'Alessio, Milena De Nicola, Silvia Nuccitelli, Andrea Magrini, Antonio Bergamaschi, and Lina Ghibelli 226

Prevention of p53 Degradation in Human MCF-7 Cells by Proteasome Inhibitors Does Not Mimic the Action of Roscovitine. *By* Carmen Ranftler, Marieta Gueorguieva, and Jòzefa Wesierska-Gadek 234

Role of ATP in Trauma-Associated Cytokine Release and Apoptosis by P2X7 Ion Channel Stimulation. *By* E. Marion Schneider, Katrin Vorlaender, Xueling Ma, Weidong Du, and Manfred Weiss 245

Experimental Sepsis: Characteristics of Activated Macrophages and Apoptotic Cells in the Rat Spleen. *By* HELLE EVI SIMOVART, ANDRES AREND, HELLE TAPFER, KERSTI KOKK, MARINA AUNAPUU, ELLE POLDOJA, GUNNAR SELSTAM, AND AADE LIIGANT 253

Insulin-Like Growth Factor-I Receptor Correlates with Connexin 26 and Bcl-xL Expression in Human Colorectal Cancer. *By* STANISLAW SULKOWSKI, LUIZA KANCZUGA-KODA, MARIUSZ KODA, ANDRZEJ WINCEWICZ, AND MARIOLA SULKOWSKA 265

Characterization of the Proapoptotic Intracellular Mechanisms Induced by a Toxic Conformer of the Recombinant Human Prion Protein Fragment 90–231. *By* VALENTINA VILLA, ALESSANDRO CORSARO, STEFANO THELLUNG, DOMENICO PALUDI, KATIA CHIOVITTI, VALENTINA VENEZIA, MARIO NIZZARI, CLAUDIO RUSSO, GENNARO SCHETTINI, ANTONIO ACETO, AND TULLIO FLORIO 276

Role of NADPH Oxidase and Calcium in Cerulein-Induced Apoptosis: Involvement of Apoptosis-Inducing Factor. *By* JI HOON YU, KYUNG HWAN KIM, AND HYEYOUNG KIM 292

Part II. Extracellular Matrix Interactions

Signaling for Integrin $\alpha 5/\beta 1$ Expression in *Helicobacter pylori*–Infected Gastric Epithelial AGS Cells. *By* SOON OK CHO, KYUNG HWAN KIM, JOO-HEON YOON, AND HYEYOUNG KIM 298

Human Recombinant Vasostatin-1 May Interfere with Cell–Extracellular Matrix Interactions. *By* VALENTINA DI FELICE, FRANCESCO CAPPELLO, ANTONELLA MONTALBANO, NELLA ARDIZZONE, CLAUDIA CAMPANELLA, ANGELA DE LUCA, DANIELA AMELIO, BRUNO TOTA, ANGELO CORTI, AND GIOVANNI ZUMMO 305

Microgravity Signal Ensnarls Cell Adhesion, Cytoskeleton, and Matrix Proteins of Rat Osteoblasts: Osteopontin, CD44, Osteonectin, and α-Tubulin. *By* YASUHIRO KUMEI, SADAO MORITA, HISAKO KATANO, HIDEO AKIYAMA, MASAHIKO HIRANO, KEI'ICHI OYHA, AND HITOYATA SHIMOKAWA 311

14-3-3 Proteins Bind Both Filamin and $\alpha_L \beta_2$ Integrin in Activated T Cells. *By* SUSANNA M. NURMI, CARL G. GAHMBERG, AND SUSANNA C. FAGERHOLM 318

MAP Kinases

Grb2-Associated Binder 1 (Gab1) Adaptor/Scaffolding Protein Regulates Erk Signal in Human B Cells. *By* ADRIENN ANGYAL, DAVID MEDGYESI, AND GABRIELLA SARMAY 326

CXC Receptor and Chemokine Expression in Human Meningioma: SDF1/CXCR4 Signaling Activates ERK1/2 and Stimulates Meningioma Cell Proliferation. *By* FEDERICA BARBIERI, ADRIANA BAJETTO, CAROLA PORCILE, ALESSANDRA PATTAROZZI, ALESSANDRO MASSA, GIANLUIGI LUNARDI, GIANLUIGI ZONA, ALESSANDRA DORCARATTO, JEAN LOUIS RAVETTI, RENATO SPAZIANTE, GENNARO SCHETTINI, AND TULLIO FLORIO 332

Reduction of Bcr-Abl Function Leads to Erythroid Differentiation of K562 Cells via Downregulation of ERK. *By* A. BRÓZIK, N.P. CASEY, CS. HEGEDŰS, A. BORS, A. KOZMA, H. ANDRIKOVICS, M. GEISZT, K. NÉMET, AND M. MAGÓCSI 344

The MAPK Pathway and HIF-1 Are Involved in the Induction of the Human PAI-1 Gene Expression by Insulin in the Human Hepatoma Cell Line HepG2. *By* ELITSA Y. DIMOVA AND THOMAS KIETZMANN 355

Role of Mitogen-Activated Protein Kinases, NF-κB, and AP-1 on Cerulein-Induced IL-8 Expression in Pancreatic Acinar Cells. *By* KYUNG DON JU, JI HOON YU, HYEYOUNG KIM, AND KYUNG HWAN KIM 368

Upregulation of VEGF by 15-Deoxy-$\Delta^{12,14}$-Prostaglandin J_2 via Heme Oxygenase-1 and ERK1/2 Signaling in MCF-7 Cells. *By* EUN-HEE KIM, HYE-KYUNG NA, AND YOUNG-JOON SURH 375

SDF-1 Controls Pituitary Cell Proliferation through the Activation of ERK1/2 and the Ca^{2+}-Dependent, Cytosolic Tyrosine Kinase Pyk2. *By* ALESSANDRO MASSA, SILVIA CASAGRANDE, ADRIANA BAJETTO, CAROLA PORCILE, FEDERICA BARBIERI, STEFANO THELLUNG, SARA ARENA, ALESSANDRA PATTAROZZI, MONICA GATTI, ALESSANDRO CORSARO, MAURO ROBELLO, GENNARO SCHETTINI, AND TULLIO FLORIO 385

Insulin Primes Human Neutrophils for CCL3-Induced Migration: Crucial Role for JNK 1/2. *By* FABRIZIO MONTECUCCO, GIORDANO BIANCHI, MARIA BERTOLOTTO, GIORGIO VIVIANI, FRANCO DALLEGRI, AND LUCIANO OTTONELLO 399

Doxorubicin-Induced MAPK Activation in Hepatocyte Cultures Is Independent of Oxidant Damage. *By* ROSAURA NAVARRO, ROSA MARTÍNEZ, IDOIA BUSNADIEGO, M. BEGOÑA RUIZ-LARREA, AND JOSÉ IGNACIO RUIZ-SANZ 408

Superoxide Anions Are Involved in Doxorubicin-Induced ERK Activation in Hepatocyte Cultures. *By* ROSAURA NAVARRO, IDOIA BUSNADIEGO, M. BEGOÑA RUIZ-LARREA, AND JOSÉ IGNACIO RUIZ-SANZ 419

MAPKinase Gene Expression, as Determined by Microarray Analysis, Distinguishes Uncomplicated from Complicated Reconstitution after Major Surgical Trauma. *By* E. MARION SCHNEIDER, MANFRED WEISS, WEIDONG DU, GERHARD LEDER, KLAUS BUTTENSCHÖN, ULRICH C. LIENER, AND UWE B. BRÜCKNER 429

Effects of Chemical Ischemia on Cerebral Cortex Slices: Focus on Mitogen-Activated Protein Kinase Cascade. *By* ANNA SINISCALCHI, SABRINA CAVALLINI, SILVIA MARINO, SOFIA FALZARANO, LARA FRANCESCHETTI, AND RITA SELVATICI 445

Amyloid Precursor Protein Modulates ERK-1 and -2 Signaling. *By* VALENTINA VENEZIA, MARIO NIZZARI, EMANUELA REPETTO, ELISABETTA VIOLANI, ALESSANDRO CORSARO, STEFANO THELLUNG, VALENTINA VILLA, PIA CARLO, GENNARO SCHETTINI, TULLIO FLORIO, AND CLAUDIO RUSSO 455

Index of Contributors 467

Contents of the Other Volumes

PART B: Stress Signaling and Transcriptional Control

Volume 1091, December 2006

CONTENTS

Preface. *By* MARC DIEDERICH xxvii

Part I. Oxidative Stress

Oxidative Upregulation of Bcl-2 in Healthy Lymphocytes. *By* SILVIA CRISTOFANON, SILVIA NUCCITELLI, MARIA D'ALESSIO, FLAVIA RADOGNA, MILENA DE NICOLA, ANTONIO BERGAMASCHI, CLAUDIA CERELLA, ANDREA MAGRINI, MARC DIEDERICH, AND LINA GHIBELLI 1

Intracellular Pro-oxidant Activity of Melatonin Deprives U937 Cells of Reduced Glutathione without Affecting Glutathione Peroxidase Activity. *By* MARIA CRISTINA ALBERTINI, FLAVIA RADOGNA, AUGUSTO ACCORSI, FRANCESCO UGUCCIONI, LAURA PATERNOSTER, CLAUDIA CERELLA, MILENA DE NICOLA, MARIA D'ALESSIO, ANTONIO BERGAMASCHI, ANDREA MAGRINI, AND LINA GHIBELLI 10

Mitochondrial "Movement" and Lens Optics following Oxidative Stress from UV-B Irradiation: Cultured Bovine Lenses and Human Retinal Pigment Epithelial Cells (ARPE-19) as Examples. *By* VLADIMIR BANTSEEV AND HYUN-YI YOUN 17

2-Methoxyestradiol Inhibits Superoxide Anion Generation while It Enhances Superoxide Dismutase Activity in Swine Granulosa Cells. *By* GIUSEPPINA BASINI, SUJEN ELEONORA SANTINI, AND FRANCESCA GRASSELLI 34

Role of Reactive Oxygen Species in Kv Channel Inhibition and Vasoconstriction Induced by TP Receptor Activation in Rat Pulmonary Arteries. *By* ANGEL COGOLLUDO, GIOVANNA FRAZZIANO, LAURA COBEÑO, LAURA MORENO, FEDERICA LODI, EDUARDO VILLAMOR, JUAN TAMARGO, AND FRANCISCO PEREZ-VIZCAINO 41

DNA Strand Breaks by Metal-Induced Oxygen Radicals in Purified *Salmonella typhimurium* DNA. *By* EZZATOLLAH KEYHANI, FATEMEH ABDI-OSKOUEI, FARNOOSH ATTAR, AND JACQUELINE KEYHANI 52

Antioxidant Enzymes during Hypoxia–Anoxia Signaling Events in *Crocus sativus* L. Corm. *By* EZZATOLLAH KEYHANI, LILA GHAMSARI, JACQUELINE KEYHANI, AND MAHNAZ HADIZADEH 65

Ataxia-Telangiectasia-Mutated-Dependent Activation of Ku in Human Fibroblasts Exposed to Hydrogen Peroxide. *By* JONG HWA LEE, JI HOON YU, KYUNG HWAN KIM, AND HYEYOUNG KIM 76

Regulation of 2-Deoxy-D-Glucose Transport, Lactate Metabolism, and MMP-2 Secretion by the Hypoxia Mimetic Cobalt Chloride in Articular Chondrocytes. *By* ALI MOBASHERI, NICOLA PLATT, COLIN THORPE, AND MEHDI SHAKIBAEI 83

Oxidative Stress Response in Telomerase-Immortalized Fibroblasts from a Centenarian. *By* CHIARA MONDELLO, MARIA GRAZIA BOTTONE, SAKON NORIKI, CRISTIANA SOLDANI, CARLO PELLICCIARI, AND ANNA IVANA SCOVASSI 94

Differential Modulation of AMPK Signaling Pathways by Low or High Levels of Exogenous Reactive Oxygen Species in Colon Cancer Cells. *By* IN-JA PARK, JIN-TAEK HWANG, YOUNG MIN KIM, JOOHUN HA, AND OCK JIN PARK 102

Alterations in Salivary Antioxidants, Nitric Oxide, and Transforming Growth Factor-β_1 in Relation to Disease Activity in Crohn's Disease Patients. *By* ALI REZAIE, FAKHTEH GHORBANI, AZADEH ESHGHTORK, MOHAMMAD J. ZAMANI, GHOLAMREZA DEHGHAN, BARDIA TAGHAVI, SHEKOUFEH NIKFAR, AZADEH MOHAMMADIRAD, NASSER E. DARYANI, AND MOHAMMAD ABDOLLAHI 110

Control of Bioamine Metabolism by 5-HT$_{2B}$ and α_{1D} Autoreceptors through Reactive Oxygen Species and Tumor Necrosis Factor-α Signaling in Neuronal Cells. *By* BENOIT SCHNEIDER, MATHÉA PIETRI, SOPHIE MOUILLET-RICHARD, MYRIAM ERMONVAL, VINCENT MUTEL, JEAN-MARIE LAUNAY, AND ODILE KELLERMANN 123

Determination of Oxidative Stress Status and Concentration of TGF-β1 in the Blood and Saliva of Osteoporotic Subjects. *By* GHOLAMREZA YOUSEFZADEH, BAGHER LARIJANI, AZADEH MOHAMMADIRAD, RAMIN HESHMAT, GHOLAMREZA DEHGHAN, ROJA RAHIMI, AND MOHAMMAD ABDOLLAHI 142

Part II. Transcriptional Control

Targeting Signal-Transducer-and-Activator-of-Transcription-3 for Prevention and Therapy of Cancer: Modern Target but Ancient Solution. *By* BHARAT B. AGGARWAL, GAUTAM SETHI, KWANG SEOK AHN, SANTOSH K. SANDUR, MANOJ K. PANDEY, AJAIKUMAR B. KUNNUMAKKARA, BOKYUNG SUNG, AND HARUYO ICHIKAWA 151

Gene Expression Modulation in A549 Human Lung Cells in Response to Combustion-Generated Nano-Sized Particles. *By* ANDREA ARENZ, CHRISTINE E. HELLWEG, NEVENA STOJICIC, CHRISTA BAUMSTARK-KHAN, AND HORST-HENNING GROTHEER 170

Multiple Levels of Control of the Expression of the Human AβH-J-J Locus Encoding Aspartyl-β-hydroxylase, Junctin, and Junctate. *By* GIORDANA FERIOTTO, ALESSIA FINOTTI, GIULIA BREVEGLIERI, SUSAN TREVES, FRANCESCO ZORZATO, AND ROBERTO GAMBARI 184

Activation of Nuclear Factor κB by Different Agents: Influence of Culture Conditions in a Cell-Based Assay. *By* CHRISTINE E. HELLWEG, ANDREA ARENZ, SUSANNE BOGNER, CLAUDIA SCHMITZ, AND CHRISTA BAUMSTARK-KHAN 191

Atrial Appendage Transcriptional Profile in Patients with Atrial Fibrillation with Structural Heart Diseases. *By* MARIA S. KHARLAP, ANGELICA V. TIMOFEEVA, LUDMILA E. GORYUNOVA, GEORGE L. KHASPEKOV, SERGEY L. DZEMESHKEVICH, VLADIMIR V. RUSKIN, RENAT S. AKCHURIN, SERGEY P. GOLITSYN, AND ROBERT SH. BEABEALASHVILLI 205

DNA Hypomethylation of *CAGE* Promotors in Squamous Cell Carcinoma of Uterine Cervix. *By* TAEK SANG LEE, JAE WEON KIM, GYEONG HOON KANG, NOH HYUN PARK, YONG SANG SONG, SOON BEOM KANG, AND HYO PYO LEE 218

The *MECP2* Gene Mutation Screening in Rett Syndrome Patients from Croatia. *By* TANJA MATIJEVI, JELENA KNEŽEVIČ, INGEBORG BARIŠIĆ, BISERKA REŠIĆ, VIDA ČULIĆ, AND JASMINKA PAVELIĆ 225

Prostaglandins Regulate Transcription by Means of Prostaglandin Response Elements Located in the Promoters of Mammalian Na, K-ATPase $\beta 1$ Subunit Genes. *By* KEIKANTSE MATLHAGELA AND MARY TAUB 233

Different Modulation of ER-Mediated Transactivation by Xenobiotic Nuclear Receptors Depending on the Estrogen Response Elements and Estrogen Target Cell Types. *By* GYESIK MIN 244

Effects of TK Promotor and Hepatocyte Nuclear Factor-4 in CAR-Mediated Transcriptional Activity of Phenobarbital Responsive Unit of *CYP2B* Gene in Monkey Kidney Epithelial-Derived Cell Line COS-7. *By* GYESIK MIN 258

Expression of the E2F Family of Transcription Factors and Its Clinical Relevance in Ovarian Cancer. *By* DANIEL REIMER, SUSANN SADR, ANNEMARIE WIEDEMAIR, GEORG GOEBEL, NICOLE CONCIN, GERDA HOFSTETTER, CHRISTIAN MARTH, AND ALAIN G. ZEIMET 270

Isolation and Characterization of the Rat SND p102 Gene Promoter: Putative Role for Nuclear Factor-Y in Regulation of Transcription. *By* LORENA RODRÍGUEZ, NEREA BARTOLOMÉ, BEGOÑA OCHOA, AND MARÍA J. MARTÍNEZ 282

The cAMP-Responsive Unit of the Human Insulin-Like Growth Factor–Binding Protein-1 Coinstitutes a Functional Insulin-Response Element. *By* GHISLAINE SCHWEIZER-GROYER, GUILLAUME FALLOT, FRANÇOISE CADEPOND, CHRISTELLE GIRARD, AND ANDRÉ GROYER 296

c-Jun and JunB Are Essential for Hypoglycemia-Mediated *VEGF* Induction. *By* BJÖRN TEXTOR, MELANIE SATOR-SCHMITT, KARL HARTMUT RICHTER, PETER ANGEL, AND MARINA SCHORPP-KISTNER 310

Altered Gene Expression Pattern in Peripheral Blood Leukocytes from Patients with Arterial Hypertension. *By* A.V. TIMOFEEVA, L.E. GORYUNOVA, G.L. KHASPEKOV, D.A. KOVALEVSKII, A.V. SCAMROV, O.S. BULKINA, YU.A. KARPOV, K.A. TALITSKII, V.V. BUZA, V.V. BRITAREVA, AND R.SH. BEABEALASHVILLI 319

Effects of AT_1 Receptor-Mediated Endocytosis of Extracellular Ang II on Activation of Nuclear Factor-κB in Proximal Tubule Cells. *By* JIA L. ZHUO, OSCAR A. CARRETERO, AND XIAO C. LI 336

Part III. Histone Deacetylase

Retinoic Acid and Histone Deacetylase Inhibitor BML-210 Inhibit
Proliferation of Human Cervical Cancer HeLa Cells. *By*
Veronika V. Borutinskaite, Ruta Navakauskiene, and
Karl-Eric Magnusson 346

Effects of Histone Deacetylase Inhibitors, Sodium Phenyl Butyrate, and
Vitamin B3, in Combination with Retinoic Acid, on Granulocytic
Differentiation of Human Promyelocytic Leukemia HL-60 Cells. *By*
Rasa Merzvinskyte, Grazina Treigyte, Jurate Savickiene,
Karl-Eric Magnusson, and Ruta Navakauskiene 356

The Histone Deacetylase Inhibitor FK228 Distinctly Sensitizes the Human
Leukemia Cells to Retinoic Acid-Induced Differentiation. *By*
Jurate Savickiene, Grazina Treigyte, Veronika Borutinskaite,
Ruta Navakauskiene, and Karl-Eric Magnusson 368

Effect of Valproic Acid, a Histone Deacetylase Inhibitor, on Cell Death
and Molecular Changes Caused by Low-Dose Irradiation. *By*
Darina Záškodová, Martina Řezáčová, Jiřina Vávrová,
Doris Vokurková, and Aleš Tichý 385

Part IV. Novel Technological and Therapeutical Approaches

Protein Folding Information in Nucleic Acids Which Is Not Present in the
Genetic Code. *By* Jan C. Biro 399

Antigens and Cytokine Genes in Antitumor Vaccines: The Importance of the
Temporal Delivery Sequence in Antitumor Signals. *By*
María José Herrero, Rafael Botella, Francisco Dasí,
Rosa Algás, María Sánchez, and Salvador F. Aliño 412

Arrest of Cancer Cell Proliferation by dsRNAs. *By* Tatyana O. Kabilova,
Albina V. Vladimirova, Elena L. Chernolovskaya, and
Valentin V. Vlassov 425

Design and Functional Activity of Phosphopeptides with Potential
Immunomodulating Capacity, Based on the Sequence of Grb2-Associated
Binder 1. *By* Akos Kertesz, Balazs Takacs, Gyorgyi Varadi,
Gabor K. Toth, and Gabriella Sarmay 437

Distinct Activity of Peptide Mimetic Intracellular Ligands (Pepducins) for
Proteinase-Activated Receptor-1 in Multiple Cells/Tissues. *By*
Satoko Kubo, Tsuyoshi Ishiki, Ichiko Doe, Fumiko Sekiguchi,
Hiroyuki Nishikawa, Kenzo Kawai, Hirofumi Matsui, and
Atsufumi Kawabata 445

A New Method to Assess Drug Sensitivity on Breast Tumor Acute Slices
Preparation. *By* Pedro Mestres, Andrea Morguet, Werner Schmidt,
Axel Kob, and Elke Thedinga 460

Process Simulation in a Mechatronic Bioreactor Device with Speed-Regulated
Motors for Growing of Three-Dimensional Cell Cultures. *By*
Mina Mihailova, Vassil Trenev, Penka Genova, and
Spiro Konstantinov 470

Animal Model of Drug-Resistant Tumor Progression. *By*
NADEZDA MIRONOVA, OLGA SHKLYAEVA, EKATERINA ANDREEVA,
NELLY POPOVA, VASILYI KALEDIN, VALERYI NIKOLIN,
VALENTIN VLASSOV, AND MARINA ZENKOVA 490

Preparation and Characterization of Recombinant Chicken Growth
Hormone (chGH) and Its Putative Antagonist chGH G119R Mutein. *By*
HELENA E. PACZOSKA-ELIASIEWICZ, GILI SALOMON, SHAY REICHER,
EUGENE E. GUSSAKOWSKY, ANNA HRABIA, AND ARIEH GERTLER 501

Induction of Apoptosis of Osteoclasts by Targeting Transcription Factors
with Decoy Molecules. *By* ROBERTA PIVA, LETIZIA PENOLAZZI,
MARGHERITA ZENNARO, ERCOLINA BIANCHINI, EROS MAGRI,
MONICA BORGATTI, ILARIA LAMPRONTI, ELISABETTA LAMBERTINI,
ELISA TAVANTI, AND ROBERTO GAMBARI 509

Competition Effects Shape the Response Sensitivity and Kinetics of
Phosphorylation Cycles in Cell Signaling. *By* CARLOS SALAZAR AND
THOMAS HÖFER 517

Preparation of Leptin Antagonists by Site-Directed Mutagenesis of Human,
Ovine, Rat, and Mouse Leptin's Site III: Implications on Blocking
Undesired Leptin Action *In Vivo*. *By* GILI SOLOMON, LEONORA
NIV-SPECTOR, DANA GONEN-BERGER, ISABELLE CALLEBAUT,
JEAN DJIANE, AND ARIEH GERTLER 531

Dual Activity of Phosphorothioate CpG Oligodeoxynucleotides on HIV:
Reactivation of Latent Provirus and Inhibition of Productive Infection
in Human T Cells. *By* CARSTEN SCHELLER, ANETT ULLRICH, STEFAN
LAMLA, ULF DITTMER, AXEL RETHWILM, AND ELENI KOUTSILIERI 540

Prostanoids with Cyclopentenone Structure as Tools for the Characterization
of Electrophilic Lipid–Protein Interactomes. *By*
KONSTANTINOS STAMATAKIS AND DOLORES PÉREZ-SALA 548

Index of Contributors 571

PART C: Cell Signaling in Health and Disease

Volume 1095, January 2007

CONTENTS

Preface. *By* MARC DIEDERICH xxvii

Part I. Cancer

Cationic Surfactants Induce Apoptosis in Normal and Cancer Cells. *By* RIYO ENOMOTO, CHIE SUZUKI, MASATAKA OHNO, TOSHINORI OHASI, RYOKO FUTAGAMI, KEIKO ISHIKAWA, MIKA KOMAE, TAKAYUKI NISHINO, YASUO KONISHI, AND EIBAI LEE 1

DMNQ S-64 Induces Apoptosis Via Caspase Activation and Cyclooxygenase-2 Inhibition in Human Nonsmall Lung Cancer Cells. *By* EU-SOO LIM, YUN-HEE RHEE, MIN-KYU PARK, BEOM-SANG SHIM, KYOO-SEOK AHN, HEE KANG, HWA-SEUNG YOO, AND SUNG-HOON KIM 7

Bcl-2 Expression in Oral Squamous Cell Carcinoma. *By* B. POPOVIĆ, B. JEKIĆ, I. NOVAKOVIĆ, L.J. LUKOVIĆ, Z. TEPAVČEVIĆ, V. JURIŠIĆ, M. VUKADINOVIĆ, AND J. MILAŠIN 19

Apoptotic Effect of Celecoxib Dependent Upon p53 Status in Human Ovarian Cancer Cells. *By* YOO-CHEOL SONG, SU-HYEONG KIM, YONG-SUNG JUHNN, AND YONG-SANG SONG 26

Breast Cancer Cells Response to the Antineoplastic Agents Cisplatin, Carboplatin, and Doxorubicin at the mRNA Expression Levels of Distinct Apoptosis-Related Genes, Including the New Member, *BCL2L12*. *By* HELLINIDA THOMADAKI AND ANDREAS SCORILAS 35

Effect of Distinct Anticancer Drugs on the Phosphorylation of p53 Protein at Serine 46 in Human MCF-7 Breast Cancer Cells. *By* JÓZEFA WĘSIERSKA-GĄDEK, MARIETA GUEORGUIEVA, IRENE HERBACEK, AND CARMEN RANFTLER 45

Significant Coexpression of GLUT-1, Bcl-xL, and Bax in Colorectal Cancer. *By* ANDRZEJ WINCEWICZ, MARIOLA SULKOWSKA, MARIUSZ KODA, LUIZA KANCZUGA-KODA, EWA WITKOWSKA, AND STANISLAW SULKOWSKI 53

Combination of Doxorubicin and Sulforaphane for Reversing Doxorubicin-Resistant Phenotype in Mouse Fibroblasts with p53^{Ser220} Mutation. *By* CARMELA FIMOGNARI, MONIA LENZI, DAVIDE SCIUSCIO, GIORGIO CANTELLI-FORTI, AND PATRIZIA HRELIA 62

Aromatase Expression Was Not Detected by Immunohistochemistry in Endometrial Cancer. *By* YONG-TARK JEON, SO YEON PARK, YONG-BEOM KIM, JAE WEON KIM, NOH-HYUN PARK, SOON-BEOM KANG, HYO-PYO LEE, AND YONG-SANG SONG 70

ER Stress Induces the Expression of Jpk, which Inhibits Cell Cycle Progression in F9 Teratocarcinoma Cell. *By* HYE SUN KIM, KYOUNG-AH KONG, HYUNJOO CHUNG, SUNGDO PARK, AND MYOUNG HEE KIM 76

Akt Involvement in Paclitaxel Chemoresistance of Human Ovarian Cancer Cells. *By* SU-HYEONG KIM, YONG-SUNG JUHNN, AND YONG-SANG SONG 82

Expression of Leptin, Leptin Receptor, and Hypoxia-Inducible Factor 1α in Human Endometrial Cancer. *By* MARIUSZ KODA, MARIOLA SULKOWSKA, ANDRZEJ WINCEWICZ, LUIZA KANCZUGA-KODA, BOGUSLAW MUSIATOWICZ, MAGDALENA SZYMANSKA, AND STANISLAW SULKOWSKI 90

The Cyclooxygenase-2 Selective Inhibitor Celecoxib Suppresses Proliferation and Invasiveness in the Human Oral Squamous Carcinoma. *By* YOUNG EUN KWAK, NAM KYEOUNG JEON, JIN KIM, AND EUN JU LEE 99

The Epidermal Growth Factor Receptor Tyrosine Kinase Inhibitor ZD1839 (Iressa) Suppresses Proliferation and Invasion of Human Oral Squamous Carcinoma Cells via p53 Independent and MMP, uPAR Dependent Mechanism. *By* EUN JU LEE, JIN HA WHANG, NAM KYEONG JEON, AND JIN KIM 113

Role of Vascular Endothelial Growth Factor-D (VEGF-D) on IL-6 Expression in Cerulein-Stimulated Pancreatic Acinar Cells. *By* JANGWON LEE, KYUNG HWAN KIM, AND HYEYOUNG KIM 129

Lack of Association of the Cyclooxygenase-2 and Inducible Nitric Oxide Synthase Gene Polymorphism with Risk of Cervical Cancer in Korean Population. *By* TAEK SANG, LEE YONG TARK JEON, JAE WEON KIM, NOH HYUN PARK, SOON BEOM KANG, HYO PYO LEE, AND YONG SANG SONG 134

Increased Cyclooxygenase-2 Expression Associated with Inflammatory Cellular Infiltration in Elderly Patients with Vulvar Cancer. *By* TAEK SANG LEE, YONG TARK JEON, JAE WEON KIM, JAE KYUNG WON, NOH HYUN PARK, IN AE PARK, YONG SUNG JUHNN, SOON BEOM KANG, HYO PYO LEE, AND YONG SANG SONG 143

Viability of a Human Melanoma Cell after Single and Combined Treatment with Fotemustine, Dacarbazine, and Proton Irradiation. *By* IVAN M. PETROVIĆ, LELA B. KORIĆANAC, DANIJELA V. TODOROVIĆ, ALEKSANDRA M. RISTIĆ-FIRA, LUCIA M. VALASTRO, GIUSEPPE PRIVITERA, AND GIACOMO CUTTONE 154

Response of a Human Melanoma Cell Line to Low and High Ionizing Radiation. *By* ALEKSANDRA M. RISTIC-FIRA, DANIJELA V. TODOROVIC, LELA B. KORICANAC, IVAN M. PETROVIC, LUCIA M. VALASTRO, PABLO G.A. CIRRONE, LUIGI RAFFAELE, AND GIACOMO CUTTONE 165

Effect of Paclitaxel on Intracellular Localization of c-Myc and P-c-Myc in Prostate Carcinoma Cell Lines. *By* ROSANNA SUPINO, ENRICA FAVINI, GIUDITTA CUCCURU, FRANCO ZUNINO, AND A. IVANA SCOVASSI 175

Erufosine: A Membrane Targeting Antineoplastic Agent with Signal Transduction Modulating Effects. *By* M.M. ZAHARIEVA, S.M. KONSTANTINOV, B. PILICHEVA, M. KARAIVANOVA, AND M.R. BERGER 182

Part II. Cell Signaling in Health and Disease

SHP-1 Tyrosine Phosphatase in Human Erythrocytes. *By* MARCANTONIO BRAGADIN, FLORINA ION-POPA, GIULIO CLARI, AND LUCIANA BORDIN 193

Is the Distribution of Selenium and Zinc in the Sublocations of Spermatozoa Regulated? *By* HOLGER BERTELSMANN, HARALD SIEME, DIETRICH BEHNE, AND ANTONIOS KYRIAKOPOULOS 204

Immune Complexes Induce Monocyte Survival through Defined Intracellular Pathways. *By* GIORDANO BIANCHI, FABRIZIO MONTECUCCO, MARIA BERTOLOTTO, FRANCO DALLEGRI, AND LUCIANO OTTONELLO 209

Inhibition of Serine–Threonine Protein Phosphatases in Monocyte Chemoattractant Protein-1 Expression in *Helicobacter pylori*-Stimulated Gastric Epithelial Cells. *By* HAE-YUN CHUNG, BORAM CHA, AND HYEYOUNG KIM 220

Cardioprotective Action of Urocortin in Early Pre- and Postconditioning. *By* BARBARA CSEREPES, GABOR JANCSO, BALAZS GASZ, BOGLARKA RACZ, ANDREA FERENC, LASZLO BENKO, BALAZS BORSICZKY, MARIA KURTHY, SANDOR FERENCZ, JANOS LANTOS, JANOS GAL, ENDRE ARATO, ATTILA MISETA, GYORGY WEBER, AND ELIZABETH ROTH 228

Novel Involvement of the Immunomodulator AS101 in IL-10 Signaling, via the Tyrosine Kinase Fer. *By* RAMI HAYUN, SALI SHPUNGIN, HANA MALOVANI, MICHAEL ALBECK, EITAN OKUN, URI NIR, AND BENJAMIN SREDNI 240

Expression and Protective Role of Heme Oxygenase-1 in Delayed Myocardial Preconditioning. *By* GÁBOR JANCSÓ, BARBARA CSEREPES, BALÁZS GASZ, LÁSZLÓ BENKŐ, BALÁZS BORSICZKY, ANDREA FERENC, MÁRIA KÜRTHY, BOGLÁRKA RÁCZ, JÁNOS LANTOS, JÁNOS GÁL, ENDRE ARATÓ, LÓSZLÓ SÍNAYC, GYÖRGY WÉBER, AND ERZSÉBET RÓTH 251

Expression of Insulin Signaling Transmitters and Glucose Transporters at the Protein Level in the Rat Testis. *By* KERSTI KOKK, ESKO VERÄJÄNKORVA, XIAO-KE WU, HELLE TAPFER, ELLE PÕLDOJA, HELLE-EVI SIMOVART, AND PASI PÖLLÄNEN 262

Effect of Endothelin on Sodium/Hydrogen Exchanger Activity of Human Monocytes and Atherosclerosis-Related Functions. *By* GEORGE KOLIAKOS, CHRISTINA BEFANI, KONSTANTINOS PALETAS, AND MARTHA KALOYIANNI 274

Small GTPase Ras and Rho Expression in Rat Osteoblasts during Spaceflight. *By* YASUHIRO KUMEI, HITOYATA SHIMOKAWA, KEI'ICHI OHYA, HISAKO KATANO, HIDEO AKIYAMA, MASAHIKO HIRANO, AND SADAO MORITA 292

Protein Expression in the Tissues of the Cardiovascular System of the Rat under Selenium Deficiency and Adequate Conditions. *By* A. KYRIAKOPOULOS, A. RICHTER, T. POHL, C. WOLF, I. GRBAVAC, A. PLOTNIKOV, M. KÜHBACHER, H. BERTELSMANN, AND D. BEHNE 300

Study on the Correlations among Disease Activity Index and Salivary Transforming Growth Factor-β1 and Nitric Oxide in Ulcerative Colitis Patients. *By* ALI REZAIE, SARA KHALAJ, MARYAM SHABIHKHANI, SHEKOUFEH NIKFAR, MOHAMMAD J. ZAMANI, AZADEH MOHAMMADIRAD, NASER E. DARYANI, AND MOHAMMAD ABDOLLAHI 305

Generation of ΔTAp73 Proteins by Translation from a Putative Internal Ribosome Entry Site. *By* A. EMRE SAYAN, JEAN-PIERRE ROPERCH, BERNA S. SAYAN, MARIO ROSSI, M.J. PINKOSKI, RICHARD A. KNIGHT, ANNE E. WILLIS, AND GERRY MELINO 315

IRF-7: New Role in the Regulation of Genes Involved in Adaptive Immunity. *By* MARCO SGARBANTI, GIULIA MARSILI, ANNA LISA REMOLI, ROBERTO ORSATTI, AND ANGELA BATTISTINI 325

A New Transcript Splice Variant of the Human Glucocorticoid Receptor: Identification and Tissue Distribution of hGRΔ313–338, an Alternative Exon 2 Transactivation Domain Isoform. *By* JONATHAN D. TURNER, ANDREA B. SCHOTE, MARC KEIPES, AND CLAUDE P. MULLER 334

Exploitation of Host Signaling Pathways by B Cell Superantigens—Potential Strategies for Developing Targeted Therapies in Systemic Autoimmunity. *By* MONCEF ZOUALI 342

Part III. Chemoprevention

Antineoplastic and Anticlastogenic Properties of Curcumin. *By* TZVETAN ALAIKOV, SPIRO M. KONSTANTINOV, TZVETOMIRA TZANOVA, KYRIL DINEV, MARGARITA TOPASHKA-ANCHEVA, AND MARTIN R. BERGER 355

Sanguinarine Inhibits VEGF-Induced Akt Phosphorylation. *By* GIUSEPPINA BASINI, SUJEN ELEONORA SANTINI, SIMONA BUSSOLATI, AND FRANCESCA GRASSELLI 371

Effect of Curcumin Treatment on Protein Phosphorylation in K562 Cells. *By* ROMAIN BLASIUS, MARIO DICATO, AND MARC DIEDERICH 377

Dosage Effects of Ginkgolide B on Ethanol-Induced Cell Death in Human Hepatoma G2 Cells. *By* WEN-HSIUNG CHAN AND YAN-DER HSUUW 388

Attenuation of Aβ-Induced Apoptosis of Plant Extract (Saengshik) Mediated by the Inhibition of Mitochondrial Dysfunction and Antioxidative Effect. *By* CHU-YUE CHEN, JUNG-HEE JANG, MI HYUN PARK, SUNG JOO HWANG, YOUNG-JOON SURH, AND OCK JIN PARK 399

Wogonin Prevents Immunosuppressive Action but Not Anti-Inflammatory Effect Induced by Glucocorticoid. *By* RIYO ENOMOTO, CHIE SUZUKI, CHIKA KOSHIBA, TAKAYUKI NISHINO, MIKIKO NAKAYAMA, HIROYUKI HIRANO, TOSHIO YOKOI, AND EIBAI LEE 412

The Analgesic Effect of *Tribulus terrestris* Extract and Comparison of Gastric Ulcerogenicity of the Extract with Indomethacine in Animal Experiments. *By* M.R. HEIDARI, M. MEHRABANI, A. PARDAKHTY, P. KHAZAELI, M.J. ZAHEDI, M. YAKHCHALI, AND M. VAHEDIAN 418

Epigallocatechin Gallate Dose-Dependently Induces Apoptosis or Necrosis in Human MCF-7 Cells. *By* YAN-DER HSUUW AND WEN-HSIUNG CHAN 428

Resveratrol Induces Apoptosis in Chemoresistant Cancer Cells via Modulation of AMPK Signaling Pathway. *By* JIN-TAEK HWANG, DONG WOOK KWAK, SUN KYO LIN, HYE MIN KIM, YOUNG MIN KIM, AND OCK JIN PARK 441

Antioxidative Effects of Plant Polyphenols: From Protection of G Protein Signaling to Prevention of Age-Related Pathologies. *By* VIKTOR JEFREMOV, MIHKEL ZILMER, KERSTI ZILMER, NENAD BOGDANOVIC, AND ELLO KARELSON 449

Jaceosidin, a Pharmacologically Active Flavone Derived from *Artemisia argyi*, Inhibits Phorbol-Ester-Induced Upregulation of COX-2 and MMP-9 by Blocking Phosphorylation of ERK-1 and -2 in Cultured Human Mammary Epithelial Cells. *By* MIN A JEONG, KI WON LEE, DO-YOUNG YOON, AND HYONG JOO LEE 458

Effects of Selenium Diet on Expression of Selenoproteins in the Lung of the Rat. *By* KATARZYNA BUKALIS, DOROTHEA ALBER, GREGOR BUKALIS, DIETRICH BEHNE, AND ANTONIOS KYRIAKOPOULOS 467

Protective Effects of Piceatannol against Beta-Amyloid–Induced Neuronal Cell Death. *By* HYO JIN KIM, KI WON LEE, AND HYONG JOO LEE 473

Jaceosidin Induces Apoptosis in *ras*-Transformed Human Breast Epithelial Cells through Generation of Reactive Oxygen Species. *By* MIN-JUNG KIM, DO-HEE KIM, KI WON LEE, DO-YOUNG YOON, AND YOUNG-JOON SURH ... 483

Involvement of AMPK Signaling Cascade in Capsaicin-Induced Apoptosis of HT-29 Colon Cancer Cells. *By* YOUNG MIN KIM, JIN-TAEK HWANG, DONG WOOK KWAK, YUN KYUNG LEE, AND OCK JIN PARK 496

Epigallocatechin Gallate Inhibits Phorbol Ester-Induced Activation of NF-κB and CREB in Mouse Skin: Role of p38 MAPK. *By* JOYDEB KUMAR KUNDU AND YOUNG-JOON SURH 504

Peonidin Inhibits Phorbol-Ester–Induced COX-2 Expression and Transformation in JB6 P$^+$ Cells by Blocking Phosphorylation of ERK-1 and -2. *By* JUNG YEON KWON, KI WON LEE, HAENG JEON HUR, AND HYONG JOO LEE 513

Wogonin, a Plant Flavone, Potentiates Etoposide-Induced Apoptosis in Cancer Cells. *By* EIBAI LEE, RIYO ENOMOTO, CHIE SUZUKI, MASATAKA OHNO, TOSHINORI OHASHI, AZUSA MIYAUCHI, ERIKO TANIMOTO, KAORI MAEDA, HIROYUKI HIRANO, TOSHIO YOKOI, AND CHIYOKO SUGAHARA 521

Inhibitory Effects of 7-Carboxymethyloxy-3′,4′,5-Trimethoxy Flavone (DA-6034) on *Helicobacter pylori*-Induced NF-κB Activation and iNOS Expression in AGS Cells. *By* JEONG-SANG LEE, HYUN-SOO KIM, KI-BAIK HAHM, MI-WON SOHN, MOOHI YOO, JEFFREY A. JOHNSON, AND YOUNG-JOON SURH 527

Phenolic Phytochemicals Derived from Red Pine (*Pinus densiflora*) Inhibit the Invasion and Migration of SK-Hep-1 Human Hepatocellular Carcinoma Cells. *By* SANG JUN LEE, KI WON LEE, HAENG JEON HUR, JI YOUNG CHUN, SEO YOUNG KIM, AND HYONG JOO LEE 536

KG-135 Inhibits COX-2 Expression by Blocking the Activation of JNK and AP-1 in Phorbol Ester–Stimulated Human Breast Epithelial Cells. *By* SIN-AYE PARK, EUN-HEE KIM, HYE-KYUNG NA, AND YOUNG-JOON SURH 545

Resveratrol Inhibits IL-1β–Induced Stimulation of Caspase-3 and Cleavage of PARP in Human Articular Chondrocytes *in vitro*. *By* MEHDI SHAKIBAEI, THILO JOHN, CLAUDIA SEIFARTH, AND ALI MOBASHERI 554

Possible Link Between NO Concentrations and COX-2 Expression in
  Systems Treated with Soy-Isoflavones. *By* JANG-IN SHIN,
  YUN-KYUNG LEE, YOUNG MIN KIM, JIN-TAEK HWANG, AND
  OCK JIN PARK 564

Assessment of the Effect of *Echinacea purpurea* (L.) Moench on Apoptotic
  and Mitotic Activity of Liver Cells during Intoxication by Cadmium. *By*
  ALINA SMALINSKIENE, VAIVA LESAUSKAITE, STANISLOVAS RYSELIS,
  OLEG ABDRAKHMANOV, RIMA KREGZDYTE, ILONA SADAUSKIENE,
  LEONID IVANOV, NIJOLE SAVICKIENE, VIRGILIJUS ZITKEVIČIUS, AND
  ARUNAS SAVICKAS 574

Influence of *Echinacea purpurea* (L.) Moench Extract on the Toxicity of
  Cadmium. *By* VIRGILIJUS ZITKEVICIUS, ALINA SMALINSKIENE,
  VAIVA LESAUSKAITE, NIJOLE SAVICKIENE, ARUNAS SAVICKAS,
  STANISLOVAS RYSELIS, RIMA KREGZDYTE, OLEG ABDRAKHMANOV,
  ILONA SADAUSKIENE, AND LEONID IVANOV 585

Index of Contributors 593

PART D: Inflammatory Signaling Pathways and Neuropathology

Volume 1096, January 2007

CONTENTS

Preface. *By* MARC DIEDERICH xxvii

Part I. Inflammatory Signaling Pathways

Targeting Bacterial Endotoxin: Two Sides of a Coin. *By* HERBERT BOSSHART
 AND MICHAEL HEINZELMANN 1

Interaction between the *Helicobacter pylori* CagA and α-Pix in Gastric
 Epithelial AGS Cells. *By* HYE YEON BAEK, JOO WEON LIM, AND
 HYEYOUNG KIM 18

Expression of Suppressors of Cytokine Signaling-3 in *Helicobacter
 pylori*-Infected Rat Gastric Mucosal RGM-1Cells. *By* BORAM CHA,
 KYUNG HWAN KIM, HIROFUMI MATSUI, AND HYEYOUNG KIM 24

Role of Proteinase-Activated Receptor-2 on Cyclooxygenase-2 Expression in
 H. pylori–Infected Gastric Epithelial Cells. *By* JI HYE SEO,
 KYUNG HWAN KIM, AND HYEYOUNG KIM 29

Control of Human Herpes Virus Type 8-Associated Diseases by NK Cells. *By*
 MARIA C. SIRIANNI, MASSIMO CAMPAGNA, DONATO SCARAMUZZI,
 MAURIZIO CARBONARI, ELENA TOSCHI, ILARIA BACIGALUPO,
 PAOLO MONINI, AND BARBARA ENSOLI 37

Analysis of Tissue Distribution of TNF-α, TNF-α-Receptors, and the
 Activating TNF-α–Converting Enzyme Suggests Activation of the
 TNF-α System in the Aging Intervertebral Disc. *By*
 BEATRICE E. BACHMEIER, ANDREAS G. NERLICH, CHRISTOPH WEILER,
 GÜNTHER PAESOLD, MARIANNE JOCHUM, AND NORBERT BOOS 44

Upregulation of Apolipoprotein B Secretion, but Not Lipid, by Tumor Necrosis
 Factor-α in Rat Hepatocyte Cultures in the Absence of Extracellular Fatty
 Acids. *By* NEREA BARTOLOMÉ, LORENA RODRÍGUEZ, MARÍA J. MARTÍNEZ,
 BEGOÑA OCHOA, AND YOLANDA CHICO 55

Gene Expression Profiling of LPS-Stimulated Murine Macrophages and Role
 of the NF-κB and PI3K/mTOR Signaling Pathways. *By* S. DOS SANTOS,
 A.-I. DELATTRE, F. DE LONGUEVILLE, H. BULT, AND M. RAES 70

Modification of Proteins by Cyclopentenone Prostaglandins is Differentially
 Modulated by GSH *in vitro*. *By* JAVIER GAYARRE, M. ISABEL AVELLANO,
 FRANCISCO J. SÁNCHEZ-GÓMEZ, M. JESÚS CARRASCO, F. JAVIER CAÑADA,
 AND DOLORES PÉREZ-SALA 78

Signaling Pathways Involved in Proteinase-Activated Receptor$_1$-Induced
 Proinflammatory and Profibrotic Mediator Release Following
 Lung Injury. *By* PAUL F. MERCER, XIAOLING DENG, AND
 RACHEL C. CHAMBERS 86

Signaling Pathway Used by HSV-1 to Induce NF-κB Activation: Possible
Role of Herpes Virus Entry Receptor A. *By* M. TERESA SCIORTINO,
M. ANTONIETTA MEDICI, FRANCESCA MARINO-MERLO, DANIELA ZACCARIA,
MARIA GIUFFRÈ, ASSUNTA VENUTI, SANDRO GRELLI, AND
ANTONIO MASTINO 89

Melphalan Reduces the Severity of Experimental Colitis in Mice by Blocking
Tumor Necrosis Factor-α Signaling Pathway. *By* GALINA SHMARINA,
ALEXANDER PUKHALSKY, VLADIMIR ALIOSHKIN, AND
ALEX SABELNIKOV 97

Part II. Neuropathology

Cellular Prion Protein Signaling in Serotonergic Neuronal Cells. *By*
SOPHIE MOUILLET-RICHARD, BENOÎT SCHNEIDER, ELODIE PRADINES,
MATHÉA PIETRI, MYRIAM ERMONVAL, JACQUES GRASSI,
J. GRAYSON RICHARDS, VINCENT MUTEL, JEAN-MARIE LAUNAY, AND
ODILE KELLERMANN 106

Strongly Reduced Number of Parvalbumin-Immunoreactive Projection
Neurons in the Mammillary Bodies in Schizophrenia: Further Evidence
for Limbic Neuropathology. *By* HANS-GERT BERNSTEIN,
STEPHANIE KRAUSE, DIETER KRELL, HENRIK DOBROWOLNY,
MARION WOLTER, RENATE STAUCH, KARIN RANFT, PETER DANOS,
GUSTAV F. JIRIKOWSKI, AND BERNHARD BOGERTS 120

Alterations of Synaptic Turnover Rate in Aging May Trigger Senile Plaque
Formation and Neurodegeneration. *By* CARLO BERTONI-FREDDARI,
PATRIZIA FATTORETTI, BELINDA GIORGETTI, YESSICA GROSSI,
MARTA BALIETTI, TIZIANA CASOLI, GIUSEPPINA DI STEFANO, AND
GEMMA PERRETTA 128

Preservation of Mitochondrial Volume Homeostasis at the Early Stages of
Age-Related Synaptic Deterioration. *By* CARLO BERTONI-FREDDARI,
PATRIZIA FATTORETTI, BELINDA GIORGETTI, YESSICA GROSSI,
MARTA BALIETTI, TIZIANA CASOLI, GIUSEPPINA DI STEFANO, AND
GEMMA PERRETTA 138

Immunohistochemical Evidence for Impaired Neuregulin-1 Signaling in the
Prefrontal Cortex in Schizophrenia and in Unipolar Depression. *By*
IRIS BERTRAM, HANS-GERT BERNSTEIN, UWE LENDECKEL,
ALICJA BUKOWSKA, HENRIK DOBROWOLNY, GERBURG KEILHOFF,
DIMITRIOS KANAKIS, CHRISTIAN MAWRIN, HENDRIK BIELAU,
PETER FALKAI, AND BERNHARD BOGERTS 147

Dysregulation of GABAergic Neurotransmission in Mood Disorders: A
Postmortem Study. *By* HENDRIK BIELAU, JOHANN STEINER,
CHRISTIAN MAWRIN, KURT TRÜBNER, RALF BRISCH,
GABRIELA MEYER-LOTZ, MICHAEL BRODHUN, HENRIK DOBROWOLNY,
BRUNO BAUMANN, TOMASZ GOS, HANS-GERT BERNSTEIN, AND
BERNHARD BOGERTS 157

Release of β-Amyloid from High-Density Platelets: Implications for
Alzheimer's Disease Pathology. *By* TIZIANA CASOLI,
GIUSEPPINA DI STEFANO, BELINDA GIORGETTI, YESSICA GROSSI,
MARTA BALIETTI, PATRIZIA FATTORETTI, AND
CARLO BERTONI-FREDDARI 170

The Role of Selenite on Microglial Migration. *By* LISA DALLA PUPPA, NICOLAI E. SAVASKAN, ANJA U. BRÄUER, DIETRICH BEHNE, AND ANTONIOS KYRIAKOPOULOS 179

Altered Subcellular Distribution of the Alzheimer's Amyloid Precursor Protein Under Stress Conditions. *By* SARA C.T.S. DOMINGUES, ANA GABRIELA HENRIQUES, WENJUAN WU, EDGAR F. DA CRUZ E. SILVA, AND ODETE A.B. DA CRUZ E. SILVA 184

Differential Distribution of Alzheimer's Amyloid Precursor Protein Family Variants in Human Sperm. *By* MARGARIDA FARDILHA, SANDRA I. VIEIRA, ALBERTO BARROS, MÁRIO SOUSA, ODETE A.B. DA CRUZ E. SILVA, AND EDGAR F. DA CRUZ E. SILVA 196

The Effect of Repeated Physical Exercise on Hippocampus and Brain Cortex in Stressed Rats. *By* DRAGANA FILIPOVIĆ, LJUBICA GAVRILOVIĆ, SLADJANA DRONJAK, AND MARIJA B. RADOJČIĆ 207

Prion Protein Aggregation and Neurotoxicity in Cortical Neurons. *By* JOANA BARBOSA MELO, PAULA AGOSTINHO, AND CATARINA RESENDE OLIVEIRA 220

Intensive Remodeling of Purkinje Cell Spines after Climbing Fibers Deafferentation Does Not Involve MAPK and Akt Activation. *By* JELENA M. MILAŠIN, ANNALISA BUFFO, DANIELA CARULLI, AND PIERGIORGIO STRATA 230

Immunomorphological Analysis of RAGE Receptor Expression and NF-κB Activation in Tissue Samples from Normal and Degenerated Intervertebral Discs of Various Ages. *By* ANDREAS G. NERLICH, BEATRICE E. BACHMEIER, ERWIN SCHLEICHER, HELMUT ROHRBACH, GUENTHER PAESOLD, AND NORBERT BOOS 239

Amyloid Precursor Protein and Presenilin 1 Interaction Studied by FRET in Human H4 Cells. *By* MARIO NIZZARI, VALENTINA VENEZIA, PAOLO BIANCHINI, VALENTINA CAORSI, ALBERTO DIASPRO, EMANUELA REPETTO, STEFANO THELLUNG, ALESSANDRO CORSARO, PIA CARLO, GENNARO SCHETTINI, TULLIO FLORIO, AND CLAUDIO RUSSO 249

Amino-Terminally Truncated Prion Protein PrP90-231 Induces Microglial Activation *in vitro*. *By* STEFANO THELLUNG, ALESSANDRO CORSARO, VALENTINA VILLA, VALENTINA VENEZIA, MARIO NIZZARI, MICHELA BISAGLIA, CLAUDIO RUSSO, GENNARO SCHETTINI, ANTONIO ACETO, AND TULLIO FLORIO 258

Activation and Endocytic Internalization of Melanocortin 3 Receptor in Neuronal Cells. *By* S.J.M. WACHIRA, B. GURUSWAMY, L. URADU, C.A. HUGHES-DARDEN, AND F.J. DENARO 271

Index of Contributors 287

The New York Academy of Sciences believes it has a responsibility to provide an open forum for discussion of scientific questions. The positions taken by the participants in the reported conferences are their own and not necessarily those of the Academy. The Academy has no intent to influence legislation by providing such forums.

Preface

In 1998, we organized the first specialized meeting in the field of signal transduction and gene expression in Luxembourg. This type of meeting was originally intended to teach doctoral students attending the cellular and molecular biology program (DEA de Pharmacologie moléculaire) of the University of Nancy I (France).

From 1998 to 2004, this teaching program became a full-size international meeting, and more than 4,300 fundamental and clinical researchers have gathered in Luxembourg to discuss therapeutic applications in the field of signal transduction, transcription, and translation. Our meetings allow new insights into a rapidly moving field. Novel antibodies against receptors, protein kinase inhibitors, and siRNA targeting both signal transduction and gene expression will certainly lead to the therapeutic approaches that will be developed and used in this new century.

This is the first of four volumes forming the proceedings of our Cell Signaling World 2006 meeting. The contributions are divided according to areas of research. Part A focuses on basic research, and the chapters are divided into the following sections: apoptotic cell signaling mechanisms, extracellular matrix interactions, and MAP kinases. Chapters on hypoxia signal transduction, phosphoserine/threonine-binding domains, targeting of polycomb repressive complexes, conserved signaling mechanisms in innate immunity, and signal transduction by stress-activated MAP kinases have been provided by the keynote speakers at the meeting. Other topics included among these reports on recent research are receptor signaling, protein kinase cascades as therapeutic targets, cell death in cancer, inflammation-specific signaling, cell signaling pathways leading to regulated chromatin modifications, and transcriptional control.

This field is moving forward so rapidly that another meeting has been set for January 23–25, 2008, at which fundamental and clinical researchers will gather again in Luxembourg for an eighth meeting, entitled Apoptosis World 2008: From Mechanisms to Applications. The details of this meeting, which will focus on the evolution of the therapeutic applications derived from the field of signal transduction, can be accessed at <http://www.transduction-meeting.lu>.

I would like to thank the editorial department of the *Annals of the New York Academy of Sciences* for its help in publishing these proceedings. I would also like to extend my special gratitude to the City of Luxembourg, as well as to the Fondation de Recherche Cancer et Sang (FRCS) for their generous contributions in support of the conference. I also thank Q8 Petroleum, Novartis, and Alexis for supporting our meetings, and I look forward to the next one in 2008.

—MARC DIEDERICH
Laboratoire de Biologie Moléculaire et Cellulaire du Cancer
Hôpital Kirchberg
L-2540 Luxembourg

Pleiotropic Effects of PI-3′ Kinase/Akt Signaling in Human Hepatoma Cell Proliferation and Drug-Induced Apoptosis

CATHERINE ALEXIA,[a] MARLÈNE BRAS,[a,b] GUILLAUME FALLOT,[a] NATHALIE VADROT,[a] FANNY DANIEL,[a] MALIKA LASFER,[a] HOUDA TAMOUZA,[a] AND ANDRÉ GROYER[a,c]

[a]*INSERM U.481, Faculté de Médecine Xavier Bichat, 75870 Paris Cédex 18, France*

[b]*Institut Pasteur, Département d'Immunologie, 75724 Paris Cédex 15, France*

[c]*INSERM U.773, Faculté de Médecine Xavier Bichat, 75870 Paris Cédex 18, France*

ABSTRACT: IGF-II and type I-IGF receptor (IGF-IR) gene expression is increased in primary liver tumors, and transgenic mice overexpressing IGF-II in the liver develop hepatocellular carcinoma (HCC) spontaneously, suggesting that alterations of IGF-IR signaling *in vivo* may play a role in the auto/paracrine control of hepatocarcinogenesis. We have addressed the contribution of PI-3′K/Akt signaling on the proliferation of HepG$_2$ human hepatoma cells and on their protection against doxorubicin-induced apoptosis. Both basal HepG$_2$ cell DNA replication and that stimulated by IGF-IR signaling were inhibited by the specific PI-3′K inhibitor Ly294002 (Ly). In the former case, PI-3′K signaling overcame cell cycle arrest in G$_1$ via increased cyclin D$_1$ protein and decreased p27^{kip1} gene expression. Doxorubicin treatment induced apoptosis in HepG$_2$ cells and was concomitant with the proteolytic cleavage of Akt-1 and -2. Drug-induced apoptosis was reversed by IGF-I and this effect was (i) dependent on Akt-1 and -2 phosphorylation and (ii) accompanied by the inhibition of initiator caspase-9 activity, suggesting that IGF-IR signaling interferes with mitochondria-dependent apoptosis. Accordingly, Ly enhanced doxorubicin-induced apoptosis and suppressed its reversal by IGF-I. Altogether, the data emphasize the crucial role of PI-3′K/Akt signaling (i) in basal as well as IGF-IR-stimulated HepG$_2$ cell proliferation and (ii) in controlling both doxorubicin-induced apoptosis

Address for correspondence: Dr. André Groyer, INSERM U.773, Faculté de Médecine Xavier Bichat, 16, rue Henri Huchard; BP 416, 75870-Paris Cédex 18, France. Voice: 33-1-44-85-61-39; fax: 33-1-42-28-87-65.

e-mail: groyer@bichat.inserm.fr

(e.g., drug-induced cleavage of Akt) and its reversal by IGF-I (protection against apoptosis parallels the extent of Akt phosphorylation). They suggest that targeting Akt activity or downstream Akt effectors (e.g., GSK3-beta, FOXO transcription factors) may help define novel therapeutic strategies of increased efficacy in the treatment of HCC-bearing patients.

KEYWORDS: human hepatoma cells; proliferation; apoptosis; insulin-like growth factors; PI-3' kinase signaling

INTRODUCTION

Hepatocellular carcinoma (HCC) is one of the most common types of fatal cancer (80–90% of liver cancer case worldwide)[1] and arises from multiple risk factors in humans.[2] HCC has a very poor prognosis, and current treatment efficacy of nonresectable tumors is limited by low response rates, severe toxicity, and high recurrence rates.[3] Thus, novel strategies are needed to improve the survival of patients with HCC. In this connection, understanding the molecular mechanisms that underlie impaired balance in proliferation/apoptosis and lead to hepatocarcinogenesis is needed to design new chemotherapeutic strategies. Such an imbalance may result from the loss of coordinated response to growth factors and cytokines,[2,4,5] among which the insulin-like growth factors (IGF-I and -II) stand as suitable candidates.

Indeed, the IGFs are synthesized and secreted in extracellular fluids by a large variety of cell types,[6] and their interaction with type-I IGF receptor (IGF-IR) plays a pivotal role in the proliferation,[7] in the control of cell cycle progression in G_1,[8] in the regulation of early phases of tumorigenicity,[9,10] in the maintenance of the tumorigenic phenotype,[11-13] and in the prevention of apoptosis.[13-15]

The following evidence supports the putative contribution of the IGF system in primary liver cancer[16] and specifically that of IGF-IR signaling to dysregulated hepatocyte proliferation *in vivo*: (1) *IGF-II* and *IGF-IR* gene expression are significantly increased in human cirrhotic liver and in primary liver cancer,[17-20] and hepatic *IGF-II* gene expression is reactivated during hepatocarcinogenesis in transgenic mice.[21] Overexpression of IGF-II in HCC correlated with increased rates of cell mitotic activity, but not with changes in apoptosis.[19,21] (2) Expression of HBx was often associated with that of IGF-II and IGF-IR in hepatitis B–infected liver and in hepatoma cells[22,23] and increased IGF-IR level in HBx-expressing hepatoma cells enhances the mitogenic effect of the IGFs.[23] (3) IGF-I gene expression is enhanced in non-burdened liver lobes of rats that had developed HCCs in one lobe after inoculation of H_4II rat hepatoma cells,[23] suggesting (i) a paracrine role for IGF-I in HCC-bearing livers and (ii) a putative synergism between IGF-I and IGF-II in hepatocarcinogenesis or in HCCs cell proliferation. (4) IRS-1, an adapter molecule in IGF-IR signaling, is overexpressed in human HCCs versus adjacent, nontumoral liver.[25] Accordingly, targeting IRS-1 overexpression in the liver of transgenic mice yields constitutive activation of MEK/extracellular signal-regulated kinases (Erk-1/2) and of phosphatidylinositol-3′ kinase (PI-3′K) signaling and increases the DNA synthesis.[26]

PI-3′K/Akt signaling plays a major role in mediating antiapoptotic signals, including those under stress conditions.[27–29] Accordingly, overexpression of Akt-1/-2 or of a constitutively active mutant of Akt confers drug resistance,[30] and conversely inactivation of Akt by caspase-dependent cleavage has been reported to occur during Fas-mediated apoptosis.[31] Finally, apoptosis-inducing agents may also activate Akt and such activation delays apoptosis.[32] Several mechanisms have been reported to convey the suppression of apoptosis by Akt: inhibition of proapoptotic Bad, caspase-9, and of Foxo-mediated transcription, stimulation of the transactivation potential of NF-κB.[28,33–35] On the other hand, PI-3′K/Akt signaling has also been shown to control cell cycle progression in G_1 and G_2.[35–39] Altogether, these observations suggest that PI-3′K/Akt signaling may serve a molecular target for anticancer drug development.[40]

Using human HepG$_2$ hepatoma cells as a model system, we report that PI-3′K/Akt signaling (i) plays a key role both in basal cell proliferation and in IGF-I-induced DNA replication and (ii) controls the G_1/S transition in correlation with decreased *p27^{kip1}* gene expression and increased cyclin D$_1$ level. In addition, we have observed that the proteolytic cleavage of Akt-1/-2 may contribute to apoptosis in doxorubicin-treated HepG$_2$ cells and, conversely, that activation of Akt-1/-2 by PI-3′K is instrumental in protecting HepG$_2$ cells against apoptosis.

MATERIALS AND METHODS

Cell Culture and Treatments

Cell Culture

HepG$_2$ cells were grown in DMEM supplemented with 10% heat-inactivated fetal calf serum, 2 mM L-glutamine, 100 U/mL penicillin, and 100 μg/mL streptomycin and were maintained in a humidified atmosphere of 5% CO$_2$ in air.

Thymidine Incorporation and Kinase Activation Studies

HepG$_2$ cells were plated at a density of 4×10^4 cells/cm^2. Twenty-four hours post seeding the culture medium was replaced by serum-free DMEM. After an additional 24-h culture, 14 nM rhIGF-I (R&D Systems, Lille, France) was added in fresh serum-free medium. When appropriate, Ly294002 (Ly, 10 μM) was added 10 min before rhIGF-I.

PARP Cleavage

HepG$_2$ cells were seeded at a density of 4×10^4 cells/cm^2. Seventy-two hours after plating, the cells were treated with 4 μg/mL doxorubicin ± 14 nM rhIGF-I for 24 h (rhIGF-I was added 1 h before the onset of drug treatment). When appropriate, vehicle (dimethylsulfoxide [DMSO]) or Ly (10 μM) or PD98059 (50 μM) was added 10 min before the start of drug ± IGF-I treatment.

Cell Proliferation

Thymidine Incorporation

DNA was labeled with [*methyl*-^{3}H] thymidine (1.5 μCi/mL) for the last 3 h of IGF-I treatment. Cells were then fixed with chilled 10% TCA, washed once with 10% TCA, lysed in 0.1 N NaOH, and incorporated radioactivity was measured by liquid scintillation counting. DNA replication measured in IGF-I ± Ly-treated cells was compared to that of vehicle-treated cells.

Crystal Violet Staining

HepG$_2$ cells were plated in 96-well plates (4×10^4 cells/cm^2). Twenty-four hours post seeding the culture medium was renewed and after an additional 72-h culture in the presence DMSO or Ly (5–40 μM), the cells were fixed in 70% ethanol and stained with 0.08% crystal violet. Cell-associated crystal violet was solubilized with 33% acetic acid and OD was measured at 540 nM.

Cell Cycle Analysis

Adherent cells were trypsinized, collected by centrifugation at 1,500 *g* (3 min; RT), and fixed in ice-cold 70% ethanol for 24 h. Fixed cells were resuspended in phosphate-buffered solution (PBS) containing 180 μg/mL of DNase-free RNase A and 50 μg/mL of propidium iodide for at least 15 min, and DNA fluorescence (λ_{ex} 540 nm/λ_{em} 625 nm) was analyzed using a Coulter Epics XL Flow Cytometer (Beckman-Coulter France, Roissy, France) and the Multicycle Software.

Preparation of Whole-Cell and Nuclear Extracts

Whole-Cell Extracts (WCEs)

Cells were lysed for 30 min in 20 mM Tris-HCl, pH 7.5, 150 mM NaCl, 10% glycerol, 1% Nonidet P-40, 10 mM NaF, 10 mM Na$_3$VO$_4$, 10 mM β-glycerophosphate, 2 μM leupeptin, 2 μM aprotinin, 1 mM PMSF. Lysates were cleared by centrifugation (20,000 g × 15 min) and the supernatants (WCEs) were stored frozen.

Crude Nuclear Extracts (CNEs)

Cells were lysed in 20 mM Hepes, pH 7.9, 20 mM NaF, 1 mM Na_3VO_4, 1 mM β-glycerophosphate, 1mM EDTA, 1 mM EGTA, 1 mM DTT, 0.5 mM PMSF. Nuclei were pelleted by centrifugation at 16,000 g (20 sec); then nuclear proteins were extracted in the same buffer supplemented with 20% glycerol, 420 mM NaCl, 1 μg/mL leupeptin, 1 μg/mL aprotinin. CNEs were cleared by centrifugation (16000 g × 20 min, 4°C) and stored frozen.

Western Blot

WCE or CNE proteins (50–100 μg) were resolved by SDS-PAGE (7–12% acrylamide) and electrotransferred to nitrocellulose membranes (BA83; Whatman Schleicher & Schuell, Dassel, Geramany). The membranes were probed with the appropriate primary antibody at 1/1,000 dilution (polyclonal anti-pAkt, anti-Akt, anti-cyclin D_1 antibodies: Cell Signaling; monoclonal anti-U_1 SnRNP$_{70}$, anti-β-actin antibodies: Santa Cruz; anti-PARP clone C2-10: Pharmigen). Immune complexes were then detected with goat anti-rabbit (1/10,000) or anti-mouse (1/4,000) or rabbit anti-goat (1/5,000) IgG antibodies coupled to horseradish peroxidase and revealed by enhanced chemiluminescence (ECL; Amersham, Buckinghamshire, UK).

Quantitative Analysis

Western blots were analyzed by laser densitometry scanning, and quantification was performed using NIH Image 1.60.

RNAse Protection Assay (RPA)

Riboprobes

Stretches of p27^{kip}, cyclin D_1 and ribosomal protein S_6 cDNAs (see TABLE 1) were amplified by polymerase chain reaction (PCR) and inserted into pBluescript II SK+ (Stratagene). [α^{32}P]-UTP-labeled antisense RNA probes were transcribed from the T7 RNA polymerase promoter of pBluescript II.

TABLE 1. cDNA fragments used for probing gene expression by RPA and characteristics of the antisense RNA probes

Gene name	Accession number	Cloned fragment	Antisense RNA probe (nt)	
			Synthesized	Protected
Cycline D_1	M64349	nt 210–480	349	270
$p27^{kip1}$	X84849	nt 33–277	323	245
S6 ribosomal	NM_001010	nt 553–683	314	130

Cells were lysed in 5 M guanidine isothiocyanate, 0.1 M EDTA, pH 7.4. Aliquots of the lysates were incubated with the appropriate ^{32}P-labeled riboprobe (annealing), and then sequentially digested with RNAses A + T_1 (digestion of nonhybridized transcripts and riboprobes) and proteinase K.[41] The protected RNA sequences were resolved by polyacrylamide gel electrophoresis and their radioactivity was measured with an Instant Imager (Packard). The abundance of each test transcript was standardized relative to that of the S_6 gene.

RESULTS

IGF-I Stimulates DNA Replication of $HepG_2$ Cells: An Effect Dependent on IGF-IR and PI-3'K Signaling

$HepG_2$ cells were always cultured for 24 h in serum-free medium before IGF-I treatment. This allowed a significant increase in the percentage of cells that accumulated in G_0/G_1 : 90.8% ± 2.2 versus 63.8% ± 2.4 in cells growing in serum-supplemented medium (mean ± SEM, $n = 6$)(not shown). Serum-depleted $HepG_2$ cells were treated with a physiological concentration of IGF-I. After a 4-h lag, [^3H]-thymidine incorporation into DNA was gradually increased, an effect, which was maximal after 16 h and onward, at least up to 22 h (1.5 ± 0.09-fold; mean ± SEM, $n = 6$; $P < 0.05$) (FIG. 1A). When IGF-I treatment was performed in the presence of αIR3, an antagonistic anti-IGF-IR antibody, IGF-I-induced DNA replication was reproducibly inhibited by ∼50% (not shown). IGF-I-induced DNA replication was almost completely blunted (89–100% inhibition) when serum-starved $HepG_2$ cells were cotreated with IGF-I and the PI-3'K inhibitor Ly (FIG. 1A).

On the other hand, no phospho-Akt (Ser^{473}, pAkt) band was detected in serum-deprived $HepG_2$ cells, suggesting no constitutive activation of Akt (FIG. 1B). IGF-I treatment yielded a *bona fide* pAkt band (activated Akt), the intensity of which was decreased in the presence of Ly (FIG. 1B). These results establish that IGF-I stimulates DNA replication in $HepG_2$ cells, an effect that is dependent on IGF-IR signaling through the PI-3'K pathway.

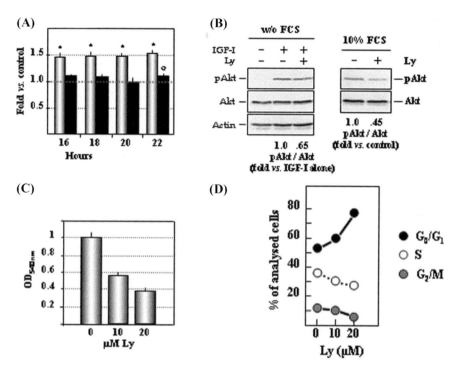

FIGURE 1. IGF-IR signaling stimulates DNA replication in HepG$_2$ human hepatoma cells. (**A**) Monolayers of serum-deprived HepG$_2$ cells were either left untreated or treated with 14 nM IGF-I in the presence (*black bars*) or absence (*shaded bars*) of 10 μM Ly. (**B**) w/o FCS : Monolayers of serum-deprived HepG$_2$ cells were either left untreated (control) or treated with IGF-I alone or cotreated with IGF-I and 10 μM Ly for 3 h. WCEs were prepared and analyzed by Western blot using anti-pAkt or anti-Akt or anti-β-actin antibodies. The pAkt/Akt ratios were calculated and computed relative to that of IGF-I-treated cells. 10% FCS : WCEs were prepared from control and Ly-treated (10 μM) HepG$_2$ cells grown in complete medium (i.e., containing 10% of heat-inactivated serum) and analyzed by Western blot using anti-pAkt or anti-Akt antibodies. The pAkt/Akt ratios were calculated and computed relative to that of control cells. Blots are representative of three independent experiments. (**C,D**) HepG$_2$ cells were seeded at a density of 4×10^4 cells/cm^2, grown in serum-containing medium, and then treated with either vehicle (DMSO) or Ly (5–40 μM) for 72 h. (**C**) Proliferation was monitored by crystal violet staining. Mean ± SD of four independent determinations. (**D**) Cell cycle was monitored by flow cytometry and the percentage of cells in each phase of the cell cycle was quantified using WinCycle and plotted as function of Ly concentration. Mean of three independent determinations; SD was always <10% of the mean value.

PI-3′K Signaling Plays a Role in IGF-I-Independent Proliferation of HepG$_2$: A G$_1$ Phase-Mediated Effect

HepG$_2$ cells were grown asynchronously in a medium containing heat-inactivated FCS (devoid of biologically active IGF-I/-II) and treated with

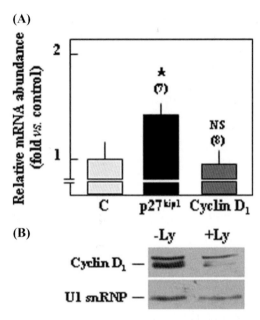

FIGURE 2. Modulation of G_1 phase gene expression by PI-3′K signaling. HepG$_2$ cells were seeded at a density of 4×10^4 cells/cm^2 grown in serum-containing medium, and then treated with either vehicle (DMSO, C) or 10 μM Ly for 72 h. (**A**) Cell lysates were prepared in GuSCN and the relative abundance of p27^{kip1} and cyclin D$_1$ transcripts was determined by RPA. Mean ± SEM ; ($n=8$), number of independent determinations in at least three separate experiments. The relative abundance of transcripts present in Ly- versus vehicle-treated cells was compared using the paired Student's t-test; *, $P <$ 0.05; NS, not significant. (**B**) WCEs were prepared, and then analyzed by Western blot using anti-cyclin D$_1$ or anti-U$_1$ snRNP antibodies (blots are representative of at least three independent experiments).

Ly (5–40 μM). A dose-dependent inhibition of cell proliferation was observed by crystal violet staining (FIG. 1C), growth inhibition being significant at 5 μM Ly and over (maximal inhibition: 75% vs. control). Inhibition of cell proliferation was concomitant with ~55% decrease in Akt phosphorylation, (i.e., attenuation of PI-3′K / Akt signaling) (FIG. 1B). To avoid cell death at high Ly concentrations (>40 μM), 10 or 20 μM were routinely used throughout the study.

A dose-dependent accumulation of HepG$_2$ cells in G_0/G_1 was observed by flow cytometry as Ly concentration was increased (FIG. 1D). It was significant at 20 μM Ly (1.3-fold, $P < 0.002$), as were the 1.3- and 2.1-fold decreases in the percentage of cells in S ($P < 0.04$) and G_2/M ($P < 0.001$) phases, respectively (FIG. 1D). Interestingly, p27^{kip1} transcripts (a cyclin-dependent kinase inhibitor that inhibits cell cycle progression through the G_1 phase) were significantly increased in Ly-treated HepG$_2$ cells, whereas the abundance of the cyclin D_1 protein was drastically reduced (80% inhibition) (FIG. 2A, B). In addition, the lack of alteration of cyclin D_1 transcripts in Ly-treated in $HepG_2$ cells (FIG. 2A) and the decrease in the half-life of the protein (not shown) suggested tight post-translational control of cyclin D_1 expression.

These results suggest that impaired G_1 to S transition through inhibition of PI-3′K signaling in unsynchronized $HepG_2$ cells correlates with increased expression of genes that impede and with decreased expression of genes that favor progression through G_1. Neither progression through the S phase nor the G_2 to M transition was altered in our experimental design.

Drug-Induced Apoptosis and Its Reversion by IGF-I are Dependent on PI-3′K Signaling

We have focused on the role of PI-3′K/Akt signaling in drug-induced apoptosis. Treatment of $HepG_2$ cells with doxorubicin, an acknowledged chemotherapeutic agent, increased apoptosis as monitored by poly [ADP-ribose] polymerase (PARP) cleavage[42] (FIG. 3A), DAPI staining, and flow cytometry (not shown). Whatever the analytical criterion used (PARP cleavage or flow cytometry), drug-induced apoptosis was reversed when the cells were cotreated with IGF-I (84% protection; $P < 0.05$) (FIG. 3A).

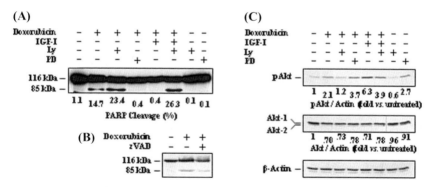

FIGURE 3. PI-3′K signaling in doxorubicin-induced apoptosis of $HepG_2$ cells: modulation of pathway's activation by doxorubicin and evidence for a protective effect of IGF-I. Monolayers of $HepG_2$ cells were either left untreated (control) or treated with 4 μg/mL doxorubicin (4 μg/mL). When required, the cells were co-treated with zVAD (30 μM) or with IGF-I (14 nM), Ly or PD or alone, or in combination. (**A, B**) CNEs were prepared, analyzed by Western blot using anti-PARP antibodies, and the percentage of cleaved PARP was calculated. Similar patterns were obtained in three independent experiments. (**C**) WCEs were prepared and analyzed by Western blot using anti-pAkt or anti-Akt or anti-β-actin antibodies. The pAkt/β-actin and Akt/β-actin ratios were calculated under each experimental condition and compared to that of untreated cells. Each experiment was repeated three times and the blots presented are representative of the overall experimental data.

As shown in FIGURE 3A, Ly by itself was unable to trigger HepG$_2$ cell apoptosis. In contrast, inhibition of PI-3'K signaling in HepG$_2$ cells cotreated with doxorubicin and Ly enhanced drug-induced apoptosis by ~1.6-fold. Moreover, Ly abolished the protection against drug-induced apoptosis elicited by IGF-I. Doxorubicin-induced PARP cleavage suggests activation of effector caspases in doxorubicin-treated cells. This conclusion was supported by inhibition of PARP cleavage by the broad-spectrum caspase inhibitor zVAD (FIG. 3B).

Dual Effects of Doxorubicin on PI-3'K/Akt Signaling in HepG$_2$ Cells

Doxorubicin treatment of HepG$_2$ cells led to a 42.5 ± 7.7 % (mean ± SEM) decrease in the relative level of total Akt when monitored by Western blot using an anti-Akt antibody. This decrease was a hallmark of doxorubicin treatment: it was always observed in the presence of doxorubicin, whatever the apoptotic status of the cell (e.g., presence or absence of IGF-I) and never observed in its absence (middle rows of FIG. 3C) and may correspond to caspase-dependent, proteolytic cleavage of Akt.

On the other hand, exposure of HepG$_2$ cells to doxorubicin alone led to a >2-fold increase in Akt activation as evidenced by enhanced levels of Ser^{473} phosphorylation, and Akt activation was further increased by ~2-fold when the cells were cotreated with doxorubicin and IGF-I (upper rows of FIG. 3C). Interestingly, Akt was also activated by the MEK-1/-2 inhibitor PD (FIG. 3C), in keeping with our previous observation that doxorubicin-induced apoptosis was dependent on the activity of Raf/MEK-1/-2 / Erk-1/Erk-2 signaling in HepG$_2$ (Ref. 16; compare also lanes doxorubicin ± PD in FIG. 3A).

Altogether, these observations point out that Akt may play a dual role in the control of drug-induced apoptosis of HepG$_2$ cells. Cleavage of Akt (e.g., by caspase-3) (see DISCUSSION, below) may be instrumental in mediating doxorubicin-induced apoptosis. In contrast, an increased pAkt/Akt ratio (e.g., after activation of PI-3'K signaling by IGF treatment) triggers protection against apoptosis.

DISCUSSION

Deregulation of proliferation and programmed cell death has been shown to contribute to tumorigenesis. Accordingly, overexpression of growth factors and aberrant growth factor signaling occur in the course of HCC pathogenesis.[43,44] Both lead to baseline activation of signaling cascades[25,26] and interfere with the tumor's response to chemotherapeutic agents.[30,45] In this connection, it should be emphasized that in primary liver cancer, IGF-I is synthesized in peritumoral liver tissue[24] and IGF-II is overexpressed in HCC cells. However, their putative autocrine/paracrine effects in hepatocarcinogenesis remain to be elucidated. Using human hepatoma cells as a model system, this study has investigated the effect of IGF-IR signaling on cell proliferation and on drug-induced apoptosis. It was carried out with IGF-I rather than with IGF-II since (i) paracrine effects of IGF-I may synergize with other growth factors or cytokines in hepatocarcinogenesis or in the dysregulation of HCC cell proliferation and apoptosis[24] and (ii) contrarily to IGF-II, IGF-I does not bind to type A insulin receptor and thus allows investigation of the effects mediated by IGF-IR *per se*.[46]

Cell Proliferation

Our results establish that a physiological concentration of IGF-I increases the rate of DNA replication of serum-starved $HepG_2$ cells, suggesting reentry in the S phase of the cell cycle (~90% of serum-starved cells accumulate in G_1) and are consistent with data obtained with H_4II rat hepatoma cells (Alexia *et al.*, in press). That the stimulation of DNA replication by IGF-I is mediated by IGF-IR signaling is supported by the following: (1) $HepG_2$ cells express a significant amount (unpublished results) of functional IGF-IR[47] this study (our unpublished results). (2) Although $HepG_2$ cells express IR-A and IR-B (unpublished observation) IGF-I does not bind to IR-A with high affinity and 14 nM is not a high enough concentration to allow low-affinity binding of IGF-I to the insulin receptor. (3) IGF-I-induced DNA replication is inhibited, albeit partially, when $HepG_2$ cells are cotreated with IGF-I and αIR3, an anti-IGF-IR monoclonal antibody which impedes IGF-I action (unpublished observation). Accordingly, neutralization of IGF-IR signaling reduces the autocrine effect of IGF-II on $HepG_2$ cell proliferation.[47] Altogether, these data support the conclusion that in primary liver tumors *in vivo*, IGF-IR mediates both the paracrine effect of IGF-I and the auto/paracrine effect of IGF-II on HCC cell proliferation.

We have previously shown that the MEK/Erk-1/Erk-2 signaling cascade is involved in IGF-I-induced $HepG_2$ and rat H_4II hepatoma cell proliferation[16,48] (Alexia *et al.*, in press). We now report that PI-3′K signaling is also instrumental in mediating IGF-I-induced proliferation of serum-starved $HepG_2$ cells, as it is the case for normal rat hepatocytes,[49] H_4II rat hepatoma (Alexia *et al.*, in press) and several cell types of diverse origin (reviewed in Reference 50). In keeping with these results, quantitative studies of Akt phosphorylation demonstrated that in $HepG_2$ cells the PI-3′K/Akt pathway (i) is not constitutively activated and (ii) is activated upon IGF-I treatment of serum-deprived cells. However, since Ly treatment of serum-deprived $HepG_2$ cells inhibits IGF-I-induced Erk-1/Erk-2 phosphorylation/activation (unpublished observation), we cannot exclude that the contribution of PI-3′K/Akt signaling on $HepG_2$ cell proliferation may at least in part rely on its cross-talk with the

MEK/Erk-1/Erk-2 signaling cascade. This contrasts with the dominant role of Akt on IGF-I-induced H_4II rat hepatoma cell proliferation (Alexia et al., in press).

On the other hand, we have shown that the proliferation of $HepG_2$ cells grown in the presence of fetal bovine serum devoid of bioactive IGF-I/-II is also dependent on PI-3'K signaling (Erk-1/Erk-2 activation was unaffected by Ly in cells grown in the presence of serum, not shown). Specifically, PI-3'K activity is needed for unsynchronized $HepG_2$ cells not to accumulate in the G_0/G_1 phase of the cell cycle, indicating that it is instrumental in overstepping the restriction point in late G_1, as already shown in A14, U87MG, lymphoblastoid cells, and NIH3T3 fibroblasts.[29,37,38,51] That PI-3'K signaling is involved in G_1 to S transition is corroborated by PTEN/MMAC1-mediated accumulation in G_1[52] and by the observation that activation of PI-3'K was sufficient to promote entry into S phase of 3Y1 rat embryo fibroblasts.[53] In addition, our data show that PI-3'K signaling overcome cell cycle arrest in G_1 via the control of G_1 phase controlling gene expression: decreased expression of genes that impede ($p27^{kip1}$) and increased expression of genes that favor (cyclin D_1) progression in G_1 phase. We have furthermore observed that posttranslational mechanisms may be involved in the regulation of cyclin D_1 expression (a $\sim$2-fold increase in the protein half-life; unpublished observation), as already reported by Diehl and co-workers in NIH3T3 cells.[36] Altogether, these data support the conclusion that PI-3'K-dependent cell cycle progression of $HepG_2$ human hepatoma cells might rely on at least two phenomena: (1) the regulation of key regulatory kinases lying downstream of Akt (e.g., glycogen synthase-3β and mTor)[39,51]; and (2) the inhibition of the Foxo family of transcription factors.[40] They further emphasize the pivotal role of cyclin D_1 in the control of hepatoma cell proliferation, as is the case for normal mouse hepatocytes in vivo.[54] Of note, this study was carried out with unsynchronized $HepG_2$ cells under experimental conditions (high percentage of cells in G_1) that do not allow pinpointing of putative regulations at the G_2/M transition. Accordingly, it has been reported that attenuation of the PI-3'K pathway is required to allow activation of Foxo transcription factors in G_2 and cell cycle termination in mammalian cells,[55] although others have demonstrated a role for PI-3'K and Akt activities in the G_2/M phase of the cell cycle.[39]

Doxorubicin-Induced Apoptosis of Hepatoma Cells: Underlying Mechanisms

We have observed that in $HepG_2$ cells treated with doxorubicin, the decreased amount in total Akt protein was solely dependent on exposure to doxorubicin: specifically, it was similar when the cells were treated with doxorubicin alone or cotreated with doxorubicin in the presence of either Ly, PD, or IGF-I. This observation allows us to hypothesize that caspase-dependent proteolytic cleavage of Akt may account for the involvement of Akt *per se* in the control of doxorubicin-induced apoptosis. Accordingly, proteolytic Akt cleavage by caspase-3 activity sensitizes RKO colon and MDA–MB breast carcinoma cells to apoptosis-inducing stimuli.[56,57] In contrast, we have observed that doxorubicin treatment triggers Akt activation in HepG$_2$ cells. This phenomenon may be due to oxidative stress-dependent activation of PI-3'K/Akt signaling (e.g., through inactivation of PTEN in cells treated with doxorubicin)[58] and may contribute to decreased doxorubicin sensitivity.

Protection of HepG$_2$ Cells against Doxorubicin-Induced Apoptosis by IGF-I Is Dependent on PI-3'K Signaling

The protection of HCC cells from apoptosis by autocrine secretion of IGF-II (overexpression of IGF-II is observed in 30–40% of human HCCs) has been suggested long ago. It has recently been reported and is mediated via IGF-IR signaling in HepG$_2$ cells.[47] Our results show that protection of HepG$_2$ cells against doxorubicin-induced apoptosis by IGF-I (i) involves inhibition of effector caspases activities (as mirrored by PARP cleavage), and (ii) appears to be solely dependent on the pAkt/Akt ratio.

We have previously shown that drug-induced activation of MEK/Erk-1/-2 signaling plays a critical role in doxorubicin-induced apoptosis of HepG$_2$cells.[16,48] Accordingly, our novel observation that the MEK-1/-2 inhibitor PD98059 enhances Akt phosphorylation in doxorubicin-treated HepG$_2$ cells may help explain its antiapoptotic effect and strongly suggests that the cross-talk between MEK/Erk-1/-2 and PI-3'K/Akt signaling controls drug-induced human hepatoma cell apoptosis and it reversion by IGF-IR signaling. PI-3'K/Akt signaling to mitochondria (e.g., through rapid accumulation of Akt in mitochondria) has recently been reported in various cell types.[59,60] Whether it is involved in the protective effect of IGF-I against drug-induced apoptosis of human hepatoma cells is currently addressed in the laboratory.

ACKNOWLEDGMENTS

We thank G. Schweizer-Groyer for critical reading of the manuscript. C. Alexia was supported by the Ministère de l'Education Nationale et de la Recherche and La Fondation de la Recherche Médicale.

REFERENCES

1. LIANG, T.J. *et al.* 1993. Viral pathogenesis of hepatocellular carcinoma in the United States. Hepatology **18:** 1326–1333.

2. MACDONALD, G.A. 2001. Pathogenesis of hepatocellular carcinoma. Clin. Liver Dis. **5:** 69–85.
3. LEVIN, B. & C. AMOS. 1995. Therapy of unresectable hepatocellular carcinoma. N. Engl. J. Med. **332:** 1294–1296.
4. PITOT, H.C. 1998. Hepatocyte death in hepatocarcinogenesis. Hepatology **28:** 1–5.
5. MICHALOPOULOS, G.K. & M.C. DEFRANCES. 1997. Liver regeneration. Science **276:** 60-66.
6. HUMBEL, R.E. 1990. Insulin-like growth factors I and II. Eur. J. Biochem. **190:** 445–462.
7. SCRIMGEOUR, A.G. et al. 1997. Mitogen-activated protein kinase and phosphatidylinositol 3-kinase pathways are not sufficient for insulin-like growth factor I-induced mitogenesis and tumorigenesis. Endocrinology **138:** 2552–2558.
8. BASERGA, R. 1994. Oncogenes and the strategy of growth factors. Cell **79:** 927–930.
9. KALEKO, M., W.J. RUTTER & A.D. MILLER. 1990. Overexpression of the human insulin-like growth factor I receptor promotes ligand-dependent neoplastic transformation. Mol. Cell. Biol. **10:** 464–473.
10. SELL, C. et al. 1993. Simian virus 40 large tumor antigen is unable to transform mouse embryonic fibroblasts lacking type 1 insulin-like growth factor receptor. Proc. Natl. Acad. Sci. USA **90:** 11217–11221.
11. RESNICOFF, M. et al. 1994. Rat glioblastoma cells expressing an antisense RNA to the insulin-like growth factor-1 (IGF-1) receptor are nontumorigenic and induce regression of wild-type tumors. Cancer Res. **54:** 2218–2222.
12. KHANDWALA, H.M. et al. 2000. The effects of insulin-like growth factors on tumorigenesis and neoplastic growth. Endocr. Rev. **21:** 215–244.
13. LEROITH, D. & C.T. ROBERTS, JR. 2003. The insulin-like growth factor system and cancer. Cancer Lett. **195:** 127–137.
14. RESNICOFF, M. et al. 1995. The insulin-like growth factor I receptor protects tumor cells from apoptosis in vivo. Cancer Res. **55:** 2463–2469.
15. SELL, C., R. BASERGA & R. RUBIN. 1995. Insulin-like growth factor I (IGF-I) and the IGF-I receptor prevent etoposide-induced apoptosis. Cancer Res. **55:** 303–306.
16. ALEXIA, C. et al. 2004. An evaluation of the role of insulin-like growth factors (IGF) and of type-I IGF receptor signaling in hepatocarcinogenesis and in the resistance of hepatocarcinoma cells against drug-induced apoptosis. Biochem. Pharmacol. **68:** 1003–1015.
17. CARO, J.F. et al. 1988. Insulin-like growth factor I binding in hepatocytes from human liver, human hepatoma, and normal, regenerating, and fetal rat liver. J. Clin. Invest. **81:** 976–981.
18. CARIANI, E. et al. 1988. Differential expression of insulin-like growth factor II mRNA in human primary liver cancers, benign liver tumors, and liver cirrhosis. Cancer Res. **48:** 6844–6849.
19. NARDONE, G. et al. 1996. Activation of fetal promoters of insulin-like growth factors II gene in hepatitis C virus-related chronic hepatitis, cirrhosis, and hepatocellular carcinoma. Hepatology **23:** 1304–1312.
20. SCHARF, J.G., G. RAMADORI & F. DOMBROWSKI. 2000. Analysis of the IGF axis in preneoplastic hepatic foci and hepatocellular neoplasms developing after low-number pancreatic islet transplantation into the livers of streptozotocin diabetic rats. Lab. Invest. **80:** 1399–1411.

21. SCHIRMACHER, P. et al. 1992. Reactivation of insulin-like growth factor II during hepatocarcinogenesis in transgenic mice suggests a role in malignant growth. Cancer Res. **52:** 2549–2556.
22. SU, Q. et al. 1994. Expression of insulin-like growth factor II in hepatitis B, cirrhosis and hepatocellular carcinoma: its relationship with hepatitis B virus antigen expression. Hepatology **20:** 788–799.
23. TAO, X. et al. 2000. Hepatitis B virus X protein activates expression of IGF-IR and VEGF in hepatocellular carcinoma cells. Zhonghua Gan Zang Bing Za Zhi **8:** 161–163.
24. PRICE, J.A. et al. 2002. Insulin-like growth factor I is a comitogen for hepatocyte growth factor in a rat model of hepatocellular carcinoma. Hepatology **36:** 1089–1097.
25. NISHIYAMA, M. & J.R. WANDS. 1992. Cloning and increased expression of an insulin receptor substrate-1-like gene in human hepatocellular carcinoma. Biochem. Biophys. Res. Commun. **183:** 280–285.
26. TANAKA, S. et al. 1997. Biological effects of human insulin receptor substrate-1 overexpression in hepatocytes. Hepatology **26:** 598–604.
27. MITSUI, H. et al. 2001. The MEK1-ERK map kinase pathway and the PI 3-kinase-Akt pathway independently mediate anti-apoptotic signals in HepG2 liver cancer cells. Int. J. Cancer **92:** 55–62.
28. KOPS, G.J. et al. 2002. Forkhead transcription factor FOXO3a protects quiescent cells from oxidative stress. Nature **419:** 316–321.
29. BURGERING, B.M. & G.J. KOPS. 2002. Cell cycle and death control: long live Forkheads. Trends Biochem. Sci. **27:** 352–360.
30. NAKASHIO, A. et al. 2000. Prevention of phosphatidylinositol 3′-kinase-Akt survival signaling pathway during topotecan-induced apoptosis. Cancer Res. **60:** 5303–5309.
31. WIDMANN, C., S. GIBSON & G.L. JOHNSON. 1998. Caspase-dependent cleavage of signaling proteins during apoptosis. A turn-off mechanism for anti-apoptotic signals. J. Biol. Chem. **273:** 7141–7147.
32. TANG, D. et al. 2001. Akt is activated in response to an apoptotic signal. J. Biol. Chem. **276:** 30461–30466. Epub 2001 June 8.
33. MADRID, L.V. et al. 2000. Akt suppresses apoptosis by stimulating the transactivation potential of the RelA/p65 subunit of NF-kappaB. Mol. Cell. Biol. **20:** 1626–1638.
34. DIJKERS, P.F. et al. 2002. FKHR-L1 can act as a critical effector of cell death induced by cytokine withdrawal: protein kinase B-enhanced cell survival through maintenance of mitochondrial integrity. J. Cell. Biol. **156:** 531–542.
35. DATTA, S.R. et al. 1997. Akt phosphorylation of BAD couples survival signals to the cell-intrinsic death machinery. Cell **91:** 231–241.
36. DIEHL, J.A. et al. 1998. Glycogen synthase kinase-3beta regulates cyclin D1 proteolysis and subcellular localization. Genes Dev. **12:** 3499–3511.
37. MEDEMA, R.H. et al. 2000. AFX-like Forkhead transcription factors mediate cell-cycle regulation by Ras and PKB through p27kip1. Nature **404:** 782–787.
38. KOPS, G.J. et al. 2002b. Control of cell cycle exit and entry by protein kinase B-regulated forkhead transcription factors. Mol. Cell. Biol. **22:** 2025–2036.
39. SHTIVELMAN, E., J. SUSSMAN & D. STOKOE. 2002. A role for PI 3-kinase and PKB activity in the G2/M phase of the cell cycle. Curr. Biol. **12:** 919–924.

40. MITSIADES, C.S., N. MITSIADES & M. KOUTSILIERIS. 2004. The Akt pathway: molecular targets for anti-cancer drug development. Curr. Cancer Drug Targets **4:** 235–256.
41. KAABACHE, T. *et al.* 1995. Direct solution hybridization of guanidine thiocyanate-solubilized cells for quantitation of mRNAs in hepatocytes. Anal. Biochem. **232:** 225–230.
42. DURIEZ, P.J. & G.M. SHAH. 1997. Cleavage of poly(ADP-ribose) polymerase: a sensitive parameter to cell death. Biochem. Cell. Biol. **75:** 337–349.
43. BUENDIA, M.A. 2000. Genetics of hepatocellular carcinoma. Semin. Cancer Biol. **10:** 185–200.
44. HUYNH, H. *et al.* 2003. Over-expression of the mitogen-activated protein kinase (MAPK) kinase (MEK)-MAPK in hepatocellular carcinoma: its role in tumor progression and apoptosis. BMC Gastroenterol. **3:** 19.
45. HAYAKAWA, J. *et al.* 1999. Inhibition of extracellular signal-regulated protein kinase or c-Jun N-terminal protein kinase cascade, differentially activated by cisplatin, sensitizes human ovarian cancer cell line. J. Biol. Chem. **274:** 31648–31654.
46. FRASCA, F. *et al.* 1999. Insulin receptor isoform A, a newly recognized, high-affinity insulin-like growth factor II receptor in fetal and cancer cells. Mol. Cell. Biol. **19:** 3278–3288.
47. LUND, P. *et al.* 2004. Autocrine inhibition of chemotherapy response in human liver tumor cells by insulin-like growth factor-II. Cancer Lett. **206:** 85–96.
48. ALEXIA, C., M. LASFER & A. GROYER. 2004b. Role of constitutively activated and insulin-like growth ERK1/2 signaling in human hepatoma cell proliferation and apoptosis: evidence for heterogeneity of tumor cell lines. Ann. N. Y. Acad. Sci. **1030:** 219–229.
49. COUTANT, A. *et al.* 2002. PI3K-FRAP/mTOR pathway is critical for hepatocyte proliferation whereas MEK/ERK supports both proliferation and survival. Hepatology **36:** 1079–1088.
50. BRADER, S. & S.A. ECCLES. 2004. Phosphoinositide 3-kinase signaling pathways in tumor progression, invasion and angiogenesis. Tumori **90:** 2–8.
51. BRENNAN, P. *et al.* 2002. Phosphatidylinositol 3-kinase is essential for the proliferation of lymphoblastoid cells. Oncogene **21:** 1263–1271.
52. CHENEY, I.W. *et al.* 1999. Adenovirus-mediated gene transfer of MMAC1/PTEN to glioblastoma cells inhibits S phase entry by the recruitment of p27Kip1 into cyclin E/CDK2 complexes. Cancer Res. **59:** 2318–2323.
53. KLIPPEL, A. *et al.* 1998. Activation of phosphatidylinositol 3-kinase is sufficient for cell cycle entry and promotes cellular changes characteristic of oncogenic transformation. Mol. Cell. Biol. **18:** 5699–5711.
54. NELSEN, C.J. *et al.* 2001. Transient expression of cyclin D1 is sufficient to promote hepatocyte replication and liver growth in vivo. Cancer Res. **61:** 8564–8568.
55. ALVAREZ, B. *et al.* 2001. Forkhead transcription factors contribute to execution of the mitotic programme in mammals. Nature **413:** 744–747.
56. BACHELDER, R.E. *et al.* 1999. p53 inhibits alpha 6 beta 4 integrin survival signaling by promoting the caspase 3-dependent cleavage of AKT/PKB. J. Cell. Biol. **147:** 1063–1072.
57. ROKUDAI, S. *et al.* 2000. Cleavage and inactivation of antiapoptotic Akt/PKB by caspases during apoptosis. J. Cell. Physiol. **182:** 290–296.

58. LESLIE, N.R. *et al*. 2003. Redox regulation of PI 3-kinase signaling via inactivation of PTEN. EMBO J. **22:** 5501–5510.
59. LAI, H.C. *et al*. 2003. Insulin-like growth factor-1 prevents loss of electrochemical gradient in cardiac muscle mitochondria via activation of PI 3 kinase/Akt pathway. Mol. Cell. Endocrinol. **205:** 99–106.
60. BIJUR, G.N. & R.S. JOPE. 2003. Rapid accumulation of Akt in mitochondria following phosphatidylinositol 3-kinase activation. J. Neurochem. **87:** 1427–1435.

Expression of Bcl-xL, Bax, and p53 in Primary Tumors and Lymph Node Metastases in Oral Squamous Cell Carcinoma

MAREK BALTAZIAK,[a] EWA DURAJ,[b] MARIUSZ KODA,[a]
ANDRZEJ WINCEWICZ,[a] MARCIN MUSIATOWICZ,[c]
LUIZA KANCZUGA-KODA,[a] MAGDALENA SZYMANSKA,[a]
TOMASZ LESNIEWICZ,[a] AND BOGUSLAW MUSIATOWICZ[a]

[a]*Department of Clinical and General Pathomorphology, Medical University of Bialystok, Waszyngtona 13, 15-269 Bialystok, Poland*

[b]*Department of Maxillofacial Surgery, Medical University of Bialystok, Waszyngtona 13, 15-269 Bialystok, Poland*

[c]*Department of Pediatric Otorhinolaryngology, Medical University of Bialystok, Waszyngtona 13, 15-269 Bialystok, Poland*

ABSTRACT: Disturbances in expression of apoptosis-associated proteins take part in the development and progression of many human malignancies. The aim of this study was the assessment of correlations among proteins involved in apoptosis—Bcl-xL, Bax, and p53—as well as relationships of these proteins with selected clinicopathological features in oral squamous cell carcinoma. Consequently, we examined by immunohistochemistry, using the avidin–biotin–peroxidase method, Bcl-xL, Bax, and p53 expression in 56 samples of primary oral squamous cell carcinoma and in 22 matched pairs of primary and metastatic tumors. The evaluation of immunostaining of Bcl-xL, Bax, and p53 was analyzed in 10 different tumor fields, and the mean percentage of tumor cells with positive staining was evaluated. The significance of the associations was determined using Spearman correlation analysis and the chi-square test. We found positive Bcl-xL, Bax, and p53 immunostaining in 44.6%, 28.6%, and 58.9% of the studied primary tumors and in 63.6%, 45.5%, and 72.7% of lymph node metastases, respectively. Analysis of associations among studied proteins revealed positive correlation between Bcl-xL and Bax in primary tumors ($P < 0.03$, $r = 0.307$). Statistically significant relationship between p53 expression in primary oral cancers and its expression in lymph node metastases ($P < 0.02$) as well as increased expression of Bcl-xL, Bax, and p53 in

Address for correspondence: Marek Baltaziak, M.D., Department of Pathomorphology, Medical University of Bialystok, Waszyngtona 13, 15-269 Bialystok, Poland. Voice: +48-85-7485945; fax: +48-85-7485944.
e-mail: sulek@zeus.amb.edu.pl

metastatic sites compared with primary tumors could indicate an association of these proteins with oral cancer progression and development of metastases. Moreover, we suppose that knowledge about heterogeneity between primary and metastatic tumor might help to understand mechanisms of oral cancer progression.

KEYWORDS: oral cancer; apoptosis; Bcl-xL; Bax; p53

INTRODUCTION

Dysregulation of apoptosis is one of the most important features of carcinogenesis.[1,2] Despite significant steps forward in understanding the events at the pathway of cancer development, apoptosis protein regulators are still insufficiently described and their role poorly understood in progression of cancer. The Bcl-2 family of proteins comprises promotors (Bax, Bak, Bad, Bcl-xS) and inhibitors (Bcl-2, Bcl-xL, Mcl-1) of the apoptotic pathway as well.[3] These proteins may form homo- or heterodimers and balance between these proteins regulates the cell cycle and apoptosis. The balance between the Bcl-2 family proteins is also regulated among other things by p53 protein.[4]

Nuclear protein p53 controls cell cycle arrest and apoptosis via regulation of the transcription of genes, such as the cyclin-dependent kinase inhibitor p21 (Waf1/Cip1), the protein GADD45, and the apoptosis proteins, such as Bax.[5] Over 50% of human cancers were found with mutations of the p53 tumor-suppressor gene.[6] The role of p53 mutations and protein accumulation in cancer progression has not been well characterized. Loss of the protective effects of its tumor-suppressor function would seem to indicate a more aggressive phenotype and correspond clinically with increased metastasis.[7,8]

The metastatic process is very complex and the metastatic cells have to possess several properties to be able to perform all actions during the neoplastic spread. Lee et al.[9] imputed an important role to hypoxia in solid tumors in increasing the metastatic potential of tumor cells by promoting tissue remodeling, inducing angiogenesis, and reducing apoptosis. Taking into account that metastasizing cells harbor different genotypes, their phenotypic appearance, including invasiveness and chemosensitivity, could strongly differ between primary tumors and metastases. Therefore metastasizing cells are a novel target for anticancer therapy. During the process of metastasis, cells are exposed to various apoptotic stimuli. Thus, in addition to genetic changes that promote proliferation, successful metastatic cells are expected to have a decreased sensitivity to apoptotic stimuli.[10] Differences were observed in apoptosis induction between primary and metastatic oral cancer cells—primary oral cancer revealed greater susceptibility to apoptosis.[11] Moreover, in other cancers there were differences in expression of proteins associated with apoptosis.[12] Gu et al.[13] hypothesized that the development of the most resistant cells during metastasis is favored by antiapoptotic proteins, leading to the acquisition of an adaptive phenotype crucial to drug resistance at the metastatic foci.

Fernandez et al.[14] suggested that antiapoptotic Bcl-xL promotes metastasis of cancer cells through induction resistance to cytokines. p53 may increase the chance of metastasis directly by affecting cell properties that function to reduce proapoptotic stress at individual steps of metastasis as well as by increasing the invasiveness of the cell.[10]

The diagnostic studies are performed in majority cases on samples obtained from the primary tumor and little is known about the biological factors taking part in lymph node metastases and oral cancer progression. An assessment of these biological factors in lymph node metastases could help us to understand the molecular mechanisms of the oral cancer progression. The purpose of the study was to evaluate the expression and correlations between proteins involved in regulation of apoptosis (Bcl-xL, Bax and p53) in the primary oral cancer and in lymph node metastases as well as to estimate the relationships between Bcl-xL, Bax, p53, and the chosen clinicopathological features of the oral cancer.

MATERIALS AND METHODS

We examined a series of 56 cases of oral squamous cell cancer and 22 matched cases of lymph node metastases. The obtained biopsy specimens were fixed in a 10% buffered formalin solution, embedded in paraffin at 56°C, and then cut into 5-μm slices and stained with hematoxylin and eosin (H+E). The studied protein expression was evaluated by immunohistochemical reaction as described previously,[15] using polyclonal antibodies for Bcl-xL and Bax from Santa Cruz Biotechnology, USA and monoclonal antibody for p53 from Dako Cytomation, Denmark. In order to visualize the antigen-antibody reaction, the LSAB technique was applied, using diaminobenzidine as a chromogen (DAB; Dako Cytomation, Denmark). We omitted primary antibodies in negative controls, whereas samples of breast cancer tissue, which showed a strong positive Bcl-xL, Bax, and p53 immunoreactions, were used as positive controls. The evaluation of immunostaining of Bcl-xL, Bax, and p53 was analyzed in 10 different tumor fields using a magnification of 20×. The mean percentage of tumor cells with positive staining was evaluated. The sections were classified as positive if at least 10% of cells expressed the studied antigens. We analyzed Bcl-xL, Bax, and p53 expression in relation to the patients' age, grading and staging, and lymph node metastases. The significance of the associations was determined using Spearman correlation analysis and the chi-square test, employing the SPSS software package v.8.0 for Windows (SPSS Inc., Chicago, IL, USA). The values at $P < 0.05$ were considered statistically significant.

RESULTS

Our study included only oral cancers, classified histopathologically as squamous cell carcinoma: 7 cases in G1, 44 in G2, and 5 in G3. There were 22

TABLE 1. Expression of Bax, Bcl-xl, and p53 in relation to different pathological advancement of primary tumors and clinical data of patients

	Bax			Bcl-xL			p53		
	−	+	P	−	+	P	−	+	P
pN (−)	23	10	0.731	17	16	0.488	13	20	0.759
(+)	17	6		14	9		10	13	
pT (1 + 2)	31	9	0.105	23	17	0.610	16	24	0.796
(3 + 4)	9	7		8	8		7	9	
G 1	5	2	0.328	7	0	<0.04	5	2	0.157
2	30	14		21	23		17	27	
3	5	0		3	2		1	4	
Age ≤ 50	14	2	0.083	8	8	0.610	8	8	0.390
> 50	26	14		23	17		15	25	

tumors in pT1, 18 in pT2, 7 in pT3, and 9 in pT4. In all studied groups, 40 patients were above 50 years old. Twenty-three of 56 (23/56) (41.1%) patients had lymph nodes involved at the time of diagnosis (TABLE 1). We found positive Bcl-xL, Bax, and p53 immunostaining in 44.6%, 28.6%, and 58.9% of studied primary tumors and in 63.6%, 45.5%, and 72.7% of lymph node metastases, respectively. Analysis of relationships between studied proteins and selected clinicopathological features revealed statistically significant association of Bcl-xL with differentiation of primary tumor ($P < 0.04$; TABLE 1).

To assess whether the expression of Bcl-xL, Bax, and p53 undergoes changes during oral cancer progression, these proteins were studied in 22 matched pairs of primary tumors and lymph node metastases (TABLE 2). Bcl-xL, Bax, and p53 expression in primary tumors and metastases to regional lymph nodes was similar in 13/22 (59.1%), 12/22 (54.5%), and 18/22 (81.8%) of cases, respectively. Seven of 13 (53.8%) Bcl-xL-negative primary tumors produced Bcl-xL-positive metastases to regional lymph nodes and 2/9 (22.2%) Bcl-xL-positive primary tumors developed Bcl-xL-negative metastases. Three out of

TABLE 2. Comparison of Bax, Bcl-xL, and p53 expressions between primary tumors (Tu) and lymph node metastases (Mts)

		Bax Tu			Bcl-xL Tu			p53 Tu		
		−	+	P	−	+	P	−	+	P
Bax Mts	(−)	9	3	0.793	6	6	0.304	3	9	0.221
	(+)	7	3		7	3		5	5	
Bcl-xL Mts	(−)	6	2	0.856	6	2	0.245	3	5	0.641
	(+)	10	4		7	7		5	9	
p53 Mts	(−)	5	1	0.493	5	1	0.178	5	1	<0.02
	(+)	11	5		8	8		3	13	

six (50.0%) Bax-positive primary tumors had Bax-negative metastases, while in 7/16 (43.7%) cases we observed Bax-positive metastases despite Bax-negative primary tumors. In only 1 of the 14 cases (7.1%) of p53-positive primary oral cancers we found p53-negative lymph node metastasis, while in 3/5 (60.0%) cases of p53-negative primary tumors, p53-positive metastases were detected. Analysis of relationships between Bcl-xL, Bax, p53 expression in primary tumors and their expression in lymph node metastases revealed statistically significant relationship only in the case of p53 ($P < 0.02$; TABLE 2).

Next we analyzed correlations among studied proteins in primary tumors and in lymph node metastases. Statistically significant positive correlation between Bcl-xL and Bax in primary tumors as well as in lymph node metastases was found ($P < 0.03, r = 0.307; P < 0.03, r = 0.500$). We did not observe significant correlations between Bcl-xL and p53 as well as between Bax and p53.

DISCUSSION

In majority of tumors, several regulatory pathways are altered during multistage tumor progression, among others, the control of proliferation, the balance between cell survival and apoptosis, tumor cell migration, invasion, and metastatic dissemination. In this study we analyzed Bcl-xL, Bax, and p53 expression in primary oral cancers and in matched specimens of primary tumors and lymph node metastases. We tried to answer whether the expression of Bcl-xL, Bax, and p53 underwent changes during metastasis from primary site to lymph nodes. We observed that Bcl-xL, Bax, and p53 were expressed in primary tumors as well as lymph node metastases, but the expression in metastatic tumors was more frequent (44.6% vs. 63.6%, 28.6% vs. 45.5%, and 58.9% vs. 72.7%, respectively). Interestingly, Bcl-xL-, Bax-, and p53-positive and Bcl-xL-, Bax-, and p53-negative primary tumors produced both positive and negative metastases. The development of Bcl-xL-, Bax-, and p53-positive metastases from negative tumors is not unusual because most of negative tumors contain a small percentage of positive cells that could give rise to a metastatic population.

It is likely that metastasis can fail at many stages of the process due to cell destruction by the immune system, necrosis, or perhaps most importantly, apoptosis.[10] Protection from apoptosis is especially important in the metastatic process, because conditions associated with this process, such as loss of intercellular contact, loss of essential growth factors and hypoxia, normally induce apoptosis. The immune system, unsuitable tissue environment, and hypoxia can also induce apoptosis and destroy metastatic malignant cells. We suppose that the increase in Bcl-xL, Bax, and p53 expression in lymph node metastases observed in this study could be a result, among others, of immune system influence.

p53 mutations generate production of indolent proteins without tumor growth–suppressing activity and accumulation of inactive p53 mutants inside cancer cells.[16] Expression of p53 appears early in the malignant transformation in the oral cavity. The presence of the protein was detected in a suprabasal layer of precancerous planoepithelial lesions.[4,17] On the other hand, Shahnavaz et al.[18] found that p53 mutations occurred at a relatively late stage of oral carcinogenesis and are involved in acquisition of metastatic potential. p53-positive metastases were noted in 3/5 cases of p53-negative primary tumors of this study. Ito et al.[19] revealed poor prognosis of patients, which was related to positive expression of p53 in primary oral and oropharyngeal carcinoma. On the other hand, Liu et al.[20] did not observe any correlation between p53 expression and patients' survival in buccal squamous cell carcinoma. Mutated p53 gene may facilitate metastasis. Adenoviral delivery of wild-type p53 gene in squamous cell carcinomas of the head and neck resulted in inhibition of production of matrix metalloproteinases, which plays an important role in cancer invasion and metastasis.[21] p53 has also been suggested to increase metastatic potential and enhance metastasis by promotion of cell survival within the circulation as well as by promotion of angiogenesis in colon cancer progression.[22]

Fernandez et al.[14] suggested that Bcl-xL expression in cancer cells could increase metastatic activity. This function could expand into inducing resistance to apoptosis against cytokines, increasing cell survival in circulation, and enhancing anchorage-independent growth. In the study of Gu et al.[13] Bcl-xL enhanced cell genetic instability, which could be a molecular mechanism by which Bcl-xL evolved in the selection process of cancer progression. By means of that, different genetic changes among metastases result.

Positive expression of Bax was inversely associated with advanced tumor stage, lymph node metastasis, clinical stage, and poor prognosis in oral and oropharyngeal carcinoma.[19] Xie et al.[23] also found in oral squamous cell carcinoma of the tongue correlation between low Bax expression and poor prognosis. On the other hand, Staibano et al.[24] revealed that high expression of Bax in oral cancer appeared to correlate with a worse prognosis. Chemotherapy has an important role in multidisciplinary treatment of head and neck cancer. Antitumor effects of chemotherapeutic agents, mainly aimed at metastases, are commonly associated with the induction of apoptosis.[25] Bax may contribute to the antitumor effects of chemotherapy in oral squamous cell carcinoma, while mutated p53 and Bcl-xL overexpression may function as the key components conferring multiple drug-resistance in squamous cell carcinomas.[26,27] In lymph node metastases we found increased Bax expression compared with primary tumors, but its significance, especially for prognosis and chemotherapy, should be studied.

In conclusion, Bcl-xL, Bax, and p53 might be associated with oral cancer progression and metastagenicity, as expression of these proteins increased in metastatic sites in comparison with primary tumors. That is why, to understand the process of oral cancer progression and metastagenicity, it is imperative to identify heterogeneity between primary and metastatic tumors.

REFERENCES

1. SOTIRIOU, C., P. LOTHAIRE, D. DEQUANTER, et al. 2004. Molecular profiling of head and neck tumors. Curr. Opin. Oncol. **16:** 211–214.
2. SULKOWSKA, M., W. FAMULSKI, S. SULKOWSKI, et al. 2003. Correlation between Bcl-2 protein expression and some clinicopathological features of oral squamous cell carcinoma. Pol. J. Pathol. **54:** 49–52.
3. REED, J.C. 1998. Bcl-2 family proteins. Oncogene **17:** 3225–3236.
4. SULKOWSKA, M., W. FAMULSKI, L. CHYCZEWSKI & S. SULKOWSKI. 2001. Evaluation of p53 and bcl-2 oncoprotein expression in precancerous lesions of the oral cavity. Neoplasma **48:** 94–98.
5. DE LAURENZI, V. & G. MELINO. 2000. Evolution of functions within the p53/p63/p73 family. Ann. N. Y. Acad. Sci. **926:** 90–100.
6. THOR, A.D., D.H. MOORE, S.M. EDGERTON, et al. 1992. Accumulation of p53 tumor suppressor gene protein: an independent marker of prognosis in breast cancers. J. Natl. Cancer Inst. **84:** 845–855.
7. SILVESTRINI, R., M.G. DAIDONE, E. BENINI, et al. 1996. Validation of p53 accumulation as a predictor of distant metastasis at 10 years of follow-up in 1400 node-negative breast cancers. Clin. Cancer Res. **2:** 2007–2013.
8. KASTRINAKIS, W.V., N. RAMCHURREN, K.M. RIEGER, et al. 1995. Increased incidence of p53 mutations is associated with hepatic metastasis in colorectal neoplastic progression. Oncogene **11:** 647–652.
9. LE, Q.T., N.C. DENKO, & A.J. GIACCIA. 2004. Hypoxic gene expression and metastasis. Cancer Metastasis. Rev. **23:** 293–310.
10. TOWNSON, J.L., G.N. NAUMOV & A.F. CHAMBERS. 2003. The role of apoptosis in tumor progression and metastasis. Curr. Mol. Med. **3:** 631–642.
11. VIGNESWARAN, N., J. WU, N. NAGARAJ, et al. 2005. Differential susceptibility of metastatic and primary oral cancer cells to TRAIL-induced apoptosis. Int. J. Oncol. **26:** 103–112.
12. VILLAR, E., M. REDONDO, I. RODRIGO, et al. 2001. Bcl-2 expression and apoptosis in primary and metastatic breast carcinomas. Tumour Biol. **22:** 137–145.
13. GU, B., L. ESPANA, O. MENDEZ, et al. 2004. Organ-selective chemoresistance in metastasis from human breast cancer cells: inhibition of apoptosis, genetic variability and microenvironment at the metastatic focus. Carcinogenesis **25:** 2293–2301.
14. FERNANDEZ, Y., L. ESPANA, S. MANAS, et al. 2000. Bcl-xL promotes metastasis of breast cancer cells by induction of cytokines resistance. Cell. Death Differ. **7:** 350–359.
15. KODA, M., J. RESZEC, M. SULKOWSKA, et al. 2004. Expression of the insulin-like growth factor-I receptor and proapoptotic Bax and Bak proteins in human colorectal cancer. Ann. N. Y. Acad. Sci. **1030:** 377–383.
16. SULKOWSKA, M., W. FAMULSKI, A. STASIAK-BARMUTA, et al. 2001. PCNA and p53 expression in relation to clinicopathological features of oral papilloma. Folia Histochem. Cytobiol. **39**(Suppl 2): 193–194.

17. Nylander, K., E. Dabelsteen & P.A. Hall. 2000. The p53 molecule and its prognostic role in squamous cell carcinoma of the head and neck. J. Oral Pathol. Med. **29:** 413–425.
18. Shahnavaz, S.A., J.A. Regezi, G. Bradley, et al. 2000. p53 gene mutations in sequential oral epithelial dysplasias and squamous cell carcinomas. J. Pathol. **190:** 417–422.
19. Ito, T., S. Fujieda, H. Tsuzuki, et al. 1999. Decreased expression of Bax is correlated with poor prognosis in oral and oropharyngeal carcinoma. Cancer Lett. **140:** 81–91.
20. Liu, C.J., K.W. Chang, S.Y. Chao, et al. 2004. The molecular markers for prognostic evaluation of areca-associated buccal squamous cell carcinoma. J. Oral Pathol. Med. **33:** 327–334.
21. Ala-aho, R., R. Grenman, P. Seth & V.M. Kahari. 2002. Adenoviral delivery of p53 gene suppresses expression of collagenase-3 (MMP-13) in squamous carcinoma cells. Oncogene **21:** 1187–1195.
22. Faviana, P., L. Boldrini, R. Spisni, et al. 2002. Neoangiogenesis in colon cancer: correlation between vascular density, vascular endothelial growth factor (VEGF) and p53 protein expression. Oncol. Rep. **9:** 617–620.
23. Xie, X., O.P. Clausen, P. De Angelis & M. Boysen. 1999. The prognostic value of spontaneous apoptosis, Bax, Bcl-2, and p53 in oral squamous cell carcinoma of the tongue. Cancer **86:** 913–920.
24. Staibano, S., M.D. Mignogna, L. Lo Muzio, et al. 1998. Overexpression of cyclin-D1, bcl-2, and bax proteins, proliferating cell nuclear antigen (PCNA), and DNA-ploidy in squamous cell carcinoma of the oral cavity. Hum. Pathol. **29:** 1189–1194.
25. Kerr, J.F., C.M. Winterford & B.V. Harmon. 1994. Apoptosis. Its significance in cancer and cancer therapy. Cancer **73:** 2013–2026.
26. Takemura, K., M. Noguchi, K. Ogi, et al. 2005. Enhanced Bax in oral SCC in relation to antitumor effects of chemotherapy. J. Oral Pathol. Med. **34:** 93–99.
27. Noutomi, T., H. Chiba, M. Itoh, et al. 2002. Bcl-x(L) confers multi-drug resistance in several squamous cell carcinoma cell lines. Oral Oncol. **38:** 41–48.

Reactive Structural Dynamics of Synaptic Mitochondria in Ischemic Delayed Neuronal Death

CARLO BERTONI-FREDDARI,[a] PATRIZIA FATTORETTI,[a] TIZIANA CASOLI,[a] GIUSEPPINA DI STEFANO,[a] MORENO SOLAZZI,[a] ELISA PERNA,[b] AND CLARA DE ANGELIS[b]

[a]*Neurobiology of Aging Laboratory, INRCA Research Department, Via Birarelli 8, 60121 Ancona, Italy*

[b]*Morphometry Laboratory, Sigma-Tau, 00040 Pomezia, Italy*

ABSTRACT: The effect of transient global ischemia on the ultrastructural features of synaptic mitochondria at the distal dendrites of CA1 hippocampal neurons was investigated in 3-month-old rats. Sham surgery was performed on age-matched controls. The number of mitochondria/μm^3 of neurophils (Nv: numeric density), the mitochondrial average size (average volume: V), and longer diameter (Fmax) as well as the overall fraction of neurophils occupied by mitochondria (volume density: Vv) were measured by computer-assisted morphometry. In ischemic rats, a 10% nonsignificant decrease of Nv was found, V increased nonsignificantly by 11%, and Fmax increased nonsignificantly by 5% versus controls. As a final outcome of these balanced changes, Vv remained unchanged between the two experimental groups investigated. In ischemic animals, the percentage distribution of V showed that the population of CA1 synaptic mitochondria was composed by an increased fraction of oversized organelles, while the Fmax distribution revealed that this enlargement was due to an increased percentage of elongated organelles. Thus, the observed increase in size should not be considered as a swelling phenomenon; on the contrary, it may represent a physiological and well-documented step in mitochondrial biogenesis. The above parameters are currently supposed to provide information on the adaptive structural reorganization of mitochondrial morphology under different environmental stimulations. Conceivably, these findings document a positive reactive response to ischemia of the mitochondrial structural dynamics at CA1 synaptic terminals and suggest consideration of these organelles as reliable targets in the development of neuroprotective therapeutic interventions to treat vascular brain diseases, for example, stroke.

Address for correspondence: C. Bertoni-Freddari, Neurobiology of Aging Laboratory, INRCA Research Department Via Birarelli 8, 60121 Ancona, Italy. Voice: +32-936-04196; fax: +39-071-206791.
e-mail: c.bertoni@inrca.it

KEYWORDS: synaptic mitochondria; morphometry; CA1 neurons; experimental ischemia-delayed cell death; mitochondrial structural dynamics

INTRODUCTION

Ischemia is a condition of reduced or blocked blood flow to a localized area of the body resulting in deprivation of vital nutritional substances, inadequate oxygen supply, accumulation of potentially injurious metabolic products, and loss of function of the affected tissues. The term *ischemia* is generally used to denote interference with arterial blood flow to a circumscribed zone. The brain is characterized by a high metabolic rate associated with low oxygen stores and small reserves of high-energy phosphates or carbohydrates. In turn, these peculiar features account for its high sensitivity to ischemia. Although an ischemic damage of the brain is the final outcome of a combination of several pathophysiological events, a very critical factor is the development of severe hypoperfusion and hypoxia, leading to energy depletion, membrane damage, and cell death. The vulnerability of the central nervous system (CNS) to ischemic damage varies from zone to zone and is different even within select neuronal populations. The hippocampus is particularly sensitive to ischemia, but while the CA1 area suffers from ischemic damage, the adjacent CA3 sector and the dentate gyrus are significantly less vulnerable.[1] Namely, it is well documented that a brief period of transient global ischemia is responsible for cell death in CA1 pyramidal neurons occurring days after reperfusion. This specific vulnerability of CA1 neurons is referred to as delayed neuronal death and has been shown to begin 3 to 4 days after the initial ischemic insult.[2]

In this article we report the results of an investigation conducted on the structural dynamics of synaptic mitochondria at the distal dendrites of CA1 pyramidal neurons (stratum lacunosum moleculare) from young rats that had undergone a 20-min global transient ischemia.[3] The rationale that led us to undertake this study was that a better understanding of the changes occurring in ischemic conditions may suggest prompt interventions on well-identified targets. Conceivably, prompt and adequate treatments of this pathological condition, particularly in old age when the prevalence of brain diseases depending on vascular changes increases, may improve the prognosis of ischemic injury and the patients' quality of life.

MATERIALS AND METHODS

Experimental Ischemia

The four-vessel occlusion model of transient forebrain ischemia was carried out on female Wistar–Kyoto rats of 3 months of age according to the method of Pulsinelli and Brierley,[3] as described also in details in our previous paper.[4] Sham-operated animals of the same age served as controls. Briefly, the rat's head was placed in stereotactic ear bars in the proper position for surgical operation, each common carotid artery (CCA) was atraumatically clasped and exteriorized through a ventral midline neck incision. A dorsal incision enabled us to identify the alar foramina of the first cervical vertebrae and to localize the two vertebral arteries. A small monopolar electrocautery needle (diameter: 0.5 mm) was then inserted through each foramen to electrocoagulate permanently the vertebral arteries. Twenty-four hours later, a 20-min bilateral occlusion of the common carotid arteries was carried out; then the clasps were removed to permit reperfusion. A prompt unresponsiveness and the loss of the righting reflex were considered signs of a successful ischemic procedure. Restoration of carotid blood flow resulted in the presence of the corneal reflex as well as the recovery of walking and climbing functions.

Mitochondrial Morphometry

Seven days after the ischemic procedure reported above, the rats were anesthetized and perfused for 10 min through the left ventricle with the following fixation solution: 2% glutaraldehyde, 4% formalin, 0.005 M CaCl2 in 0.1 M cacodylate buffer. The left hippocampus was excised and histopathological analysis was carried out in the hippocampal CA1 pyramidal cell bodies to confirm that the ischemic procedure was successfully performed[5] on the basis of the number of neurons damaged. The hippocampi were processed for conventional electron microscopy. Ultrathin serial sections were prepared from each sample. By adopting the recently introduced disector counting and sampling method,[6] we counted the number of mitochondria/μm^3 of tissue (numeric density: Nv). In detail, mitochondrial profiles were counted in a known volume of tissue by analyzing pairs of sections: a reference section and a serial section, the "look-up" section (which is not necessarily the adjacent serial section). The profiles of the structures of interest (in our case the mitochondria) are located in sample areas of the reference section and are then checked in the same areas of the look-up section. The mitochondria present in the reference section, but not in the look-up section, are counted. The formula applied is:

$$Nv = \Sigma Q^- / \Sigma V_{(dis)}$$

where Q^- is the number of mitochondria present in the reference section, but not in the look-up section, and $V_{(dis)}$ is the volume of the disector obtained by $a_{dis} \cdot h$, where a_{dis} is the area of the section analyzed and h is the distance between the reference and the look-up planes of the disector. A computer-assisted image analyzer (Kontron KS300) was used in semiautomatic mode to calculate the mitochondrial volume fraction, that is, the volume occupied by mitochondria/μm^3 of tissue (volume density: Vv) and the average mitochondrial volume (V) by applying conventional morphometric formulas as reported in our previous paper.[7] Further, the longer diameter (Fmax) of each organelle counted by the disector was automatically calculated by the computer program of the image analyzer. Statistical comparisons were performed by the Student's t-test and the data were considered significant for P-values < 0.05.

RESULTS

At variance with classical morphometric methods, the disector sampling and counting method[6] enables the unbiased quantitative estimation of structures within a well-defined volume of tissue. This procedure is particularly suitable to perform measurements of the tissue numeric density of a given object (in this study, the mitochondria) without taking into account an anatomical reference structure, for example, an ordered row of cells.[8] Thus, the present data must be considered as referring to a selected volume of neurophil in the stratum lacunosum moleculare of CA1 neurons.

In ischemic animals versus controls, we found a 10% nonsignificant decrease of Nv, an 11% nonsignificant increase of V, a 5% nonsignificant increase of Fmax, and no significant difference between the two groups investigated as regards Vv (FIG. 1). In ischemic animals, a percentage distribution of V (FIG. 2) showed that the population of synaptic mitochondria is composed of a higher fraction of enlarged organelles, while the percentage distribution of Fmax revealed an increased percentage of elongated mitochondria (FIG. 2).

DISCUSSION

Hippocampus and CA1 Neurons as Anatomical and Cellular Models of Ischemia

The hippocampal formation constitutes an anatomical model particularly suitable to perform studies on the effects of an ischemic injury since its vascular anatomy is unlike that of the rest of the brain. That is, the hippocampal arteries are arranged in a rake-like fashion with branches arising at right angles to the main trunks penetrating the hippocampus as end arteries. Because of this peculiar anatomical feature, while a drop in blood pressure is distributed equally in blood vessels branching dichotomously, in the rake-like pattern of the hippocampus, the blood pressure drops critically before it occurs in other zones of the brain. As a consequence, the hippocampus is affected by ischemic damage earlier and more severely than elsewhere in the brain. In addition to this peculiar organization of the vascular bed of the hippocampus, the remarkable correlation between the arrangement of cell and fiber layers and the vascular architecture constitutes another critical determinant of the vulnerability of the hippocampal formation to ischemia. Namely, it has been clearly documented

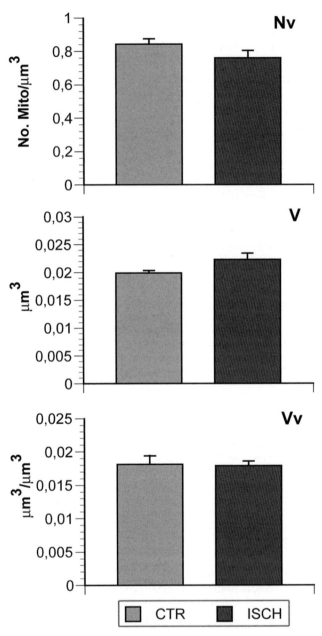

FIGURE 1. Nv, V, and Vv of synaptic mitochondria at distal dendrites of hippocampal CA1 pyramidal cells. No significant difference was found comparing control, sham-operated 3-month-old rats (CTR) versus age-matched animals undergoing transient global ischemia (ISCH) according to Pulsinelli's paradigm.[3,4]

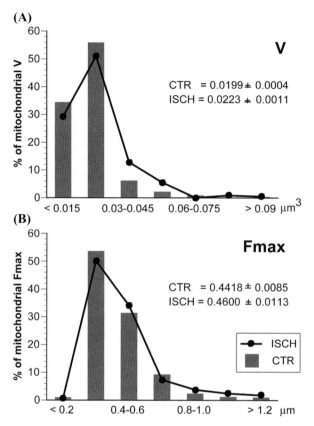

FIGURE 2. Percentage distribution of the mitochondrial V and Fmax in 3-month-old control (CTR) rats and age-matched ischemic animals (ISCH). Mean average values of V and Fmax are reported on right. Taken together, both distributions show that the increased percentage of enlarged mitochondria in ischemic rats is due to an increased fraction of elongated mitochondria.

that in elderly individuals and in patients affected by Alzheimer's disease, the severity of granulovacuolar degenerative phenomena and the density of neurofibrillary tangles is similar in distribution to the neural damage caused by oxygen and blood insufficiency.[9] Thus, changes in microcirculation may be responsible for acute and chronic lesions in the hippocampus. As mentioned above, within the hippocampus, the vulnerability to ischemia of CA1 pyramidal neurons is documented to be remarkably higher than that of other adjacent neurons. Specifically, CA1 neurons are characterized by a delayed cell death phenomenon supposed to be triggered by the deleterious events occurring during the early steps of an ischemic injury.[10] A peculiar feature of CA1 neurons is that they use glutamate as neurotransmitter,[11] and thus their vulnerability may rely on the toxic action of this excitatory molecule when its homeostatic control is impaired. Moreover, the capillary density in this zone of the hippocampus has been reported to be lower than in other hippocampal regions and this may significantly contribute to the drop in blood pressure occurring during an ischemic injury.[12] Although several hypotheses have been proposed, the clear-cut mechanisms(s) or causal event(s) responsible for the selective vulnerability of CA1 pyramidal neurons to ischemia are still unknown. Many potential bases for this phenomenon have been identified and all together participate either as initiating and activating events or (at later stages) as perpetrating circumstances in the induction of long-term functional damage due to ischemia. The many detrimental changes affecting more severely CA1 neurons than other hippocampal cells have been documented to occur at the various stages of progression of the ischemic injury and include: depression of protein synthesis, slow recovery of pH, selective vulnerability to superoxide radical, maintained elevation of free fatty acids, and others.[13] In consideration of these well-documented features, both the hippocampus and the CA1 neurons appear to represent anatomical and cellular models to be adopted to obtain reliable data in experimental brain injury.

Reactive Morphology of Synaptic Mitochondria in Ischemic Delayed Neuron Death

Although no significant differences have been found between sham-operated and ischemic rats, the results of this study show clear trends that can be summarized as follows. At the distal dendrites of CA1 neurons of ischemic rats, a decrease in the number of synaptic mitochondria (Nv) and an increase of the average volume of the organelles (V) were found. The final outcome of these balanced changes was that the volume fraction of synaptic mitochondria (Vv) was the same in both the groups investigated. These findings are in agreement with previous data obtained in our laboratory from the same ischemic and sham-operated rats and showing that, in the same CA1 subfield, the numeric density of synapses was decreased at a nonsignificant extent, but the persisting junctions were enlarged in size (average area): this resulted in a substantial preservation of the overall area of the synaptic contact zones per cubic micron of tissue (surface density).[4] It must be stressed that our present and previous study[4] was carried out 7 days after the ischemic/reperfusion intervention; thus our data specifically provide information on recovery strategies carried out by the CNS plastic potential following ischemic damage.

Reasonable interpretations of these findings must take into account the temporal sequence and severity of the many deteriorative events occurring during ischemia. It is well documented that in severe ischemic injury no aspect of cellular metabolism is spared. With specific reference to nerve cells, their functions and metabolism are critically dependent on sufficient oxygen supply, and any reduction in blood flow causes a proportional decrease in oxygen consumption. In these conditions, the ability of ionic pumps to maintain cellular integrity is impaired and anaerobic glycolysis is used to produce ATP. This leads to extracellular and intracellular acidosis, membrane depolarization and damage, uncoupling of the oxidative phosphorylation process, and energy depletion. A significant swelling of mitochondria is reported to occur during the time course (seconds to minutes)[14] of the above sequence of steps, probably as a consequence of the early imbalance of Ca^+ ions homeostasis triggered by a large increase in glutamate-mediated activation of N-methyl-D-aspartate (NMDA) receptors. Thus, just from the beginning of the ischemic injury, mitochondrial functions are seriously impaired since swelling is followed by rupture of the outer membrane and permeability transition pore (PTP) activation.[15] In addition to these alterations, neuronal mitochondria are reported to show dramatic signs of injury following exposure to ischemia–reperfusion conditions and these include increased matrix density, deposits of electron-dense material followed by disintegration by 24 h.[16] Considering the very dynamic condition of mitochondrial ultrastructure, the results of the present investigation obtained 7 days after ischemia document a remarkable plastic response of synaptic organelles in ischemic animals. Namely, although not significant, the increase in V appears to represent a compensating reaction able to counteract efficiently the decrease in Nv, as confirmed by the unchanged value of Vv between sham-operated and ischemic rats. Moreover, the higher percentage of oversized organelles found in ischemic animals is constituted by elongated mitochondria (FIG. 2 B) and this suggests that their enlarged size must not be considered as due to a swelling phenomenon, but rather as an adaptive structural reorganization of the morphology of synaptic mitochondria. Although from the present data it cannot be inferred whether these ultrastructural changes result in a functional improvement of the mitochondrial metabolic competence, it can be reasonably hypothesized that an increase in average size (V) constitutes an extension of the organelle's area potentially involved in cellular respiration, while an increase in elongation is a well-documented step in the process of mitochondria biogenesis. Accordingly, our findings can be reliably interpreted as a positive reaction of the mitochondrial ultrastructural dynamics to an ischemic injury. Although this study was conducted on young rats, it is currently reported that the plasticity of mitochondrial ultrastructure is partially maintained and can be functionally modulated also in old age,[7,17,18] when the prevalence of brain diseases depending on vascular changes increases and ischemia is reported to be responsible for atrophy, degeneration, and gliosis.[13] Thus, the present findings suggest that mitochondria may represent reliable targets for the development of neuroprotective pharmacological interventions to be associated with currently adopted thrombolytic therapies to treat patients affected by stroke.

REFERENCES

1. SCHMIDT-KASTNER, R. & T.F. FREUND. 1991. Selective vulnerability of the hippocampus in brain ischemia. Neuroscience **40:** 599–636.
2. KIRINO, T. 1982. Delayed neuronal death in the gerbil hippocampus following ischemia. Brain Res. **239:** 57–69.
3. PULSINELLI, W.A. & J.B. BRIERLEY. 1979. A new model of bilateral hemispheric ischemia in the unanesthetized rat. Stroke **10:** 267–272.
4. FATTORETTI, P., C. BERTONI-FREDDARI, U. CASELLI, *et al*. 2002. Adaptive capacities of the synaptic contact zones in hypertensive and ischemic young rats. Ann. N.Y. Acad. Sci. **977:** 109–114.
5. HASKER, P.D., J.R. BARANOWSKY, W.A. PULSINELLI & B.T. VOLPE. 1987. Retention of reference memory following ischemic hippocampal damage. Physiol. Behav. **39:** 783–786.
6. WEST, M.J. 1999. Stereological methods for estimating the total number of neurons and synapses: issues of precision and bias. Trends Neurol. Sci. **22:** 51–61.
7. BERTONI-FREDDARI, C., P. FATTORETTI, T. CASOLI, *et al*. 1993. Morphological plasticity of synaptic mitochondria during aging. Brain Res. **628:** 193–200.
8. BERTONI-FREDDARI, C., P. FATTORETTI, R. RICCIUTI, *et al*. 2002. Morphometry of E-PTA stained synapses at the periphery of pathological lesions. Micron **33:** 447–451.
9. BALL, M.J. 1978. Topographic distribution of neurofibrillary tangles and granulovacuolar degeneration of hippocampal cortex of ageing and demented patients: a quantitative study. Acta Neuropathol. **42:** 73–80.
10. ABE, K., M. AOKI, J. KAWAGOE, *et al*. 1995. Ischemic delayed neuronal death. A mitochondrial hypothesis. Stroke **26:** 1478–1489.
11. ZILLES, K. 1988. Receptor autoradiography in the hippocampus of man and rat. Adv. Anat. Embr. Cell Biol. **111:** 61–80.
12. COYLE, P. 1987. Spatial features of rat hippocampal vascular system. Exp. Neurol. **58:** 549–561.
13. LIPTON, J. 1999. Ischemic cell death in brain neurons. Physiol. Rev. **79:** 1431–1568.
14. FLYNN, C.J., A.A. FAROOQUI & L.A. HORROCKS. 1989. Ischemia and hypoxia. *In* Basic Neurochemistry: Molecular, Cellular and Medical Aspects. G.J. Siegel, B. Agranoff, R.W. Albers & P. Molinoff, Eds.: 783–795. Raven Press. New York.
15. BEALS, M.F. 2000. Energetics in the pathogenesis of neurodegenerative diseases. Trends Neurosci. **23:** 298–304.
16. SOLENSKY, N., C.G. DIPIERRO, P.A. TRIMMER, *et al*. 2002. Ultrastructural changes of neuronal mitochondria after transient and permanent cerebral ischemia. Stroke **33:** 816–824.
17. WALTER, P.B., K.B. BECKMAN & B.N. AMES. 1999. The role of iron and mitochondria in aging. *In* Understanding the Process of Aging: The Role of Mitochondria, Free Radicals and Antioxidants. E. Cadenas & L. Packer, Eds.: 203–227. M. Dekker. New York.
18. BERTONI-FREDDARI, C., P. FATTORETTI, B. GIORGETTI, *et al*. 2004. Role of mitochondrial deterioration in physiological and pathological brain aging. Gerontology **50:** 187–192.

Selective Resistance of Tetraploid Cancer Cells against DNA Damage-Induced Apoptosis

MARIA CASTEDO,[a] ARNAUD COQUELLE,[a] ILIO VITALE,[a] SONIA VIVET,[a] SHAHUL MOUHAMAD,[a] SOPHIE VIAUD,[b] LAURENCE ZITVOGEL,[b] AND GUIDO KROEMER[a]

[a]*Centre National de la Recherche Scientifique, UMR8125, Camille-Desmoulins, F-94805 Villejuif, France*

[b]*Unité d'Immunologie, ERM0208 INSERM, Camille-Desmoulins, F-94805 Villejuif, France*

Institut Gustave Roussy, 39 rue Camille-Desmoulins, F-94805 Villejuif, France

ABSTRACT: Aneuploidy and chromosomal instability, which are frequent in cancer, can result from the asymmetric division of tetraploid precursors. Genomic instability may favor the generation of more aggressive tumor cells with a reduced propensity for undergoing apoptosis. To assess the impact of tetraploidization on apoptosis regulation, we generated a series of stable tetraploid HCT116 and RKO colon carcinoma cell lines. When comparing diploid parental cells with tetraploid clones, we found that such cells were equally sensitive to a series of cytotoxic agents (staurosporine [STS], hydroxyurea, etoposide), as well as to the lysis by natural killer cells. In strict contrast, tetraploid cells were found to be relatively resistant against a series of DNA-damaging agents, namely cisplatin, oxaliplatin, camptothecin, and γ- and UVC-irradiation. This increased resistance correlated with a reduced manifestation of apoptotic parameters (such as the dissipation of the mitochondrial transmembrane potential and the degradation of nuclear DNA) in tetraploid as compared to diploid cells subjected to DNA damage. Moreover, tetraploid cells manifested an enhanced baseline level of p53 activation. Inhibition of p53 abolished the difference in the susceptibility of diploid and tetraploid cancer cells to DNA damage-induced apoptosis. These data point to an intrinsic resistance of tetraploid cells against radiotherapy and DNA-targeted chemotherapy that may be linked to the status of the p53 system.

KEYWORDS: cell death; chemoresistance; irradiation; p53

Address for correspondence: Dr. Guido Kroemer, CNRS-UMR8125, Institut Gustave Roussy, PR1 38 rue Camille Desmoulins, F-94805 Villejuif, France. Voice: 33-1-42-11-60-46; fax: 33-1-42-11-60-47.
e-mail: kroemer@igr.fr

INTRODUCTION

Cancer results from the accumulation of genetic and epigenetic alterations in which genomic instability dictates the progressive acquisition of an ever more aggressive phenotype. One of the mechanisms of genomic instability involves a transient phase of polyploidization (in most cases a tetraploidization), followed by asymmetric cell divisions and/or chromosome loss, resulting in aneuploidization and chromosomal instability.[1,2]

Polyploidization may result from two distinct processes, namely cell fusion[3,4] or by duplication of the normal chromosomal number in the absence of nuclear and cellular division. Fusion between cells from the same type (homotypic fusion) can be induced *in vitro* and can lead to the generation of pseudodiploid, aneuploid cells generated by an asymmetric reduction-mitosis.[5] Fusion between distinct cell types (heterotopic fusion) has been documented *in vivo*, for instance between tumor cells and infiltrating cells from myeloid origin[6] and may contribute to the plasticity of cancer. Duplication of chromosomes within the same cell occur physiologically through endoreplication (DNA replication without mitosis) or endomitosis (karyokinesis without cytokinesis) or, pathologically, through mitotic failure.[1] Thus, activation of the spindle assembly checkpoint usually arrests mitosis during the metaphase until the problems accounting for checkpoint activation have been solved and mitotic division can ensue correctly. However, a prolonged arrest due to the impossibility to satisfy the spindle assembly checkpoint leads to checkpoint "adaptation," "slippage," or "leakage" with the consequent exit of mitosis and fixation of a tetraploid state.[7,8] Recently, it has also been reported that chromosome dysjunction may give rise to binucleated, tetraploid cells that are formed by bipolar mitosis followed by regression of the cleavage furrow hours later.[9]

Experimental tetraploidization of p53-negative mammary epithelial cells can be employed as a method to generate transformed, tumorigenic cells, thus providing a proof-of-principle that tetraploidization may constitute an important intermediate step in carcinogenesis.[2] Moreover, a fusogenic retrovirus can induce the transformation of human cells *in vitro* through cell fusion.[10] Accordingly, two precancerous lesions in humans—oral leukoplakia[11] and Barrett's esophagus[12]— are characterized by the accumulation of tetraploid cells with premalignant or malignant properties.

One of the major hallmarks of cancer is disabled apoptosis.[13-16] A reduced probability of undergoing cell death increases the cancer cell's resistance to chemotherapy and radiotherapy and thus constitutes a major medical problem. In addition, it appears plausible that apoptosis may constitute a mechanism through which potentially malignant cells are aborted,[17,18] meaning that a reduction of the apoptotic potential may enhance the probability of survival of transformed cells.[19] Thus, polyploid cells formed by fusion[10,20] or by endoreplication[2,21,22] are usually aborted by apoptosis in a p53-dependent fashion.

On the basis of these premises, we decided to investigate whether experimental tetraploidization affects the probability of cells to undergo apoptosis.[22] We found that tetraploid clones, once established, demonstrate an intrinsic resistance against DNA-damaging chemotherapy as well as radiotherapy.

MATERIALS AND METHODS

Cell Lines and Culture

HCT116 and RKO colon cancer cells were grown in McCoy's 5A medium supplemented with 10% fetal calf serum (FCS). To generate tetraploid HCT116 derivatives,[22] cells were cultured for 48 h in the presence of nocodazole (100 nM; Sigma, St. Louis, MO, USA) or cytochalasin D (600 nM; Sigma), washed and cultured for 30 days in standard culture conditions, in the absence of these agents, followed by staining with Hoechst 33342(2 μM; Molecular Probes, Paris, France), which marks chromatin, for 30 min at 37°C.[23] Then, the cells were separated on a fluorescence-activated cell sorting (FACSVantage; Becton Dickinson Le Pont de Claix, France), setting the gate on the ~8N population (which corresponds to tetraploid cells in the G2/M phase of the cell cycle), followed by cloning by limiting dilution (FIG. 1A). To generate tetraploid and diploid RKO derivatives,[22,24,25] the parental cell line RKO, which contains ~5% tetraploid cells, was subcloned by FACS separation, limiting dilution into diploid and tetraploid clones (FIG. 1B). One diploid clone was transfected with a cDNA encoding H2B-GFP (Pharmingen, Paris, France), selected in blasticidine (20 μg/mL; Invitrogene, Paris, France), and FACS-separated into subsets of cells enriched in a diploid or tetraploid DNA content to generate diploid and tetraploid H2B-GFP-expressing clones (FIG. 1C).

Assessment of Cytotoxic Responses

A colorimetric assay for quantification of cell viability based on the cleavage of the tetrazolium salt WST-1 (Roche Diagnostics, Germany) was used to measure the IC_{50} for cytotoxic agents. In a standard assay, the cells were exposed to distinct doses of agents such as staurosporine (STS), hydroxyurea, etoposide, camptothecin, cisplatin, or oxaliplatin (all from Sigma), followed by the measurement of tetrazolium reduction, 72 h later.[26] Some experiments were performed in the presence of 30 μM of cyclic pifithrin-α (Calbiochem). In a further series of experiments, human $CD3^-CD56^+$ NK (natural killer) cells purified from healthy individuals were activated overnight with 1,000 IU/mL of recombinant human interleukin-2 and added to ^{51}Cr-labeled tetraploid clones at different effector/target cell ratios, following standard protocols.[27–29]

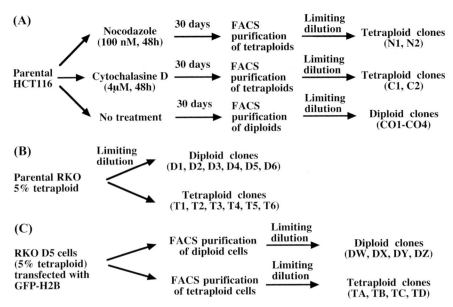

FIGURE 1. Genealogy of the cell lines used in this study. (**A**) Genealogy of diploid and tetraploid HCT116 cells. After transient culture in nocodazole or cytochalasin D, tetraploid cells were enriched by FACS, and tetraploid clones (N1, N2, C1, C2) were isolated by limiting dilution. Diploid controls (CO1-4) were obtained by sham treatment followed by a similar procedure. (**B**) Genealogy of diploid and tetraploid RKO cells. Parental RKO cells, which contain a minority of tetraploid cells, were FACS-separated into diploid and tetraploid populations, and tetraploid (T1-7) and diploid (D1-6) were derived. (**C**) Genealogy of histone H2B-GFP-expressing diploid and tetraploid RKO cells. The diploid RKO clone D5 (see B) was stably transfected with histone H2B-GFP and then subjected to a similar procedure as in B to generate diploid (DW, DX, DY, DZ) and tetraploid (Ta, TB, TC, TC) subclones.

Life Cell Staining, Cytofluorimetric Analyses, and Immunofluorescence

For the simultaneous assessment of mitochondrial apoptosis and plasma membrane permeabilization, live cells were stained with the $\Delta\Psi_m$-sensitive dye 3 tetramethylrhodamine methylester (TMRM, 150nM; Molecular Probes) and Hoechst 33342 (2 μM) which measures DNA content, for 30 min at 37°C.[23,30–32] Cytofluorometric analyses were performed on a FACSVantage (Becton Dickinson) equipped with a 70 μM nozzle and CellQuest software.[33,34] The ploidy status of cells was periodically checked by cell cycle analyses on ethanol-permeabilized cells stained with propidium iodide, by means of standard procedures.[35] For immunofluorescence analyses, cells were fixed with paraformaldehyde (4% w:v) and then stained with a rabbit antiserum specific for p53 phophorylated on serine 15 (Cell Signaling Technology, Beverly, MA), revealed with a goat anti-rabbit IgG conjugated to Alexa 568 (red) fluorochrome from Molecular Probes.[20,36,37]

RESULTS AND DISCUSSION

Normal Susceptibility of Tetraploids to Cell Death Induction by Cytotoxic Agents and NK Cells

We generated a series of tetraploid colon carcinoma clones from two parental cell lines that are characterized by microsatellite instability but no signs of chromosomal instability, HCT116 and RKO.[38,39] These tetraploids were obtained by several distinct experimental strategies. HCT116 cells were transiently (2 days) exposed to nocodazol, a reversible microtubule inhibitor that causes metaphase arrest, or to cytochalasin D, an inhibitor of the actin cytoskeleton that blocks cytokinesis. In these conditions, tetraploidy was induced and tetraploids survived at a low frequency (<1%).[22] However, tetraploids could be enriched by cytofluorometric purification, and tetraploid clones could be isolated by limiting dilution (FIG. 1A). RKO cells were observed to undergo tetraploidization spontaneously at a low rate, and 2–5% of the parental cells were found to be tetraploid. Tetraploids were isolated either from the parental RKO cell (FIG. 1B) or from a diploid RKO clone that was transfected with a GFP-histone H2B fusion construct, allowing for the determination of chromatin content without external dyes (FIG. 1C). Although tetraploids isolated from HCT116 cells had survived a crisis of apoptosis induced by nocodazole or cytochalasin D,[22] we found no indication that such cells would be particularly resistant against apoptosis inducers such as the general tyrosine kinase inhibitor STS or the topoisomerase-2 inhibitor etoposide (FIGS. 2A and 2D). Similarly, tetraploid RKO cells succumbed to STS and hydroxyurea in a similar fashion as their diploid GFP-H2B-negative or GFP-H2B-expressing controls (FIGS. 2B–2F). Moreover, diploid and tetraploid HCT116 cells were equally susceptible to lysis by primary human NK cells (FIG. 3). Altogether, these data suggest the absence of a general cell death blockade in tetraploids.

Reduced Susceptibility of Tetraploids to Cell Death Induction by Cisplatin, Oxaliplatin, and Camptothecin

Although tetraploids do not manifest a general apoptosis defect, they did exhibit a relative resistance against cisplatin, as compared to diploid controls. This was observed for tetraploid HCT116 cells (FIG. 4A) and GFP-H2B-negative RKO cells (FIG. 4B) as well as for GFP-H2B-expressing RKO cells (FIG. 4C), when cell survival was measured by means of a tetrazolium reduction assay. Similar results indicating an increased cisplatin resistance of tetraploid versus diploid cells were obtained when cell death was measured by assessing typical apoptosis-associated parameters such as the loss of the mitochondrial transmembrane potential ($\Delta\Psi$), leakage of cytochrome c from mitochondria, activation of caspase-3, exposure of phosphatidylserine on the cell surface, or apoptotic DNA degradation (not shown). Tetraploids were not only resistant

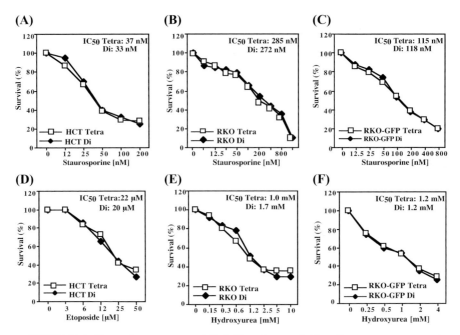

FIGURE 2. Normal responses of diploid and tetraploid cells to some apoptosis inducers. Diploid or tetraploid HCT116 cells (**A, D**), nontransfected RKO cells (**B, E**) or H2B-GFP-transfected RKO cells (**C, F**) were exposed for 48 h to the indicated concentrations of STS, etoposide or hydroxyurea, followed by dtermination of the drug effects by a tetrazolium reduction assay. The experiments were performed on all diploid and tetraploid clones described in FIGURE 1 and the results are expressed as arithmetic mean ± standard deviation (SD).

to cisplatin, but also died less with another platinum compound, oxaliplatin (FIG. 4F), as well as a unrelated compound, camptothecin, which is a topoisomerase-1 inhibitor (FIGS. 4D and 4E). Thus, although the half-lethal doses of STS were similar for diploid and tetraploid HCT116 cells, GFP-H2B-negative RKO cells as well as GFP-H2B-expressing RKO cells (FIG. 2), the half-lethal dose of DNA-damaging chemicals required to kill tetraploids was elevated in all three models included in this study (FIG. 4). Thus, tetraploid cells are intrinsically resistant against chemical DNA damage.

Enhanced Resistance of Tetraploid Cells against γ- and UVC-Irradiation-Induced Apoptosis

Intrigued by the theoretical possibility that increased resistance against DNA-targeted chemicals might be related to issues of drug influx/efflux or the altered cellular and nuclear geometry, we comparatively assessed the effect of

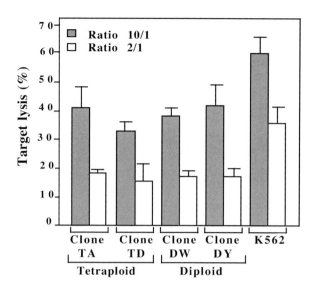

FIGURE 3. Normal lysis of tetraploid RKO cells by human NK cells. Tetraploid clones (TA and TD) or diploid controls (DW and DY) were labeled with ^{51}Cr and then cocultured with IL-2-activated, purified CD3-CD56+ NK cells from healthy donors at an effector: target cell ratio of 10:1 or 2:1. The ^{51}Cr release was then determined.

physical DNA damage on diploid and tetraploid cells. Tetraploid GFP-H2B-expressing RKO cells exhibited fewer signs of apoptosis than their diploid counterparts when exposed to γ-irradiation or UVC light. This applies both to the $\Delta\Psi$ loss, as determined with a $\Delta\Psi$-sensitive fluorochrome (FIG. 5A) and to the apoptotic DNA loss, as measured with the chromatin stain Hoechst 33343 (FIG. 5B). Irrespective of the apoptotic parameter that was assessed, tetraploid RKO cells underwent less cell death than diploid controls. Similar results were obtained for diploid versus tetraploid HCT116 cells (not shown). In conclusion, tetraploid cells are relatively resistant to apoptosis induction by chemical and physical DNA damage.

Decisive Implication of p53 in the Differential Resistance of Diploid and Tetraploid Cells to Apoptosis Induction

DNA-damaging agents can signal for apoptosis induction by activating p53 and the consequent transcription of proapoptotic genes.[15,40,41] Hence, we investigated whether inhibition of p53 by the chemical agent cyclic pifithrin-α[42] would abolish the difference in the apoptosis susceptibility of diploid versus tetraploid cells exposed to DNA-damaging agents. Diploid GFP-H2B-expressing and tetraploid GFP-H2B-negative cells were admixed at a ratio of 1:1 and then cultured in the absence or presence of cisplatin or oxaliplatin, alone or in combination with cyclic pifithrin-α Using the $\Delta\Psi$-sensitive dye

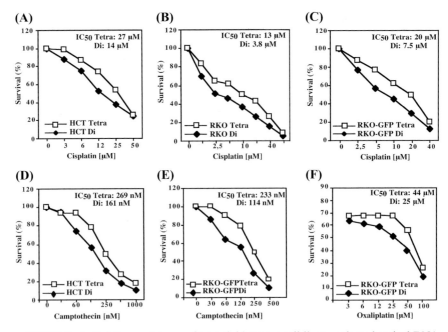

FIGURE 4. Relative resistance of tetraploid cancer cell lines against chemical DNA-damaging agents. Diploid or tetraploid HCT116 cells (**A, D**), nontransfected RKO cells (**B**) or H2B-GFP-transfected RKO cells (**C, E, F**) were cultured for 48 h to the indicated concentrations of cisplatin, oxaliplatin or hydroxyurea. Then the viability of cells was assessed by means of a tetrazolium reduction assay. The data are mean ± SD obtained for all diploid and tetraploid clones described in FIGURE 1.

TMRM, we simultaneously monitored the $\Delta\Psi$ in diploid (GFP$^+$) and tetraploid (GFP^{--}) cells in the same test tube. In response to DNA-damaging agents, diploid cells exhibited mitochondrial apoptosis at a higher frequency than tetraploid cells, and this differential response was blunted by cyclic pifithrin-α. (FIG. 6A). Accordingly, treatment with cisplatin or oxaliplatin induced a shift in favor of the selective survival of tetraploid cells (GFP$^-$), detectable 5 days after initiation of the cultures, and this shift was abolished by addition of cyclic pifithrin-α (FIG. 6B). In addition, we found that p53 exhibited the activating phosphorylation of serine 15 in a fraction of untreated tetraploid cells that was higher than in untreated diploid cells (FIG. 7). This confirms the increase in the baseline expression of p53-inducible genes detected in tetraploid cells by microrrays.[22] Cisplatin induced p53 phosphorylation on serine 15 at a similar frequency in diploid and tetraploid colon carcinoma cells (FIG. 7). These data point to a decisive role of p53 in the differential resistance of diploid and tetraploid cells to DNA damage-induced apoptosis.

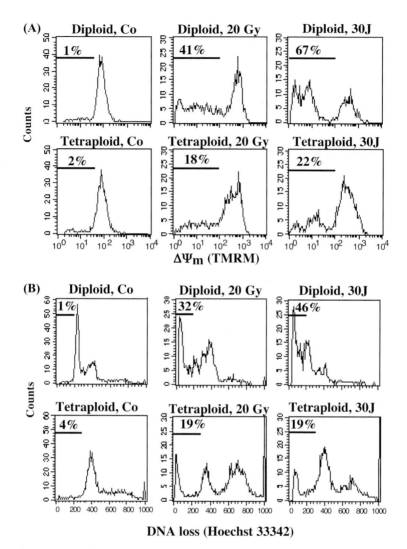

FIGURE 5. Resistance of tetraploids against UVC and γ-irradiation. Diploid and tetraploid RKO cells were subjected to the indicated treatment and were analyzed 3 days later for loss of the ΔΨm (with TMRM staining in **A**) and DNA loss (with Hoechst 33342 staining in **B**). Percentage values in each graph indicate the fraction of cells with a low ΔΨm or a subnormal (sub-G1) DNA content. The results shown are representative for all four diploid and tetraploid H2B-GFP-expressing RKO clones.

CONCLUDING REMARKS

The results presented here and in Ref. 22 indicate that tetraploid cancer cell lines are relatively resistant against DNA damage-induced apoptosis. This

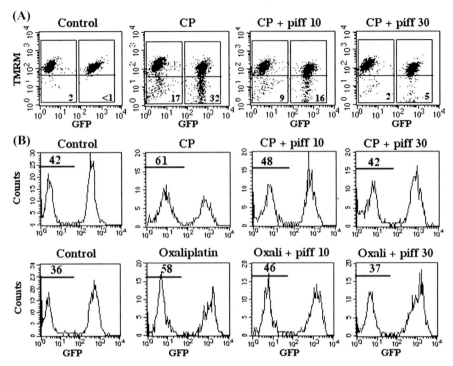

FIGURE 6. Effect of cyclic pifithrin-α on the differential apoptotic response of diploid and tetraploid cells. The H2B-GFP-expressing diploid RKO clone DW was mixed with the H2B-GFP-negative tetraploid clone T4, cultured for 2 days (**A**) or 5 days (**B**) in the absence or presence of 50 nM cisplatin 100 nM oxaliplatin and/or 30μM cyclic pifithrin-α. Effect of cyclic pifithrin-α on the apoptotic $\Delta\Psi m$ dissipation. The $\Delta\Psi m$ was measured by TMRM staining, 2 days after initiation of the treatment with cisplatin and/or cyclic pifithrin-α. The percentages in each of the lower quadrants indicates the fraction of cells manifesting a $\Delta\Psi m$ dissipation, considering the entire diploid or tetraploid population as 100%. (**B**) Relative abundance of diploid and tetraploid cells 5 days after initiation of cultures. Similar results were obtained for other combinations of diploid and tetraploid clones differing in GFP expression.

applies to two distinct colon carcinoma cell lines (HCT116 and RKO) that were rendered tetraploid by three distinct methods (spontaneous tetraploidization, blockade of mitosis with consequent mitotic slippage, or inhibition of cell division) (FIG. 1), as well as to primary mouse embryonic fibroblasts rendered tetraploid by treatment with an inhibitor of heat-shock protein 90.[22] In all cases, apoptosis resistance was not general (FIGS. 2 and 3) and rather specific for the DNA damage inflicted by ionizing irradiation (FIG. 5) or by chemotherapeutic agents such as cisplatin, oxaliplatin, or camptothecin (FIG. 4). Moreover, in all cases, the differential apoptosis regulation of diploid and tetraploid cells could be linked to the activation of the p53 system in tetraploid cells (FIGURES 6 and 7).[22]

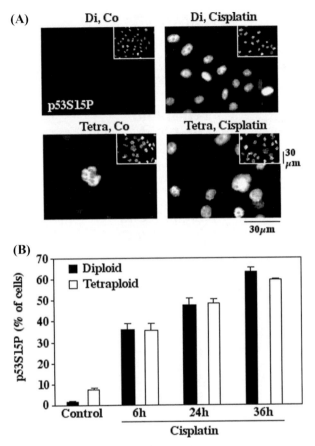

FIGURE 7. Enhanced activation of the p53 system in tetraploid cells. Diploid or tetraploid HCT116 cells were either left untreated or treated for 36 h with 20 μM cisplatin, followed by fixation, permeabilization, and staining with an antibody recognizing a phospho-neoepitope created by phosphorylation of p53 on serine 15. Representative immunofluorescence pictures of diploid and tetraploid HCT116 cells are shown in (**A**). *Insets* show the counterstaining with Hoechst 33342. The frequency of cells exhibiting a clear nuclear staining for serine 15-phosphorylated p53 (mean of three independent experiments performed on two distinct diploid and tetraploid clones each) is shown in (**B**) for distinct time points. Note the increased baseline phosphorylation of p53 in untreated cells.

Our data may be inscribed in a hypothetical scenario describing the relationship between ploidy and apoptosis (FIG. 8). Shortly after illicit tetraploidization, cells tend to undergo apoptosis in a p53-dependent fashion, through a mitochondrial pathway that relies on the proapoptotic Bcl-2 family members Bax and Bak[22]. This is most likely the default pathway, meaning that the vast majority of tetraploid cells are aborted by apoptosis immediately after their formation. However, a small portion (<1%) of tetraploid cells survive, and these cells exhibit, according to our data, an exact duplication of the genome,

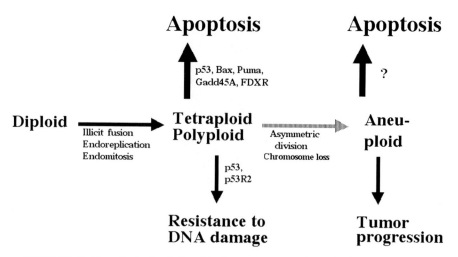

FIGURE 8. Hypothetical relationship between apoptosis and ploidy. The scheme depicts the transition from euploidy to aneuploidy via tetraploidy. The thickness of errors symbolizes the relative importance of default pathways versus the rarity of exceptional events. For details and comments consult main text.

with no major disequilibrium in the transcriptome.[22] However, tetraploid cells manifest the constitutive activation of p53, at a low subapoptotic level, and overexpress a panel of p53 target genes including RRM2B, the gene encoding p53R2, the p53-inducible ribonucleotide reductase-2(Ref. 22). The inhibition of p53 (FIG. 7) as well as the knock-down of p53R2 (a protein required for the conversion of ribonucleotides into desoxyribonucleotides necessary for DNA repair)[22] abolishes the difference of diploid and tetraploid cells in the DNA damage response, thus providing a molecular explanation for the increased resistance of tetraploid cells against DNA damage (FIG. 8).

Chromosomal loss from the tetraploid genome and/or asymmetric cell division may generate an aneuploid offspring. We have obtained evidence that aneuploid daughter cells produced by the division of polyploid mother cells die from apoptosis shortly after their generation.[5,43,44] At present, we ignore the question of whether the death of such aneuploid cells is a result of proapoptotic signals received during the asymmetric division signals or whether it results from acutely lethal metabolic deficiencies, for example, on account of a nullisomy of autosomes or gene dosage disequilibria. We suspect that at least a fraction of the rare cells that survive aneuploidization would possess disabled apoptotic pathways (FIG. 8), although this hypothesis remains to be proven.

It is a still-unsolved conundrum as to which among the biological properties of tumors are acquired and in which order during multistep oncogenesis, and the molecular history of genomic instability vis-à-vis that of apoptosis suppression remains elusive. However, on the basis of the present data it appears plausible that disabled apoptosis could be permissive for the generation of tetraploid cells (which are *bona fide* precursors of aneuploid cells), while tetraploidization has an intrinsic effect on apoptosis regulation.

ACKNOWLEDGMENTS

Guido Kroemer is supported by Ligue Nationale contre le Cancer, European Community (Active p53, RIGHT), and Institut Gustave Roussy.

REFERENCES

1. STORCHOVA, Z. & D. PELLMAN. 2004. From polyploidy to aneuploidy, genome instability and cancer. Nat. Rev. Mol. Cell. Biol. **5:** 45–54.
2. FUJIWARA, T. *et al.* 2005. Cytokinesis failure generating tetraploids promotes tumorigenesis in p53 null mice. Nature **437:** 1043–1047.
3. DUELLI, D. & Y. LAZEBNIK. 2003. Cell fusion: a hidden enemy? Cancer Cell **3:** 445–448.
4. OGLE, B.M., M. CASCALHO & J.L. PLATT. 2005. Biological implications of cell fusion. Nat. Rev. Mol. Cell. Biol. **6:** 567–575.
5. ROUMIER, T. *et al.* 2005. A cellular machine generating apoptosis-prone aneuploid cells. Cell Death Differ. **12:** 91–93.
6. PAWELEK, J. 2005. Tumour-cell fusion as a source of myeloid traits in cancer. Lancet Oncol. **6:** 988–993.
7. RIEDER, C.L. & H. MAIATO. 2004. Stuck in division or passing through: what happens when cells cannot satisfy the spindle assembly checkpoint. Dev. Cell. **7:** 637–651.
8. WEAVER, B.A. & D.W. CLEVELAND. 2005. Decoding the links between mitosis, cancer, and chemotherapy: the mitotic checkpoint, adaptation, and cell death. Cancer Cell **8:** 7–12.
9. SHI, Q. & R.W. KING. 2005. Chromosome nondisjunction yields tetraploid rather than aneuploid cells in human cell lines. Nature **437:** 1038–1042.
10. DUELLI, D.M. *et al.* 2005. A primate virus generates transformed human cells by fusion. J. Cell. Biol. **171:** 493–503.
11. SUDBO, J. *et al.* 2004. The influence of resection and aneuploidy on mortality in oral leukoplakia. N. Engl. J. Med. **350:** 1405–1413.
12. MALEY, C.C. *et al.* 2004. The combination of genetic instability and clonal expansion predicts progression to esophageal adenocarcinoma. Cancer Res. **64:** 7629–7633.
13. ZAMZAMI, N. *et al.* 2000. Bid acts on the permeability transition pore complex to induce apoptosis. Oncogene **19:** 6342–6350.
14. GREEN, D.R. & G. KROEMER. 2004. The pathophysiology of mitochondrial cell death. Science **305:** 626–629.
15. ZHIVOTOVSKY, B. & G. KROEMER. 2004. Apoptosis and genomic instability. Nat. Rev. Mol. Cell. Biol. **117:** 4461–4468.

16. GREEN, D.R. & G. KROEMER. 2005. Pharmacological manipulation of cell death: clinical applications in sight? J. Clin. Invest. **115:** 2610–2617.
17. BARTKOVA, J. *et al.* 2005. DNA damage response as a candidate anti-cancer barrier in early human tumorigenesis. Nature **434:** 864–870.
18. BRAUN, T. *et al.* 2005. NF-{kappa}B constitutes a potential therapeutic target in high-risk myelodysplastic syndromes. Blood. **107:** 1156–1165.
19. LOWE, S.W., E. CEPERO & G. EVAN. 2004. Intrinsic tumour suppression. Nature **432:** 307–315.
20. CASTEDO, M. *et al.* 2002. Sequential involvement of Cdk1, mTOR and p53 in apoptosis induced by the human immunodeficiency virus-1 envelope. EMBO J. **21:** 4070–4080.
21. MERALDI, P., R. HONDA & E.A. NIGG. 2002. Aurora-A overexpression reveals tetraploidization as a major route to centrosome amplification in p53(−/−) cells. EMBO J. **21:** 483–492.
22. CASTEDO, M. *et al.* 2006. Apoptosis regulation in tetraploid cancer cells. EMBO J. In press.
23. CASTEDO, M. *et al.* 2002. Quantitation of mitochondrial alterations associated with apoptosis. J. Immunol. Methods **265:** 39–47.
24. COQUELLE, A. *et al.* 2006. Enrichment of non-synchronized cells in the G1, S, and G2 phases of the cell cycle for the study of apoptosis. Biochem. Pharmacol. In press.
25. COQUELLE, A. *et al.* 2006. Cell cycle-dependent cytotoxic and cytostatic effects of bortezomib on colon cancer cells. Cell Death Differ. **25:** 2584–2595.
26. GOURDIER, I. *et al.* 2004. The oxaliplatin-induced mitochondrial apoptotic response of colon carcinoma cells does not require nuclear DNA. Oncogene **23:** 7449–7457.
27. GHIRINGHELLI, F. *et al.* 2005. Tumor cells convert immature myeloid dendritic cells into TGF-beta-secreting cells inducing CD4+CD25+ regulatory T cell proliferation. J. Exp. Med. **202:** 919–929.
28. CASARES, N. *et al.* 2005. Caspase-dependent immunogenicity of doxorubicin-induced tumor cell death. J. Exp. Med. **202:** 1691–1701.
29. TAIEB, J. *et al.* 2006. A novel dendritic cell subset specialized in tumor immunosurveillance. Nat. Med. **12:** 214–219.
30. CASTEDO, M. *et al.* 1996. Sequential acquisition of mitochondrial and plasma membrane alterations during early lymphocyte apoptosis. J. Immunol. **157:** 512–521.
31. METIVIER, D. *et al.* 1998. Cytofluorometric detection of mitochondrial alterations in early CD95/Fas/APO-1-triggered apoptosis of Jurkat T lymphoma cells. Comparison of seven mitochondrion-specific fluorochromes. Immunol. Lett. **61:** 157–164.
32. ANDREAU, K. *et al.* 2004. Pre-apoptotic chromatin condensation upstream of the mitochondrial checkpoint. J. Biol. Chem. **279:** 55937–55945.
33. RAVAGNAN, L. *et al.* 1999. Lonidamine triggers apoptosis via a direct, Bcl-2-inhibited effect on the mitochondrial permeability transition pore. Oncogene **18:** 2537–2546.
34. MARCHETTI, P. *et al.* 1999. The novel retinoid 6-[3-(1-adamantyl)-4-hydroxyphenyl]-2-naphtalene carboxylic acid can trigger apoptosis through a mitochondrial pathway independent of the nucleus. Cancer Res. **54:** 6257–6275.
35. ZAMZAMI, N. *et al.* 1995. Reduction in mitochondrial potential constitutes an early irreversible step of programmed lymphocyte death in vivo. J. Exp. Med. **181:** 1661–1672.

36. PERFETTINI, J.-L. et al. 2004. NF-kB and p53 are the dominant apoptosis-inducing transcription factors elicited by the HIV-1 envelope. J. Exp. Med. **199:** 629–640.
37. PERFETTINI, J.-L. et al. 2005. Essential role of p53 phosphorylation by p38 MAPK in apoptosis induction by the HIV-1 envelope. J. Exp. Med. **201:** 279–289.
38. BHATTACHARYYA, N.P. et al. 1994. Mutator phenotypes in human colorectal carcinoma cell lines. Proc. Natl. Acad. Sci. USA **91:** 6319–6323.
39. ESHLEMAN, J.R. et al. 1996. Diverse hypermutability of multiple expressed sequence motifs present in a cancer with microsatellite instability. Oncogene **12:** 1425–1432.
40. VOGELSTEIN, B., D. LANE & A.J. LEVINE. 2000. Surfing the p53 network. Nature **408:** 307–310.
41. VOUSDEN, K.H. & X. LU. 2002. Live or let die: the cell's response to p53. Nat. Rev. Cancer **2:** 594–604.
42. KOMAROV, P.G. et al. 1999. A chemical inhibitor of p53 that protects mice from the side effects of cancer therapy. Science **285:** 1733–1737.
43. CASTEDO, M. et al. 2004. Mitotic catastrophe. A special case of apoptosis preventing aneuploidy. Oncogene **23:** 4362–4370.
44. CASTEDO, M. et al. 2004. Cell death by mitotic catastrophe: a molecular definition. Oncogene **23:** 2825–2837.

Molecular Determinants Involved in the Increase of Damage-Induced Apoptosis and Delay of Secondary Necrosis due to Inhibition of Mono(ADP-Ribosyl)ation

CLAUDIA CERELLA,[a] CRISTINA MEARELLI,[a] SERGIO AMMENDOLA,[b] MILENA DE NICOLA,[a] MARIA D'ALESSIO,[a] ANDREA MAGRINI,[c] ANTONIO BERGAMASCHI,[c] AND LINA GHIBELLI[a]

[a]*Dipartimento di Biologia, Università di Roma "Tor Vergata," via Ricerca Scientifica 1, 00133 Rome, Italy*

[b]*Ambiotec SS, via Fosso dell'Acqua Mariana 125, 00040 Roma, Italy*

[c]*Cattedra di Medicina del Lavoro, Università di Roma "Tor Vergata," 00133 Rome, Italy*

ABSTRACT: ADP-ribosylations are reversible posttranslational modifications that regulate the activity of target proteins, catalyzed by two different classes of enzymes, namely poly(ADP-ribosyl)polymerases (PARPs) and mono(ADP-ribosyl)transferases (ADPRTs). It is now emerging that ADP-ribosylation reactions control signal transduction pathways, mostly as a response to cell damage, aimed at both cell repair and apoptosis. Inhibition of ADPRTs, but not PARPs, increases the extent of apoptosis induced by cytocidal treatments, at the same time delaying secondary necrosis, the process leading to plasma membrane collapse in apoptotic cells, and responsible for apoptosis-related inflammation *in vivo*. Thus, ADPRT inhibitors may be ideal as adjuvants to cytocidal therapies; to this purpose, we investigated the molecular determinant(s) for such effects by probing a set of molecules with similar structures. We found that the apoptosis-modulating effects were mimicked by those compounds possessing an amidic group in the same position as two of the most popular ADPRT inhibitors, namely, 3-aminobenzamide and nicotinamide. This study may provide useful suggestions in designing molecules with therapeutic potential to be used as adjuvant in cytocidal therapies.

KEYWORDS: apoptosis; secondary necrosis; ADP-ribosylation

Address for correspondence: Claudia Cerella, Dipartimento di Biologia, Università di Roma "Tor Vergata," via Ricerca Scientifica 1, 00133 Rome, Italy. Voice: +39-06-7259-4323; fax: +39-06-2023500.

e-mail: cerella@uniroma2.it

INTRODUCTION

Cells may be exposed to many kinds of stress-causing damage. Based on the extent of damage, cells may have different fate. If damage is very great, cells get passively killed, rapidly losing plasma membrane integrity (primary necrosis). When damage is milder, cells may have the time to choose to autoeliminate by activating the apoptotic program. Apoptotic cells, unlike necrotic ones, possess a still-functional energy metabolism, maintaining plasma membrane integrity. As a last step, apoptotic cells lose energy and plasma membrane integrity by a process known as secondary necrosis. In physiological conditions, secondary necrosis does not take place *in vivo*, since apoptosizing cells are removed earlier by phagocytes, that is, before energetic collapse. Secondary necrosis was considered as a phenomenon occurring essentially in *in vitro* experimental conditions, that is, in the absence of phagocytes. Evidence is, however, emerging that in many pathological conditions, such as phagocytic deficit or massive production of apoptotic cells, apoptosizing cells may accumulate and saturate the removal system, thus reaching the secondary necrosis stage, and leading to inflammation due to the liberation of highly inflammogenic cell debris. The best-known case is the massive apoptosis due to cytocidal therapies. The mechanisms responsible for plasma membrane collapse in apoptotic cells are not known, although they may consist of a block of the ion transport mechanism in plasma membrane.[1] Also, caspase-3[2] seems to be involved in plasma membrane permeabilization.[2] The detailed knowledge of such an event may help to define a line of research committed to find/design novel drugs able to limit/reduce the serious side effects of secondary necrosis/inflammation.

ADP-ribosylations are posttranslational modifications that modulate the activity of target proteins.[3,4] They are catalyzed by two classes of enzymes, such as a set of poly(ADP-ribosyl)polymerases (PARPs) and a number of mono(ADP-ribosyl)transferases (ADPRTs), both present in many cell compartments. It is becoming known that ADP-ribosylation reactions can control pathways of signal transduction, mostly as response to cellular stress, including apoptosis.

The role of PARP in apoptosis has been extensively analyzed,[5,6] and has been shown to be quite a complex one. Indeed, PARP was shown to consume cellular NAD, thus carrying on the so-called "NAD depletion suicide."[7] NAD suicide was shown to be a cell death by necrosis,[8] proposing PARP as a necrogenic enzyme. Accordingly, PARP is cleaved and inactivated early in apoptosis, because its activity as a NAD-consuming enzyme, strongly stimulated by the huge number of PARP-stimulatory DNA breaks, would very soon lead to energetic collapse, thus turning apoptosis into primary necrosis.[9] PARP may also mediate apoptosis by producing a turnover in chromatin proteins,[10] perhaps activating p53, a known target of PARP.[11] Unlike PARP, ADPRTs were poorly studied with respect to apoptosis. It has been shown that ADPRT inhibition[8,12] increases stress-induced apoptosis by a mechanism possibly involving glycolysis.[13]

It is known that the ADP-ribosylation inhibitor 3-aminobenzamide (3-ABA) is able to shift the mode of cell death from primary necrosis to apoptosis[14–17]; though 3ABA inhibits both PARPs and ADPRTs,[18] current evidence points to this effect of PARP inhibition.[17] We have shown that ADPRTs play a role in modulating secondary necrosis as well as the extent of damage-induced apoptosis (Cerella *et al.*, in preparation). The potential use of these combined effects as adjuvant for cytocidal therapies pushed us to explore the molecular determinants of such effects, with the goal of understanding whether the determinants of the two effects reside on the same determinants.

MATERIALS AND METHODS

Cell Culture

U937 human tumor monocytes were cultured in RPMI 1640 medium supplemented with 10% FCS, 2 mM L-glutamine, 100 IU/mL penicillin and streptomycin, and kept in a controlled atmosphere (5% CO_2) incubator at 37°C. All experiments were performed in cells at the log phase of cell growth, with viability (as results from propidium iodide (PI) staining) $\geq 98\%$ and spontaneous apoptosis $\leq 2\%$.

Induction of Apoptosis

Apoptosis was induced with 10 μg/mL puromycin (PMC) or 50 μM etoposide (VP16). Oxidative stress-induced apoptosis was carried out according to a two-step procedure, consisting of a stress phase and a recovery phase: 8×10^5/mL U937 cells were treated for 1 h with 1 mM H_2O_2. Then cells were washed and resuspended at the same density in fresh medium for recovery to allow apoptosis to develop. Apoptosis was evaluated at the times indicated.

ADP-Ribosylation Reaction Inhibition

3-Aminobenzamide (3-ABA; 5 mM, unless otherwise specified) and nicotinamide (NA; 10 mM) were added 30 min before induction of apoptosis.

Other Treatments

Isobutyramide (IBA), malonamide (MA), or trimethylacetamide (TMA), all at the concentrations of 5 mM, were added 30 min before induction of apoptosis.

Analysis of Apoptosis

Apoptosis was characterized by nuclear fragmentation detectable by fluorescence microscopy on cells stained 15 min before with DNA-specific cell-permeable dye Hoechst 33342 (1 μg/mL).[3–6]

Analysis of Plasma Membrane Integrity

Five minutes before the analysis by fluorescence microscopy, cells were stained with 5 μg/mL PI, which permeates only cells with damaged plasma membrane.

Quantification of Apoptosis-Secondary Necrosis

The fraction of U937 cells with fragmented nuclei among the total cell population was calculated, counting at least 300 cells in at least 10 randomly selected fields.[3–6] Apoptotic cells was further distinguished in fresh apoptotic cells (PI-negative) and cells in secondary necrosis (PI-positive) as revealed by double Hoechst–PI staining.

Statistical Analysis

Statistical analysis was performed using Student's t-test and P-values < 0.05 were considered significant. Data are presented as mean $\pm$ SD.

RESULTS

FIGURE 1 shows the effects of 3-ABA and NA, two known ADP-ribosylation inhibitors, on apoptosis and secondary necrosis in U937 cells, a human tumor promonocytic cell line.

U937 cells were induced to apoptosis with PMC, VP16, and H_2O_2, which act with different mechanisms. A 5-mM 3-ABA and 10-mM NA increase apoptosis independently on the apoptogenic agent (FIG. 1 A).

In the same samples, in contrast to increased apoptosis, 3-ABA and NA sensibly delay secondary necrosis, as revealed by the ratio between "fresh" apoptotic cells (PI-negative) and cells in secondary necrosis (PI-positive) (FIG. 1B). This indicates that the inhibition of ADP-ribosylation reactions is involved in modulating apoptosis and secondary necrosis.

The important implications that the duplex proapoptotic and antisecondary necrosis role of 3-ABA and NA might have for the clinic and therapeutics

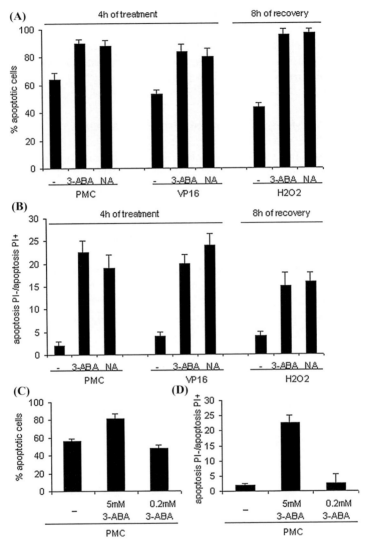

FIGURE 1. The inhibition of mono(ADP-ribosyl)ation reactions increases apoptosis and slows down the shift to secondary necrosis. The extent of apoptosis (**A**) induced with 10 μg/mL PMC, 50 μM VP16, and 1 mM H_2O_2, which operate with different mechanisms, and apoptosis PI−/apoptosis PI+ ratio as measure of secondary necrosis incidence (**B**) were analyzed in the presence/absence of 5 mM 3-ABA or 10 mM NA at the times indicated for treatment/recovery (see the MATERIALS AND METHODS section). At these doses, 3-ABA and NA increase apoptosis with all apoptogenic agents and strongly reduce the extent of secondary necrosis. Similar results were obtained by inducing apoptosis with VP16 (data not shown). Panels (**C**) and (**D**) show that a low dose of 3-ABA (0.2 mM) that specifically inhibits PARP, slightly reduces apoptosis but does not affect secondary necrosis. The results are the average of three experiments ± SD.

prompted us to further study the mechanisms responsible for these effects, thus following two lines of investigation.

First, at the concentrations used in our experiments, 3-ABA and NA might inhibit both ADPRTs and PARP.[18] To discriminate between the two, we specifically inhibited only PARP with low doses of 3-ABA (0.2 mM, known to inhibit only PARP[4,18]). We found that low 3-ABA doses are ineffective (FIG. 1C and D).

Second, since both 3-ABA and NA are aromatic compounds belonging to the family of amides, we wanted to analyze whether amidic group might be essential for determining 3-ABA and NA effects on apoptosis/secondary necrosis. In order to do this, we tested the ability to affect apoptosis/secondary necrosis extent of three linear chain (aliphatic) amides, IBA, MA, and TMA (see FIG. 2A), all used at the concentration of 5 mM, which is equimolar to the high 3-ABA doses. FIGURE 2B shows that all the three compounds tested increase apoptosis to a similar extent as 3-ABA and NA. The analysis of PI−/PI+ apoptotic cells ratio revealed that IBA, MA, and TMA are able to prevent secondary necrosis, but are slightly (about twofold) less effective with respect to 3-ABA and NA (FIG. 2C).

DISCUSSION

In this study, we report that the inhibition of mono(ADP-ribosyl)ation reactions slows down the process of secondary necrosis, though increasing the extent of apoptosis.

For several years, it was assumed that the inhibition of PARP (essentially PARP-1) is related to the prevention of primary necrosis.[14–17] Thus, research has been focused on inhibiting PARP with multiple strategies in order to prevent necrosis, whereas, in the meantime, mono(ADP-ribosyl)ation reactions and their role on necrosis have been practically ignored.

Secondary necrosis, that is, the loss of plasma membrane integrity in late apoptotic cells, is a phenomenon typical of apoptosis induced *in vitro* as the consequence of the absence of phagocytosis. But this phenomenon may take place also in pathophysiological conditions *in vivo* when phagocytes response is lacking or not sufficient. One of these cases is the massive therapeutic administration of apoptogenic treatments, such as chemo/radiotherapics in tumor treatments. The action of all chemotherapics, indeed, is based on massive cytocidal effects on tumor cells; but this leads also to a phagocytosis engulfment, thus leading to the permanence of apoptotic cells until they lose their plasma membrane integrity, which, in turn, determines a large inflammation. Our studies have shown that high doses of 3-ABA and NA that inhibit mono(ADP-ribosyl)ations increase damage-induced apoptosis (interestingly, also by etoposide, a widely used chemotherapic agent) and strongly contrasts with the appearance of secondary necrosis. These two features are of paramount

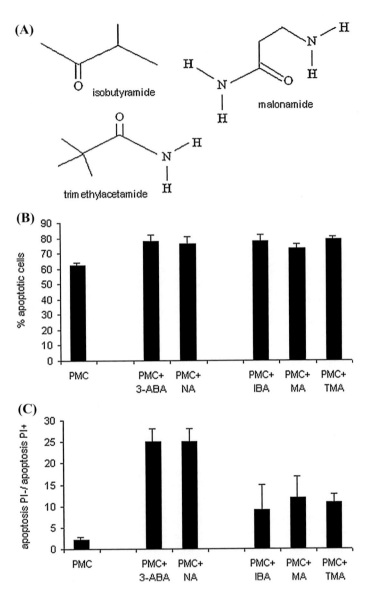

FIGURE 2. The amidic groups of 3-ABA and NA are probably responsible for the apoptosis-modulating effects. IBA, MA, and TMA are three linear chain (aliphatic) amides (**A**), all used at 5 mM (equimolar to high doses of 3-ABA). U937 cells were induced to apoptosis with 10 μg/mL PMCs in the presence of IBA, MA, or TMA. The extent of apoptosis (**B**) and the ratio between apoptosis PI−/apoptosis PI+ (**C**) were analyzed. All the three compounds increase apoptosis to an extent similar to 5 mM 3-ABA (**B**) and prevent secondary necrosis, although less effectively than 3-ABA and NA (**C**). The results are the average of two experiments ±SD.

importance in the development of drugs to be used as adjuvant to cytocidal therapies, with the goal of potentiating the efficacy of apoptogenic treatments and, at the same time, reducing the side effect of inflammation, which often accompanies massive apoptosis.

A parallel study (Cerella *et al.*, in preparation) allowed us to show that the modulation of apoptosis and the effect on secondary necrosis do depend on mono(ADP-ribosyl)ations, but could be uncoupled, that is, they are due to different and independent mechanisms. Here we wish to give molecular support to this evidence. Indeed, in searching for other compounds with pharmacological activity, we have shown that IBA, MA, and TMA, three amides that differ from aromatic 3-ABA and NA for their aliphatic structure, increase apoptosis with similar extent to 3-ABA and NA, but their ability to reduce the shift to secondary necrosis is more moderate.

REFERENCES

1. BARROS, L.F., J. CASTRO & C.X. BITTNER. 2002. Ion movements in cell death: from protection to execution. Biol. Res. **35:** 209–214.
2. SCHWAB, B.L., D. GUERINI & C. DIDSZUN, *et al.* 2002. Cleavage of plasma membrane calcium pumps by caspases: a link between apoptosis and necrosis. Cell Death Differ. **9:** 818–831.
3. BURKLE, A. 2005. Poly(ADP-ribose). The most elaborate metabolite of NAD. FEBS J. **272:** 4576–4589.
4. OKAZAKI, I.J., & J. MOSS 1996. Mono-ADP-ribosylation: a reversible posttranslational modification of proteins. Adv. Pharmacol. **35:** 247–280.
5. SCOVASSI, I. & G.G. POIRIER. 1999. Poly(ADP-ribosylation) and apoptosis. Mol. Cell. Biochem. **199:** 125–137.
6. DE MURCIA, G. & S. SHALL. 2000. Poly(ADP-ribosyl)ation reactions. From DNA Damage and Stress Signalling to Cell Death. Oxford University Press. Oxford, England.
7. BERGER, N.A. 1985. Poly(ADP-ribose) in the cellular response to DNA damage. Rad. Res. **101:** 4–15.
8. COPPOLA, S., C. NOSSERI, V. MARESCA, *et al.* 1995. Different basal NAD levels determine opposite effects of poly(ADP-ribosyl)polymerase inhibitors on H_2O_2-induced apoptosis. Exp. Cell Res. **221:** 462–469.
9. BOUCHARD, V.J., M. ROULEAU & G.G. POIRIER. 2003. PARP-1, a determinant of cell survival in response to DNA damage. Exp. Hematol. **31:** 446–454.
10. NOSSERI, C., S. COPPOLA & L. GHIBELLI. 1994. Possible involvement of poly(ADP-ribosyl)polymerase in triggering stress-induced apoptosis. Exp. Cell Res. **212:** 367–373.
11. VALENZUELA, M.T., R. GUERRERO, M.I. NUMEZ, *et al.* 1990. PARP-1 modifies the effectiveness of p53-mediated DNA damage response. Oncogene **21:** 1108–1116.
12. GHIBELLI, L., C. NOSSERI, S. COPPOLA, *et al.* 1995. The increase in H2O2-induced apoptosis by ADP-ribosylation inhibitors is related to cell blebbing. Exp. Cell Res. **221:** 470–477.

13. COLUSSI, C., M.C. ALBERTINI, S. COPPOLA, et al. 2000. H_2O_2-induced block of glycolysis as an active ADP-ribosylation reaction protecting cells from apoptosis. FASEB J. **14:** 2266–2276.
14. PALOMBA, L., P. SESTILI, F. CATTABENI, et al. 1996. Prevention of necrosis and activation of apoptosis in oxidatively injured human myeloid leukemia U937 cells. FEBS Lett. **390:** 91–94.
15. HA, H.C. & S.H. SNYDER. 1999. Poly(ADP-ribose)polymerase is a mediator of necrotic cell death by ATP depletion. Proc. Natl. Acad. Sci. USA **96:** 13978–13982.
16. HERCEG, Z. & Z.Q. WANG. 1999. Failure of poly(ADP-ribose)polymerase cleavage by caspases leads to induction of necrosis and enhanced apoptosis. Mol. Cell. Biol. **19:** 5124–5133.
17. FILIPOVIC, D.M., X. MENG & W.B. REEVES. 1999. Inhibition of PARP prevents oxidant-induced necrosis but not apoptosis in LLC-PK1 cells. AJP Renal Physiol. **277:** 428–432.
18. RANKIN, P.W., E.L. JACOBSON, R.C. BENJAMIN, et al. 1989. Quantitative studies of inhibitors of ADP-ribosylation *in vitro* and *in vivo*. J. Biol. Chem. **264:** 4312–4317.

Magnetic Fields Protect from Apoptosis via Redox Alteration

M. DE NICOLA,[a] S. CORDISCO,[a] C. CERELLA,[a] M.C. ALBERTINI,[b]
M. D'ALESSIO,[a] A. ACCORSI,[b] A. BERGAMASCHI,[c] A. MAGRINI,[c]
AND L. GHIBELLI[a]

[a] *Dipartimento di Biologia, Università di Roma Tor Vergata, via della Ricerca Scientifica, 00133 Roma, Italy*

[b] *Istituto di Chimica Biologica Giorgio Fornaini, Università degli Studi di Urbino, via Saffi, 61029 Urbino, Italy*

[c] *Cattedra di Medicina del Lavoro, Università di Roma Tor Vergata, via della Ricerca Scientifica, 00133 Roma, Italy*

> ABSTRACT: Magnetic fields (MFs) are receiving much attention in basic research due to their emerging ability to alter intracellular signaling. We show here that static MFs with intensity of 6 mT significantly alter the intracellular redox balance of U937 cells. A strong increase of reactive oxygen species (ROS) and a decrease of glutathione (GSH) intracellular levels were found after 2 h of MF exposure and maintained thereafter. We found that also other types of MFs, such as extremely-low-frequency (ELF) MFs affect intracellular GSH starting from a threshold at 0.09 mT. We previously reported that static MFs in the intensity range of 0.3–60 mT reduce apoptosis induced by damaging agents (Fanelli *et al.*, 1998). Here, we show that ELF-MFs are also able to protect U937 from apoptosis. Interestingly, this ability is limited to the ELF intensities able to alter redox equilibrium, indicating a link between MF's antiapoptotic effect and the MF alteration of intracellular redox balance. This suggests that MF-produced redox alterations may be part of the signaling pathway leading to apoptosis antagonism. Thus, we tested whether MFs may still exert an antiapoptotic action in cells where the redox state was artificially altered in both directions, that is, by creating an oxidative (via GSH depletion with BSO) or a reducing (with DTT) cellular environment. In both instances, MFs fail to affect apoptosis. Thus, a correct intracellular redox state is required in order for MFs to exert their antiapoptotic effect.
>
> KEYWORDS: static magnetic fields; extremely-low-frequency magnetic fields; ROS; glutathione; U937; apoptosis

Address for correspondence: Milena De Nicola, Dipartimento di Biologia, Università di Roma Tor Vergata, via della Ricerca Scientifica, 00133 Roma, Italy. Voice: +39-06-72594335; fax: +39-06-2023500.
  e-mail: milena.de.nicola@uniroma2.it

INTRODUCTION

In recent years, a growing interest has developed concerning the biological effects of magnetic fields (MFs) and the possible hazards for human health. Several epidemiological studies indicated a correlation between exposure to magnetic fields and the elevated frequency of various forms of cancer, including childhood leukemia, breast cancer, brain tumors, and lymphomas[1-3]; other investigations failed to find similar associations, and the involvement of electromagnetic field exposure in tumor incidence is still open to debate.[4-6] Although heterogeneous in terms of MF intensity, field types, exposure protocols, cell types, and data from the literature indicate a link between exposure to magnetic fields and tumorigenicity. The mechanism of MF tumorigenic effect remains to be established. Since no direct tumorigenic or mutagenic effect has ever been attributed to MFs, it has been hypothesized that MFs play the role of a tumor promoter in the presence of a primary initiator.[7]

Apoptosis is possibly the most potent defense against cancer because it is the mechanism used by metazoans to eliminate deleterious cells. Defects in apoptosis promote tumor formation. Therefore, the ability to understand and manipulate this cell death machinery is an obvious goal of medical research.

Previously, we demonstrated that the exposure of human monocyte tumor cells, U937, to static MFs with intensities starting from 0.6 mT, decrease in an intensity-dependent fashion, reaching a plateau at 6 mT, the extent of cell death by apoptosis.[8] MFs reduce apoptosis induced by several agents by interfering with the apoptotic process.[8] The presence of MFs allows the indefinite survival and replication of a substantial fraction of cells hit by apoptogenic agents that would have been eliminated by apoptosis in the absence of MFs. The protective effect was found to be mediated by the ability of the fields to enhance Ca^{2+} influx from the extracellular medium.

The recent literature provides convincing evidence of many effects of MFs on intracellular targets, showing that cell behavior, if not cell viability, may be deeply affected by exposure to different types of MFs.[9] Several *in vivo* and *in vitro* experiments documented that the basic harmful effect of static or ELF-MFs involves different cellular perturbations. Among them, reactive oxygen species (ROS) production has a prominent role, whereby an increase in their production has been reported for a number of cells and organs upon ELF exposure.[10] ROS are known for their cellular reactivity and toxic effects, including those on intracellular homeostasis of calcium, which has significant implications for signal transduction pathways, cell proliferation, oncogene expression,[11] and apoptosis.[12]

Glutathione (GSH), the most abundant intracellular thiol, acts as a major antioxidant by protecting against the damaging effects of free radicals and ROS.[13] Its important role is due to its complex, multifaceted detoxification system, which includes the transformation of the reduced to the oxidized glutathione disulfide form. We have shown that GSH is a key regulator of apoptosis, since cells actively extrude intracellular GSH[14] through specific carriers[15]; the consequent redox disequilibrium[16] triggers the activation of Bax[17] and cytochrome *c* release.[18]

The aim of this study was to investigate the potential effect of static and ELF MFs exposure on redox equilibrium, and the eventual relationship between MF exposure, oxidative alteration, and apoptosis.

MATERIALS AND METHODS

Cell Culture and Treatments

U937 cells were kept in RPMI 1640 supplemented with 10% inactivated fetal calf serum (FCS), 2 mM L-glutamine, 100 IU/mL penicillin and streptomycin, and kept in a controlled atmosphere (5% CO_2) incubator at 37°C. Experiments were performed at a concentration of 10^6 cells/mL.

Apoptosis was induced with 10 μg/mL puromycin (PMC), and was quantified as the fraction of apoptotic nuclei upon staining with the DNA-specific dye Hoechst 33342 (1μg/mL). GSH depletion was induced by inhibiting glutathione neosynthesis with 1 mM buthionine sulfoximine (BSO), which depletes U937 of GSH in ~20 h as described.[15] BSO was added 20 h before the apoptogenic treatment and kept throughout the experiment.

Reduced intracellular environment was induced with 10 μM dithiothreitol (DTT). DTT is a reducing agent that disrupts disulfide bridges (S-S). DTT was added 30 min before the apoptogenic treatment and kept throughout the experiment.

PMC, BSO, and DTT were purchased from Sigma Chemical Co. (St. Louis, MO, USA); Hoechst 33342 from Calbiochem (San Diego, CA, USA).

Analysis of Intracellular GSH and ROS

Cells were loaded, respectively, with 10 μM dichlorodihydrofluoresceindiacetate (DHCFDA) or 10 nM chloromethylfluoresceindiacetate (CMFDA) by incubation at 37°C for 20 min. Cells were then washed and immediately analyzed using FACScan (Becton Dickinson, San Jose, CA, USA); data (10,000 events) were elaborated with the Cell Quest software. DHCFDA and CMFDA were purchased from Molecular Probes.

MF Application

Static MFs

MFs were produced by metal magnetic disks of known intensity; magnets produce static MFs without producing alternating MF or any temperature increase. Magnets were placed under the culture petri dish. Since MF intensity decreases according to the square of the distance, the actual field intensity on the cells was calculated taking the thickness of the bottom of the petri dish (1.2 mm) into account to give an actual intensity of 0.6 mT on the cells.

ELF MFs

Exposure to ELF magnetic fields was performed as described in previous studies[19] in which cells were exposed only to the magnetic component of ELF inside a Faraday cage. Intensity of magnetic field was measured with a specific sensor placed perpendicularly to the field force-line. Cells were exposed to a range of intensity of magnetic fields from 0.07 mT to 0.1 mT. Since no metal objects, such as thermostatic equipment, may be placed in the Faraday cage, experiments were performed with a system of circulating warm water, to guarantee that the water bath with the cells multiwell plate was stably maintained at 37°C (continually monitored by an alcohol thermometer). Control cells were placed in the thermostatic water bath outside the Faraday cage in a magnetic field-null area, and were considered as sham-exposed.

Statistical Analysis

Statistical analysis was performed using Student's *t*-test for impaired data and *P*-values < 0.05 were considered significant. Data are presented as mean ± SD.

RESULTS

MFs Increase Intracellular ROS and Decrease GSH Level

MFs were proved to be able to increase intracellular ROS level in U937 cells. Magnetic disks were placed under the culture multiwell plates for 2 h and then ROS generation was analyzed by means of a flow cytometric analysis of cells stained with DHCFDA, a dye that becomes fluorescent only when oxidized. As shown in FIGURE 1A, MF exposure causes a strong increase in ROS intracellular level.

Reduced GSH is present in all cellular compartments. It acts as a cell protectant, especially able in scavenging radicals through a cycling redox process. Strong oxidative stress can use intracellular GSH, causing its depletion. Thus, we investigated the fate of GSH in the same exposure conditions. FIGURE 1B shows the values of intracellular GSH at the single cell level by means of flow cytometric analysis of cells stained with CMFDA, a probe that fluoresces when complexed with GSH by glutathione-S-transferase. After 2 h of MF

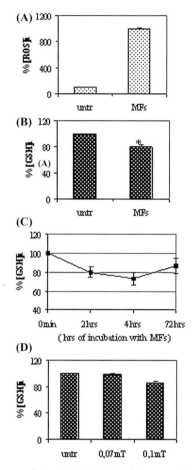

FIGURE 1. MFs increase intracellular ROS and decrease GSH level. (**A**) Effect of MFs on ROS level. After 2 h of exposure with MFs U937 cells were stained with DHCFDA and fluorescence intensity was measured by flow cytometric analysis. Values indicate differences with respect to control, posed = 100. MFs exposure causes a strong increase in ROS intracellular level. (**B**) Effect of MFs on intracellular GSH level. After 2 h of exposure to MFs, U937 cells were stained with CMFDA and fluorescence intensity was measured by flow cytometric analysis. Values indicate differences with respect to control, posed = 100. MFs exposure causes a significant ($P < 0.05$) decrease in GSH intracellular level. The time course analysis of the changes in CMFDA fluorescence intensity (**C**) revealed that the decrement in GSH level is maintained during several days of MFs exposure. Values indicate differences with respect to control, posed = 100. (**D**) Effect of ELF on intracellular GSH level. After 2 h of exposure with ELF-MFs U937 cells were stained with CMFDA and fluorescence intensity was measured by flow cytometric analysis. Values indicate differences with respect to control, posed = 100. The GSH decrease by 0.1 mT ELF-MFs is significant ($P < 0.05$). Values are the average of at least five experiments for each treatment ($\pm$ SEM).

exposure, a significant decrease in intracellular GSH was found. A similar decrease in GSH was found analyzing the GSH content by different methods, such as DTNB (5,5'-dithiobis(2-nitrobenzoic acid)[20] and the canonical high-performance liquid chromatography (HPLC) separation[21] (data not shown). In order to understand whether GSH intracellular loss was maintained during MF exposure, we performed a time-course analysis of the GSH content (FIG. 1C). The decrement observed after 2 h of exposure was maintained thereafter, up to 72 h. The loss of intracellular GSH was not paralleled by an increase of GSH in the extracellular medium (data not shown). This indicates that GSH is not extruded upon MF exposure, but rather that MF exposure induces a GSH usage.

In order to investigate the effect of different types of magnetic fields on intracellular redox equilibrium, we exposed U937 to two different intensities of ELF-MFs and analyzed their effect on intracellular GSH. We show that ELF with intensity of 0.07 mT does not alter GSH content with respect to control, whereas ELF-MFs with intensity of 0.1 mT cause a significant decrease in intracellular GSH (FIG. 1D). Thus GSH decrease depends on ELF-MF intensity.

A Correct Intracellular Redox State Is Required for MFs to Exert Their Antiapoptotic Effect

We had previously demonstrated that exposure of U937 to static MFs decreases the extent of apoptosis.[8] We show here that ELF-MFs with intensity 0.1 mT, though not exerting any toxic or apoptogenic effect on U937 cells, is able to reduce apoptosis (FIG. 2A; see also Nuccitelli *et al.*, submitted). This protection occurs with the same Ca^{2+} requirement as the one produced by static MFs, thus possibly occurring with similar mechanisms (Albertini *et al.*, in preparation). Also for ELF, we were able to detect a threshold effect, since intensities <0.08 mT were ineffective (see FIG. 2A for 0.07 mT). Interestingly, the threshold value for the antiapoptotic effect coincides with the threshold necessary to induce redox imbalance. This suggests the existence of a possible cause–effect relationship between MF-induced redox imbalance and MFs' antiapoptotic effect. Indeed, we have deeply investigated the role of GSH alterations in apoptosis.[14–18]

This observation gave us the logical basis that allowed us to ask the question of whether MF alteration of GSH content might be the cause of the antiapoptotic effect of MFs. The link between the MF antiapoptotic effect and the MF alteration of intracellular redox balance was investigated by testing whether MFs may still exert an antiapoptotic action in cells artificially deprived of GSH. We have already reported that inhibition of GSH neosynthesis with BSO in U937 cells leads to complete GSH depletion with 24 hours.[15]

In cells depleted of GSH by BSO, MFs fail to reduce apoptosis (FIG. 2B), showing that the early redox imbalance caused by MFs is part of the signaling pathway that allows MFs to antagonize apoptosis.

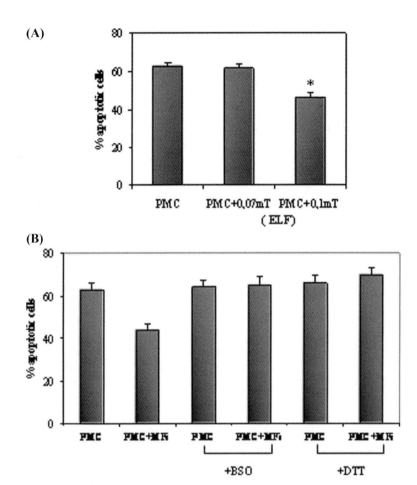

FIGURE 2. Effect of oxidant and reducing environment on MFs' antiapoptotic effect. (**A**) Effect of ELF on apoptosis. Apoptosis was induced with puromycin (PMC) on cells placed with 0.007 mT, 0.1 mT ELF, and sham-exposed, and was measured at 4 h of treatment. The protection by 0.1 mT is significant ($P < 0.5$). (**B**) Effect of MFs on apoptosis of redox-altered cells. BSO and DTT abrogate the antiapoptotic effect of MFs. BSO was added 24 h and DTT 30 min before MF exposure and apoptogenic treatment and maintained through the experiment. Apoptosis was measured at 4 h of treatment with PMC. Values are the average of at least five experiments for each treatment ($\pm$ SEM).

Next, we artificially modulated redox intracellular environment in the other direction, that is, by reducing it with DTT. Intriguingly, 10 μM DTT equally abrogated MFs' antiapoptotic action (FIG. 2B). These data indicate that MFs need a correct intracellular redox state in order to exert their antiapoptotic effect.

DISCUSSION

In this study we show that static MFs with intensity of 6 mT significantly alter the intracellular redox balance of U937 cells. In particular, MF exposure causes a large increase of ROS. Dysregulated intracellular production of ROS is harmful to the cell and is counteracted by various antioxidant defenses. GSH is the most rapid and abundant weapon against ROS. Its protective action is based on oxidation of the thiol group of its cysteine residue with the formation of GSSG, which in turn is catalytically reduced back to the thiol form (GSH) by glutathione reductase.[22] As expected, we found that the increases in ROS due to MF exposure is concomitant with a drop in GSH intracellular levels. What is the mechanism responsible for MF-induced GSH decrease? The decrement in GSH intracellular level can have different causes on account of its central role in protecting organisms against toxicity and disease. GSH may be linked to nucleophilic molecules by glutathione-S-transferase, and leave cells as a conjugated adduct.[23] Upon oxidative stress, GSSG may either recycle to GSH or exit from the cells; or GSH can be bound to proteins, leading to the formation of glutathionylated proteins.[24] In addition, recent evidences show that GSH may conjugate to nitric oxide (NO) to form an S-nitroso-glutathione (GSNO) adduct.[25] GSH can also be extruded from cells, via specific carriers,[26] thus regulating extracellular redox state,[22] promoting apoptosis[15] or GGT-mediated intracellular signaling.[27] Since a concomitant production of ROS is not a discriminant between these hugely different types of depletion, we can only speculate about the possible mechanism through which MFs decrease intracellular GSH. Preliminary results suggest that the type of ROS produced upon MF exposure may be radicals derived from nitric oxide (NO•; NOO-, which can be detected by DHCFDA); thus the decrement in GSH level may be due to the formation of S-nitroso-glutathione adducts (in preparation).

We have shown that cells deprived of GSH are not rescued by MFs from stress-induced apoptosis. Thus, it seems that the possibility of using GSH as an early response to MF exposure is required in order to build up an antiapoptotic pathway. A possible explanation may be the possible production of GSNO, which is known to be a modulator of apoptosis.[28] Alternatively, we may hypothesize that GSH is required in order to maintain a correct conformation of Ca^{2+} channels/pumps, which are known to require a correct redox equilibrium in order to work,[29] and are essential for MFs' antiapoptotic effect.[8] This latter interpretation may also explain why the reducing environment created by DTT leads to the same results, that is, inhibition of MFs' antiapoptotic effect.

We have shown that the decrement of intracellular GSH level was maintained for a long time (72 h) of static MF exposure. Thus, U937 cells present changes in GSH homeostasis for several days. The lower efficiency in neutralizing ROS can be predictive of increased susceptibility toward oxidative stress and onset of cellular damages through oxidative pathways.[30] This finding suggests the need to deeply investigate the long-term effects of oxidative responses to magnetic fields and the risk that chronic oxidative impairment might enhance susceptibility to other forms of pathologies or diseases.

It is well known that oxygen free radicals and other activated oxygen species are implicated in various physiological and pathological conditions, namely, the aging process, inflammatory injury, atherosclerosis, arthritis, etc.[31] GSH is a critical factor in protecting organisms against toxicity and disease. Changes in GSH homeostasis were implicated in the etiology and progression of a range of human disorders, including cancer and neurodegenerative and cardiovascular diseases.[32] The finding of the alterations exerted by MF on intracellular redox state might be of great importance for human health, since ROS and GSH alterations may negatively affect general homeostasis.

As a last point, we want to note that ELF and static MFs seem to exert similar effect in terms of apoptosis; also the mechanisms investigated so far (Ca2$^+$ influx; GSH loss) are shared. So far, we notice a clear difference of the threshold—0.6 mT for static MFs and 0.09 mT for ELF-MFs. It will be important to understand whether their effects may totally overlap.

REFERENCES

1. FEYCHTING, M., U. FORSSEN & B. FLODERUS. 1997. Occupational and residential magnetic field exposure and leukemia and central nervous system tumors. Epidemiology **8:** 384–389.
2. LI, C.Y., G. THERIAULT & R.S. LIN. 1997. Residential exposure to 60 Hz magnetic fields and adult cancers in Taiwan. Epidemiology **8:** 25–30.
3. UK CHILDHOOD CANCER STUDY INVESTIGATORS. 1999. Lancet **354:** 1925–1931.
4. MCCANN, J., F. DIETRICH & C. RAFFERTY. 1998. The genotoxic potential of electric and magnetic fields: an update. Mutat. Res. **411:** 45–86.
5. AHLBOM, I.C. et al. 2001. ICNIRP (International Commission for Non-Ionizing Radiation Protection) Standing Committee on Epidemiology. Review of the epidemiologic literature on EMF and health. Environ. Health Perspect. **109:** 911–933.
6. CRUMPTON, M.J. & A.R. COLLINS. 2004. Are environmental electromagnetic fields genotoxic? DNA Repair **3:** 1385–1387.
7. LACY-HULBERT, A., J.C. METCALFE & R. HESKETH. 1998. Biological responses to magnetic fields. FASEB J. **12:** 395–420.
8. FANELLI, C. et al. 1999. Magnetic fields increase cell survival by inhibiting apoptosis via modulation of Ca^{2+} influx. FASEB J. **13:** 95–102.
9. NAARALA, J., A. HOYTO & A. MARKKANEN. 2004. Cellular effects of electromagnetic fields. Altern. Lab. Anim. **32:** 355–360.
10. WOLF, F.I. et al. 2005. 50-Hz extremely low frequency electromagnetic fields enhance cell proliferation and DNA damage: possible involvement of a redox mechanism. Biochim. Biophys. Acta **1743:** 120–129.
11. LÖSCHER, W. & R.P. LIBURDY. 1998. Animal and cellular studies on carcinogenic effects of low frequency (50/60-Hz) magnetic fields. Mutat. Res. **410:** 185–220.

12. SIMON, H.U., A. HAJ-YEHIA & F. LEVI-SCHAFFER. 2000. Role of reactive oxygen species (ROS) in apoptosis induction. Apoptosis **5:** 415–418.
13. HEDLEY, D. & S. CHOW. 1994. Glutathione and cellular resistance to anti-cancer drugs. Methods Cell. Biol. **42:** 31–44.
14. GHIBELLI, L. *et al.* 1995. Non-oxidative loss of glutathione in apoptosis via GSH extrusion. Biochem. Biophys. Res. Comm. **216:** 313–320.
15. GHIBELLI, L. *et al.* 1998. Rescue of cells from apoptosis by inhibition of active GSH extrusion. FASEB J. **12:** 479–486.
16. D'ALESSIO, M. *et al.* 2003. Apoptotic GSH extrusion is associated to free radical generation. Ann. N.Y. Acad. Sci. **1010:** 449–452.
17. D'ALESSIO, M. *et al.* 2005. Oxidative Bax dimerization promotes its translocation to mitochondria independently of apoptosis. FASEB J. **19:** 1504–1506.
18. GHIBELLI, L. *et al.* 1999. Glutathione depletion causes cytochrome c release even in the absence of cell commitment to apoptosis. FASEB J. **13:** 2031–2036.
19. FIORANI, M. *et al.* 1992. Electric and/or magnetic field effect on DNA structure and function in cultured human cells. Mutat. Res. **282:** 25–29.
20. VANDEPUTTE, C. *et al.* 1994. A microtiter plate assay for total glutathione and glutathione disulfide contents in cultured/isolated cells: performance study of a new miniaturized protocol. Cell. Biol. Toxicol. **10:** 415–421.
21. REED, D.J. *et al.* 1980. High performance liquid chromatography analysis of nanomole levels of glutathione, glutathione disulphide and related thiols and disulphides. Anal. Biochem. **106:** 55–62.
22. MEISTER, A. & M.E. ANDERSON. 1983. Glutathione. Annu. Rev. Biochem. **52:** 711–760.
23. JEFFERIES, H. *et al.* 2003. Glutathione. ANZ J. Surg. **73:** 517–522.
24. POMPELLA, A. *et al.* 2003. The changing faces of glutathione, a cellular protagonist. Biochem. Pharmacol. **66:** 1499–1503.
25. ANDRE, M. & E. FELLEY-BOSCO. 2003. Heme oxygenase-1 induction by endogenous nitric oxide: influence of intracellular glutathione. FEBS Lett. **546:** 223–227.
26. AW, T.Y., M. OOKTENS & N. KAPLOWITZ. 1984. Inhibition of glutathione efflux from isolated rat hepatocytes by methionine. J. Biol. Chem. **259:** 9355–9358.
27. PAOLICCHI, A. *et al.* 2002. Glutathione catabolism as a signaling mechanism. Biochem. Pharmacol. **64:** 1027–1035.
28. SANDAU, K. & B. BRUNE. 1996. The dual role of S-nitrosoglutathione (GSNO) during thymocyte apoptosis. Cell. Signal. **8:** 173–177.
29. WARING, P. 2005. Redox active calcium ion channels and cell death. Arch. Biochem. Biophys. **434:** 33–42.
30. HALLIWELL, B. & J.M.C. GUTTERIDGE. 1990. Role of free radicals and catalytic metal ions in human disease: an overview. *In* Methods in Enzymology, Vol. 186. L. Packer & A.N. Glazer, Eds.: 1–85. Academic Press. New York.
31. STADTMAN, E.R. 1992. Protein oxidation and aging. Science **257:** 1220–1224.
32. TOWNSEND, D.M., K.D. TEW & H. TAPIERO. 2003. The importance of glutathione in human disease. Biomed. Pharmacother. **57:** 145–155.

The Cleavage Mode of Apoptotic Nuclear Vesiculation Is Related to Plasma Membrane Blebbing and Depends on Actin Reorganization

M. DE NICOLA,[a] C. CERELLA,[a] M. D'ALESSIO,[a] S. COPPOLA,[a]
A. MAGRINI,[b] A. BERGAMASCHI,[b] AND L. GHIBELLI[a]

[a]*Dipartimento di Biologia, Università di Roma Tor Vergata, via della Ricerca Scientifica, 00133 Roma, Italy*

[b]*Cattedra di Medicina del Lavoro, Università di Roma Tor Vergata, via della Ricerca Scientifica, 00133 Roma, Italy*

ABSTRACT: In U937 monocytic cells induced to apoptosis, plasma membrane blebbing of different intensities appears, before the development of nuclear alterations; this latter phenomenon can occur through two major pathways, namely the cleavage and the budding mode (Dini *et al.*, 1996). Strongly blebbing cells develop deep nuclear constrictions leading to nuclear fragmentation according to the cleavage mode, while cells with milder forms of blebbing, or no blebbing at all, undergo nuclear fragmentation along the budding mode. Compounds interfering with different cytoskeletal components affect blebbing, which is completely inhibited by the actin polymerization inhibitors, cytochalasins, while disturbance of tubulin network with taxol limits blebbing to milder forms. At the same time, the cytoskeletal poisons affect the type of nuclear fragmentation, abolishing the cleavage mode, shifting all events into the budding pathway. Adherent cells, which possess a more structured cytoskeleton, do not develop strong blebs and undergo nuclear fragmentation via budding. These observations suggest that the deep cytoskeletal movements that cause the strongest forms of plasma membrane blebbing strangle the nucleus, leading to the constrictions that later evolve into nuclear fragmentation by cleavage. The trigger for the cytoskeletal movements, known to be redox-sensitive, is probably the apoptotic GSH extrusion.

KEYWORDS: cleavage; budding; apoptotic morphology

Address for correspondence: Lina Ghibelli, Dipartimento di Biologia, Università di Roma Tor Vergata, via della Ricerca Scientifica, 00133 Roma, Italy. Voice: +39-06-7259-4323; fax: +39-062023500.
e-mail: ghibelli@uniroma2.it

INTRODUCTION

Plasma membrane blebbing and nuclear fragmentation are two dispensable, though quite common, events of apoptosis. Apoptotic blebbing is frequent among cells growing in suspension, and requires a deep cytoskeletal reorganization.[1] It is a very striking phenomenon, which turns a spheroid cell into a set of bubbles, connected through their cytoplasm, within a few minutes; these bubbles may detach, originating the so-called apoptotic bodies. Blebbing is not exclusively related to apoptosis, but can occur, with different and less striking modalities, in several cytopathological conditions, such as oxidative stress. Though not clarified yet as to its genesis or biochemical mechanisms, blebbing has been shown to depend on Ca^{2+} and/or thiol disturbances, which interfere with the actin cytoskeletal network.[2–5] Indeed, the involvement of cytoskeleton in apoptotic blebbing was shown by experiments indicating that inhibition of actin polymerization inhibits blebbing[6]; however, this does not affect the development of other apoptotic features, such as DNA digestion or chromatin condensation, and the loss of cell viability.[6]

Nuclear fragmentation is a striking alteration occurring in apoptosis as an alternative to nuclear shrinkage,[7] the choice between the two events being cell type–specific rather than depending on the type of apoptogenic inducer. An ultrastructural study of apoptotic nuclear fragmentation in U937 cells indicated that it occurs through multiple, alternative pathways, the most frequent being the budding and the cleavage modes.[7] Nuclear fragmentation by cleavage occurs when a cleft in the nuclear sap forms, at a point in the nuclear periphery, which is free of condensed chromatin, growing inward and eventually cleaving the nucleus; chromatin condenses in tiny, regularly shaped crescents. Nuclear fragmentation by budding involves chromatin condensation to form large clumps, which press against the nuclear envelope, protruding, and eventually budding, from the nuclear membrane; the budding mode is the one that is most commonly observed and shown in the specific literature.[7]

Apoptosizing cells get rid of reduced glutathione prior to apoptosis[8] by promoting its efflux via specific carriers: this is a *sine qua non* for damage-induced apoptosis.[9] The mechanism through which GSH efflux is necessary for the progression of apoptosis may involve the oxidative dimerization and translocation of Bax[10] and cytochrome *c* release.[11]

The assembly and remodeling of actin cytoskeleton is regulated by, and very sensitive to, redox alterations,[12] and is involved in the deep morphological alterations of blebbing.[13]

In this study, we demonstrate a relationship between cytoplasmic and nuclear events in apoptosis, possibly mediated by cytoskeletal-associated contractile forces, by showing that the cleavage mode of nuclear fragmentation depends on the development of plasma membrane blebbing, which is able to "strangle" the nucleus, thus triggering the cleavage pathway. Reshaping of the actin cytoskeleton may be a consequence of apoptotic GSH loss.

MATERIALS AND METHODS

Cell Culture and Treatments

U937, CEM, and HepG2 cells were kept in log phase in RPMI 1640 supplemented with 10% (5% for CEM) inactivated FCS; the experiments were performed at the concentration 10^6 cells/mL (for HepG2, 2×10^6 cells/25 cm^2 flask). Macrophages from human peripheral blood were isolated as described by Ghibelli et al.[14] Apoptosis was induced with 10 μg/mL puromycin (PMC); 100 μM etoposide (VP16)[9]; 1 mM H2O2 for 1 h followed by recovery in fresh medium.[15] Inhibition of tubulin assembly was induced with 100 nM taxol. Taxol was added 1 h before the apoptogenic treatment and kept throughout the experiment.

Analysis of Blebbing and Nuclear Fragmentation

Nuclear fragmentation was detected after staining with the specific DNA dye Hoechst (1μg/mL), and analyzed by the fluorescence microscope according to the nuclear morphological ultrastructural features described by Ghibelli et al.[9] For simultaneous detection of nuclear features and cytoplasmic shape, Hoechst-stained cells were also analyzed under the phase-contrast microscope. In order to avoid blebs, disappearance on account of centrifugation of nonfixed cells, cells were directly examined without fixation upon staining with the supravital Hoechst dye (type 33342).

Quantitation of Blebbing and Nuclear Fragmentation

The fraction of cells with fragmented, crescent-shaped, or shrunken nuclei was evaluated among the Hoechst-stained cells by counting at least 300 cells in at least three random selected fields.

Inhibition of Actin Polymerization

A 5 μg/mL cytochalasin D was added to U937 cells 1h before the apoptogenic treatment and kept throughout the experiment.

Fluorescence Microscopy and Digital Photomicrography

For fluorescence microscopy analysis, stained cells were observed with a Nikon Eclipse TE 200 microscope equipped with a 100-W mercury lamp. Images were recorded with a CoolSNAP digital camera.

Statistical Analysis

Statistical analysis was performed using Student's t-test for unpaired data and P-values < 0.05 were considered significant. Data are presented as mean $\pm$ SD.

RESULTS

Types of Apoptotic Blebbing

We followed the development of plasma membrane blebbing in U937 cells induced to apoptosis by puromycin, a protein synthesis inhibitor that produces almost 100% apoptotic cells in 5 h of continuous incubation. The direct observation of living cells at the phase contrast reveals that blebbing is a strikingly rapid phenomenon: the round shape of the cells suddenly disappears to give rise to a set of spherically shaped, easily recognizable protrusions (FIG. 1 A). Each blebbing cell has its own peculiar shape; however, three main categories can be distinguished, as depicted in FIGURE 1B: total blebbing is the most frequent; it involves the whole cell body with very deep constrictions, giving a berry-like shape; milder forms of blebbing involve only a region or just the cortical part of the cytoplasm.

Blebbing Precedes Nuclear Alterations in Apoptosis

To study the temporal relationship between blebbing and nuclear fragmentation, the simultaneous observation of Hoechst-stained nuclei at the fluorescence microscope and cell contour at the phase-contrast microscope is required. The direct observation of living cells, stained with the supravital DNA dye Hoechst 33342, allowed us to establish a precise order of events. Indeed, the time-course analysis of the fraction of blebbing cells and cells with apoptotically altered nuclei upon puromycin treatment, shown in FIGURE 1C, reveals that blebbing cells appear before the appearance of cells with altered nuclei. The observation that many blebbing cells have a still normally shaped nucleus confirms that plasma membrane blebbing precedes nuclear alterations.

Relationship between Blebbing and Type of Nuclear Fragmentation

In the already mentioned TEM study we had described multiple pathways for apoptotic nuclear vesiculation occurring in U937 cells as a response to different apoptogenic agents; the identification of intermediate stages allowed us to demonstrate that the different morphologies do not interconvert one into

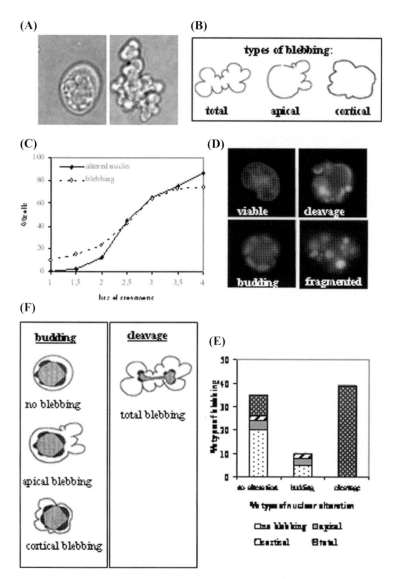

FIGURE 1. Relationship between plasma membrane blebbing and type of nuclear fragmentation (**A**) shows a direct living cell and blebbing cells at the phase-contrast microscope as described in MATERIALS AND METHODS. (**B**) Different types of blebbing. (**C**) Time-course analysis of blebbing and nuclear apoptosis in U937 cells induced to apoptosis with PMC showing that blebbing precedes nuclear fragmentation. Values represent mean ± SD ($n = 4$). (**D**) Nuclear apoptotic morphology unambiguously revealed by Hoechst 33342 staining in PMC-treated U937 cells as described in MATERIALS AND METHODS. (**E**) Relationship between blebbing and type of nuclear fragmentation. (**F**) Extent of different types of blebbing with respect to nuclear alteration in U937 cells induced to apoptosis with PMC at 3 h of treatment. Values represent mean ± SD ($n = 4$).

the other, but are the result of independent pathways.[7] The two main interphase morphologies, budding and cleavage, are distinguishable for the nuclear membrane structure and the peculiar shape of chromatin clamps condensing at the borders, being either a protruding, budding mass in the budding cells, or a nonprotruding crescent in cells in cleavage. Nuclear vesiculation is then completed with separation of the vesicles, by budding or by invagination and sealing of the inner and outer nuclear membrane, respectively, to reach the final fragmented morphology.

The shape of condensed chromatin is easily recognizable by fluorescent microscopy after staining with the DNA-specific dye Hoechst 33342. In FIGURE 1D we show U937 induced to apoptosis with puromycin, where different apoptotic morphologies are recognizable; budding nuclei have protruding chromatin masses, whereas nuclei in cleavage have chromatin condensed in crescent shapes. Budding and cleavage decline at later times to give rise to the final fragmented nuclei.

We wanted to investigate whether a specific type of blebbing is associated with one or the other pathway of nuclear fragmentation. To study the temporal relationship between blebbing and nuclear fragmentation, we induced U937 cells to apoptosis with PMC and simultaneously observed Hoechst-stained nuclei at the fluorescence microscope and cell contour at the phase-contrast microscope. It was found that totally blebbing cells have either a normal, a cleaved, or a fragmented nucleus. Cells with apical or cortical blebbing, or without blebbing, have either a normal, or a budding, or a fragmented nucleus (FIG. 1E). FIGURE 1F shows the frequencies of the different nuclear shapes for each cytoplasmatic morphology. This analysis demonstrates that total blebbing is associated with the cleavage pathway of nuclear fragmentation, while mildly blebbing or non-blebbing cells undergo apoptotic nuclear fragmentation by budding. Moreover, this analysis shows that blebbing precedes nuclear alterations in cells that are to take the cleavage route; another implication is that with this analysis we confirm that total fragmentation is the final converging point of both types of nuclear vesiculation.

Effects of Cytoskeletal Poisons on Plasma Membrane Blebbing and Nuclear Fragmentation

In order to confirm the latest assumption, we analyzed the effects that compounds that alter in different ways the cytoskeletal network exert both on plasma membrane blebbing and type of nuclear fragmentation. It is known indeed that the inhibitors of actin polymerization, cytochalasins B and D, are able to inhibit blebbing. In our experiments, both cytochalasins inhibited all types of blebbing (FIG. 2A). This is accompanied by the complete shift of the nuclear fragmentation pathway toward the budding mode (FIG. 2A, B).

We then investigated the role of compounds interfering with non-actin cytoskeleton, such as taxol, an inhibitor of tubulin assembly, on apoptotic bleb

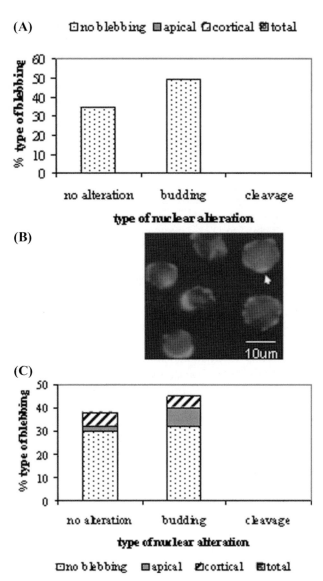

FIGURE 2. Effects of cytoskeletal poisons on plasma membrane blebbing and nuclear fragmentation. (**A**) The extent of different type of blebbing with respect to nuclear alteration, in U937 cells induced to apoptosis with PMC at 3 h of treatment, evaluated in the presence of CCD (actin polymerization inhibitor) Values represent mean ± SD ($n = 4$). (**B**) The nuclear apoptotic morphology revealed by Hoechst 33342 staining in PMC-treated U937 cells in the presence of CCD. (**C**) The extent of different types of blebbing with respect to nuclear alteration, in U937 cells induced to apoptosis with PMC at 3 h of treatment, evaluated in the presence of taxol (inhibitor of tubulin assembly). Values represent mean ± SD ($n = 4$).

formation. It turned out that taxol affected cell blebbing, inhibiting the appearance of totally blebbing cells, although allowing cortical or apical blebbing to take place (FIG. 2C). These results also indicate that microtubules contribute to the change of cell shape occurring in apoptosis. As occurs in apoptosizing U937 cells naturally developing mild blebbing as well as in taxol-treated cells, cortical or apical blebbing is accompanied by nuclear fragmentation by budding.

Analysis of Blebbing and Nuclear Fragmentation in Different Cell Systems

We extended the analysis of the relationship between blebbing and type of nuclear fragmentation both on U937 cells induced to apoptosis by other agents and on other cell types. On U937, etoposide-induced apoptosis elicits strong blebbing and mostly produces cleaved nuclei, while upon treatment with hydrogen peroxide, apoptotic cells do not develop strong blebs, and nuclei fragment according to the budding mode (not shown). In other cell systems, we observed that suspended lymphoid CEM cells behave as U937 cells, undergoing either strong blebbing associated with nuclear cleavage, or nuclear budding in the absence of blebbing (not shown), even though a slight nuclear shrinkage overlaps and partially obscures the detection of the different types of nuclear fragmentation (not shown). On adherent cells, we analyzed the hepatoma cell line HepG2 and macrophages from human peripheral blood, both induced to apoptosis by puromycin: in these cells no blebbing occurs, and nuclei fragment according to the budding mode, whereas the cells are still attached (not shown); cells then detach and a mild form of plasma membrane blebbing may take place definitely after the onset of nuclear fragmentation.

DISCUSSION

In this study we show that one of the major modes of apoptotic nuclear fragmentation occurring in cells in suspension, namely the cleavage pathway, is dependent on the previous development of plasma membrane blebbing. Most studies addressing questions on the evolution of the morphological aspects of apoptosis describe in detail the type of nuclear fragmentation we call the budding mode, while little attention has been paid so far to the steps and mechanisms of the cleavage mode. This may be due to the different frequency of the two pathways: indeed, our observations suggest that nuclear fragmentation by budding might be the default mode; the events that accompany plasma membrane blebbing may create overlapping effects, which shift nuclear fragmentation toward the cleavage mode. These effects are most likely related to the reorganization of the actin cytoskeleton, since they are completely inhibited by cytochalasins. Data showing the involvement of myosin in apoptotic blebbing[16] indicate that actin-related contractile forces may indeed develop in the cytoplasm of apoptosizing cells, thus possibly explaining the mechanism of nuclear strangling observed in our systems.

We know that redox alterations induced by GSH efflux cause apoptosizing cells to undertake the cleavage morphological route (De Nicola, in preparation). The mechanism through which this occurs is a matter of further studies; we begin to address this question here, showing that reshaping of the actin cytoskeleton is essential for nuclear vesiculation by cleavage. The notion that the usually cytosolic actin may play a role in determining nuclear apoptotic morphology is in line with recent findings indicating that actin fibers form in the nucleus after an apoptogenic stimulus.[17] Actin cytoskeleton is very sensitive to redox modulation,[12] and the deep actin reshaping occurring in apoptosis (i.e., blebbing) has been attributed to redox alterations.[5] We have evidence that the very rapid GSH loss occurring in apoptosis may alter intracellular redox equilibrium to such an extent as to remodel the actin cytoskeleton and to shift the nuclear morphology toward cleavage. According to this view, budding might be the default mode, and the redox alterations consequent to GSH loss superimpose additional effects that turn budding into cleavage.

REFERENCES

1. BELLOMO, G. *et al.* 1988. Oxidative stress-induced plasma membrane blebbing and cytoskeletal alterations in normal and cancer cells. Ann. N. Y. Acad. Sci. **551:** 128–130.
2. BELLOMO, G.G. *et al.* 1990. The cytoskeleton as a target in quinone toxicity. Free Radic. Res. Commun. **8:** 391–399.
3. MALORNI, W. *et al.* 1994. Both UVA and UVB induce cytoskeleton-dependent surface blebbing in epidermoid cells. J. Photochem. Photobiol. **26:** 265–270.
4. GABAI, V.L., A.E. KABAKOV & A.F. MOSIN. 1992. Association of blebbing with assembly of cytoskeletal proteins in ATP-depleted EL-4 ascites tumour cells. Tissue Cell **24:** 171–177.
5. JEWELL, S.A. *et al.* 1982. Bleb formation in hepatocytes during drug metabolism is caused by disturbances in thiol and calcium ion homeostasis. Science **217:** 1257–1259.
6. GHIBELLI, L. *et al.* 1992. Cycloheximide can rescue heat-shocked L cells from death by blocking stress-induced apoptosis. Exp. Cell Res. **201:** 436–443.
7. DINI, L. *et al.* 1996. Multiple pathways for apoptotic nuclear fragmentation. Exp. Cell Res. **223:** 340–347.
8. GHIBELLI, L. *et al.* 1995. Non-oxidative loss of glutathione in apoptosis via GSH extrusion. Biochem. Biophys. Res. Commun. **216:** 313–320.
9. GHIBELLI, L. *et al.* 1998. Rescue of cells from apoptosis by inhibition of active GSH extrusion. FASEB J. **12:** 479–486.
10. D'ALESSIO, M. *et al.* 2005. Oxidative Bax dimerization promotes its translocation to mitochondria independently of apoptosis. FASEB J. **19:** 1504–1506.
11. GHIBELLI, L. *et al.* 1999. Glutathione depletion causes cytochrome c release even in the absence of cell commitment to apoptosis. FASEB J. **13:** 2031–2036.

12. DALLE-DONNE, I.R. *et al.* 2001. The actin cytoskeleton response to oxidants: from small heat shock protein phosphorylation to changes in the redox state of actin itself. Free Radic. Biol. Med. **31:** 1624–1632.
13. MIRABELLI, F. *et al.* 1988. Menadione-induced bleb formation in hepatocytes is associated with the oxidation of thiol groups in actin. Arch. Biochem. Biophys. **264:** 261–269.
14. GHIBELLI, L. *et al.* 2003. Anti-apoptotic effect of HIV protease inhibitors via direct inhibition of calpain. Biochem. Pharmacol. **66:** 1505–1512.
15. NOSSERI, C., S. COPPOLA & L. GHIBELLI. 1994. Possible involvement of poly(ADP-ribosyl) polymerase in triggering stress-induced apoptosis. Exp. Cell Res. **212:** 367–373.
16. MILLS, J.C. *et al.* 1998. Apoptotic membrane blebbing is regulated by myosin light chain phosphorylation. J. Cell. Biol. **140:** 627–636.
17. LUCHETTI, F. *et al.* 2002. Actin involvement in apoptotic chromatin changes of hemopoietic cells undergoing hyperthermia. Apoptosis **7:** 143–152.

Experimental Apoptosis Provides Clues about the Role of Mitochondrial Changes in Neuronal Death

PATRIZIA FATTORETTI,[a] CARLO BERTONI-FREDDARI,[a] RINA RECCHIONI,[b] BELINDA GIORGETTI,[a] MARTA BALIETTI,[a] YESSICA GROSSI,[a] MORENO SOLAZZI,[a] TIZIANA CASOLI,[a] GIUSEPPINA DI STEFANO,[a] AND FIORELLA MARCHESELLI[b]

[a]*Neurobiology of Aging Laboratory, INRCA Research Department, 60121 Ancona, Italy*

[b]*Centre of Cytology, INRCA Research Department, 60121 Ancona, Italy*

ABSTRACT: A quantitative morphometric study has been carried out in human neuroblastoma SK-N-BE cells to evaluate the ultrastructural features and the metabolic efficiency of mitochondria involved in the early steps of apoptosis. In mitochondria from control and apoptotic cells cytochrome oxidase (COX) activity was estimated by preferential cytochemistry. Number of mitochondria (numeric density: Nv), volume fraction occupied by mitochondria/μm^3 of cytoplasm (volume density: Vv), and average mitochondrial volume (V) were calculated for both COX-positive and -negative organelles. The ratio (R) of the cytochemical precipitate area to the overall area of each mitochondrion was evaluated on COX-positive organelles to estimate the inner mitochondrial membrane fraction actively involved in cellular respiration. Following apoptotic stimulus, the whole mitochondrial population showed a significant increase of Nv and Vv, while V was significantly decreased. In COX-positive organelles higher values of Nv were found, V appeared significantly reduced, and Vv was unchanged. R was increased at a nonsignificant extent in apoptotic cells. COX-positive mitochondria accounted for 21% and 35% of the whole population in control and in apoptotic cells, respectively. These findings document that in the early stages of apoptosis the increased fraction of small mitochondria provides an adequate amount of ATP for progression of the programmed cell death and these more efficient organelles appear to represent a reactive response to the loss of metabolically impaired mitochondria. A better understanding of the mitochondrial role in neuronal apoptosis may suggest potential interventions to prevent the extensive nerve cell death typical of neurodegenerative diseases.

Address for correspondence: Patrizia Fattoretti, Neurobiology of Aging Laboratory, INRCA Research Department, Via Birarelli 8, 60121 Ancona, Italy. Voice: +39-071-800-4502; fax: +39-071-206791.
e-mail: p.fattoretti@inrca.it

Ann. N.Y. Acad. Sci. 1090: 79–88 (2006). © 2006 New York Academy of Sciences.
doi: 10.1196/annals.1378.008

KEYWORDS: morphometry; mitochondrial structural dynamics; apoptosis; cytochrome oxidase activity; mitochondrial metabolic competence; SK-N-BE cells

INTRODUCTION

Apoptosis is a programmed, physiological cell death that plays an important role in the normal development and differentiation of multicellular organisms as well as in the degeneration of many mammalian tissues.[1-3] Programmed cell death is frequently defined by characteristic morphological changes including fragmentation of nuclei with aggregation of chromatin and formation of surface protuberances that separate as membrane-bound globules (apoptotic bodies), whereas organelles in the cytoplasm appear to be intact.[4] The subcellular events that lead to cell death are fundamental to understanding several human pathological conditions as diverse as cancer, diabetes, obesity, ischemia–reperfusion injury, and neurodegenerative disorders, such as Alzheimer's, Parkinson's, and Huntington's diseases. Recent studies have suggested that mitochondria are involved in apoptotic signal transduction,[5,6] thus playing a critical role in determining whether a cell lives or dies. Current findings have been reported that fragmentation of mitochondria occurs upon induction of apoptosis,[7] and it has been suggested that activation of the mitochondrial fission machinery is one of the primary triggering events of this process. Moreover, the observation that during apoptosis within a discrete population of mitochondria not all the organelles may depolarize at the same time in response to a proapoptotic stimulus suggests that some mitochondria are involved in signaling during apoptosis, while others may continue to provide ATP during the early critical stages of apoptosis.[8] To get more information on the role of mitochondria in apoptosis, by means of computer-assisted morphometry and preferential cytochemistry, we carried out an investigation on the metabolic competence of these organelles in SK-N-BE cells following an apoptotic stimulus.

MATERIALS AND METHODS

Cell Culture

The human neuroblastoma SK-N-BE cells were seeded at a density of 10^4 cells/cm^2 in plastic culture flasks and grown to confluence in RPMI 1640 medium (ICN) containing 10% heat-inactivated fetal calf serum (Bio Whittaker), 2 mM glutamine, 100 units/mL penicillin, and 100 μg/mL streptomycin (ICN) at 37°C in a 5% CO_2 humidified atmosphere. Growth medium was changed three times a week and upon confluence the cells were dispersed with trypsin, splitted and subcultured at a density of 10^5 cells/cm^2 in chamber slides for electron microscopy experiments and in 6-well plates for flow cytometry detection of apoptosis.

Apoptosis Induction

The bacterial alkaloid staurosporine (STS) was used as inducer of apoptosis. To determine the time scale of apoptosis, SK-N-BE human neuroblastoma cells, 24 h after seeding, were treated with 0.1, 0.5, and 1 μM STS. The effect of STS was investigated 3, 6, 9, 18, and 24 h following treatment. The aim of these experiments was to find the optimal concentration of STS with which to perform the electron microscopic experiments.

Detection of Apoptosis by Flow Cytometry

Cytofluorimetric analysis was performed using an Epics XL flow cytometer (Coulter Corporation Miami, FL, USA) fitted with an air-cooled argon laser emitting at 488 nm. Typically, forward and orthogonal scatter signals were used to gate out live cells; in each experiment 10,000 events were collected by log amplification.

Measurement of Phosphatidylserine (PS) Exposure

PS redistribution from the inner to the outer leaflet is an early and common event during apoptosis. Surface exposure of PS by apoptotic cells was measured by Annexin V-FITC, which allows detecting and discriminating between apoptotic and dead cells. Apoptotic cells are stained positively for Annexin V-FITC, but are negative for staining with propidium iodide (PI). Dead cells stained positively for Annexin V-FITC and PI, whereas viable cells are negative for both Annexin V-FITC and PI. After trypsinization, cells were resuspended in binding buffer (10 mM Hepes/NaOH, pH 7.4, 140 mM NaCl, 2.5 mM $CaCl_2$) to a final concentration of 5×10^5 cells/mL. Annexin V-FITC was added to cell suspension in a final concentration of 5 μg/mL. The cells were subsequently incubated for 10 min in the dark at room temperature, then washed twice and resuspended in binding buffer. Prior to flow cytometric analysis, PI was added to a final concentration of 1 μg/mL.

Analysis of Mitochondrial Membrane Potential

The mitochondrial permeability transition is an important step in the induction of cellular apoptosis. During this process, the electrochemical gradient across the mitochondrial membrane (referred to as mitochondrial membrane potential or $\Delta\psi_m$) collapses. JC-1 is a cationic dye able to signal the loss of mitochondrial membrane potential. SK-N-BE adherent cells, after STS treatment, were incubated with the molecular probe JC-1 at a final concentration of 1 μg/mL for 15 min in 5% CO_2 at 37°C. The cells were then washed and harvested by trypsinization. Stained cells were washed twice and resuspended in 1 mL of phosphate-buffered solution (PBS) prior to flow cytometric analysis with excitation of 488 nM. The emission was between 525 and 590 nm.[9]

DNA Fragmentation Analysis

DNA fragmentation was measured by quantification of hypoploid nuclei after DNA staining with PI. SK-N-BE cells treated with STS were harvested by trypsinization and resuspended in 1 mL PI staining solution (50 μg/mL PI in 0.1% sodium citrate + 0.1% Triton X-100). The cell suspension was then incubated at 4°C. After washing in PBS, stained nuclei were analyzed on a FACS and cell debris was gated out based on light-scatter measurements before histograms of each parameter were drawn (excitation: 488 nM; emission: 620 nM). Hypoploid nuclei appear as a Sub G_0/G_1 peak.

Cytochrome Oxidase (COX) Cytochemistry

Control and STS-treated SK-N-BE cells were processed for electron microscopy according to a COX preferential cytochemical technique.[10,11] The cells were washed with 0.1 M phosphate buffer pH 7.4 (PB) and fixed with 1.5% glutaraldheyde in PB plus 4% sucrose for 20°C at room temperature. After a rinse with PB, cells were incubated for 2 h at 37 °C in the dark in the following solution (100 mL): 50 mg diaminobenzidine (DAB), 27 mg cytochrome c, and 4 g sucrose. Then cells were post fixed in 1% osmium tetroxide for 30°C at room temperature, processed by dehydration, and embedded, in epoxy resin according to conventional electron microscopic procedures. Ultrathin sections were stained in uranyl acetate and lead citrate and examined by an electron microscope (FIG. 1).

Electron Microscopy

In ultrathin sections (thickness: 60 nm), 700 μm² of cytoplasm were sampled. Electron microscopic images, at a magnification of 12,000×, of COX-positive and -negative mitochondria were acquired by a TV camera directly connected to the image analysis system (Kontron KS300). The following mitochondrial morphometric parameters were calculated: the number of

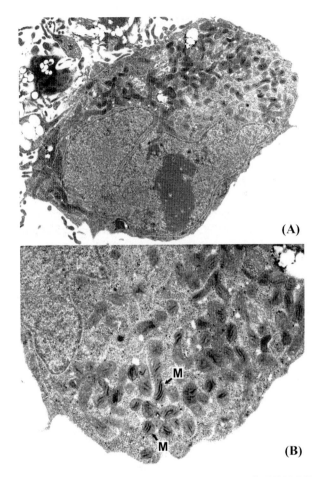

FIGURE 1. (**A**) Electron microscopic picture of an apoptotic SK-N-BE cell (0.5 μM STS for 6 h) stained by DAB method to evidence COX activity. (**B**) Detail of COX-positive mitochondria (M) in the SK-N-BE apoptotic cell. The dark cytochemical precipitate due to the DAB-COX reaction is sharply evidenced at the inner mitochondrial membrane.

mitochondria/μm^3 of cytoplasm (numeric density: Nv), the volume fraction occupied by mitochondria/μm^3 of cytoplasm (volume density: Vv), and the average volume (V). In COX-positive mitochondria, the total area of the cytochemical precipitates due to COX activity (CPA)/mitochondrion and the area (MA) of each COX-positive organelle were semiautomatically measured by the computer program of our image analyzer. The ratio (R) between CPA and MA was also calculated and referred to as the percentage of mitochondrial inner membrane area involved in COX activity.[12] The percentage of COX-positive organelles in control and apoptotic cells was also calculated. Statistical comparisons were performed by Student's *t*-test.

RESULTS

STS Induces Significant Apoptosis in SK-N-BE Human Neuroblastoma Cells Detected with PI Staining

A 0.1 μM STS induced nonsignificant effect on cellular apoptosis, whereas incubation with either 0.5 μM or 1 μM STS over 24 h was found to produce apoptotic death in a dose-dependent manner; therefore the dose-response curve, using PI staining, showed that both concentrations of 0.5 μM and 1 μM STS are adequate to induce apoptosis. The first apoptotic effect was detected about 6 h following STS treatment. STS (0.5 μM) resulted in an increase of apoptotic nuclei from $1.62\% \pm 0.35$, in the control cells, to $5.35\% \pm 0.58$ in the treated ones, reaching a percentage of $12.74\% \pm 0.8$, $22.99\% \pm 1.01$, $47.41\% \pm 1.07$, respectively after 9, 18, and 24 h. The dose-dependent effect was also observed with 1 μM STS, where the percentage of apoptotic nuclei increased drastically from $15.42\% \pm 0.67$, $28.95\% \pm 0.64$, $51.95\% \pm 0.71$ after 6, 9, and 18 h, respectively, up to $90.65\% \pm 1.81$ after 24 h. It is well known that PI assay is based on events that occur late in apoptosis, for example, development of a "leaky" plasma membrane, nuclear breakdown, and chromosomal fragmentation. To that end, we were interested in obtaining cells in early stages of apoptosis, and so we used Annexin V-FITC assay to detect apoptotic cells significantly earlier than would have been possible by DNA-based assay, and to choose the right dose and incubation time of STS.

Early Apoptosis of SK-N-BE Cells Detected by Annexin V-FITC Binding

We performed studies of Annexin V binding only on SK-N-BE cells treated with 0.5 μM STS after 6 and 9 h. In fact, we considered these points as the border between the early and late stages of apoptosis, where membrane integrity is still well preserved, as demonstrated by the very low percentage of cells that is able to uptake PI as a dye related to membrane impairment. Early apoptotic cells appeared in the Annexin V+/PI- fraction and remained constant between 6 h and 9 h while the number of Annexin V+/PI+ (reporting on the late apoptotic or necrotic cells), that is, those that lost their membrane integrity, increased. At 6 h, in the presence of STS, the percentage of Annexin V+/PI- cells increased from $4.1\% \pm 0.3$ to $28.9\% \pm 1.15$. At 9 h the number of Annexin V+/PI- cells was of $31.4\% \pm 0.85$, but Annexin V+/PI+ cells started to increase from $1.9\% \pm 0.9$, following 6-h treatment, to $8.2\% \pm 0.6$ after 9 h. Considering these preliminary dose- and time-effect studies, the concentration of 0.5 μM STS for 6-h incubation was chosen for the following experiments of electron microscopy because it offered the best compromise between early and late apoptosis during a period of 24 h.

STS Induces Elevation of the $\Delta\psi_m$ in SK-N-BE Human Neuroblastoma Cells

Following exposure to 0.5 μM STS, there was a significant increase of cell percentage with a higher $\Delta\psi_m$ compared to the control cells and this is maintained over the entire incubation period (data not shown). Based on forward and side scatter analysis, the cells with high membrane potential represent the living population; however, at 24 h, the population with high membrane potential includes also those cells that underwent apoptotic cell death. Before apoptosis is detectable, the increase in $\Delta\psi_m$ occurs after 3-h treatment with 0.5 μM STS. These results clearly demonstrate that SK-N-BE cells are committed to apoptosis by 0.5 μM STS after 3 h of incubation, whereas the early apoptotic phase is detectable after 6 h.

Ultrastructural Features and Metabolic Competence of Mitochondria in Control and SK-N-BE-Treated Cells

Following apoptotic stimulus, the whole mitochondrial population showed a significant increase of Nv (+83.75%) and Vv (+40.95%), whereas V was significantly decreased by 29.4%. In COX-positive organelles, higher values of Nv (+96.5%) were found, V appeared significantly reduced (−38.3%) and Vv was unchanged (FIG. 2). R was increased at a nonsignificant extent (+34.57%) in apoptotic cells. COX-positive mitochondria accounted for 21% and 35% of the whole population in control and in apoptotic cells, respectively. A percentage distribution of V of the whole mitochondrial population showed a higher fraction of smaller organelles (<0.1 μm^3) (66.9%) in apoptotic cells than in control ones (51%). COX-positive mitochondria exhibit the same trend with 45.9% of small organelles compared with control ones (30.5%).

DISCUSSION

A marked structural remodeling of the mitochondria of neuroblastoma SK-N-BE cells at the early stages of programmed cell death due to STS treatment constitutes the main finding of this study. The morphological rearrangements appear to be of functional significance since they are paired by significant increases of COX activity, a marker of neuronal energetic metabolism. Namely, the increase of Nv and Vv as well as the decrease of V in the whole population of mitochondria (FIG. 2A) can be found also in FIGURE 2B reporting the data of COX-positive organelles. Considering that the value of R was also increased (+34.5%) in COX-positive mitochondria of apoptotic cells versus controls, these data suggest that, at the early apoptotic stages, mitochondria are actively participating in this program by providing more amounts of ATP than in normal functional conditions. Moreover, the clear evidence that not all SK-N-BE

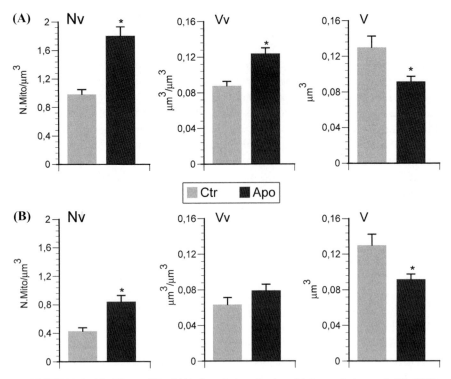

FIGURE 2. Nv, Vv, and V of (**A**) the whole mitochondrial population and (**B**) COX-positive organelles in control and apoptotic SK-N-BE cells. $*P < 0.05$

mitochondria are positive to the COX cytochemical staining procedure (as shown by comparing the data of Nv reported in FIG. 2A vs. B) suggests that a fraction of the population of these organelles is actively promoting the apoptotic process. This suggestion finds support in the many data in the literature documenting a role of mitochondria in the regulation of apoptosis.[8,13,14]

In interpreting these findings in terms of mitochondrial structural dynamics, it must be mentioned that the current knowledge on the morphological status of mitochondria within a given cell documents that these organelles may (1) constitute a unique network or (2) be separated units of the cell's energetic machinery. At a morphological inspection, the mitochondria may appear in either of the two conditions according to the actual requests of energy for an adequate functional response. In this context, an ongoing mechanism of fusion and fission is postulated to take place[7,15,16] and is supposed to account for the well-documented plasticity of selected pools of organelles. On the basis of these concepts, our present findings can be interpreted as a marked shift toward fission of mitochondria in SK-N-BE cells undergoing programmed cell deathing as a consequence of an STS stimulus. It must be stressed that a percent distribution of the V values of COX-positive organelles has shown that in apoptotic cells the fraction of smaller mitochondria is higher than in controls (45.9% vs. 30.5%) and this can be interpreted as a precocious reaction of the mitochondrial metabolic competence to sustain energetically the cell death program. A recent report from our laboratory has clearly shown that the small and medium-sized organelles are much more efficient in providing adequate ATP amounts than are the organelles of larger size.[17,18]

Considering our findings altogether, the present investigation confirms and strengthens the notion that mitochondria play a central role in regulating the processes involved in programmed cell death. In turn, this suggests that mitochondria may be considered reliable targets for intervention strategies in several pathological conditions (such as the neurodegenerative diseases typical of aging), for which cell death constitutes a well-documented step in the progression of the pathology, that is, Alzheimer's and Parkinson's disease.

ACKNOWLEDGMENT

The authors are particularly grateful to Mr. G. Bernardini for his skillful technical assistance.

REFERENCES

1. KERR, J.F.R., A.H. WYLLIE & A.R. CURRIE. 1972. Apoptosis: a basic biological phenomenon with wide-ranging implications in tissue kinetics. Br. J. Cancer **26:** 239–257.
2. WYLLIE, A.H., J.F.R. KERR & A.R. CURRIE. 1980. Cell death: the significance of apoptosis. Int. Rev. Cytol. **68:** 251–306.
3. MACAYA, A. 1996. Apoptosis in the nervous system. Rev. Neurol. **24:** 1356–1360.
4. ARENDS, M.J. & A.H. WYLLIE. 1991. Apoptosis: mechanisms and roles in pathology. Int. Rev. Exp. Pathol. **32:** 223–254.
5. KROEMER, G., B. DALLAPORTA & M. RESCHE-RIGON. 1998. The mitochondrial death/life regulator in apoptosis and necrosis. Annu. Rev. Physiol. **60:** 619–641.
6. GREEN, D.R. & J.C. REED. 1998. Mitochondria and apoptosis. Science **281:** 1309–1312.
7. KARBOWSKI, M. & R.J. YOULE. 2003. The dynamics of mitochondrial morphology in healthy cells and during apoptosis. Cell Death Differ. **10:** 870–880.
8. LEIST, M., B. SINGLE, A.F. CASTOLDI, et al. 1997. Intracellular adenosine triphosphate (ATP) concentration: a switch in the decision between apoptosis and necrosis. J. Exp. Med. **185:** 1481–1486.
9. COSSARIZZA, A., G. KALASHNIKOVA, E. GRASSELLI, et al. 1994. Mitochondrial modifications during rat thymocyte apoptosis: a study at the single cell level. Exp. Cell. Res. **214:** 323–330.
10. SELIGMAN, A.-M., M.J. KARNOVSKY, H.L. WASSERKRUG, et al. 1968. Nondroplet ultrastructural demonstration of cytochrome oxidase activity with a polymerising osmiophilic reagent, diaminobenzidina (DAB). J. Cell. Biol. **38:** 1–14.

11. WONG-RILEY, M.T.T. 1989. Cytochrome oxidase: an endogenous metabolic marker for neuronal activity. Trends Neurosci. **12:** 94–101.
12. BERTONI-FREDDARI, C., P. FATTORETTI, T. CASOLI, *et al.* 2001. Quantitative cytochemical mapping of mitochondrial enzymes in rat cerebella. Micron **35:** 405–410.
13. NICOTERA, P., M. LEIST & E. FERRANDO-MAY. 1998. Intracellular ATP, a switch in the decision between apoptosis and necrosis. Toxicol. Lett. **102–103:** 139–142.
14. FINKEL, E. 2001. The mitochondrion: is it central to apoptosis? Science **292:** 624–626.
15. BOSSY-WETZEL, E., M.J. BARSOUM, A. GODZIK, *et al.* 2003. Mitochondrial fission in apoptosis, neurodegeneration and aging. Curr. Opin. Cell. Biol. **15:** 706–716.
16. YOULE, R.J. & M. KARBOWSKI. 2005. Mitochondrial fission in apoptosis. Nat. Rev. **6:** 657–663.
17. BERTONI-FREDDARI, C., P. FATTORETTI, R. PAOLONI, *et al.* 2003. Inverse correlation between mitochondrial size and metabolic competence: a quantitative cytochemical study of cytochrome oxidase activity. Naturwissenschaften **90:** 68–71.
18. BERTONI-FREDDARI, C., P. FATTORETTI, B. GIORGETTI, *et al.* 2005. Age-related decline in metabolic competence of small and medium-sized synaptic mitochondria. Naturwissenschaften **92:** 82–85.

Alterations in mRNA Expression of Apoptosis-Related Genes *BCL2*, *BAX*, *FAS*, *Caspase-3*, and the Novel Member *BCL2L12* after Treatment of Human Leukemic Cell Line HL60 with the Antineoplastic Agent Etoposide

KOSTAS V. FLOROS,[a] HELLINIDA THOMADAKI,[a] DIMITRA FLOROU,[a] MAROULIO TALIERI,[b] AND ANDREAS SCORILAS[a]

[a]*Department of Biochemistry and Molecular Biology, Faculty of Biology, University of Athens, Panepistimiopolis, 15701 Athens, Greece*

[b]*"G. Papanicolaou" Research Center for Oncology, Saint Sava's Hospital, 11522 Athens, Greece*

ABSTRACT: Apoptotic cell death is a highly regulated process, which plays a crucial role in many biological events. Etoposide is an antineoplastic drug, which targets the DNA unwinding enzyme, topoisomerase II. The aim of the present research approach to investigate the expression of the apoptosis-related genes *BCL2* (Bcl-2), *FAS*, *Caspase-3*, *BAX* and the new member *BCL2L12*, cloned by our group, along with treatment of HL-60 leukemia cells with etoposide. The kinetics of apoptosis induction and cell toxicity was evaluated by DNA laddering and MTT method, respectively. The mRNA expression levels of the genes were analyzed by RT-PCR using gene-specific primers. β-*Actin* was used as a control gene. An important downregulation of *BCL2L12* was observed at 4 h of drug treatment, whereas *BAX* was upregulated at the same time point. No alteration in the expression pattern of the other apoptosis-related genes was detected. Since, the main anticarcinogenic effect of etoposide is due to the induction of apoptosis, these changes observed in the mRNA expression levels of the genes may be an underlying mechanism.

KEYWORDS: apoptosis; *BCL2L12*; BCL2-family; Bcl-2; HL-60; etoposide; caspases

Address for correspondence: Andreas Scorilas, Department of Biochemistry and Molecular Biology, Faculty of Biology, University of Athens, Panepistimiopolis, 15701 Athens, Greece. Voice: +30-210-727-4306; fax: +30-210-727-4158.
e-mail: ascorilas@biol.uoa.gr; scorilas@netscape.net

INTRODUCTION

Apoptosis is a complex, highly regulated process, characterized by cell shrinkage, chromatin condensation, and internucleosomal DNA fragmentation. The BCL2-family includes a growing number of proteins that serve as critical regulators of pathways involved in apoptosis, acting to either inhibit or promote programmed cell death. These proteins contain at least one of the four homology domains BH1, BH2, BH3, or BH4. The *BCL2L12* gene, the novel member of the BCL2-family, encodes a BCL2-like, proline-rich protein.[1] The BCL2L12 protein contains one BH2 homology domain, which is present in most antiapoptotic proteins.[1] PCR screening for *BCL2L12* transcripts demonstrated the presence of two bands. The upper band represents the classical form of the gene and the lower is a splice variant.[1]

The *BCL2* (B cell lymphoma gene-2) gene was the first oncogene to be implicated in the regulation of apoptosis. It is the most characteristic member of the family that inhibits apoptosis. Caspases are the executioners in the process of apoptosis. They are cysteine proteases with a specificity for aspartic acid and are divided in two subgroups: the upstream or initiator caspases and the downstream or effector caspases. Activation of the caspases proceeds by proteolytic cleavage of the expressed pro-form. The role of initiator caspases is to function as signaling molecules activating the effector caspases via proteolytic cleavages. The effector caspases are directly implicated in the execution of apoptotic cells. One of the downstream caspases is caspase-3. Once activated, it is thought to be one of these molecules that are responsible for the actual demolition of the cell during apoptosis.[2]

BAX (Bcl-2-associated protein X) is the most characteristic death-promoting member of the BCL2-family. The *BAX* gene encodes a protein that is primarily localized to the cytosol and after apoptotic stimulation is translocated to the mitochondria. At the mitochondria it triggers the release of cytochrome *c* and forms a complex with other cofactors that triggers the activation of caspase-9, initiating a downstream caspase cascade leading finally to cell death.[3]

The *Fas* gene (also called *Apo-1* or *CD95*) encodes a protein that has a central role in the regulation of programmed cell death and is the first of a series of apoptosis-inducing receptors. It is a death domain–containing member of the tumor necrosis factor receptor (TNFR) superfamily.[4] It is well known that one of the main pathways leading to the induction of apoptosis, the extrinsic cell death pathway, can be initiated upon Fas ligand–Fas receptor interactions at the cell surface.

Etoposide (VP-16) is an antineoplastic agent that targets the DNA unwinding enzyme, topoisomerase II. It was the first agent recognized as a topoisomerase II–inhibiting anticancer drug. When used at specific concentrations, it causes death of myeloid leukemia cells by apoptosis.[5,6] At concentrations achieved *in vivo*, etoposide causes dose-dependent single-strand and double-strand DNA breaks.[7] Like other topo II inhibitors, it does not kill cells by blocking the catalytic function of topoisomerase. VP-16 poisons the enzyme by increasing the steady-state concentration of its covalent DNA cleavage complex. This disrupts the cleavage complex and converts single- or double-strand breaks into permanent double-stranded fractures which are no longer held together. When these permanent DNA breaks are present at sufficient concentration, they trigger a number of events that ultimately lead to cell death by apoptosis.[8]

HL-60 is a human promyelocytic cell line that is very sensitive to a number of apoptosis-inducing agents.[9] It was recently shown that etoposide alone is able to cause cell death through the induction of apoptosis in HL-60.[10] Here, we investigate the alterations in the mRNA expression of *BCL2L12* in relation to the expression of the other apoptotis-related genes (*BCL2, BAX, Fas, Caspase-3*) during etoposide-induced apoptosis in HL-60 cells.

MATERIALS AND METHODS

Drugs and Cell Culture

Etoposide was obtained from Ebewe (Pharma m.b.H. Nfg.KG) as a solution for injection. The vial of the drug (20 mg/mL) was stored at 4°C.

The human promyelocytic leukemia cell line, HL-60, was maintained in RPMI 1640 medium supplemented with 10% fetal bovine serum. A 200 KU/l benzopenicillin, 0.1 g/L streptomycin, 0.3 g/L L-glutamine, and 0.85 g/L NaHCO$_3$, in a humidified atmosphere containing 5% CO$_2$. Cells were plated at 4×10^5 cells/mL and incubated at 37°C for 48 h before any treatment.

Cytotoxicity Assay

Cell viability was assayed using a standard MTT (3-[4,5-dimethylthiazol-2-yl]-2,5-diphenyltetrazoliumbromide thiazolyl blue indicator dye) chemosensitivity assay.[11] The HL-60 cell line was assessed for its sensitivity to etoposide in order to determine the concentration that was toxic to at least 50% (IC$_{50}$) of the cells after the indicated time periods of drug treatment. Cells in quadruplicate were treated with each drug for the indicated time periods prior to addition of MTT, in the presence of which they were further incubated for 4 h at 37°C. The absorbance of the cell lysate solution was measured at 550 nM, with a reference absorbance at 690 nM, and the results were expressed as the percentage (%) of treated cells versus untreated cells.

Electrophoretic Analysis of DNA Fragmentation

Internucleosomal DNA fragmentation was detected by electrophoresis, using DNA from 10^6 HL-60 cells/well, in a 2% agarose gel according to the Eastman protocol.[12] High-molecular-weight DNA fragments were trapped in or near the well, whereas the DNA fragments of low molecular weight were run and separated through the gel. The gel was stained with ethidium bromide and photographed with a Nikon F-801.

RNA Isolation

Total RNA was extracted from HL-60 cells using the Trizol reagent (Gibco-BRL, Gaithersburg, MD, USA).), using the manufacturer's instructions. The isolated RNAs had an A_{260}/A_{280} ratio of 1.6–1.8.

RT-PCR

A 2-μg sample of total RNA was reverse-transcribed into first-strand cDNA using the Superscript preamplification system (Gibco-BRL). The final volume was 20 μL. Gene-specific primers were designed and PCR was carried out in a reaction mixture containing 1 μL of cDNA, 10 mM Tris-HCl (pH 8.3), 50 mM KCl, 1.5 mM $MgCl_2$, 200 mM dNTPs, 150 ng of primers, and 1.5 U of GoTaq DNA polymerase (Promega Corporation, Madison, WI) in a LabNet thermal cycler. The cycling conditions were: a denaturation step at a 95°C for 2 min, followed by 35–39 cycles of 95°C for 30 sec, 62–65°C for 30 sec (for β-*Actin, BCL2L12, BCL2, BAX, FA*S and *Caspase-3*), 72°C for 1 min, and a final extension step at 72°C for 5 min. Equal amounts of PCR products were electrophoresed on 2% agarose gels and visualized by ethidium bromide staining. β-actin was used as an internal control for the integrity of the mRNA. The sequence of primers used in the RT-PCR was as followed: β-*Actin*: 5'- ATC TGG CAC CAC ACC TTC TA-3' (sense) and 5'- CGT CAT ACT CCT GCT TGC TG-3' (antisense); *BCL2L12*: 5'- GGA GAC CGC AAG TTG AGT GG-3' (sense) and 5'- GTC ATC CCG GCT ACA GAA CA-3' (antisense); *BCL2*: 5'- TTT GAG TTC GGT GGG GTC AT-3' (sense) and 5'- TGA CTT CAC TTG TGG CCC AG-3' (antisense); *Fas*: 5'-ATG CTG GGC ATC TGG ACC CT-3' (sense) and 5'- GCC ATG TCC TTC ATC ACA CAA-3' (antisense); *Caspase-3*: 5'- GCA GCA AAC CTC AGG GAA AC-3' (sense) and 5' -TGT CGG CAT ACT GTT TCA GCA -3' (antisense); *BAX*: 5'- TGG CAG CTG ACA TGT TTT CTG AC-3' (sense) and 5'- TCA CCC AAC CAC CCT GGT CTT-3' (antisense).

RESULTS AND DISCUSSION

In the present study, we first measured the cytotoxicity and apoptotic effect of etoposide on HL-60 cells. FIGURE 1 shows the percentage of viable cells

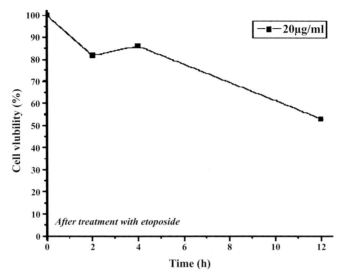

FIGURE 1. Cell viability of HL-60 cells after treatment with 34 μM etoposide (20 μg/mL) for 0, 2 h, 4 h, 12 h, assessed by MTT assay.

after treatment, for the time periods indicated, with etoposide. Assessment of cell viability by MTT assay showed that toxic phenomena (cell viability–50%) did not appear with the drug concentration used (34 μM), demonstrating that the variation in cell viability assessed by MTT mainly reflected programmed cell death. We also analyzed the appearance of DNA laddering (apoptosis) (FIG. 2). We noticed that DNA fragments were not evident until 1h of etoposide treatment. Low levels of DNA fragmentation were observed when leukemia cells were treated with 34 μM etoposide for 2 h and it was clearly visible at 4 h of cell exposure to the drug. Finally, we noticed an extensive, nonspecific degradation of DNA 12 h after drug treatment and so we did not study the expression of the apoptosis-related genes at this time point.

mRNA expression of *BCL2L12, BCL2, BAX, Fas,* and *Caspase-3* were investigated at the concentration of 34 μM (FIG. 3). A sharp downregulation of *BCL2L12* was observed at 4 h of exposure to the drug. The band of the splice variant of the novel gene seems to almost disappear at this time point. We have recently identified an upregulation of the mRNA expression levels of the gene after treatment of HL-60 cells with doxorubicin, carboplatin, and cisplatin.[13,14] In addition, preliminary experiments showed that *BCL2L12* is overexpressed more often in breast tumors with a low degree of differentiation,[15] as well as in patients in the initial stages of the disease.[15] Therefore, the mRNA expression levels of *BCL2L12* change in response to drug treatment and further experiments are required to verify its pro- or antiapoptotic behavior.

FIGURE 2. Assessment of DNA cleavage in HL-60 cells by agarose gel electrophoresis after treatment with 34 μM etoposide for the time periods indicated: 1, control; 2, 1 h; 3, 2 h; 4, 4 h; 5, 12 h.

The expression of the proapoptotic member of the BCL2 family, *BAX,* was increased at 4 h of cell treatment with the anticancer drug. In other recent studies with HL-60 cells, *BAX* also showed an increasing tendency after treatment with doxorubicin and amifostine,[16] but a gradual downregulation in its mRNA expression after exposure to carboplatin.[13]

Fas-, Caspase-3- and *BCL2-* transcripts did not show any modulation. Expression of *Fas* receptor on tumor cells has been shown to be enhanced by anticancer drugs. In the human monocytic leukemia cell line U937, it was shown that *Fas* expression was induced by low doses of etoposide.[17] In a more recent study with HL-60 cells, the expression level of *Fas*, among other cell death receptors, remained mostly unchanged during etoposide-induced apoptosis,[18] supporting our results. In addition, apoptosis, considered to involve mitochondrial dysfunction with subsequent release of cytochrome *c* from mitochondria, results in activation of *Caspase-3*. Experimental evidence with P39 cells, a myelodysplastic syndrome-derived cell line, demonstrated an etoposide-induced apoptosis and *Caspase-3* activation in this cell line, but without cytochrome *c* release.[19] Other experiments demonstrated an upregulation in the mRNA expression of *Caspase-3* after doxorubicin-induced apoptosis in HL-60 cells.[13] These modulations in the expression of *BCL2L12* and the other apoptosis-related genes, after treatment of the leukemia cells with etoposide, may be an underlining mechanism through as yet unknown apoptotic pathways or the result of a drug-dependent response.

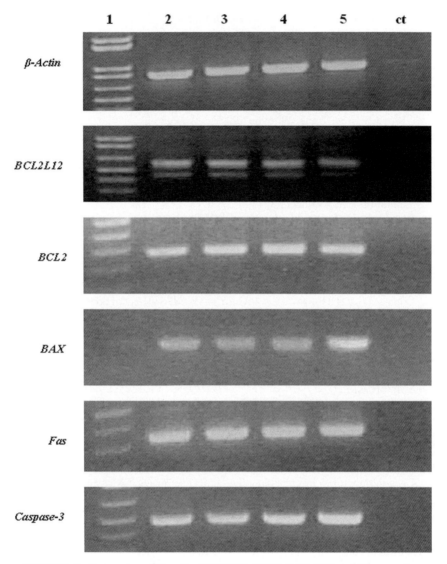

FIGURE 3. Expression of β-*actin, BCL2L12, BCL2, BAX, FAS,* and *Caspase-3* genes after treatment of the HL60 cells with etoposide (34 μM). Lane 1, DNA Markers; lane 2, control; lane 3, 1 h (34 μM); lane 4, 2 h (34 μM); lane 5, 4 h (34 μM); ct, H_2O (negative control).

ACKNOWLEDGMENTS

This work was supported by a Greek-Spanish joint research and technology grant (EPAN.M.4.3.6.1 No. 6014) co-funded by the European Regional Development Fund, the General Secretariat for Research & Technology of Greece, and by the Biopaths Co., Athens, Greece.

REFERENCES

1. SCORILAS, A., L. KYRIAKOPOULOU, G.M. YOUSEF, et al. 2001. Molecular cloning, physical mapping, and expression analysis of a novel gene, *BCL2L12*, encoding a proline-rich protein with a highly conserved BH2 domain of the *BCL2* family. Genomics **72:** 217–221.
2. CREAGH, E.M. & S.J. MARTIN. 2001. Caspases: cellular demolition experts. Biochem. Soc. Trans. **29:** 696–702.
3. ANTONSSON, B. 2001. Bax and other pro-apoptotic Bcl-2 family "killer-proteins" and their victim, the mitochondrion. Cell Tissue Res. **306:** 347–361
4. TRAUTH, B.C., C. KLAS, A.M. PETERS, et al.1989. Monoclonal antibody-mediated tumor regression by induction of apoptosis. Science **245:** 301–305.
5. HANDE, K.R. 1998. Etoposide: four decades of development of a topoisomerase II inhibitor. Eur. J. Cancer **34:** 1514–1521.
6. BERTRAND, R., E. SOLARY, J. JENKINS & Y. POMMIER 1993. Apoptosis and its modulation in human promyelocytic HL-60 cells treated with DNA topoisomerase I and II inhibitors. Exp. Cell Res. **207:** 388–397.
7. GARCÍA-BERMEJO, L., C. PÉREZ, N.E. VILABOA, et al. 1998. cAMP increasing agents attenuate the generation of apoptosis by etoposide in promonocytic leukemia cells. J. Cell. Sci. **111:** 637–644.
8. LOIKE, J.D., S.B. HORWITZ & A.P. GROLLMAN. 1976. Effect of podophyllotoxin and VP-16 on microtubule assembly *in vitro* and nucleoside transport in HeLa cells. Biochemistry **15:** 5435–5442.
9. LENNON, S.V., S.J. MARTIN & T.G. COTTER. 1991. Dose dependent induction of apoptosis in human tumour cell lines by widely diverging stimuli. Cell Prolif. **24:** 203–214.
10. DEBORAH, M.F., J.W. NIGEL, P.A.-M. GUSTAVO, et al. 1999. Collapse of the inner mitochondrial transmembrane potential is not required for apoptosis of HL60 cells. Exp. Cell Res. **251:** 166–174.
11. MOSMANN, T. 1983. Rapid colorimetric assay for the cellular growth and survival: application to proliferation and cyctotoxic assays. J. Immunol. Methods **65:** 55–63.
12. BARRY, M.A., C.A. BEHNKE & A. EASTMAN. 1990. Activation of programmed cell death (apoptosis) by cisplatin, other anticancer drugs, toxins and hyperthermia. Biochem. Pharmacol. **40:** 2353–2362.
13. FLOROS, K.V., H. THOMADAKI, N. KATSAROS, et al. 2004. mRNA expression analysis of a variety of apoptosis-related genes, including the novel gene of the BCL2-family, *BCL2L12*, in HL-60 leukemia cells after treatment with carboplatin and doxorubicin. Biol. Chem. **385:** 1099–1103.
14. FLOROS, K.V., H. THOMADAKI, G. LALLAS, et al. 2003. Cisplatin-induced apoptosis in HL-60 human promyelocytic leukaemia cells: differential expression of BCL2 and novel apoptosis-related gene *BCL2L12*. Ann. N.Y. Acad. Sci. **1010:** 153–158.
15. TALIERI, M., E.P. DIAMANDIS, N. KATSAROS, et al. 2003. Expression of *BCL2L12*, a new member of apoptosis-related genes, in breast tumors. Thromb. Haemost. **89:** 1081–1088.
16. RÓZALSKI, M., M. MIROWSKI, E. BALCERCZAK, et al. 2005. Induction of caspase 3 activity, bcl-2 bax and p65 gene expression modulation in human acute promyelocytic leukemia HL-60 cells by doxorubicin with amifostine. Pharmacol. Rep. **57:** 360–366.

17. AKIYAMA, H., T. INO, E. TOKUNAGA, *et al.* 2003. A synergistic increase of apoptosis utilizing Fas antigen expression induced by low doses of anticancer drug. Rinsho Byori **51:** 733–739.
18. BERGERON, S., M. BEAUCHEMIN & R. BERTRAND. 2004. Camptothecin- and etoposide-induced apoptosis in human leukemia cells is independent of cell death receptor-3 and -4 aggregation but accelerates tumor necrosis factor–related apoptosis-inducing ligand–mediated cell death. Mol. Cancer Ther. **3:** 1659–1669.
19. T. HISHITA, S. TADA-OIKAWA, K. TOHYAMA, *et al.* 2001. Caspase-3 activation by lysosomal enzymes in cytochrome *c*-independent apoptosis in myelodysplastic syndrome-derived cell line P39. Cancer Res. **61:** 2878–2884.

Using Janus Green B to Study Paraquat Toxicity in Rat Liver Mitochondria

Role of ACE Inhibitors (Thiol and Nonthiol ACEi)

M. GHAZI-KHANSARI,[a] A. MOHAMMADI-BARDBORI,[a,b] AND M-J. HOSSEINI[a]

[a]*Department of Pharmacology, School of Medicine, Tehran University of Medical Sciences, Tehran, Iran*

[b]*Faculty of Pharmacy, Shiraz University of Medical Sciences, Shiraz, Iran.*

ABSTRACT: Janus green B (JG-B) dye is used for vital staining of mitochondria and its reduction and oxidation shows the electron transfer chain alteration. The defect in electron transfer chain of mitochondria by paraquat is linked to free radical formation. In this present study we compared the abilities of different angiotensin-converting enzyme inhibitors, captopril (a thiol ACEi), enalapril, and lisinopril (two nonthiol ACEi) on mitochondria toxicity due to paraquat. The rat liver mitochondria were first isolated by centrifuge (at 4°C at a speed of 7,000 g) in a mixture of 0.25 M saccharose solution and 0.05 M Tris buffer. Various concentrations of paraquat (1, 5, 10 mM), enalapril (0.25, 0.5, 1 mM), lisinopril (0.01, 0.05, 0.1 mM), and captopril (0.08, 0.1, 1 mM) on the mitochondria isolated from the liver with respect to time were investigated. Paraquat at a concentration of 5 mM was determined to be significantly different compared to control values ($P < 0.05$) and captopril at a concentration of 0.08 mM, lisinopril (0.01 mM), and enalapril (0.25 mM) were found not to be significantly different from controls as found by spectroscopy at wavelength of 607 nm. Simultaneous treatment of mitochondria with captopril (0.08 mM) and paraquat (5 mM) significantly ameliorates the mitochondria toxicity of paraquat (5 mM) alone ($P < 0.05$). Our results show that captopril is a more effective antioxidant than the nonthiol ACEi. Lisinopril (0.01 mM) and enalapril (0.25 mM) did not significantly change the mitochondrial toxicity by paraquat (5 mM) ($P > 0.05$). The antioxidative

Address for correspondence: M. Ghazi-Khansari, Ph.D., Department of Pharmacology, School of Medicine, Tehran University of Medical Sciences, P.O. Box 13145-784, Tehran, Iran. Voice/fax: +9821-6640-2569.
 e-mail: ghazikha@sina.tums.ac.ir; khansagm@yahoo.com

action of captopril appears to be attributable to the sulfahydryl group (SH) in the compound. This effect may be due to captopril's abilities to scavenge reactive oxygen species.

KEYWORDS: paraquat; Janus green B; rat liver mitochondria; angiotensin-converting enzyme inhibitor (ACEi); ACE inhibitors (ACEi); rat liver

INTRODUCTION

Lewis (1923) used Janus green staining in a study on the role of the mitochondria in the development of the visual cells of chick embryos.[1] JG-B has been used as a method for studying the oxidation-reduction potential of various tissues and organisms.[2] It is probable that the cytochrome oxidase system plays a role in the super vital staining reaction.[3] J-G B dye is prepared by conjugating diethyl safranine to dimethyl aniline through an azo linkage. Diethyl safranine is a red dye and JG-B is blue. Thus the conjugation with dimethyl aniline alters the resonating structure of the diethyl safranine and shifts the position of the absorption spectrum maximum toward the longer wavelengths[4] with visible absorption spectrum maximum at 605 nm at pH 7. It has been shown that cytochrome c is reduced by leukosafranine and the leukosafranine is in turn oxidized to diethyl safranine. The cytochrome oxidase system, an enzyme within mitochondria, prevents the reduction of JG-B. This enzyme system is oxygen-dependent and cyanide-sensitive.[3]

Paraquat (N, N'-dimethyl 4, 4'-bipyridium) is very toxic to animals, including humans, with putative toxicity mechanisms associated with mitochondrial redox systems.[5] Mitochondria are candidate targets of paraquat toxicity in animal tissues and plants.[6] The mechanisms of paraquat toxicity are frequently related to the generation of the superoxide anion, which can lead to the formation of more toxic reactive oxygen species, for example, hydrogen peroxide, often taken as the main toxicant.[7,8] Palmeira et al.[9] reported that paraquat at a dose of 1mM, through its effect on uncoupling system, causes respiratory system depression by means of inhibition of mitochondrial complexes (I) and (V).

Angiotensin-converting enzyme (ACE) inhibitors, which have both an antihypertensive and a cardioprotective action, are commonly used in the treatment of hypertension and most forms of heart failure.[10–12] The beneficial effects of ACE inhibitors were thought to be primarily due to the inhibition of angiotensin II formation. However, a number of studies have shown improvement of oxidant stress and fibrosis. Treatment with enalapril or captopril was shown to increase antioxidant enzymes and nonenzymatic antioxidant defenses in several mouse tissues.[13]

In this report, we studied the role of three ACE inhibitors—captopril, enalapril, and lisinopril—to protect against mitochondria toxicity by paraquat.

MATERIALS AND METHODS

Compounds

Paraquat dichloride salt, captopril, lisinopril, and enalapril were purchased from the Sigma Chemical Co. (St. Louis, MO, USA). Janus green was obtained from Merck (Germany). All other chemicals were obtained from the Sigma Chemical Co.

Animals and Experimental Groups

This study was performed on 36 male Wister albino rats (weighing 80–120 g). They were kept in individual cages in a controlled room (temperature, 20–25°C humidity, 70% to 80%, exposed to 12 h of daylight). The rats were fed with standard rat food and tap water until experimentation. Twelve hours before the experiment the rats were stopped from feeding, but were allowed free access to tap water. Limitation of food and water was not applied to the animals that were put into their cages after the experiments.

Preparation of Mitochondria

The liver was removed and minced with small scissors in a cold manitol solution containing 0.225 M D-manitol, 75 mM sucrose, and 0.2 mM ethylenediaminetetraacetic acid (EDTA).[14] The minced liver (30 g) was gently homogenized in a glass homogenizer with a Teflon pestle and then centrifuged at 700 × g for 10 min at 4°C at remove nuclei, unbroken cells, and other non-subcellular tissue. The supernatants were centrifuged at 7,000 × g for 20 min. These second supernatants were pooled as the crude microsomal fraction and the pale loose upper layer, which was rich in swollen or broken mitochondria, lysosomes, and some microsomes, of sediments was washed away. The dark packed lower layer (heavy mitochondrial fraction) was resuspended in the manitol solution and recentrifuged twice at 7,000 × g for 20 min. The heavy mitochondrial sediments were suspended in Tris solution containing 0.05 M Tris-HCl buffer (pH 7.4), 0.25 M sucrose, 20 mM KCl, 2.0 mM $MgCl_2$, and 1.0 mM Na_2HPO_4 at 4°C before assay.

Mitochondrial Staining Procedure

A 1.0 mL of mitochondrial (10 mg protein) suspension treated with different concentrations of paraquat and ACEi were added with 1mL of JG-B (1 ppm). The blank solution was prepared in the same way without addition of mitochondrial suspension. Each sample was then measured via spectrophotometry at 607 nm within 40 min at every 10-min interval.

Viability Percentage Calculation

The percentage of mitochondrial viability of each test sample was calculated as described by Heidari *et al.* (2001) as follows:[15,16]

$$\%\text{Mitochondrial toxicity} = 1 - \frac{\text{Mean absorbance of toxicant}}{\text{Mean absorbance of negative control}} \times 100$$

$$\%\text{viability} = 100 - \%\text{Mitochondrial toxicity}$$

Experimental Design

Mitochondria suspension was divided into 13 groups for dose–response determination. Group 1 contained control, groups 2–4 contained enalapril (0.25, 0.5, 1 mM), groups 5–7 contained captopril (0.1, 0.08, 1 mM), groups 8–10 contained lisinopril (0.01, 0.05, 0.1 mM), and groups 11–13 contained paraquat (1, 5, 10 mM).

For treatment study, mitochondria suspension was divided into eight groups. Group 1 contained control, group 2 contained paraquat only (5 mM),

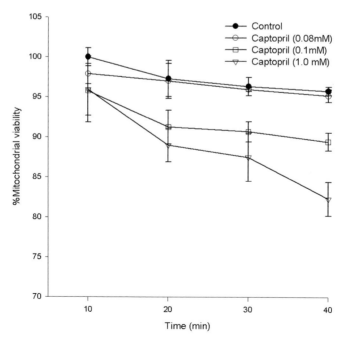

FIGURE 1. Effect of different concentrations of captopril (0.1, 0.08, 1 mM) on percentage viabilities of rat liver mitochondria suspension. There was no significant decrease in viability at 0.08 mM captopril dose levels as compared to control groups. All data are given as mean ± SEM, 3–4 mitochondria suspensions per group.

group 3–5 contained enalapril (0.25 mM) or captopril (0.08 mM) or lisinopril (0.01 mM), and group 5–8 contained paraquat (5 mM) together with enalapril (0.25 mM) or captopril (0.08 mM) or lisinopril (0.01 mM).

Statistical Analysis

All values were expressed as mean ± standard error (SEM) of 3–4 samples. Analysis of variance (ANOVA) followed by Student–Newmans–Keuls test was used to evaluate the significance of the results obtained. All computations were analyzed by computer using SPSS software.

RESULTS

Dose Response of Angiotensin-Converting Enzyme Inhibitors and Paraquat

Angiotensin-converting enzyme inhibitors were shown to decrease viability dose-dependently (FIGS. 1–3). Enalapril at a dose of 0.25 mM, captopril at a

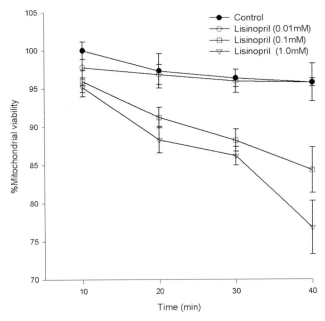

FIGURE 2. Effect of different concentrations of lisinopril (0.01, 0.05, 0.1 mM) on percentage viabilities of rat liver mitochondria suspension. There was no significant decrease in viability at 0.01 mM lisinopril dose levels as compared to control groups. All data are given as mean ± SEM, 3–4 mitochondria suspensions per group.

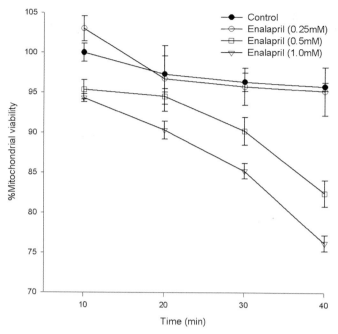

FIGURE 3. Effect of different concentrations of enalapril (0.25, 0.5, 1 mM) on percentage viabilities of rat liver mitochondria suspension. There was no significant decrease in viability at 0.25 mM enalapril as compared to control groups. All data are given as mean ± SEM, 3–4 mitochondria suspensions per group.

dose of 0.08 mM, and lisinopril at a dose of 0.01 mM were shown not to be significantly different from control ($P > 0.05$). Therefore, the above doses were used for the treatment study. Paraquat at a concentration of 5 mM (FIG. 4) was determined to be significantly different compared to control ($P < 0.05$).

Effect of Angiotensin-Converting Enzyme Inhibitors on Mitochondria Toxicity of Paraquat

Simultaneous treatment of mitochondria with captopril (0.08 mM) and paraquat (5 mM) (FIG. 5A) significantly ameliorates the mitochondria toxicity of paraquat (5 mM) alone ($P < 0.05$). Lisinopril (0.01 mM) (FIG. 5B) and enalapril (0.25 mM) (FIG. 5C) did not significantly change the mitochondrial toxicity by paraquat (5 mM) ($P > 0.05$).

DISCUSSION

There are two suggested mechanisms for paraquat mitochondrial toxicity.

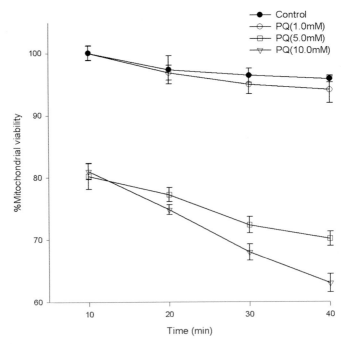

FIGURE 4. Effect of different concentrations of paraquat (1, 5, 10 mM) on percentage viabilities of rat liver mitochondria suspension. There was a significant decrease in viability at 5 mM paraquat as compared with control groups. All data are given as mean ± SEM, 3–4 mitochondria suspensions per group.

Alternate Electron Acceptors

Alternate electron acceptors are substances capable of extracting electrons from intermediates in the respiratory chain, competing with the natural substrates. These substances may also affect redox cycle, passing electrons back to the respiratory chain at a later point, bypassing sites in the chain essential for energy generation.

Inhibition of the Respiratory Chain

The respiratory chain can be inhibited at any of the four protein complexes in the respiratory chain, although effects on complex IV (cytochrome *c* oxidase) are the most severe because this is the step where oxygen is reduced to water. Inhibition at complex III can result in the generation of reactive oxygen species as the consequence of the inherent instability of the electron transfer process to this complex from reduced ubiquinone. JG-B in this study was not only used for vital staining of mitochondria, but also shows the alteration of the

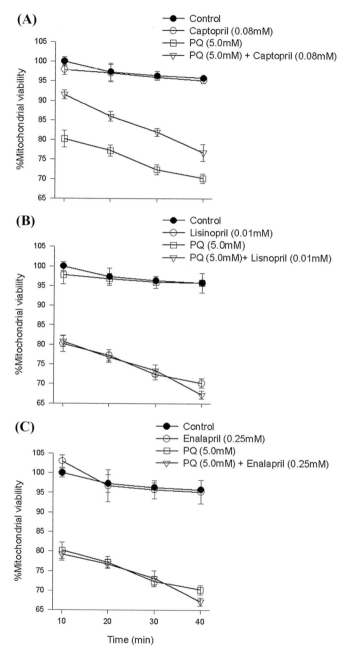

FIGURE 5. Effect of captopril (0.08 mm), lisinopril (0.01mM), and enalapril (0.25 mM) on paraquat toxicity (5 mM) isolated rat liver mitochondria. All dose levels compared with control groups. All data are given as mean ± SEM, 3–4 mitochondria suspensions per group.

electron transfer chain in mitochondria. Since paraquat also affects the electron transfer chain, JG-B is a good marker for its toxicity. FIGURE 4 shows paraquat dose-dependently increased the mitochondria toxicity. FIGURE 1–5 also shows that JG-B is reliable to study the effect of ACEi on paraquat mitochondria toxicities.

We found significant differences among three assayed ACEi in mitochondria toxicity induced by paraquat (FIG. 5A, B, C). Captopril was the most effective in decreasing mitochondria toxicity by paraquat. Our observation suggests that the presence of a thiol group in the ACEi structure may be a determinant for the antioxidant properties.[18] Under normal physiological conditions, antioxidants protect against reactive oxygen species (ROS) generated in the mitochondria. Numerous past studies have employed antioxidants, such as vitamin E[17] and melatonin, as treatments for excessive ROS production. Captopril was shown to partially prevent the decrease of coenzyme Q10 level, dimension of cytochrome oxidase activity state 3, oxidative phosphorylation rate (OPR), and the enhancement of mitochondrial F_1ATPase protein concentration. CoQ10 transfers electrons from complex I and II to complex III, and stabilizes respiratory complexes at the level of the inner mitochondrial membrane.[19] It seems that ACEi may protect tissues from oxidative damage by increased mitochondrial coenzyme Q10 level, improved respiratory chain function, and energy production. These effects may be due to captopril's abilities to scavenge reactive oxygen species as evaluated *in vitro* by Janus green B, a vital marker of mitochondria.

Of the angiotensin-converting enzyme inhibitor we used only pretreatment with thiols containing compounds (captopril) that were able to quench ROS generation from isolated mitochondria. Our data demonstrate that treatment with captopril can prevent damage induced by paraquat and therefore may be considered a therapeutic agent in the prevention and treatment of environmental toxins such as paraquat. Our results also suggest that JG-B dye is a simple, rapid, safe, cost-effective, reproducible, and reliable method to study mitochondria toxicity.

ACKNOWLEDGMENTS

This study was supported by a grant from the Vice Chancellor of Research of Tehran University of Medical Sciences (132/11737, March 18, 2006).

REFERENCES

1. ELFVIN, L.G. 1953. The supravital staining with Janus green B, a of mitochondria in the retinal rods of the guinea-pig eye. Exp. Cell Res. **5:** 554–556.
2. COOPERSTEIN, S.J., A. LAZAROW & J.W. PATERSON. 1953. II. Reactions and properties of Janus green B and its derivatives. Exp. Cell Res. **5:** 69–82.

3. COOPERSTIN, S.J. & A. LAZAROW. 1953. Studies on the mechanism of Janus green B staining of mitochondria. Exp. Cell Res. **5:** 82–97.
4. SATA, T., K. TAKESHIGE, R. TAKAYANAGI & S. MINAKAMI. 1983. Lipid peroxidation by bovine heart submitochondrial particles stimulated by 1, 1-dimethylbipyridylium. Biochem. Pharmacol. **32:** 13–19.
5. LAMBERT, C. & E.S.C. BONDY. 1989. Effects of MPTP, MPP and paraquat on mitochondrial potential and oxidative stress. Life Sci. **44:** 1277–1284.
6. TAYLOR, N.L., D.A. DAY & A.H. MILLAR. 2002. Environmental stress causes oxidative damage to plant mitochondria leading to inhibition of glycine decarboxylase. J. Biol. Chem. **277:** 42663–42668.
7. FARRINGTON, J.A., M. EBERT, E.J. LAND & K. FLETCHER. 1973. Bipyridylium salts and related compounds. V. Pulse radiolysis studies of the reaction of paraquat. Biochim. Biophys. Acta **314:** 372–381.
8. GHAZI-KHANSARI, M., G. NASIRI & M. HONARJOO. 2005. Decreasing the oxidant stress from paraquat in isolated perfused rat lung using captopril and niacin. Arch. Toxicol. **79:** 341–345.
9. PALMEIRA, C.M. A.J. MORENO & V.M. MADEIRA. 1995. Mitochondrial bioenergetics is affected by the herbicide paraquat. Biochim. Biophys. Acta **1229:** 187–192.
10. BRUNNER, H.R., H. GAVRAS, B. WAEBER, et al. 1979. Oral angiotensin-converting enzyme inhibitor in long-term treatment of hypertensive patients. Ann. Intern. Med. 19–23.
11. KIOWSKI, W., M. ZUBER, S. ELSASSER, et al. 1991. Coronary vasodilatation and improved myocardial lactate metabolism after angiotensin converting enzyme inhibition with cilazapril in patients with congestive heart failure. Am. Heart J. **122:** 1382–1388.
12. KONSTAM, M.A., M.F. ROUSSEAU, M.W. KRONENBERG, et al. 1992. Effects of the angiotensin converting enzyme inhibitor enalapril on the long-term progression of left ventricular dysfunction in patients with heart failure. SOLVD Investigators. Circulation **86:** 431–438.
13. CAVANAGH, E.M., C.G. FRAGA, L. FERDER & F. INSERRA. 1997. Enalapril and captopril enhance antioxidant defenses in mouse tissues. Am. J. Physiol. **272:** 514–518.
14. RAMASARMA, T. 1982. Generation of H2O2 in biomembranes. Biochim. Biophys. Acta **694:** 69–93.
15. HEIDARI, M., M. GHAZI-KHANSARI, M. KHOSHNOODI & F. SHOKRI. 2003. Comparative measurement of in vitro T-2 toxin cytotoxicity using three different cytotoxicity assays. Toxicol. Mech. Methods **13:** 153–157.
16. SHOKRI, F., M. HEIDARI, S. GHARAGOZLOO & M. GHAZI-KHANSARI. 2000. In vitro inhibitory effects of antioxidants on cytotoxicity of T-2 toxin. Toxicology **146:** 171–176.
17. GARCIA-ESTRADA, J., O. GONZALEZ-PEREZ, R.E. GONZALEZ-CASTANEDA, et al. 2003. An alpha-lipoic acid-vitamin E mixture reduces post-embolism lipid peroxidation, cerebral infarction, and neurological deficit in rats. Neurosci. Res. **47:** 219–224.
18. GHAZI-KHANSARI, M., G. NASIRI & M. HONARJOO. 2005. Decreasing the oxidant stress from paraquat in isolated perfused rat lung using captopril and niacin. Arch. Toxicol. **79:** 341–345.
19. GVOZDJAKOVA, A., F. SIMKO, J. KUCHARSKA, et al. 1999. Captopril increased mitochondrial coenzyme Q10 level, improved respiratory chain function and energy production in the left ventricle in rabbits with smoke mitochondrial cardiomyopathy. Biofactors **10:** 61–65.

Membrane Fluidity Changes Are Associated with Benzo[a]Pyrene-Induced Apoptosis in F258 Cells

Protection by Exogenous Cholesterol

MORGANE GORRIA,[a] XAVIER TEKPLI,[a] ODILE SERGENT,[b]
LAURENCE HUC,[a] FRANÇOIS GABORIAU,[c] MARY RISSEL,[a]
MARTINE CHEVANNE,[b] MARIE-THÉRÈSE DIMANCHE-BOITREL,[a]
AND DOMINIQUE LAGADIC-GOSSMANN[a]

[a]*INSERM U620, Université de Rennes 1, Rennes, France*

[b]*UPRES EA 3891, Université de Rennes 1, Rennes, France*

[c]*INSERM U522, Hôpital Ponchaillou, Rennes, France*

ABSTRACT: Polycyclic aromatic hydrocarbons (PAHs) such as benzo[a]yrene (B[a]P) constitute a widely distributed class of environmental pollutants, responsible for highly toxic effects. Elucidating the intracellular mechanisms of this cytotoxicity thus remains a major challenge. Besides the activation of the p53 apoptotic pathway, we have previously found in F258 hepatic cells that the B[a]P (50 nM)-induced apoptosis was also dependent upon the transmembrane transporter NHE1, whose activation might result from membrane alterations in our model. We here demonstrate that: (1) B[a]P induces a membrane fluidization surprisingly linked to NHE1 activation; (2) membrane stabilization by exogenous cholesterol protects cells from B[a]P-induced apoptosis, via an effect on late acidification and iron uptake.

KEYWORDS: benzo[a]pyrene; apoptosis; membrane fluidity; NHE1; cholesterol

INTRODUCTION

Polycyclic aromatic hydrocarbons (PAHs) are important environmental pollutants to which humans are largely exposed. Besides their well-known carcinogenic effects, PAHs also induce apoptosis in different cell types, for example, hepatic cells.[1,2] Although activation of p53 plays a primordial role in this

Address for correspondence: Dominique Lagadic-Gossmann, INSERM U620, Faculté de Pharmacie, Université de Rennes 1, 2 Av. Du. Prof Léon Bernard, 35043 Rennes cedex, France. Voice: +33-0-2-23-23-48-37; fax: +33-0-2-23-23-47-94.
e-mail: dominique.lagadic@rennes.inserm.fr

apoptosis, permissive pathways might also occur; indeed, we have recently evidenced a parallel activation of Na^+/H^+ exchanger (NHE1) with consequences on apoptosis occurrence.[1] Such an activation might result from changes in membrane characteristics, especially as PAHs are highly lipophilic molecules capable of interacting with biological membranes.[3] Increases in membrane fluidity can be involved in apoptosis, since inhibition of membrane fluidization has been shown to be protective in different apoptotic models.[4] An important role for reactive oxygen species (ROS) production has also been demonstrated in cell death induced by various chemicals.[5] As we recently demonstrated strong interactions between changes in membrane fluidity and ROS production in hepatocytes exposed to ethanol[4] and as PAHs induce ROS production,[1] we wanted to test here the possible involvement of membrane fluidity changes in B[*a*]P-induced apoptosis.

MATERIALS AND METHODS

Rat hepatic F258 epithelial cells were cultured and treated as previously described.[1] Membrane fluidity was measured either by quantification of the polarization fluorescent ratio of 1,6-diphenyl-1,3,5-hexatriene (DPH, 5 μM) or by analysis of electron paramagnetic resonance of 12-doxyl stearic acid (12-DSA, 50 μg/mL) spin label.

Apoptosis was evaluated either by Hoechst 33342 staining of chromatin or measurement of caspase 3/7 activity, and the intracellular pH (pH_i) was monitored by microspectrofluorimetry using the pH-sensitive fluorescent probe, carboxy-SNARF-1.[1]

Iron uptake was measured from cell lysates following a 6-h incubation with [^{55}Fe]-iron chloride (10 mM) performed during the late hours of each cell treatment. Lipid peroxidation was measured using the fluoroprobe C_{11}-BODIPY$^{581/591}$ (10 μM).[4]

RESULTS AND DISCUSSION

Using the fluorescence polarization technique and DPH, we first demonstrated that B[a]P (50 nM) induces a significant increase in bulk membrane fluidity following 48 h of exposure, as pointed out by the decrease of polarization ratio (FIG. 1A). Under our experimental conditions, we have previously found that B[*a*]P is metabolized by cytochromes of the CYP1 family within the first 24 h, indicating that B[*a*]P has already entered the cells through the plasma membrane at this time. We showed here that the B[*a*]P-induced membrane fluidization remained undetectable after a 24-h treatment, thus suggesting an indirect action of B[*a*]P on the membrane's physical state. After a 72-h treatment, an increase in fluidity was still detected. This latter

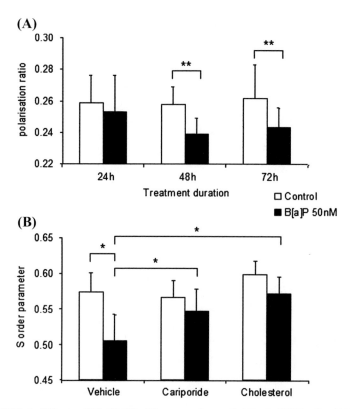

FIGURE 1. Effects of B[*a*]P (50 nM) on membrane fluidity in F258 cells. (**A**) Kinetics of B[*a*]P-induced bulk membrane fluidization. The fluorescence polarization ratio measured from DPH-labeled cells is conversely proportional to membrane fluidity. (**B**) Effects of cariporide (30 μM; B) and exogenous cholesterol (30 μg/mL;B) on B[*a*]P (48 h)-induced membrane fluidization, measured using EPR and 12-DSA spin label. S order parameter is conversely proportional to membrane fluidity. Data are given as mean ± SEM of at least three independent experiments ($^*P < 0.05$; $^{**}P < 0.01$, *t*-test).

result further reinforces the idea that a direct interaction of B[*a*]P with plasma membrane was likely not responsible for this sustained increase of membrane fluidity.

As we previously demonstrated that the transmembrane protein NHE1 was involved in our model of apoptosis, we then tested the possible involvement of this exchanger in the B[*a*]P-induced increase of membrane fluidity. Using cariporide (30 μM) to inhibit NHE1 activation observed at 48 h,[1] we then showed that the B[*a*]P-induced change in fluidity was partly related to this transporter, as pointed out by the absence of a significant decrease of the membrane order parameter estimated using 12-DSA and RPE (FIG. 1B).

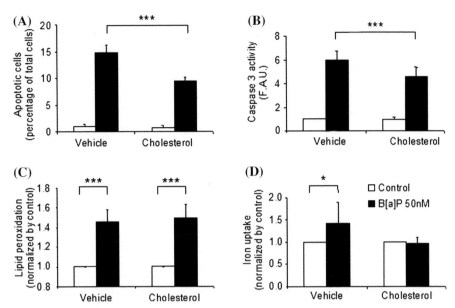

FIGURE 2. (A, B) Cotreatment with exogenous cholesterol affects B[a]P-induced apoptosis. Apoptosis was analyzed following a 72-h treatment with B[a]P by (A) counting the percentage of cells with apoptotic nuclei after Hoechst staining, and by (B) estimating caspase 3/7 activity. (C) Oxidative stress was analyzed by lipid peroxidation measurements after a 72-h treatment. (D) Iron uptake in F258 cells was analyzed using ^{55}Fe after a 72-h treatment. Results are representative of at least three independent experiments in each case ($*P < 0.05$; $***P < 0.001$, t-test).

Exogenous application of cholesterol (30 μg/mL), a well-known membrane stabilizer, was then used to test the involvement of membrane fluidization in B[a]P-induced apoptosis. Our data demonstrated that cotreatment with this compound, besides inhibiting any membrane fluidization (FIG. 1B), significantly reduced apoptosis, as evidenced by a decrease of cell population exhibiting nuclear fragmentation (FIG. 2A) and of caspase 3/7 activity (FIG. 2B). The inhibition by ~30% of B[a]P-induced apoptosis by cholesterol was in the range of inhibitions observed when the permissive apoptotic pathway relying upon NHE1 activation was targeted.[1]

In order to find out which event in the B[a]P-induced apoptotic cascade was related to membrane fluidization, we decided to test the effects of exogenous cholesterol application on the apoptosis-related intracellular acidification and the oxidative stress previously shown to be associated with B[a]P-induced apoptosis. Our results clearly demonstrated that cotreatment of F258 cells with cholesterol led to a significant reduction of B[a]P-induced acidification ($\Delta pH_i = -0.08 \pm 0.04$ [+ cholesterol] vs. $\Delta pH_i = -0.17 \pm 0.04$ pH units [− cholesterol], $n = 10$, $P < 0.01$, t-test), thus suggesting an inhibition of

B[*a*]P-induced mitochondrial damages; indeed, the observed acidification has been shown to be due to mitochondrial dysfunction.[1] We then focused on the possible involvement of membrane fluidization in B[*a*]P-induced oxidative stress. To do so, oxidative stress was evaluated by measuring lipid peroxidation. As shown in FIGURE 2C, cholesterol treatment remained ineffective on B[*a*]P-induced lipid peroxidation. On the basis of our recent observation showing that ethanol-induced membrane fluidization enhanced oxidative stress in rat hepatocytes via interactions with iron metabolism[4] and knowing that iron was involved in caspase 3/7 activation in our model (unpublished data), we finally investigated the possible action of cholesterol on B[*a*]P-elicited iron uptake. Analyzing ^{55}Fe uptake under our experimental conditions, we showed that the protective effect of cholesterol was likely through an inhibition of B[*a*]P-induced iron uptake (FIG. 2D). Altogether, our results indicate that cholesterol might reduce B[*a*]P-induced apoptosis by inhibiting both iron uptake and mitochondria-dependent acidification, thereby possibly interfering with caspase 3/7 activation.

In conclusion, this work suggests a role for NHE1 activation in B[*a*]P-induced membrane fluidization and an involvement of fluidization in the early events of B[*a*]P-induced apoptotic cascade. This, along with our previous work,[4] also points to a role for membrane fluidization in chemically induced cell death likely through interactions with iron transport.

REFERENCES

1. HUC, L. *et al.* 2004. Identification of Na^+/H^+ exchange as a new target for toxic polycyclic aromatic hydrocarbons. FASEB J. **18:** 344–346.
2. SOLHAUG, A. *et al.* 2004. Polycyclic aromatic hydrocarbons induce both apoptotic and anti-apoptotic signals in Hepa1c1c7 cells. Carcinogenesis **25:** 809–819.
3. JIMENEZ, M. *et al.* 2002. The chemical toxic benzo[a]pyrene perturbs the physical organization of phosphatidylcholine membranes. Environ. Toxicol. Chem. **21:** 787–793.
4. SERGENT, O. *et al.* 2005. Role for membrane fluidity in ethanol-induced oxidative stress of primary rat hepatocytes. J. Pharmacol. Exp. Ther. **313:** 104–111.
5. FLEURY, C., B. MIGNOTTE & J.L. VAYSSIERE. 2002. Mitochondrial reactive oxygen species in cell death signaling. Biochimie **84:** 131–141.

Metal-Containing Proteins in the Apoptosis and Redox Processes in the Rat Prostate and Human Prostate Cells

I. GRBAVAC, C. WOLF, N. WENDA, D. ALBER, M. KÜHBACHER, D. BEHNE, AND A. KYRIAKOPOULOS

Department for Molecular Trace Element Research in the Life Sciences, Hahn-Meitner-Institut, Glienicker Str. 100, 14109 Berlin, Germany

ABSTRACT: Several trace elements, such as Se, Cu, Mn, and Zn are bound to proteins (metallo- and metalloidproteins) in the prostate gland. Currently, it is known that some of those elements play a role in the apoptosis of different cells and redox processes. For the detection of such proteins, analytical and biochemical procedures were combined. SEC and ICP-MS were used to detect some trace elements, which are bound to proteins in the prostate cytosol and /or in the human prostate cell lines. Several antibodies against specific proteins were tested. By means of some of these antibodies several trace element–containing proteins, such as selenoproteins and Cu- and Cu-Zn-proteins, could be identified in the prostate. In addition, the localization of such metal- and metalloid-containing proteins in the micro organelles and cytosol of the prostate indicates specific functions of these proteins because, as it is known, such metal- and metalloid proteins play a role in the apoptosis and especially in the redox processes.

KEYWORDS: prostate; rat; metalloproteins; selenoproteins; redox process

INTRODUCTION

Several structural and regulatory functions can only be implemented in the eukaryotic cells if certain trace elements are bound to proteins. The significant role of such elements is reflected by a large number of diseases. Epidemiological studies have indicated the relationship between increased prostate cancer risk and trace elements, such as zinc and selenium.[1,2] Several metallo- and metalloidproteins are involved in the metabolic processes, such as oxidative stress

Address for correspondence: I. Grbavac, Department for Molecular Trace Element Research in the Life Sciences, Hahn-Meitner-Institut, Glienicker Str. 100, 14109 Berlin, Germany. Voice: +493080622294; fax: +493080622781.
 e-mail: Grbavac@hmi.de

Ann. N.Y. Acad. Sci. 1090: 113–119 (2006). © 2006 New York Academy of Sciences.
doi: 10.1196/annals.1378.012

and antioxidant defense system. Superoxide dismutase (SOD) and selenoenzymes are important antioxidant enzymes for cellular protection from reactive oxygen species (ROS).[3,4] The role of trace elements before the development and during the cancer process and its inhibition has been connected with many questions.[5] For example, the real effect of the trace elements and their relationship with cancer has not been reported in detail. It is not yet clear if the elements exert their effect independently or only in connection with a protein. In the case of the element Zn, it is reported that benign prostatic obstruction (BPO) is characterized by high Zn concentrations and prostate cancer is characterized by low Zn concentrations.[6] In order to obtain more information about the trace element and trace element–containing proteins, ratios in the prostate microanalytical techniques and biochemical methods were combined.

MATERIALS AND METHODS

Experiments on 6-Month-Old Animals

Two groups of Wistar rats (Charles River, Sulzfeld, Germany) were used in the study. The first group of rats was fed with a low Se diet with the Se concentration of 5–7 µg/kg, and the second was fed with the same diet and a high Se concentration (300µg/kg).

Labeling of Rats in Vivo with ^{75}Se and Recovery of Organs

Rats were labeled with ^{75}Se-selenite with a specific activity of approximately 10 MBq/µg Se. The labeling and sampling procedures has been described elsewhere.[10]

Analysis of ^{75}Se

The distribution of the ^{75}Se activity in the tissue, tissue homogenates, and tissue fractions was measured by means of a 3 × 3 inch NaI (Tl) well-type detector coupled to a multichannel analyzer (Campera, Germany).

Differential Centrifugation

The prostate samples were homogenized in a solution containing 5 mM $MgCl_2$, 0.02 M Tris- HNO_3, and several protease inhibitors (0.5% v/v aprotinin, chymostatin, 0.1% leupeptin, 0.1% v/v pepstatin, and 0.1% v/v phenylmethylfluoride). The subcellular fractionation of the tissue homogenates was carried out by differential centrifugation as previously described.[8]

Electrophoresis and Autoradiography

The proteins of the homogenates and subcellular fractions have been separated by SDS-PAGE. The SDS-PAGE was performed in a gel system according to the method of Laemmli.[12] The tracer distribution in the proteins after electrophoretic separation was determined autoradiographically using a phosphor plate, in connection with an imaging analyzer (FLA 3000, Fuji, Raytest GmbH, Straubenhardt, Germany).

Immunoblotting

After electrophoretic separation of the different prostate samples, the proteins were transferred on NC-membranes and tested with specific antibodies. After testing, the antigen–antibody complex was visualized by the IR imaging-analyzing system (LI-COR)

Analysis of the Cytosolic Samples by Means of SEC–ICP-MS

The cytosol was prepared by homogenization of a whole prostate in buffer containing a protease inhibitory cocktail, and a subsequent ultracentrifugation step. The obtained supernatant was stored at -20°C for further measurements. All vessels used were acid-cleaned to avoid metal contamination. Proteins in the cytosolic fraction were analyzed by means of size-exclusion chromatography (SEC), which was coupled to the ICP-MS system.

RESULTS AND DISCUSSION

In earlier studies the prostatic epithelial selenoprotein (PES) with the molecular mass of 15 kDa and its possible role in the prostate was reported.[7] In addition, a selenoenzyme (cGPx) and two further metalloenzymes, which are involved in redox systems and in apoptosis, were identified by the use of antibodies against these proteins in the homogenate and subcellular fractions.

The positive signal, even though it is weak after immunoblotting, of the antibody with the 15 kDa prostate protein in the homogenate and cytosol can be seen in FIGURE 2B. Positive signal was obtained after testing with the cGPx-antibodies only in the cytosolic fraction, which was to be expected because cGPx is a cellular protein and can be seen in FIGURE 1A. In the case of SOD, as seen in FIGURE 1B, the enzyme was clearly found in the homogenate and in each subcellular fraction. In the cell lines, the proteins in the molecular mass range of 18 kDa instead of 16 kDa were also found. The protein LOX could be detected in almost every sample fraction except that of the prostate nuclei and soma, which are provided in FIGURE 2A.

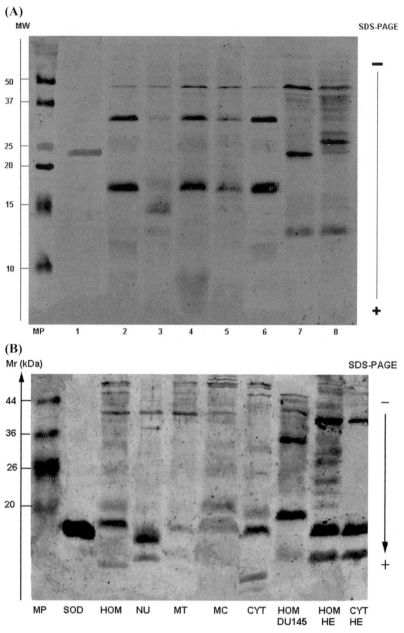

FIGURE 1. Immunotest of the cGPx and SOD in several fractions of the rat prostate. (**A**) MP, marker proteins; 1, cGPx (antigene); 2, homogenate; 3, nuclei; 4, mitochondria, 5, microsoma; 6, kidney homogenate (hom); 7, cytosol (pro); 8, liver homogenate (hom). (**B**) Marker proteins, SOD antigen, homogenate, nucleus, mitochondria, microsoma, prostate-cytosol (cyt), cell line homogenate, heart homogenate, heart cytosol.

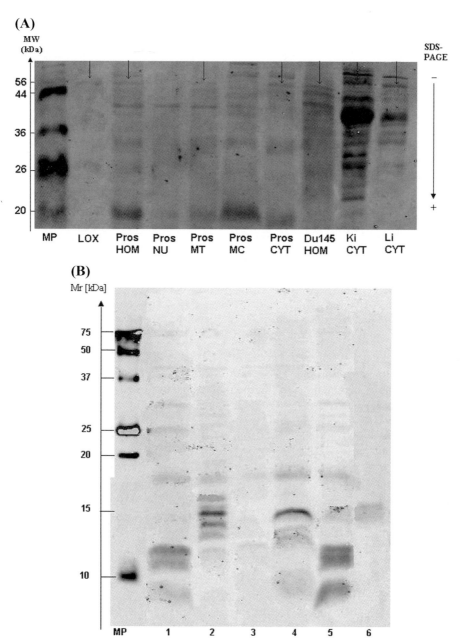

FIGURE 2. The NC-membrane shows the tested antibody against the enzyme lysyl oxidase (LOX) and the 15-kDa selenoprotein (B) in several fractions. (**A**) MP: marker protein, LOX, homogenate, nuclei, mitochondria, microsoma, cytosole, prostate cell lines, kidney cytosol, liver cytosol. (**B**) MP, marker protein; 1, homogenate; 2, nuclei; 3, mitochondria; 4, microsoma; 5, cytosole; 6, 15-kDa antigene.

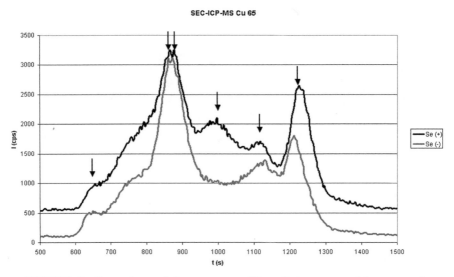

FIGURE 3. Comparison of the copper profiles of chromatographic separated rat prostate cytosols. Shown are the cytosols of a selenium-sufficient (*dark*) and a selenium-deficient (*light*) rat. The *arrows* indicate the copper-containing protein fractions.

Further investigation was carried out by means of SEC. The aim of this investigation, shown in FIGURE 3, was to determine the distribution pattern of protein-bound elements regarding the molecular weight (size) of the proteins. The molecular masses of the fractions were estimated by means of a column calibration with known molecule markers.

After the measurement of selenium-sufficient and selenium-deficient prostate cytosols by means of SEC–ICP-MS, two partially similar profiles are obtained. With exception of one peak, which indicates a Cu-binding protein of approximately 16 kDa (peak 3), four peaks were found showing Cu-binding proteins of approximately <6 kDa (1), 9 kDa (2), 29 kDa (4), and >70 kDa (5) (*black arrows*) in both samples. The 16-kDa protein occurs only contained in the selenium-sufficient cytosol.

On the basis of the implemented studies it can be said that several selenoproteins and metalloproteins, such as SOD and LOX, which are involved in the redox processes, are present in the prostate. In addition, the results of the investigations by SEC–ICP-MS indicate that copper is bound or attached to proteins or peptides with molecular masses between 2 and 72 kDa. In this study the molecular masses were estimated and therefore it is possible that the native molecular masses deviate from the masses determined by chromatography.

The identification of small copper-bound proteins or peptides is a very interesting result because small copper-bound peptides were found for the first time in the cytosol of the prostate. The characterization of those and their possible role in the apoptosis and redox processes in the rat prostate and human cell lines is the next task of our investigation.

ACKNOWLEDGMENTS

This study was partly supported by the Deutsche Forschungsgemeinschaft (DFG). We would like to thank Dr. B. Niggemann and Mr. J. Franke of the Klinikum Benjamin Franklin der Freien Universität Berlin for the help with the animal experiments.

REFERENCES

1. SCHRAUZER, G.N. 1980. The role of trace elements in the etiology of cancer. Trace Elem. Anal. Chem. Med. Biol. 183–185.
2. YAMAN, M., D. ATICI, S. BAKIRDERE & I. AKDENIZ. 2005. Comparison of trace metal concentrations in malign and benign human prostate, J. Med. Chem. **48**: 630–634.
3. HAI-NING, Y.U., Y.I.N. JUN-JIE & S.H.E.N SHENG-RONG. 2004. Growth inhibition of prostate cells by epigallocatechin gallate in the presence of Cu^{2+}. J. Agric. Food Chem. **52**: 462–466.
4. BARNES, P.J. 1990. Reactive oxygen species and airway inflammation. Free Radic. Biol. Med. **9**: 235–243.
5. GYORKEY, F., K.W. MIN, J.A. HUFF & P. GYORKEY. 1967. Zinc and magnesium in human prostate gland: normal, hyperplastic and neoplastic. Cancer Res. **27**: 1349–1353.
6. VENKATARAMAN, P., M. SIRDHAR, S. DHANAMMAL, et al. 2004. Oct antioxidant role of zinc in PCB (aroclor 1254) exposed ventral prostate of albino rats, J. Nutr. Biochem. **15**: 608–613.
7. BEHNE, D., A. KYRIAKOPOULOS, C. WEISS-NOWAK, et al. 1996. Newly found selenium-containing proteins in the tissues of the rat. Biol. Trace Element Res. **55**: 99–110.
8. CHAMBERS, J.A.A. & D. RICKWOOD 1978. Fractionation of subcellular organelles by differential centrifugation. In Centrifugation: A Practical Approach, Information Retrieval. D. Rickwood, Ed.: 33–47. London.
9. LAEMMLI, U.K. 1970. Cleavage of structural proteins during the assembly of the head of bacteriophage t4. Nature **227**: 680–685.
10. KALCKLÖSCH, M., A. KYRIAKOPOULOS, H. GESSNER & D. BEHNE 1995. A new selenoprotein found in the glandular epithelial cells of the rat prostate. Biochem. Biophys. Res. Commun. **217**: 162–170.

Does Transduced p27 Induce Apoptosis in Human Tumor Cell Lines?

MIRA GRDIŠA, ANA-MATEA MIKECIN, AND MIROSLAV POZNIC

Laboratory of Molecular Oncology, Division of Molecular Medicine, Rudjer Bošković Institute, 10 000 Zagreb, Croatia

ABSTRACT: p27 is a cyclin-dependent kinase inhibitor involved in the negative regulation of G1 progression in response to a number of antiproliferative signals. In this study, we examined the transduction of full-length Tat-p27, pt-mutated Tat-p27, and N'- Tat-p27 (truncated p27 on the C-terminal end) fusion proteins into human tumor cell lines and whether these transduced proteins induced apoptosis in the cells. Protein transduction can be described as the direct uptake by the cell of exogenous proteins/peptides as a result of a specific property of the protein/peptide component. The basic domain of human immunodeficiency virus type 1 (HIV-1) transactivator of transcription (Tat) protein possesses the ability to traverse biological membranes efficiently in a process termed *protein transduction*. Although the mechanism is unknown, transduction occurs in receptor/transporter-independent manner that appears to target the lipid bilayer directly. Thus, HIV-1 Tat proteins have tremendous potential to deliver large-sized compounds into the cells. Transduction of TAT-fusion proteins affected the proliferation of human tumor cell lines, depending on the type of protein and cell line. By Western blot analysis it was shown that some cell cycle regulatory proteins were affected, and that some proteins were responsible for the induction of apoptosis.

KEYWORDS: p27; apoptosis; cell lines

INTRODUCTION

Tumor development and progression have been shown to be dependent on cellular accumulation of various genetic and epigenetic events, including alterations in the cell cycle machinery at the G1/S checkpoint.[1,2] It is characterized by deregulated proliferation and aberrant cell cycle control. The identification of new prognostic parameters and therapeutic targets should be

Address for correspondence: Mira Grdiša, Ph.D., Division of Molecular Medicine, "Rudjer Bošković" Institute, Bijenička 54, 10 000 Zagreb, Croatia. Voice: +385-1-456-1110; fax: +385-1-456-1010.
e-mail: grdisa@rudjer.irb.hr

possible by analyzing cell cycle regulation in cancer cells. p27, a member of Kip/Cip family of CDK inhibitors, is involved in the negative regulation of G1 progression. It acts as a tumor-suppressor gene[3] that causes G1 arrest by inhibiting cyc E/cdk2 complex.[4]

The proliferating cell nuclear antigen (PCNA) is a 36-kDa molecular weight protein, synthesized in early G1 and S phases of the cell cycle. Deregulation of the mechanism controlling cell cycle progression is a hallmark of neoplasia.[5] Oncogenic transformation can occur from the dysregulation of normal cell cycle progression, abnormal activation, or inhibition of signal transduction cascades, failure to respond to apoptotic stimuli, or defective cell cycle check point activation.[6]

Apoptosis, a morphologically and biochemically defined form of cell death, plays a role in a wide variety of biological systems. The process is a highly synchronized cellular pathway, leading to activation of the downstream death machinery. Process is fundamental to normal development and tissue homeostasis, and its deregulation is associated with a variety of diseases from cancer to autoimmunity. There are efforts to manipulate the machinery that drives the apoptotic process for therapeutic gain. Central to the death machinery is a family of cysteine proteases called caspases. Caspases are cysteine proteases and potent inducers of apoptosis. These proteases are expressed as inactive precursors (zymogens) that are activated by proteolytic cleavage.[7,8] Caspases can cleave and thereby activate other caspases, but inactive caspases can also undergo autocatalytic activation when recruited into multiprotein complexes.[9] Their activation and activity is tightly regulated. There are several mechanisms by which caspases can be activated but one key pathway involves release of holocytochrome c from mitochondria into the cytoplasm. Cytoplasmic cytochrome c binds to apoptotic protease-activating factor-1 (Apaf-1), a 130- kDa protein, driving the formation of an Apaf-1 oligomer (the apoptosome), which in turn binds and activates caspase-9.[10,11,12]

Cytochrome c is normally localized with mitochondria but it is released into the cytosol by apoptotic stimuli in a process regulated by proteins of the Bcl-2 family (Bax and Bak).[13,14]

Poly(ADP-ribose) polymerases (PARPs) constitute a family of enzymes involved in the regulation of many cellular processes, such as DNA repair, gene transcription, cell cycle progression, cell death, chromatin functions, and genomic stability. PARP is defined as a family of enzymes that cleaves NAD+ to nicotinamide and ADP-ribose to form long and branched (ADP-ribose) polymers on glutamic acid residues of a number of acceptor proteins usually associated with chromatin, including PARP itself. The addition of negatively charged polymers profoundly alters the properties and functions of the target proteins.[15] Besides being involved in DNA repair, PARPs may also act as a mediator of cell death. In fact, extensive DNA damage, which saturates cell repair ability, is known to trigger PARP overactivation with consequent extensive NAD consumption during the synthesis of ADP polymers, which leads to ATP depletion and induction of necrosis.[16] Extensive PARP activation may also result in caspase-independent programmed cell death, mediated by the translocation of apoptosis-inducing factor to the nucleus.[17]

The major goal of anticancer therapy is to specifically kill tumor cells. Certain proteins and/or protein–protein interactions are specifically altered or deregulated in cancer cells and a very attractive strategy for anticancer drug is to develop molecules that target and modulate tumor-specific proteins and protein–protein interactions. An effort to modulate the biology of cancers involves the direct introduction of peptides, full-length proteins, and/or protein functional domains into tumor cells. However, to overcome cell membrane–mediated permeability barriers, methods that enable unrestricted delivery of biologically active molecules into cells are needed. It seems that protein transduction method is one of them.

MATERIALS AND METHODS

Cell Culture

Human cell lines (Molt, Nalm, Raji, SuDHL, K562, RKO, MiaPaCa2, HCT and MCF) were maintained in DMEM or RPMI (Gibco, Austria) medium plus 10% fetal bovine serum (FBS). Cells were transduced by addition of purified TAT fusion proteins (TAT-p27, TAT-pt-p27, TAT-N'-p27) at a concentration of 100–150 nM directly to cell culture medium. Transduction efficiency was verified by immunoblotting for the HA epitope contained in all TAT fusion proteins. Percentage of live cells was determined by WST-1 or MTT test. Each experiment was performed in quadriplicate and repeated 3–5 times.

Purification of TAT Fusion Proteins

The purification of TAT fusion proteins was described earlier.[18] It was performed with slight modification. In brief, bacterial lysates containing recombinant TAT fusion proteins were sonicated in 8 M urea, 100 mM NaCl, 20 mM Hepes, pH 8.0, and passed over a Ni-NTA resin (Amersham-Pharmacia Biotech AB, Uppsala, Sweden). All TAT fusion proteins contain an N-terminal 6x-His, which specifically binds on Ni-NTA resin. The proteins were eluted with imidazole. The fractions with TAT fusion proteins were pooled, diluted with 20 mM Hepes, pH 8.0, and desalted into phosphate-buffered saline (PBS) on a PD-10 column (Amersham-Pharmacia Biotech AB). All TAT fusion proteins were sterile filtered and stored in 10% glycerol at –80°C.

Western Blot Analysis

In brief, 50 μg of each lysate were separated on gradient gel (5–10%) SDS-PAGE[19] and transferred to Immobilon-P nitrocellulose membranes (Millipore,

Bedford, MA, USA). Control of loading was by staining of transferred proteins with naphthol blue-black (Sigma, St. Louis, MO, USA). After blocking with 5% nonfat milk in phosphate-buffered saline (PBS) for 30 min, membranes were cut up and the resulting subsections incubated with the following specific antibody for p27 (N-20, 0.5 μg/mL), cyclin D1 (HD11, 0.5 μg/mL), cyclin D3 (C-16, 0.5 μg/mL), cyclin E (HE-12, 0.5 μg/mL), Apaf-1 (K-20, 0.5 μg/mL), Aif (H-300, 0.5 μg/mL), caspase-3 (H-277, 0.5 μg/mL), and PCNA (0.5 μg/mL) (Santa Cruz Biotechnology Inc., Santa Cruz, CA, USA) and cyclin D2 (2 μg/mL) and PARP (dil. 1:2,000) (Pharmingen, San Diego, CA, USA). Subsequently, the membranes were washed twice in TBS-Tween buffer (10 mM Tris-HCl pH 8.0, 200 mM NaCl, 1% Tween 20) and incubated for 1 h with an appropriate horseradish peroxidase–linked secondary antibody (Dako, Denmark). Then the membranes were washed three times with TBS-Tween buffer, and proteins detected by chemiluminescence (ECL™ reagents, Amersham Pharmacia Biotech, Buckingamshire, UK) using X-ray film (Biomax™ film, Kodak, USA).

RESULTS

Influence of TAT-Fusion Proteins on Proliferation of the Cells

Different cell lines were treated with TAT-fusion proteins (p27, pt-p27, and N′-p27) and by WST or MTT test the number of cells was determined. All these results are shown in FIGURE 1. There was no difference between the effect of transduced TAT-fusion proteins on the cells grown in suspension or attached to the plastic. The influence of TAT-fusion proteins varied among the type of the cells as well as among the transduced proteins. Three days after starting the treatment, TAT-p27 decreased the cell proliferation up to 20%, TAT-pt-p27 to 35%, and TAT-N′-p27 (truncated p27) to 30–60%, on the cells grown in suspension. TAT fusion proteins were added every 12 h. In the experiments with attached cells, the effect of TAT fusion proteins decreased. The proliferation of RKO and MiaPaCa2 cells was decreased in the presence of TAT-p27 up to 20% and TAT-pt-p27 up to 15%. Whether transduced proteins caused cell cycle arrest or activation of some other signal transduction pathway was examined.

As is shown in FIGURES 2 and 3, Tat-fusion proteins (TAT-p27, TAT-pt-p27, TAT-N′-p27) were cleaved, after penetration into the cells, with some variations among the cell lines. In Molt-4 cells (T cell leukemia), TAT-fusion proteins were cleaved shortly after transduction. They were detectable at 3 h, but not at 23 h after transduction. In Nalm1 (B-myeloid leukemia) and Raji (B-lymphoma) TAT-fusion proteins have longer half-life. At 24-h posttransduction there was a remarkable amount of TAT-p27 detected by Western blot. By adding an additional amount of TAT-p27, its level increased into the cells (results are not shown).

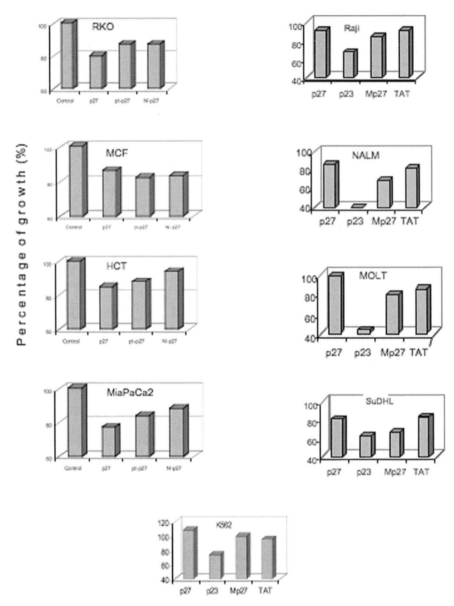

FIGURE 1. Percentage of growth of different cell lines 72 h after transduction of TAT-p27 (100 nM), TAT-pt-p27 (50 nM), and TAT-N′-p27 (50 nM). Viability of the cells was determined by MTT or WST-1. The results are expressed as a mean of 3–5 experiments, performing in quadriplicates (see details in MATERIALS AND METHODS).

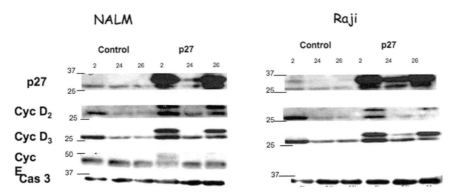

FIGURE 2. Western blot analysis of *NALM* and *Raji* cell lines, after transduction of TAT-p27 (150 nM) (details in MATERIALS AND METHODS).

Expression of Cell Cycle Regulatory Proteins

Among the cell cycle regulatory proteins, the expression of the cyclins (D1, D2, D3, and E) was determined. The results were shown in FIGURES 2 and 3. Similar to proliferation, the expression of cell cycle regulatory proteins was affected moderately, depending on the type of cells and transduced protein. It were found that levels of cyc D1 and E, the cancer biomarkers, were changed in response to transduced proteins measured in RKO cells. The levels of cyc D1 was reduced in comparison with control cells. Expression of PCNA, a marker of proliferating cells, was remarkably decreased in the presence of TAT-pt-p27. On the Molt cells the expression of cyc E and D3 were without significant changes among the TAT-proteins and control cells. Transduced TAT-p27 into Nalm and Raji cell lines causes some changes in the expression of cyc D2 and D3. An additional band appeared.

Expression of the Proapoptotic Proteins

To find out whether some of the transduced proteins induced apoptosis in cultured cell lines, the expression of the inductor of apoptosis was examined. For that purpose the expression of Apaf-1, Cas3, Aif, and PARP proteins was determined. The results are shown in FIGURES 2 and 3. TAT-fusion proteins induce an increase of Cas3 protein (pro-form). A cleavage of PARP protein on RKO cells was determined. Aif and Apaf-1 proteins were not significantly affected. Because of cleaving PARP protein, it is possible to predict an apoptotic pathway through the activation of caspase.

DISCUSSION

Molecular analysis of human tumors has demonstrated that p27 is functionally inactivated by different means in a majority of neoplasias of several

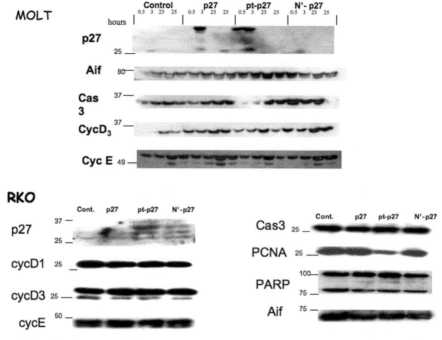

FIGURE 3. Western blot analysis of *MOLT* and *RKO* cells, 24 h after transduction with 150 nM of TAT-p27, TAT-pt-p27, and TAT-N′-p27 (details in "Methods").

different origins, suggesting that p27 levels represent an important determinant in cell transformation and cancer development. The abundance of p27 is regulated by multiple extracellular stimuli and functions as a sensor of external signals to cell cycle regulation. In normal cells p27 is expressed at high levels and decreases rapidly after mitogen stimuli.[20] In response to various stimuli, which inhibit cell proliferation, p27 levels increment.[17,22]

It is well known that the overexpression of CKIs in proliferating cells causes cell cycle arrest and inhibition of proliferation, making them good candidates for use as cytostatic agents. Numerous approaches have been tested to introduce CKIs in different cell types and the use of modified recombinant proteins seems to be the most promising one. A domain of HIV-TAT protein fused to heterologous proteins could confer the ability to efficiently transduce several cell types and introduce the CKI p27.[18]

Because of a pivotal role of p27 in cell cycle regulation and its involvement in proliferation, differentiation, and apoptosis, we examined it as a potential target in the cells, where its regular functions have failed. For that purpose we transduced p27, point mutated p27 (pt-p27), and truncated the form of p27 (N′-p27) into different tumor cell lines. As results show (FIG. 1), these proteins influenced the proliferation of examined cell lines, depending on the type of cells and protein. A different effect of the transduced proteins could be a result of different half-lives among them, which was shown by Western blot (FIGS. 2 and 3). Similar to this finding, other studies have shown that the pt-p27 (T187A mutant) was significantly more resistant to ubiquitination and subsequent degradation than was wild-type p27.[22,23] In our hands, pt-p27 causes a significant decrease of PCNA expression and PARP cleavage (FIG. 3), which point to induction of apoptosis.

Loss of cell cycle control is important in carcinogenesis. Although genetic alterations of p27 have rarely been found in cancer, downregulation of p27 is frequently observed in human cancer and has been associated with increased cell proliferation and tumor aggressiveness.[24–27] The relationship between p27 and induction of apoptosis is still unclear. p27 may be indirectly associated with apoptosis through CDK inhibition, and may be able to regulate the cell cycle or apoptosis to protect cells from overgrowth and apoptosis-inducing stimulation. By Western blot analysis we detected the changes in the expression of proteins responsible for induction of apoptosis (PARP). PARP (116 kDa) protein is cleaved by caspase-3 into 85-kDa and 25-kDa fragments.[28] We also detected a fragment of 85 kDa in RKO cells (FIG. 3). Other proteins on that signal transduction pathway (Apaf and cas 3) were not significantly changed. Perhaps the pathway through Aif did not activate after the transduction of TAT-p27 proteins, because there was no change in the expression of Aif (FIG. 3). There is some indication that activation of caspase is involved.

We also found that levels of cyc D1 and E were changed in the presence of transduced proteins, with slight variations among the cell types. Other proteins responsible for the regulation of cell cycle were not remarkably affected with transduced proteins.

Perhaps some of the TAT-fusion proteins would be useful in the treatment of tumors, but the possible antitumor effect must be additionally confirmed.

ACKNOWLEDGMENTS

This work was supported from the Grant 0098093 from the Ministry of Science, Education, and Sports, Republic of Croatia.

REFERENCES

1. MICHALIDES, R. 1999. Prognosis for G1 cell cycle regulators: useful for predicting course of disease and for assessment of therapy in cancer. J. Pathol. **188:** 341–343.
2. SHERR, C.J. 2000. The Pezcoller Lecture: Cancer cell cycles revisited. Cancer Res. **60:** 3689–3695.
3. FERO, M.L., E. RANDEL, K.E. GURLEY, *et al.* 1998. The murine gene p27kip1 is haplo-insufficient for tumor suppression. Nature **396:** 177–180.

4. POLYAK, K., M.H. LEE, H. ERDJUMENT-BROMAGE, *et al.* 1994. Cloning of p27Kip1, a cyclin dependent kinase inhibitor and a potential mediator of extracellular antimitogenic signals. Cell **78:** 59–66.
5. MORGAN, D.O. 1995. Principles of CDK regulation. Nature **374:** 131–134.
6. PRAY, R.T., F. PARLATI, J. HUANG, *et al.* 2002. Cell cycle regulatory E3 ubiquitin ligases as anticancer targets. Drug Res. Updates **5:** 249–252.
7. THORNBERRY, N.A. & Y. LAZEBNIK. 1998. Caspases–enemies within. Science **281:** 1312–1316.
8. BUDIHARDJO, I., H. OLIVER, M. LUTTER, *et al.* 1999. Biochemical pathways of caspase activation during apoptosis. Annu. Rev. Cell Dev. Biol. **15:** 269–290.
9. HERR, I. & K.M. DEBATIN. 2001. Cellular stress response and apoptosis in cancer therapy. Blood **98:** 2603–2614.
10. ANGEL, G.M., J. NAGUYEN, J.A. WELLS & H.O. FEARNHEAD. 2004. Apo cytochrome c inhibits caspases by preventing apoptosome formation. Biochem. Biophys. Res. Commun. **319:** 944–950.
11. ZOU, H., W.J. HENZEL, X. LIU, *et al.* 1997. Apaf-1, a human protein homologous to *C. elegans* ced-4, participates in cytochrome c-dependent activation of caspase-3. Cell **90:** 405–413.
12. TWIDDY, D., D.G. BROWN, C. ADRAIN, *et al.* 2004. Pro-apoptotic proteins released from the mitochondria regulate the protein composition and caspase-processing activity of the native Apaf-1/caspase-9 apoptosome complex. J. Biol. Chem. **279:** 19665–19682.
13. BORNER, C. 2003. The Bcl-2 protein family: sensors and checkpoints for life-or-death decisions. Mol. Immunol. **39:** 615–647.
14. SCORRANO, L. & S.J. KORSMEYER. 2003. Mechanism of cytochrome c release by proapoptotic Bcl-2 family members. Biochem. Biophys. Res. Commun. **304:** 437–444.
15. TENTORI, L. & G. GRAZIANI. 2005. Chemopotentiation by PARP inhibitors in cancer therapy. Pharmacol. Res. **52:** 25–33.
16. HA, H.C. & S.H. SNYDER. 1999. Poly(ADP-ribose) polymerase is a mediator of necrotic cell death by ATP depletion. Proc. Nat. Acad. Sci. USA **96:** 13978–13982.
17. YU, S.W., H. WANG M.F. POITRAS, *et al.* 2002. Mediation of poly(ADP-ribose) polymerase-1-dependent cell death by apoptosis inducing factor. Science **297:** 259–263.
18. NAGAHARA, H., A.M. VOCERO-AKBANI, E.L. SNYDER, *et al.* 1998. Transduction of full length TAT fusion proteins into mammalian cells: TAT-p27Kip1 induces cell migration. Nat. Med. **4:** 1449–1451.
19. GRDIŠA, M. 2003. Influence of CD40 ligation on survival and apoptosis of B-CLL cells *in vitro*. Leuk. Res. **1690:** 1–6.
20. REYNISDOTTIR, I., K. PLOYAK, A. IAVARONE & J. MASSAGUE. 1998. Kip/Cip and Ink4 CDK inhibitors cooperate to induce cell cycle arrest in response to TGF-beta. Genes Dev. **9:** 555–563.
21. HENGST, L., V. DULIC, J.M. SLINGERLAND, *et al.* 1994. A cell cycle regulated inhibitor of cyclin-dependent kinases. Proc. Natl. Acad. Sci. USA **91:** 5291–5295.
22. MONTAGNOLI, A., F. FIORE, E. EYTAN, *et al.* 1999. Ubiquitination of p27 is regulated by Cdk-dependent phosphorylation and trimeric complex formation. Genes Dev. **13:** 1181–1189.

23. ZANG, Q., L. TIAN, A. MANSOURI, et al. 2005. Inducible expression of a degradation-resistant form of p27Kip1 causes growth arrest and apoptosis in breast cancer cells. FEBS Lett. **579:** 3932–3940.
24. CATZAVELOS, C., N. BHATTACHARYA & Y.C. UNG. 1997. Decreased levels of the cell cycle inhibitor p27Kip1 protein: prognostic implications in primary breast cancer. Nat. Med. **3:** 227–230.
25. MORI, M., K. MIMORI, T. SHIRAISHI, et al. 1997. p27 expression and gastric carcinoma. Nat. Med. **3:** 593.
26. LODA, M., B. CUKOR & S.W. TAM. 1997. Increased proteosome-dependent degradation of the cyclin-dependent kinase inhibitor p27 in aggressive colorectal carcinomas. Nat. Med. **3:** 231–234.
27. ESPOSITO, V., A. BALDI, A. DE LKUCA, et al. 1997. Prognostic role of the cyclin-dependent kinase inhibitor p27 in non-small cell lung cancer. Cancer Res. **57:** 3381–3385.
28. PATEL, T., G. J. GORES & S.H. KAUFMANN. 1996. The role of proteases during apoptosis. FASEB J. **10:** 587–597.

Apoptotic Cell Signaling in Lymphocytes from HIV+ Patients during Successful Therapy

SANDRO GRELLI,[a] EMANUELA BALESTRIERI,[b]
CLAUDIA MATTEUCCI,[a] ANTONELLA MINUTOLO,[a]
GABRIELLA D'ETTORRE,[c] FILIPPO LAURIA,[d]
FRANCESCO MONTELLA,[d] VINCENZO VULLO,[c] STEFANO VELLA,[e]
CARTESIO FAVALLI,[a] ANTONIO MASTINO,[f]
AND BEATRICE MACCHI[b,g]

[a]*Department of Experimental Medicine and Biochemical Science, University of Rome "Tor Vergata," via Montpellier 1, 00133 Rome, Italy*

[b]*Department of Neuroscience, University of Rome "Tor Vergata," via Montpellier 1, 00133 Rome, Italy*

[c]*Department of Infectious and Tropical Diseases, University of Rome "La Sapienza," 00161 Rome, Italy*

[d]*Clinical Immunology Unit, S. Giovanni Hospital, 00184 Rome, Italy*

[e]*Istituto Superiore di Sanità, 00161 Rome, Italy*

[f]*Department of Microbiological, Genetic and Molecular Sciences, University of Messina, 98166 Messina, Italy*

[g]*IRCCS, S.Lucia, 00179 Rome, Italy*

ABSTRACT: The impact of antiretroviral therapy (ART) on immune-reconstitution and its relationship with the complex scenario of multiple cell signaling associated with apoptosis in HIV infection has not yet been fully elucidated. Here we report the results of the analysis of the expression of 13 genes involved in the apoptotic pathway, simultaneously detected by RNA-protection assay in peripheral blood mononuclear cells (PBMCs) of 12 HIV-1-infected responder patients before and during successful ART. In particular, we calculated the correlations among apoptosis and viral load (VL) levels versus the quantitative expression of genes associated with death receptors or to Bcl-2 pathways. Nonparametric bivariate Spearman's analysis of significant correlations showed that apoptosis was directly correlated with mRNA levels for caspase-8, FasL, and TRAIL. Conversely, apoptosis levels were inversely correlated with mRNA levels for Bcl-xl, Bcl-2, and Mcl-1, respectively. In addition,

Address for correspondence: Dr. Beatrice Macchi, Department of Neuroscience, Building F, Floor 0, Corridor South, Room F36, University of Rome "Tor Vergata," Via Montpellier 1, 00133 Roma, Italy. Voice: 0039-06-72596392; fax: 0039-06-72596323.
e-mail: macchi@med.uniroma2.it

Ann. N.Y. Acad. Sci. 1090: 130–137 (2006). © 2006 New York Academy of Sciences.
doi: 10.1196/annals.1378.014

while VL was directly correlated with the expression of caspase 8, it was inversely correlated with mRNA levels for Bcl-2 and Mcl-1. These results, although worthy of further investigation, show that variations of apoptosis levels in PBMCs of HIV-1+ patients during ART are strictly related to the modulation of a complex network of signaling involving both death and survival of lymphocytes.

KEYWORDS: HIV-1; apoptosis; RNA-protection assay; antiretroviral therapy

INTRODUCTION

Our understanding of the mechanisms underlying cell death in T lymphocytes from HIV-infected individuals is, at present, still incomplete. Existing data suggest that abnormal apoptosis of lymphocytes depends on several independent events occurring during infection, such as HIV-induced syncitia formation, direct HIV-protein-induced cell death, activation-induced cell death (AICD) and bystander cell killing. Independently of its origin, these events could involve both the extrinsic pathway, which is initiated by signaling through death receptors, and the intrinsic pathway of apoptosis, which is mainly regulated by mitochondrial proteins.[1,2] In fact, gene profile analysis has shown that a number of genes associated with both pathways[3,4] are modulated after infection of cell lines with HIV *in vitro*, while data *ex vivo* from patients have only partially revealed which genes could be involved in the apoptotic response to HIV infection. In particular, investigations in patients focused on single or strictly related apoptosis pathways showed decreased Fas and Fas-ligand (FasL) expression, the inhibition of apoptosis induced by Fas[5,6] or by TNFR1 and TNFR2,[7] and inhibited expression of TNF-related apoptosis-inducing ligand (TRAIL)[8] in cells or in sera from individuals undergoing antiretroviral therapy (ART). In addition, it has been reported that virological suppression during ART was associated with high expression of Bcl-2 and IL-15.[9] Thus, one of the key questions is whether changes in apoptosis during ART are more dependent on the modulation of pro- or of antiapoptotic genes. Here we present the results of a study in which apoptosis or viral load (VL) levels have been correlated with mRNA levels of several apoptosis-related genes evaluated at the same time points in the peripheral blood mononuclear cells (PBMCs) of twelve drug-naive HIV-1-infected patients undergoing ART.

In this investigation, a well-characterized cohort of 12 HIV-1-infected patients was enrolled in a longitudinal open study at the Department of Infectious and Tropical Disease, University of Rome "La Sapienza," or at the "AIDS Center," S. Giovanni Hospital, Rome. CDC classification at enrollment was as follows: class A included eight patients (two A1, four A2, two A3) and class B four (two B2, two B3). The eligibility criteria at enrollment were decided as previously described.[6] No patient lost eligibility during the course of the study. Most of the patients remained on their assigned treatment with the exception of one patient who changed therapy, without modification of the overall response. The study received the ethical approval of the institutions participating in the study and the informed consent was obtained from all participating patients. Spontaneous apoptosis and VL were measured before (time 0, T0) and regularly after 3, 6, and 12 months of therapy. PBMCs from HIV$^+$ or from a group of healthy HIV$^-$ individuals were cultured with 10% fetal calf serum for 48 h before detection of spontaneous apoptosis. Apoptosis was evaluated by flow cytometry analysis of nuclei isolated from PBMCs by detergent treatment and stained with propidium iodide. HIV-1 RNA VL levels were determined using a quantitative ultrasensitive reverse transcription-polymerase chain reaction assay (limit of detection was 50 HIV RNA copies/mL). To evaluate gene expression, total RNA was prepared from 48-h PHA-stimulated PBMCs, and the expression of multiple genes within the same RNA samples was analyzed by RNA-protection assay (RPA). In particular, the expression of different genes within the same sample was analyzed using the "multiprobe ribonuclease protection assay system" (Ambion, Austin, TX, USA) using ^{32}P-labeled antisense RNA probes generated from two commercial sets of single-strand DNA templates (hStress-1 and hApo-3 template sets [Pharmingen BD, San Diego, CA, USA]). Free probes and other single-stranded RNA were digested with RNases and the remaining hybridization-protected probes were resolved on denaturing polyacrylamide gel 5%. The intensity of bands of RNA duplexes were quantified by phosphor-imaging analysis (Molecular Dynamics, Sunnyvale, CA, USA) using Image Quant software, and normalized to L32 housekeeping mRNA levels after subtraction of the background.

Statistical analyses were done using SPSS (version 10.0, SPSS). For statistical comparisons of the laboratory data from HIV$^+$ patients at different times of therapy the paired samples Student's t-test analysis was performed. For nonparametric analysis of bivariate correlations, the Spearman's rho correlation coefficient and corresponding P values for each couple of parameters were calculated using all couples of data available at all time points tested for each patient. For statistical and graphical reasons, an arbitrary value of 25 HIV RNA copies/mL was assigned to samples under the limit of HIV-RNA detection and the numbers of HIV copies/mL were transformed as decimal logarithms.

On the basis of clinical, virological, and immunological response to therapy, patients were classified into one homogeneous cohort with a successful response. Results reported in FIGURE 1 show that VL levels were clearly decreased at 2 months from the initiation of ART. Similarly, mean values of apoptosis promptly decreased, as expected, and levels detected after 12 months of therapy were similar to those detected in a group of healthy controls (around 10%). In parallel, mean percentages of CD4$^+$ cells progressively and consistently

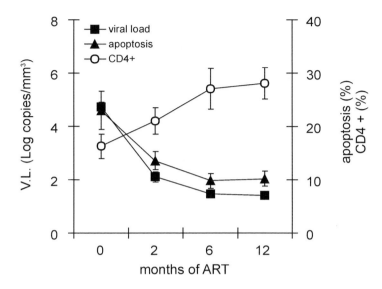

FIGURE 1. Apoptosis, VL, and CD4+ cell percentages in a group of HIV+ patients during successful ART. *Top*: levels of PBMC apoptosis, plasmatic VL, and PBMC CD4+ percentages in a group of 12 HIV+ patients before initiating ART (0) and at 2, 6, and 12 months of therapy (means ± SE). *Bottom*: significance values (P) obtained following comparison of means using the paired Student's t-test at the indicated times for each analyzed parameter.

increased during therapy. Statistical analysis showed that apoptotic changes were significant or highly significant at all times tested during ART in comparison to the baseline value. VL levels were highly significantly different at all times tested in comparison with time 0. Conversely, the percentage of CD4+ cells was significantly different only after 12 months of therapy with respect to time 0. Statistical analysis of correlations suggests that these variations were not incidental events. In fact, VL changes were significantly correlated with both apoptosis levels (rho = 564, $P < 0.0001$) and percentages of CD4+ cells (rho = −637, $P < 0.0001$).

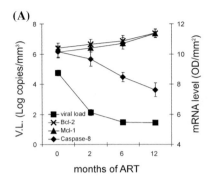

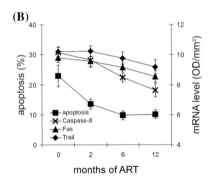

FIGURE 2. Statistically significant correlations between VL or apoptosis and expression of apoptosis-related genes analyzed by RNA-protection assay. (**A**) *Top*: Plasmatic VL levels and mRNA levels for Bcl-2, Mcl-1, and caspase-8 of stimulated PBMCs in the same group of the 12 HIV$^+$ patients in FIGURE 1, before initiating ART (0) and at 2, 6, and 12 months of therapy (means ± SE). *Bottom*: Spearman's analysis of bivariate correlations, conducted using all available couples of data. The calculated Spearman's rho and the corresponding significance values (*P*) are reported. (**B**) *Top*: PBMC apoptosis levels and mRNA levels for caspase-8, FasL, and TRAIL of stimulated PBMCs in the same group of the 12 HIV$^+$ patients in FIGURE 1, before initiating ART (0) and at 2, 6, and 12 months of therapy (means ± SE). *Bottom*: Spearman's analysis of bivariate correlations, conducted using all available couples of data. The calculated Spearman's rho and the corresponding significance values (*P*) are reported. (**C**) *Top*: PBMC apoptosis levels and mRNA levels for Bcl-xl, Bcl-2, and Mcl-1 of stimulated PBMCs in the same group of the 12 HIV$^+$ patients in FIGURE 1, before initiating ART (0) and at 2, 6, and 12 months of therapy (means ± SE). *Bottom*: Spearman's analysis of bivariate correlations, conducted using all available couples of data. The calculated Spearman's rho and the corresponding significance values (*P*) are reported.

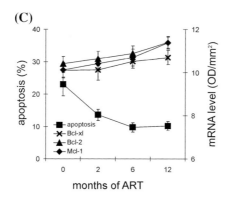

FIGURE 2. Continued.

At the same time points at which apoptosis was measured, RNA samples from PHA-stimulated PBMCs of the 12 HIV-1-infected individuals were subjected to quantitative evaluation by RPA of the following apoptosis-related genes: *FAS, FASL, FAF1, FADD, CASPASE-8, DR3, TRAIL, TNFR1, TRADD, BCL-XL, BCL-2, MCL-1,* and *BAX*. Quantitative evaluations of mRNA levels for genes associated with apoptosis were utilized to calculate the correlations with apoptosis and VL. Nonparametric bivariate Spearman's analysis of correlations showed that VL was highly significantly inversely correlated with mRNA levels for Bcl-2 (rho = 0.413, $P = 0.005$, FIG. 2A) and Mcl-1 (rho = 0.450, $P = 0.002$, FIG. 2A) and directly correlated with mRNA levels for caspase 8 (rho = 0.568, $P < 0.001$, FIG. 2A). Apoptosis levels were significantly directly correlated with mRNA levels for TRAIL (rho = 0.334, $P = 0.031$, FIG. 2B), FasL (rho = 0.335, $P = 0.030$, FIG. 2B), and caspase-8 (rho = 0.390, $P = 0.011$, FIG. 2B). Conversely, apoptosis was highly significantly inversely correlated with mRNA levels for Bcl-xl (rho = -0.436, $P = 0.004$, FIG. 2C), Bcl-2 (rho = $-0.463, P = 0.002$, FIG. 2C), and Mcl-1 (rho = -0.479, $P = 0.001$, FIG. 2C). All other correlations among mRNA levels for apoptotic genes and apoptosis or VL levels were not statistically significant.

Our results, showing that apoptosis was highly significantly inversely correlated with mRNA levels for Bcl-2 in PBMCs from HIV-1-infected patients during the first 12 months of therapy, indicate the importance of Bcl-2 in preventing lymphocytes from undergoing cell death during HIV therapy. Similarly, the finding that expression of the antiapoptotic genes *BCL-XL* and *MCL-1* were highly significantly inversely correlated to apoptosis further supports the hypothesis that antiapoptotic genes of the Bcl-2 family play a major role in the modulation of apoptosis during ART.[10] In addition, the important involvement of this group of genes in the response to therapy is confirmed by the observation that mRNA levels for Bcl-2 and Mcl-1 were also significantly inversely correlated with VL levels. Thus, these results clearly suggest that the mitochondrial pathway is preferentially involved in the evident changes in cell-death fate of immune cells in HIV-1$^+$ patients with successful response to therapy. Moreover, our data on proapoptotic *CASPASE-8*, *FASL*, and *TRAIL* gene expression also highlight that inhibition of apoptosis during ART is evidently related to the diminished activity of genes playing a key role in death-receptor mediated apoptosis, even if apparently less significantly. In addition, our analysis of correlations was carried out using a limited number of patients and, consequently, of paired observations to utilize for calculations. Thus, also correlations that did not reach a significant level in our analyses could have real relevance regarding mechanisms controlling apoptosis in PBMCs of patients during ART.

In conclusion, the results of our study, which are novel in terms of the wide range of apoptotic gene expression simultaneously analyzed during a longitudinal investigation in HIV-1$^+$ patients, indicate that both mitochondrial and death-receptor signaling seem to be involved in the long-term successful response to ART, suggesting that inhibition of cell death in HIV$^+$ patients under therapy is the result of a complex, but coherent network of multifactor signaling.

ACKNOWLEDGMENTS

We wish to thank Alison Inglis for her linguistic assistance. This work was supported by grants from the Istituto Superiore di Sanità, AIDS Project No. 30F.32 (3) and from the Italian Ministry of Education, University and Research, Research Projects of National Interest Coordinated by Prof. E. Garaci.

REFERENCES

1. MIURA, Y. & Y. KOYANAGY. 2005. Death ligand-mediated apoptosis in HIV infection. Rev. Med. Virol. **15:** 169–178.
2. GUGEON, M.L. 2003. Apoptosis as an HIV strategy to escape immune attack. Nat. Rev. Immunol. **3:** 392–404.
3. WEN, W., S. CHEN, Y. CAO, *et al.* 2005. HIV-1 infection initiates changes in the expression of a wide array of genes in U937 promonocytes and HUT78 T cells. Virus Res. **113:** 26–35.
4. CORBEIL, J., D. SHEETER, D. GENINI, *et al.* 2001. Temporal gene regulation during HIV-1 infection of human CD4$^+$ T cells. Genomic Res. **11:** 1198–1204.
5. BADLEY, A.B., D.H. DOCKRELL, A. ALGECIRAS, *et al.* 1998. *In vivo* analysis of Fas/FasL interactions in HIV-infected patients. J. Clin. Invest. **102:** 79–87.

6. GRELLI, S., S. CAMPAGNA, M. LICHTNER, et al. 2000. Spontaneous and anti-fas-induced apoptosis in lymphocytes from HIV-infected patients undergoing highly active anti-retroviral therapy. AIDS **14:** 939–949.
7. DE OLIVERA PINTO, L.M., S. GARCIA, H. LECOEUR, et al. 2002. Increased sensitivity of T lymphocytes to tumor necrosis factor receptor 1 (TNFR1)- and TNFR2-mediated apoptosis in HIV infection: relation to expression of Bcl-2 and active caspase 8 and caspase 3. Blood **99:** 1666–1675.
8. HERBEUVAL, J.P., A. BOASSO, J.C. GRIVEL, et al. 2005. TNF-related apoptosis-inducing ligand (TRAIL) in HIV-1- infected patients and its *in vitro* production by antigen-presenting cells. Blood **105:** 2458–2464.
9. TORTI, C., G. COLOGNI, M.C. UCCELLI, et al. 2004. Immune correlates of virological response in HIV-positive patients after highly active antiretroviral therapy (HAART). Viral Immunol. **17:** 279–286.
10. REGAMEY, N., T. HARR, M. BATTEGAY & P. ERB. 1999. Downregulation of Bcl-2, but not of Bax or Bcl-x, is associated with T lymphocyte apoptosis in HIV infection and restored by antiretroviral therapy or by Interleukin 2. AIDS Res. Hum. Retroviruses **9:** 803–810.

Capacitation and Acrosome Reaction in Nonapoptotic Human Spermatozoa

SONJA GRUNEWALD, THOMAS BAUMANN, UWE PAASCH, AND HANS-JUERGEN GLANDER

European Academy of Andrology Center, University of Leipzig, Philipp-Rosenthal-Str. 23-25, 04103 Leipzig, Germany

ABSTRACT: Capacitation and acrosome reaction (AR) of human spermatozoa are prerequisites for fertilization. Annexin-V-MACS is able to separate apoptotic from nonapoptotic sperm on the basis of their externalization of phosphatidylserine (EPS). The nonapoptotic (EPS$^-$) fraction is characterized by the lowest amounts of membrane alterations, caspase activation, disrupted mitochondrial potential, and DNA fragmentation. The aim of our study was to investigate the separation effect of Annexin-V-MACS on capacitation and AR in nonapoptotic sperm. Semen specimens from 10 healthy donors were separated into 2 samples each, one was left untreated (control) and the second was subjected to Annexin-V-MACS. Two aliquots of both, the control as well as the EPS$^-$ fraction after Annexin-V-MACS, were incubated in HTF at 37°C, 5% CO_2 for 3 h either with 3% BSA (capacitation) or without additives. Capacitation was monitored by tyrosine phosphorylation (TyrP) using Western blot technique. AR was determined by labeling with CD46-FITC before and after stimulation with calcium-ionophore A23187, followed by flow cytometric evaluation of the percentage of CD46$^+$ sperm. Densitometric analyses of the 105-kDa and 80-kDa bands of the TyrP Western blots demonstrated highest TyrP in the capacitated EPS$^-$ aliquots. There was no difference in spontaneous AR in all groups. AR was best inducible in EPS-negative sperm after capacitation. Nonapoptotic human spermatozoa with intact plasma membranes are characterized by superior ability to capacitate and consequently by maximum potential to perform AR after stimulation. Selection of EPS-negative sperm may be of advantage for assisted reproduction in order to prepare the sperm subpopulation with the highest fertilizing potential.

KEYWORDS: MACS; human spermatozoa; apoptosis; capacitation; acrosome reaction

Address for correspondence: Sonja Grunewald, European Academy of Andrology Center, University of Leipzig, Philipp-Rosenthal-Str. 23-25, 04103 Leipzig, Germany. Voice: +49-341-9718740; fax: +49-341-9718749.
  e-mail: sonja.grunewald@medizin.uni-leipzig.de

INTRODUCTION

Before fertilization can occur, spermatozoa undergo a series of changes to acquire the ability to bind and penetrate the oocyte. Those prerequisites for fertilization include capacitation and acrosome reaction (AR).[1] However, the exact molecular mechanism and signal transduction of capacitation and the following AR are not clearly understood. Capacitation comprises an increase in membrane fluidity, cholesterol efflux and ion fluxes. Major changes during capacitation result from tyrosine phosphorylation of proteins, a process regulated by activation of at least three intracellular signaling pathways involving cAMP/ protein kinase A, receptor tyrosine kinases, and nonreceptor tyrosine kinases.[2] Calcium, bicarbonate,[3] ROS,[4] GABA, progesterone,[5] angiotensin, and cytokines are known to modulate the signaling pathway. Tyrosine phosphorylations in human spermatozoa correlate to other findings specific for capacitation: hyperactivated motility, zona pellucida binding and AR.[6]

Ejaculated human spermatozoa have been shown to display characteristics that are typical of apoptosis such as the externalization of phosphatidylserine (EPS) from the inner to the outer leaflet of the plasma membrane, the disruption of the mitochondrial membrane potential, caspase activation and DNA fragmentation.[7–9]

On account of the high and selective affinity of annexin-V to EPS,[10] spermatozoa can be subpopulated by colloidal superparamagnetic annexin-V microbeads (50 nm in diameter) in EPS$^+$ (dead and apoptotic spermatozoa) and EPS$^-$ (nonapoptotic) fractions in a magnetic cell sorter (MACS). Placed onto a column containing iron balls followed by a passage through a strong magnetic field, EPS$^+$ cells are attached to the iron balls within the separation column. Nonapoptotic EPS$^-$ cells remain unlabeled and pass freely through the column.[11,12] Annexin-V MACS was previously proven to separate spermatozoa with activated caspases—key initiator and executor enzymes of apoptosis, disrupted MMP and DNA fragmentations preferentially in the EPS$^+$ fraction—while EPS$^-$ sperm was characterized by lowest activation of the apoptosis signaling cascade.[13,14] The MACS system has no detrimental effects on sperm motility, viability, and morphology.[15]

The aim of our study was to investigate the separation effect of Annexin-V-MACS on capacitation-related tyrosine phosphorylation and AR.

MATERIAL AND METHODS

Selection Criteria of the Semen Samples

Following a period of 3–5 days of abstinence, a total of 10 semen specimens from 10 healthy donors were collected by masturbation into sterile containers. After liquefaction for 20 min at room temperature an aliquot was examined for sperm concentration, motility and morphology according to the WHO standard guidelines. Each ejaculate exceeded the WHO reference ranges for the normal fertile population.[16]

Experimental Design

The effect of preparation of nonapoptotic human spermatozoa by Annexin-V-MACS on capacitation as well as on spontaneous and induced AR was investigated. The semen specimens were separated into two samples each: one was left untreated (control) and the second was subjected to Annexin-V-MACS. Two aliquots of both, the control as well as the EPS-negative fraction after Annexin-V-MACS, were incubated in human tubal fluid media (HTF; Irvine Scientific, Santa Ana, CA) at 37°C, 5% CO_2 for 3 h either with 3% BSA (capacitation) or without additives. Capacitation was monitored by Western blot analyses of tyrosine phosphorylation. AR was determined by detection of CD46 on the spermatozoal surface before (spontaneous AR) and after stimulation with calcium ionophore A23187 (induced AR), followed by flow cytometric evaluation of the percentage of $CD46^+$ sperm. The study was approved by the Institution Review Board of the University of Leipzig.

Sperm Preparation

After liquefaction for 20 min and dilution in 2 mL HTF, washing was performed by centrifugation at $400 \times g$ for 5 min. The supernatants were discarded and the pellets were diluted in HTF for further experiments. One portion of the sperm suspension was kept as a control aliquot, while the other portion was subjected to MACS.

Isolation of Spermatozoa with Deteriorated Membranes by MACS

Spermatozoa were incubated with Annexin-V-conjugated microbeads (Miltenyi Biotec, Bergisch-Gladbach, Germany) for 15 min at room temperature. One hundred µL of microbeads were used for each 10 million separated cells. The sperm/microbead suspension was loaded in a separation column containing iron balls, which was fitted in a magnet (MiniMACS; Miltenyi Biotec, Bergisch-Gladbach, Germany). The apoptotic spermatozoa were retained in the separation column and labeled as EPS^+, whereas the nonapoptotic spermatozoa with intact membranes passed through the column and were labeled EPS^-. The power of the magnetic field was measured as 0.5 tesla between the poles of the magnet and up to 1.5 tesla within the iron globes of the column. After the column was removed from the magnetic field, the retained EPS^+

fraction was eluted using Annexin-binding buffer (Miltenyi Biotec, Bergisch-Gladbach, Germany).

Capacitation of Human Spermatozoa

Spermatozoa were capacitated by incubation in HTF media containing 3% bovine serum albumin (BSA) for 3.0 h at 37°C and 5% carbon dioxide. The non-capacitated control aliquots were incubated under identical conditions without additives in the HTF media. Capacitation was evaluated by Western blot analyses of tyrosine phosphorylation.

Western Blot Analysis of Tyrosine Phosphorylations

The semen aliquots were washed in HTF and centrifuged at $400 \times g$ for 5 min. Laemmli buffer pH 6.7 (50 mM Tris-Cl, 2% SDS, 2% mercaptoethanol, 1% 100 mM Na_3VO) was added to the sperm pellet for cell lysis. Nucleic acids were degraded and viscosity was reduced by 2% benzonase (Calbiochem, Schwalbach am Taunus, Germany). Lysis was completed by heating for 5 min at 90°C, cooling down to room temperature and addition of an equal volume of sample buffer (87.5 mL aqua dest., 10 mL glycerol, 2.5 mL buffer [6.06 g Tris, 0.4 g SDS, 0.01 g NaAcid, 100 mL aqua dest., pH 6.8], 1 g SDS, 0.01 g NaAcid).

The protein concentration of the samples was determined by amido black.[17] A total of 15% SDS-PAGE was carried out with 15 μg sperm protein lysate per lane under reducing conditions at 200 V for 60 min on a vertical slab gel apparatus. The method applied corresponded to that of Laemmli.[18] EGF-stimulated A431 cell lysate (Upstate Biotechnology, Inc, Lake Placid, NY, USA) served as positive control and BenchMarks™ (Life Technologies, Karlsruhe, Germany) was used for the evaluation of the molecular weights.

After protein transfer to nitrocellulose membranes for 120 min at 200 mA, membranes were blocked with 2% BSA-Tris-Triton, pH 7.5 (50 mM Tris, 150 mM NaCl, 0.2% Triton-X-100) for 1 h followed by an incubation with mouse-anti-human-phosphotyrosine clone 4G10 (1:500, Upstate, Biotechnology) in 2% BSA-Tris-Triton-solution and washing steps and incubation for 1 h with peroxidase-conjugated goat-anti-mouse secondary antibody (1:10,000, Dianova, Hamburg, Germany). For negative controls the primary antibody was replaced by nonimmune mouse serum. All incubations were performed at +4°C.

Relative differences between amounts of tyrosine phosphorylations were examined by luminol-H_2O_2-detection using the ChemImageTM 4400-System (Alpha Innotech Corporation, San Leandro, CA, USA) according to Faulkner et al.[19] Bound secondary antibodies were visualized by X-ray films (Konica, Type 3A4, BW Plus Roentgen, Kamp-Crutfort, Germany).

Induction and Measurement of AR

Induction of AR was performed using the calcium ionophore A23187 (Sigma, St. Louis, MO). On the day of use, a frozen aliquot of stock solution of calcium ionophore A23187 in dimethyl sulfoxide (DMSO, Sigma) was diluted 1:5 with HTF medium, and 20 μL of the solution were added to 100 μL of the sperm suspension. As a control, an aliquot of the same sperm suspension was left untreated. Both test and control tubes were incubated for 1 h at 37°C in 5% CO_2 in air before the acrosomes were assessed.

The amount of spontaneous and ionophore-induced AR was detected using a monoclonal FITC-labeled mouse anti-human CD46 antibody (IgG2a, Biomeda, Foster City, CA). All aliquots were washed in 900 μL PBS for 4 min at 400 × g. The resulting sperm pellet was resuspended in 100 μL phosphate-buffered saline (PBS), pH 7.4, containing 4 μL of CD46-FITC and incubated for 30 min under light protection. Two more washing steps in PBS (4 min at 400 × g) were performed before the pellet was resuspended in 1 mL PBS. The extent of CD46 on spermatozoal surface was evaluated by flow cytometric analyses using Coulter® Epics® XL™ (Coulter, Germany). The percentage of positive cells and the mean fluorescence was calculated on a 1,023-channel scale by software Expo32ADC (Coulter, Germany).

Statistical Analysis

Evaluation of differences and correlation data was performed by nonparametric tests (Wilcoxon test and Spearman's rank correlation), as appropriate for data type and distribution (investigated by Shapiro-Wilk-test). All calculations were done using the computer program STATISTICA 6.0 StatSoft, Inc. (Tulsa, OK, USA). P values < 0.05 were considered as statistically significant. All values are given as mean ± standard deviation (X ± SD).

RESULTS

Tyrosine Phosphorylation in Spermatozoal Subpopulations

Capacitation was monitored by semiquantitative densitometric analyses of the 105-kDa and 80-kDa bands of the tyrosine phosphorylation Western blots (see FIG. 1). The amount of tyrosine phosphorylation in the nonseparated, noncapacitated aliquot was taken as control and set as 100%.

Serving as an internal quality control for the induction of capacitation, the capacitated aliquots always showed significant higher levels of tyrosine phosphorylation compared to the noncapacitated aliquots ($P < 0.01$). Furthermore, EPS⁻ spermatozoa separated by Annexin-V-MACS presented

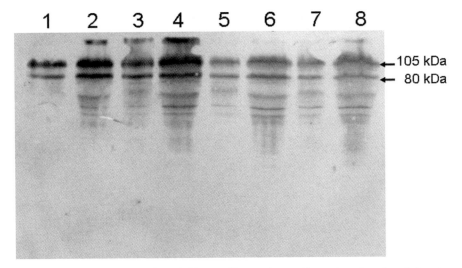

FIGURE 1. Two examples of Western blot analyses of tyrosine phosphorylation in the four subgroups evaluated semiquantitatively with the chemiluminimeter: *lanes* 1 + 5: noncapacitated and nonseparated aliquots; *lanes* 2 + 6: capacitated and nonseparated aliquots; *lanes* 3 + 7: noncapacitated and MACS-separated, EPS-negative aliquots; *lanes* 4 + 8: capacitated and MACS-separated, EPS-negative aliquots. The noncapacitated and nonseparated aliquots (*lanes* 1 + 5) were set as 100%. The highest amount of tyrosine phosphorylations was always detected in the capacitated, EPS-negative aliquots (*lanes* 4 + 8).

with significantly higher amounts of tyrosine phosphorylation compared to the nonseparated semen samples ($P < 0.01$). This was true for the capacitated as well as for the noncapacitated aliquots: control/noncapacitated, 100%; control/capacitated, $136 \pm 30\%$; EPS$^-$/noncapacitated, $115 \pm 13\%$; EPS$^-$/capacitated, $165 \pm 30\%$ (mean $\pm$ SD; see FIG. 1).

Spontaneous and Calcium-Ionophore-Induced AR in Spermatozoal Subpopulations

There was no difference in spontaneous AR as measured by detection of CD46 on the spermatozoal surface in the different study groups: $3.5 \pm 0.7\%$ of the control/noncapacitated, $4.8 \pm 1.2\%$ of the control/capacitated, $3.4 \pm 1.6\%$ of the EPS$^-$/noncapacitated, and $4.9 \pm 1.9\%$ of the EPS$^-$/capacitated spermatozoa were CD46-positive ($P > 0.05$, all values are expressed as mean $\pm$ SD, FIG. 2).

In contrast, there were considerable variances concerning the ability to induce AR in the various spermatozoal subpopulations. Because *in vivo* capacitation precedes the AR, the portion of CD46-positive, acrosome-reacted sperm was significantly higher in the capacitated aliquots compared to the

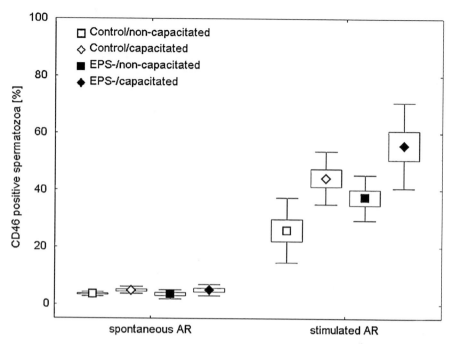

FIGURE 2. Results of CD46 detection in the different study groups before (spontaneous AR) and after induction: spontaneous AR was not influenced by capacitation and Annexin-V MACS separation. The highest potential to perform AR was found following capacitation ($P < 0.01$) compared to all other groups. Graphs are displayed as box plots; box: mean ± SEM, whisker: mean ± SD. AR = acrosome reaction.

noncapacitated aliquots (control/noncapacitated vs. control/capacitated spermatozoa: 25.9 ± 11.4% vs. 44.3 ± 9.3% CD46-positive cells, $P < 0.01$). The Annexin-V-MACS separation of EPS-negative sperm leads, in addition, to significantly increased amounts of CD46-positive cells. AR was best inducible in EPS$^-$ sperm after capacitation, while EPS$^-$/noncapacitated aliquots showed 37.4 ± 8.0% and the EPS$^-$/capacitated aliquots contained 55.7 ± 15.1% CD46-positive sperm ($P < 0.01$, FIG. 2).

Within the differently treated groups there was no association between the relative amount of tyrosine phosphorylation and spontaneous AR measured as percentage of CD46-positive spermatozoa. In contrast, the susceptibility of spermatozoa to induction of AR by calcium ionophore A23187 depended on the quality of capacitation: the more tyrosine phosphorylation was detected, the higher the percentage of CD46-positive spermatozoa after incubation with calcium ionophore A23187 (Spearmans rank correlation, $R = 0.35$, $P < 0.05$).

DISCUSSION

Capacitation and succeeding AR are indispensable steps towards full fertilization capacity of mammalian sperm. To date signaling pathways of capacitation and AR are not fully understood. Capacitation induces apoptosis-unrelated exposure of aminophospholipids such as phosphatidylserine at the plasma membrane.[20,21] Surface exposure of phosphatidylserine on sperm cell activation *in vitro* was found to be restricted to the apical area of the head plasma membrane and unrelated to apoptosis.[22] On the other hand exposure of phosphatidylserine and its specific binding to Annexin-V has been used to characterize and to select apoptotic spermatozoa.[7,13]

The aim of our study was to see whether spermatozoa free of EPS at the surface prior to capacitation are more capable of achieving full fertile potential. To address this question, induction of capacitation and AR was tested in unselected sperm and in enriched nonapoptotic (EPS$^-$) sperm prepared by the Annexin-V MACS technique.

Our results indicate that in nonapoptotic spermatozoa, selected by Annexin-V-MACS, both physiologic processes are significantly facilitated. EPS-negative human spermatozoa are characterized by a superior ability to capacitate. This superior ability to capacitate directly leads in a linear-dependent manner to a maximum potential to undergo AR after stimulation.

These findings correspond with previous studies demonstrating higher fertilizing potential of EPS-negative spermatozoa in the hamster oocyte-penetration assay.[23] Selection of EPS-negative sperm by Annexin-V-MACS may be of advantage for assisted reproduction in order to prepare the sperm subpopulation with the highest fertilizing potential.

ACKNOWLEDGMENT

Support for this study was available from the Junior Research Grant from the Faculty of Medicine, University of Leipzig (formel.1–54).

REFERENCES

1. HARRISSON, R.A.P. 1996. Capacitation mechanisms, and the role of capacitation as seen in eutherian mammals. Reprod. Fertil. Dev. **8:** 581–594.
2. NAZ, R.K. & P.B. RAJESH. 2004. Role of tyrosine phosphorylation in sperm capacitation/acrosome reaction. Reprod. Biol. Endocrinol. **2:** 75.
3. HARRISON, R.A. & B.M. GADELLA. 2005. Bicarbonate-induced membrane processing in sperm capacitation. Theriogenology **63:** 342–351.
4. FORD, W.C. 2004. Regulation of sperm function by reactive oxygen species. Hum. Reprod. Update **10:** 387–399.

5. EMILIOZZI, C., H. CORDONIER, J.F. GUERIN, et al. 1996. Effects of progesterone on human spermatozoa prepared for in vitro fertilization. Int. J. Androl. **19:** 39–47.
6. FRASER, L.R. 1998. Sperm capacitation and the acrosome reaction. Hum. Reprod. **13**(Suppl. 1): 9–19.
7. GLANDER, H.J. & J. SCHALLER. 1999. Binding of annexin V to plasma membranes of human spermatozoa: a rapid assay for detection of membrane changes after cryostorage. Mol. Hum. Reprod. **5:** 109–115.
8. PAASCH, U., S. GRUNEWALD, A. AGARWAL & H.J. GLANDERA. 2004. Activation pattern of caspases in human spermatozoa. Fertil. Steril. **81**(Suppl. 1): 802–809.
9. LOPES, S., J.G. SUN, A. JURISICOVA, et al. 1998. Sperm deoxyribonucleic acid fragmentation is increased in poor-quality semen samples and correlates with failed fertilization in intracytoplasmic sperm injection. Fertil. Steril. **69:** 528–532.
10. VERMES, I., C. HAANEN, H. STEFFENS-NAKKEN & C.P. REUTELINGSPERGER. 1995. A novel assay for apoptosis: flow cytometric detection of phosphatidylserine expression of early apoptotic cells using fluorescein labelled Annexin V. J. Immunol. Meth. **184:** 39–51.
11. GRUNEWALD, S., U. PAASCH & H.J. GLANDER. 2001. Enrichment of non-apoptotic human spermatozoa after cryopreservation by immunomagnetic cell sorting. Cell Tissue Bank. **2:** 127–133.
12. MILTENYI, S., W. MULLER, W. WEICHEL & A. RADBRUCH. 1990. High gradient magnetic cell separation with MACS. Cytometry **11:** 231–238.
13. PAASCH, U., A. AGARWAL, A.K. GUPTA, et al. 2003. Apoptosis signal transduction and the maturity status of human spermatozoa. Ann. N.Y. Acad. Sci. **1010:** 486–489.
14. PAASCH, U., S. GRUNEWALD, A. AGARWAL & H.J. GLANDER. 2004. The activation pattern of caspases in human spermatozoa. Fertil. Steril. **81:** 802–809.
15. GRUNEWALD, S., U. PAASCH & H.J. GLANDER. 2001. Enrichment of non-apoptotic human spermatozoa after cryopreservation by immunomagnetic cell sorting. Cell Tissue Bank. **2:** 127–133.
16. WHO. 1999. WHO Laboratory Manual for the Examination of Human Semen and Sperm-Cervical Mucus Interaction. Cambridge University Press. Cambridge, UK.
17. HENKEL, A.W. & S.C. BIEGER. 1994. Quantification of proteins dissolved in an electrophoresis sample buffer. Anal. Biochem. **223:** 329–331.
18. LAEMMLI, U.K. 1970. Cleavage of structural proteins during the assembly of the head of bacteriophage T4. Nature **227:** 680–685.
19. FAULKNER, K. & I. FRIDOVICH. 1993. Luminol and lucigenin as detectors for O2.- Free. Radic. Biol. Med. **15:** 447–451.
20. MURATORI, M., I. PORAZZI, M. LUCONI, et al. 2004. AnnexinV binding and merocyanine staining fail to detect human sperm capacitation. J. Androl. **25:** 797–810.
21. GADELLA, B.M. & R.A. HARRISON. 2002. Capacitation induces cyclic adenosine $3',5'$-monophosphate-dependent, but apoptosis-unrelated, exposure of aminophospholipids at the apical head plasma membrane of boar sperm cells. Biol. Reprod. **67:** 340–350.
22. DE VRIES, K.J., T. WIEDMER, P.J. SIMS & B.M. GADELLA. 2003. Caspase-independent exposure of aminophospholipids and tyrosine phosphorylation in bicarbonate responsive human sperm cells. Biol. Reprod. **68:** 2122–2134.
23. SAID, T.M., A. AGARWAL, S. GRUNEWALD, et al. 2006. Selection of non-apoptotic spermatozoa as a new tool for enhancing assisted reproduction outcomes: an in vitro model. Biol. Reprod. **74:** 530–537.

Caspase Activation and Extracellular Signal-Regulated Kinase/Akt Inhibition Were Involved in Luteolin-Induced Apoptosis in Lewis Lung Carcinoma Cells

JIN-HYUNG KIM,[a] EUN-OK LEE,[a] HYO-JUNG LEE,[b] JIN-SOOK KU,[a] MIN-HO LEE,[c] DEOK-CHUN YANG,[d] AND SUNG-HOON KIM[a,b]

[a]*Department of Oncology, Graduate School of East-West Medical Science, Kyung Hee University, 1 Seochunri, Kiheungeup, Yongin 449-701, Republic of Korea*

[b]*Laboratory of Angrogenesis and Chemoprevention, College of Oriental Medicine, 1 Hoegidong, Dongdae Mungu, Seoul 131-701, Republic of Korea*

[c]*Kyungee University-Based Enterprise of Oriental Medicinal Materials and Processing, Kyung Hee University, 1 Seochunri, Kiheungeup, Yongin 449-701, Republic of Korea*

[d]*Department of Oriental Materials and Processing, Kyung Hee University, 1 Seochuri, Kiheungeup, Yongin, Kyunggi 449-701, Republic of Korea*

ABSTRACT: Luteolin was isolated from *Scutellaria barbata* D. Don (*S. barbata*). In the present study, we examined the underlying molecular mechanism of luteolin and its effect on *in vivo* tumor growth of Lewis lung carcinoma (LLC) cells. Luteolin exhibited antiproliferative activity against LLC cells with IC_{50} of 12 μM. Luteolin effectively increased Annexin-V-positive cells as well as sub G_1 DNA portion as seen on flow cytometric analysis. Western blotting has revealed that luteolin effectively activates caspase 9 and 3, cleaves poly (ADP-ribose) polymerase (PARP), and increases the ratio of Bax/Bcl-2. Furthermore, mitochondrial membrane potential was reduced by luteolin as seen on fluorescence microscopy. Luteolin downregulated the expression of extracellular signal-regulated kinase (ERK) and Akt in a concentration-dependent manner. In addition, luteolin significantly inhibited the growth of LLC cells implanted on the flank of mice to 40% and 60% of untreated control group values at 2 mg/kg and 10 mg/kg, respectively. Similarly, luteolin significantly reduced the expression of proliferating cell nuclear antigen (PCNA) as well as increased the expression of terminal deoxynucleotidyl transferase biotin-dUTP nick end labeling (TUNEL) in tumor section

Address for correspondence: Sung-Hoon Kim, O.M.D., Ph.D., Department of Oncology, Graduate school of East-West Medical Science, Kyung Hee University, 1 Seochunri, Yongin 449-701, Republic of Korea. Voice: +82-31-201-2179; fax: +82-31-205-1074.
 e-mail: sungkim7@khu.ac.kr

of LLC-bearing mice as determined by immunohistochemistry. Taken together, these results suggest that luteolin exerts antitumor activity by caspase activation and ERK/Akt inhibition.

KEYWORDS: luteolin; apoptosis; caspase; extracellular signal-regulated kinase; Akt; Lewis lung carcinoma

INTRODUCTION

Scutellaria barbata D. Don (*S. barbata*) is a perennial herb which is natively distributed throughout southern China and Korea. It has been used in folk medicine as a remedy for cancer, inflammation and urinary disease for years.[1] Recently, there were reports of antitumor activity of *S. barbata* via regulation of calcium flow in cervical cancer cells,[2] on inhibitory effect on tumor growth by macrophage-enhancing activity,[1] antimutagenic activity[3] and apoptosis.[4,5] *S. barbata* contains alkaloids and flavonoids, which are prominent plant secondary metabolites that are consumed by humans as dietary constituents. A large number of biological activities including anticancer activity have been attributed to these compounds.[6] On the basis of our previous report,[7] we isolated luteolin from ethylacetate fraction of *S. barbata* as a leading compound.[8] Luteolin was reported to inhibit angiogenesis,[9] suppress androgen-independent prostate cancer[10] and DNA topoisomerase I and induce apoptosis in different cancer cells.[11–13] In the present study, we examined the underlying apoptotic mechanism of luteolin and its effect on *in vivo* tumor growth of LLC cells.

MATERIALS AND METHODS

Isolation of Luteolin from Ethylacetate Fraction of S. barbata

One kilogram of dried *S. barbata* was extracted three times with methanol (MeOH). Methanol solvent was then removed by evaporation, and dried MeOH extracts (122 g) were obtained. Methanol extracts were dissolved in water

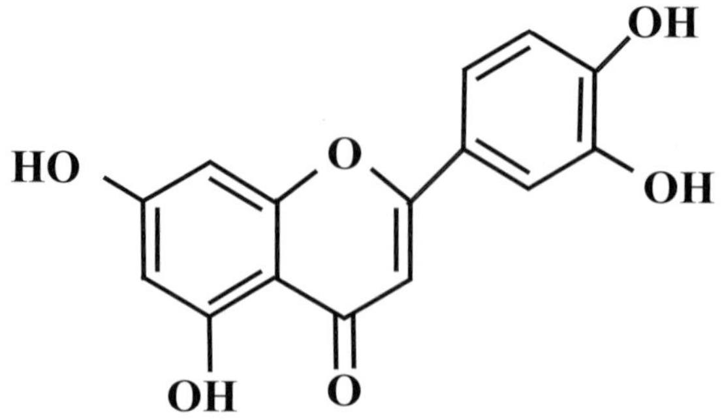

FIGURE 1. Structure of luteolin.

and separated with equal volume of hexane (Hx), methylene chloride (MC), ethyl acetate (EA), and butanol (BuOH) in sequence. Luteolin (FIG. 1) was isolated from EA fraction of *S. barbata* and purified by high-performance liquid chromatography (HPLC) with chloroform-methanol, 9:1, to afford SB-2 (25 mg). Purity was monitored by TLC and by HPLC on LiChrosorb Si 60 (4.0 mm × 25 cm; Merck KGaA, Darmstadt, Germany) with *n*-hexane-ethyl acetate, 1:1, as the mobile phase with a modified method based on Sato's protocol.[8]

Cell Culture

Lewis lung carcinoma (LLC) cell line was provided by the American Type Culture Collection. The cells were cultured in DMEM (Gibco, Grand Island, NY) supplemented with 10% fetal bovine serum, 100 units/mL antibiotic-antimycotic, 1.5 g/L sodium bicarbonate, 4.5 g/L glucose, and 4 mM L-glutamine and incubated in a humidified atmosphere of 5% CO_2 at 37°C.

Cell Proliferation Assay

The effect of luteolin on cell viability was assessed by XTT assay.[14] In brief, the cells were seeded onto 96-well microplates at a density of 1×10^4 cells per well in 100 μL of DMEM. After incubation at 37°C in a humidified incubator for 24 h, cells were treated with various concentrations of luteolin in serum-free DMEM for 24 h and 48 h. After incubation, XTT (Sigma Chemical Co., St. Louis, MO) working solution was added to each well. Cells were incubated at 37°C for 2 h and the optical density was measured using microplate reader (Molecular Devices Co., Sunnyvale, CA) at 450 nm. Cell survival was determined as a percentage of viable cells in the various concentrations of luteolin-treated group versus vehicle-treated control.

Flow Cytometric Analysis

After treatment without or with luteolin for 24 h, cells were collected and washed with cold phosphate-buffered saline solution (PBS). To confirm early-stage apoptosis, Annexin V and propidium iodide (PI) double staining was performed using Annexin V-FITC apoptosis detection kit I (BD Biosciences, San Diego, CA) as described in the manufacturer's instruction. To do cell cycle analysis, cell pellets were fixed in 70% cold ethanol overnight at −20°C. Fixed cells were centrifuged, washed and resuspended in 100 μL of PBS containing 10 μL of RNase A (10 mg/mL) and incubated for 1 h at 37°C. The cells were stained by adding 900 μL of PI (50 μg/mL) for 30 min at room temperature in the dark, and the sub G_1 DNA contents were measured using CellQuest

Software with a FACSVANTAGE SE system (Beckton Dickinson, Heidelberg, Germany).

Western Blotting

After treatment without or with luteolin for 24 h, cells were collected and cell lysates were prepared in lysis buffer (50 mM Tris-HCl [pH 7.4], 150 mM NaCl, 1% Triton X-100, 0.1% SDS, and 1 mM EDTA). The lysates containing 20 μg of protein were separated through gel and proteins were electrophoretically transferred to nitrocellulose membranes. The nitrocellulose membranes were incubated with primary antibody against Bcl-2, Bax (Santa Cruz Biotechnology, Santa Cruz, CA), caspase-9, cleaved caspase-3, PARP, extracellular signal-regulated kinase (ERK), pERK, AKT, pAKT (Cell Signaling Technology, Beverly, MA), and β-actin (Sigma Chemical Co.) at 4°C overnight and incubated with secondary antibodies (Zymed, San Francisco, CA). Signals were developed using an enhanced chemiluminescence (ECL) Western blotting detection kit (Amersham Pharmacia, Arlington Heights, IL) and exposed to X-ray films.

Measurement of Mitochondrial Membrane Potential

LLC cells (1×10^6/mL) were treated with luteolin for 24 h. The cells were washed with cold PBS and stained by adding 150 nM of the fluorescent potential-dependent indicator, tetramethylrhodamine ethyl ester (TMRE, Molecular Probes) for 30 min at 37°C. Then mitochondrial membrane potential was photographed under Axiovert S 100 microscope (Carl Zeiss, Inc., Thornwood, NJ, USA).

Evaluation of in Vivo Tumor Growth in LLC-Bearing Mice

Male C57BL/6 mice, 5 weeks old, were purchased from Daehan Biolink Co., LTD (Chungbuk, Korea) and given food and water *ad libitum*. Mice were housed in a room maintained at $25 \pm 1°C$ with 55% relative humidity. LLC cells (5×10^5/100 μL PBS) were subcutaneously implanted on the right flank of C57BL/6 mice. Luteolin was dissolved in PBS containing 0.1% dimethylsulfoxide (DMSO) and 0.5% Tween 80. Six days later, mice were daily injected i.p. with luteolin at doses of 2 mg/kg/day and 10 mg/kg/day for 7 days. Tumor volumes were measured every 2–3 days with a caliper, and calculated according to the formula ([length $\times$ width^2]/2),[15–17] where length represents the largest tumor diameter and width. All mice were sacrificed 13 days after inoculation of LLC cells and tumors were removed and weighed.

Immunohistochemical Analyses of PCNA and TUNEL

The tumors were fixed in 10% neutral buffered formalin overnight, paraffin-embedded, and sectioned at 4 μm. The thin sections were heat-immobilized, deparaffinized by xylene, rehydrated in a graded series of ethanol, and washed with distilled water. Then the sections were stained with proliferating cell nuclear antigen (PCNA) (Dako A/S, Glostrup, Denmark) using ABC and DAB kits (Vector Lab., Inc., Burlingame, CA) according to the manufacturers' protocols and counterstained with Mayer's hematoxylin solution. Terminal deoxynucleotidyl transferase biotin-dUTP nick end labeling (TUNEL) staining was performed using a TdT-FragEL™ DNA fragmentation kit (Oncogene, Boston, MA) according to the manufacturer's protocols and counterstained with methyl green. The sections were photographed under an Axiovert S 100 light microscope (Carl Zeiss, Inc.) at 400× magnification. Proliferating index was expressed as percentage of PCNA-positive cells by the following equation: proliferating cell (%) = (number of PCNA-positive cells/total number of cells) × 100. Apoptotic index was calculated as percentage of TUNEL-positive cells by the following equation: apoptotic cells (%) = (number of TUNEL-positive cells/total number of cells) × 100.

Statistical Analysis

All values represent means ±SD. The statistically significant differences between control and sample groups were calculated by the Student's t-test.

RESULTS

Luteolin Inhibited the Proliferation of LLC Cells

Inhibitory effect of luteolin on the cell proliferation of LLC cells was investigated by XTT assay. The cells were treated with various concentrations of luteolin for the indicated times. After incubation for 24 h at 37°C, 50 μL of XTT working solution was added to each well. Cell viability was then determined compared with control. Luteolin exhibited antiproliferative activity against LLC cells with IC_{50} of 12 μM by 2,3-bis[2-4-nitro-5-sulphophenyl]2H-tetrazolium-5-carboxanilide (XTT) assay (FIG. 2).

Luteolin Increased sub G_1 Portion and Membrane Alteration in LLC Cells

To examine whether the inhibitory effect of luteolin on LLC cell proliferation was due to induction of apoptosis, cell cycle analysis was carried out *in vitro*. Cells were treated with various concentrations of luteolin for 24 h,

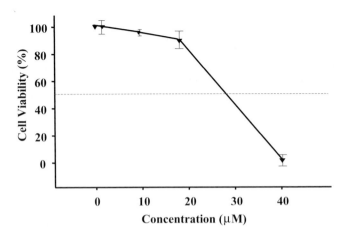

FIGURE 2. Luteolin inhibited the proliferation of LLC cells. Cell viability was evaluated by XTT assay. The cells were treated with various concentrations of luteolin for 24 h or 48 h. After incubation at 37°C, 50 μL of XTT working solution was added to each well. Cell viability was then determined using microplate reader (Molecular Devices Co.) at 450 nm. All data are expressed as means ±SE.

and the sub G_1 DNA content was analyzed by flow cytometry. Luteolin significantly increased sub G_1 peak (FIG. 3A) by PI staining, suggesting apoptosis as well as induced membrane alteration by Annexin V staining, suggesting early apoptosis in a concentration-dependent manner (FIG. 3B). These results indicate that luteolin can induce apoptosis in LLC cells.

Luteolin-Induced Apoptosis in LLC Cells via Mitochondrial Pathway

To investigate the molecular mechanism of luteolin-induced apoptosis, we examined the expression levels of Bax, Bcl-2, caspase 9 and 3, and PARP cleavage by Western blot analysis. Luteolin effectively activated caspase 9 and 3 expression (FIG. 4). In addition, PARP cleavage was also observed from a concentration of 10 μg/mL (FIG. 4). The level of Bcl-2, an antiapoptotic protein, was dramatically decreased after 24 h treatment of 5 μg/mL luteolin, while the expression of proapoptotic Bax was almost steady. Thus, the ratio of Bax to Bcl-2 was gradually increased in a concentration-dependent manner (FIG. 5). Furthermore, mitochondrial potential, which triggers mitochondrial apoptotic pathway, was remarkably reduced by luteolin (FIG. 6). These data suggest that luteolin can induce apoptosis via mitochondrial pathway.

ERK and Akt were Involved in Luteolin-Induced Apoptosis in LLC Cells

The ERK and Akt survival pathways are two such cascades, and their activation is often found to be a critical step for a favorable cellular outcome

DMNQ S-64 (µM)

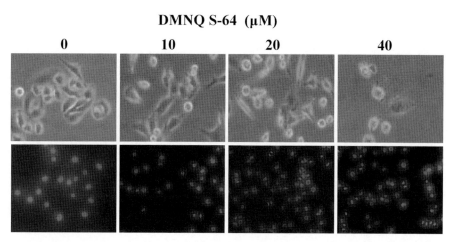

FIGURE 3. Luteolin increased the sub G_1 portion and membrane alteration in LLC cells. Cells were treated with 0 µM (A), 5 µM (B), 10 µM (C) and 20 µM(D) of luteolin for 24 h. After fixing in 75% ethanol, cells were stained by adding PI (A) and Annexin V (B), respectively. Sub G_1 DNA contents and cell membrane alteration were analyzed by flow cytometry system.

upon stressful stimulation.[18,19] In this study, the expression of ERK and Akt was downregulated in a concentration-dependent manner as shown in FIGURE 7, suggesting ERK and Akt signaling pathways were involved in luteolin-induced apoptosis.

Luteolin Suppressed the in Vivo Growth of LLC Cells

Luteolin significantly suppressed the growth of LLC cells up to 40% and 60% of control values at doses of 2 mg/kg and 10 mg/kg, respectively (FIG. 8). We also confirmed by immunohistochemical studies that luteolin can exert antitumor activity via inhibition of PCNA for survival index, along with increase of TUNEL for apoptosis index (FIG. 9).

DISCUSSION

Apoptosis is a strictly controlled mechanism of cell suicide triggered by certain internal or external signals.[20] Malfunction of apoptosis can cause a variety of diseases including cancer, neurodegenerative disorders, and autoimmune disease.[21] Because many chemotherapeutic drugs have induced apoptosis in malignant cells, apoptosis has currently been a target for developing antitumor drugs.[22] Moreover, there is a growing interest in the use of natural materials or naturally occurring substances for the treatment or prevention of cancer

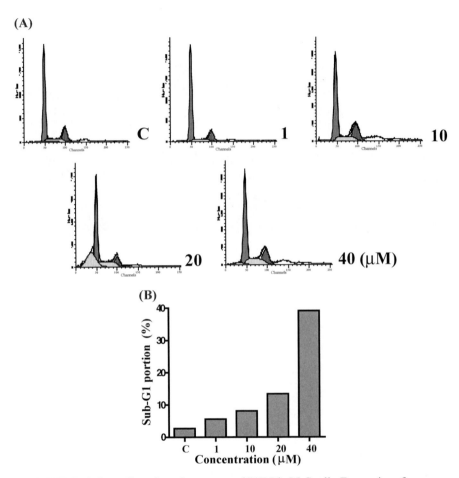

FIGURE 4. Luteolin activated caspases and PARP in LLC cells. Expression of caspases and PARP was investigated by Western blotting. Cells were treated with various concentrations of luteolin for 24 h. The cell lysates containing 20 g of proteins were separated on gels, transferred to Hybond ECL transfer membrane, and probed with specific antibodies. The proteins were developed using an ECL Western blotting detection kit (Amersham Pharmacia) and exposed to X-ray films.

because of the few side effects.[23,24] Luteolin, a naturally occurring flavonoid, was isolated from *S. barbata* D. Don by activity-based fractionation. Our study was carried out to elucidate the apoptotic mechanism of luteolin in LLC cells. Luteolin exerted antiproliferative activity in LLC cells with IC_{50} of 12 μM as seen on XTT assay. In addition, flow cytometric analysis showed that luteolin can increase sub G_1 portion in LLC cells, suggesting apoptosis. Similarly, the increase of membrane alteration in cells positively stained by Annexin V implies the early stages of apoptosis of LLC cells. These data suggest that the antiproliferative effect of luteolin in LLC cells may be due to apoptosis.

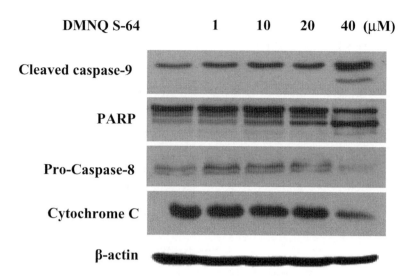

FIGURE 5. Luteolin increased the ratio of BAX to Bcl-2 protein in LLC cells. Cells were treated with various concentrations of luteolin for 24 h. The cell lysates containing 20 μg of proteins were separated on gels, transferred to Hybond ECL transfer membrane, and probed with specific antibodies. The proteins were developed using an ECL Western blotting detection kit (Amersham Pharmacia) and exposed to X-ray films.

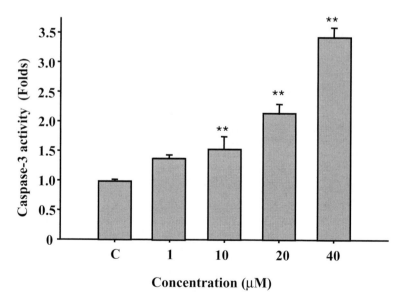

FIGURE 6. Luteolin reduced mitochondrial membrane potential in LLC cells. Cells were treated with 0 μM (**A**) and 10 μM (**B**) of luteolin for 24 h. After washing with PBS, cells were stained by adding TMRE solution (150 nM) for 30 min at 37°C. Then, randomly chosen fields were photographed under an Axiovert S 100 microscope (Carl Zeiss, Inc.).

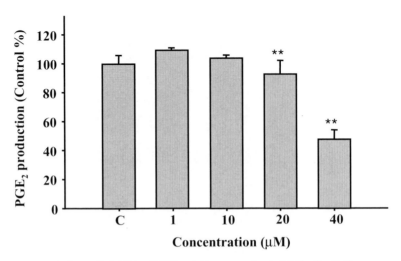

FIGURE 7. Luteolin inhibited ERK and Akt expression in LLC cells. Cells were treated with various concentrations of luteolin for 24 h. The cell lysates containing 20 μg of proteins were separated on gels, transferred to Hybond ECL transfer membrane, and probed with specific antibodies. The proteins were developed using an ECL Western blotting detection kit (Amersham Pharmacia) and exposed to X-ray films.

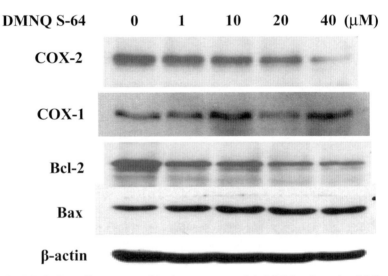

FIGURE 8. Luteolin suppressed *in vivo* tumor growth in LLC-bearing mice. LLC cells ($5 \times 10^5 / 100$ μL PBS) were subcutaneously implanted on the right flank of C57BL6 mice. Luteolin was dissolved in PBS containing 0.1% DMSO and 0.5% Tween 80. Six days later, mice were daily injected i.p. by luteolin at doses of 2 mg/kg/day and 10 mg/kg/day for 7 days. Tumor volumes were measured every 2–3 days with a caliper. All mice were sacrificed 13 days after inoculation of LLC cells and tumors were removed and weighed (**B**). All data are expressed as means ±SE., ∗: $P < 0.05$ versus untreated control.

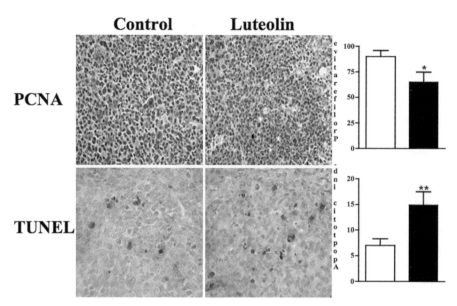

FIGURE 9. Luteolin attenuated the expression of PCNA and increased the expression of TUNEL in tumor sections as determined by immunohistochemistry. LLC-bearing mice were sacrificed on day 13, and tumors were immediately removed, fixed in 10% neutral buffered formalin overnight, paraffin-embedded, and sectioned by 4 μM. The sections were immunostained with PCNA and TUNEL based on DAB substrate staining (brown) and counterstained with Mayer's hematoxylin solution (blue), or methyl green (green). The sections were photographed under an Axiovert S 100 light microscope (Carl Zeiss, Inc.) at 400 × magnification.

There are two distinct pathways that initiate apoptosis designated as mitochondrial and death-receptor pathways. The intrinsic pathway or mitochondrial pathway is triggered by release of cytochrome c from mitochondria into the cytosol. Release of cytochrome c is regulated by Bcl-2 family. Antiapoptotic members, such as Bcl-2 and Bcl-xL, prevent cytochrome c release, whereas proapoptotic members, such as Bax and Bak promote cytochrome c release.[25] Bax forms heterodimers with multiple antiapoptotic members and also induces apoptosis by direct cytochrome c release.[26] In this study, luteolin effectively activated the expression of caspase 9 and 3, and cleaved poly (ADP-ribose) polymerase (PARP). Luteolin also significantly decreased the level of Bcl-2 in LLC cells, resulting in alteration of the Bax to Bcl-2 ratio in favor of apoptosis. In addition, the reduction of mitochondrial membrane potential by luteolin makes sure that the mitochondrial pathway was involved in luteolin-induced apoptosis. There is accumulating evidence that the ERK and Akt activity are required to modulate apoptosis as prosurvival pathways.[27,28] The downregulation of expression of ERK and Akt in this study strongly indicate that luteolin-induced apoptosis can be mediated through ERK and Akt pathways.

Furthermore, luteolin significantly inhibited the growth of LLC cells implanted on the flank of mice to ~40% and 60% greater, respectively, than control values at doses of 2 mg/kg and 10 mg/kg. Immunohistochemical studies showed that luteolin also suppressed the expression of PCNA for survival index and as increased TUNEL for apoptosis index in tumor section, suggesting that the inhibition of tumor growth by luteolin may be mediated by survival inhibition and apoptosis induction.

In summary, luteolin exerted antiproliferative activity against LLC cells. Luteolin-induced apoptosis can be mediated via mitochondrial pathway by casapse 9 and 3 activation, PARP cleavage, and decrease of mitochondrial potential. Luteolin also increased sub G_1 apoptotic portion and membrane alteration induced by Annexin V. In addition, inhibition of ERK and Akt pathways were involved in luteolin-induced apoptosis. Finally, luteolin inhibited the growth of LLC cells *in vivo* in a dose-dependent manner. Taken together, these results indicate that luteolin exerts antiproliferative activity and induces apoptosis via caspase activation and ERK/Akt inhibition in LLC cells and that luteolin can be used as a potential cancer chemopreventive agent.

ACKNOWLEDGMENT

This work was supported by Biogreen project (Grant B050007) from the Korean Ministry of Health and Welfare, and grants of SRC (R11-2005-014) and BRP (R01-2-005-000-10993-0) from KOSEF.

REFERENCES

1. WONG, B.Y., B.H. LAU, T.Y. JIA, et al. 1996. Oldenlandia diffusa and Scutellaria barbata augment macrophage oxidative burst and inhibit tumor growth. Cancer Biother. Radiopharm. **11:** 51–56.
2. GAO, D., Y. GAO & P. BAI. 2003. Influence of Scutellaria barbata on calcium beaconage of cervix cancer cells. Zhong. Yao. Cai. **10:** 730–733.
3. WONG, B.Y., B.H. LAU, T. YAMASAKI, et al. 1993. Inhibition of dexamethasone-induced cytochrome P450-mediated mutagenicity and metabolism of aflatoxin B1 by Chinese medicinal herbs. Eur. J. Cancer Prev. **24:** 351–356.
4. SHI, R.X., C.N. ONG & H.M. SHEN. 2004. Luteolin sensitizes tumor necrosis factor-alpha-induced apoptosis in human tumor cells. Oncogene **23:** 7712–7721.
5. KO, W.G., T.H. KANG, S.J. LEE, et al. 2002. Effects of luteolin on the inhibition of proliferation and induction of apoptosis in human myeloid leukaemia cells. Phytother. Res. **16:** 295–298.
6. MIDDLETON, E. JR., C. KANDASWAMI & T.C. THEOLARIDES. 2000. The effects of plant flavonoids on mammalian cells: implications for inflammation, heart disease, and cancer. Pharmacol. Rev. **52:** 673–751.

7. CHA, Y.Y., E.O. LEE, H.J. LEE, *et al.* 2004. Methylene chloride fraction of *Scutellaria barbata* induces apoptosis in human U937 leukemia cells via the mitochondrial signaling pathway. Clin. Chim. Acta **348:** 41–48.
8. SATO, Y., S. SUZAKI, T. NISHIKAWA, T, *et al.* 2000. Phytochemical flavones isolated from *Scutellaria barbata* and anti-bacterial activity against methicillin-resistant *Staphylococcus aureus*. J. Ethnopharmacol. **72:** 483–488.
9. BAGLI, E., M. STEFANIOTOU, L. MORBIDELLI, *et al.* 2004. Luteolin inhibits vascular endothelial growth factor-induced angiogenesis; inhibition of endothelial cell survival and proliferation by targeting phosphatidylinositol 3′-kinase activity. Cancer Res. **64:** 7936–7946.
10. KNOWLES, L.M., D.A. ZIGROSSI, R.A. TAUBER, *et al.* 2000. Flavonoids suppress androgen-independent human prostate tumor proliferation. Nutr. Cancer **38:** 116–122.
11. CHOWDHURY, A.R., S. SHARMA, S. MANDAL, *et al.* 2002. Luteolin, an emerging anti-cancer flavonoid, poisons eukaryotic DNA topoisomerase I. Biochem. J. **366:** 653–661.
12. HORINAKA, M., T. YOSHIDA, T. SHIRAISHI, *et al.* 2005. Luteolin induces apoptosis via death receptor 5 upregulation in human malignant tumor cells. Oncogene **24:** 7180–7189.
13. CHENG, A.C., T.C. HUANG, C.S. LAI, *et al.* 2005. Induction of apoptosis by luteolin through cleavage of Bcl-2 family in human leukemia HL-60 cells. Eur. J. Pharmacol. **509:** 1–10.
14. JOST, L.M., J.M. KIRKWOOD, T.L. WHITESIDE. 1992. Improved short- and long-term XTT-based colorimetric cellular cytotoxicity assay for melanoma and other tumor cells. J. Immunol. Methods **147:** 153–165.
15. GIAVAZZI, R., D.E. CAMPBELL, J.M. JESSUP, *et al.* 1986. Metastatic behavior of tumor cells isolated from primary and metastatic human colorectal carcinomas implanted into different sites in nude mice. Cancer Res. **46:** 1928–1933.
16. ALESSANDRIA, G., S. FILIPPESCHI, P. SINIBALDI, *et al.* 1987. Influence of gangliosides on primary and metastatic neoplastic growth in human and murine cells. Cancer Res. **47:** 4243–4247.
17. CHIRIVI, R.G., A. GAROFALO, M.J. CRIMMIN, *et al.* 1994. Inhibition of the metastatic spread and growth of B16-BL6 murine melanoma by a synthetic matrix metalloproteinase inhibitor. Int. J. Cancer **58:** 460–464.
18. DATTA, S.R., A. BRUNET & M.E. GREENBERG. 1999. Cellular survival: a play in three Akts. Genes Dev. **13:** 2905–2927.
19. WANG, X., J.L. MARTINDALE, Y. LIU, *et al.* 1998. The cellular response to oxidative stress: influences of mitogen-activated protein kinase signalling pathways on cell survival. Biochem. J. **333:** 291–300.
20. LA PORTA, C.A. 2004. Cellular targets for anti-cancer strategies. Drug Targets **5:** 347–355.
21. THOMPSON, C.B. 1995. Apoptosis in the pathogenesis and treatment of disease. Science **267:** 1456–1462.
22. KAUFMANN, S.H. & W.C. EARNSHAW. 2000. Induction of apoptosis by cancer chemotherapy. Exp. Cell Res. **256:** 42–49.
23. THATTE, U., S. BAGADEY & S. DAHANUKAR. 2000. Modulation of programmed cell death by medicinal plants. Cell. Mol. Biol. **46:** 199–214.
24. CHEN, Q., G.W. YANG & L.G. AN. 2000. Apoptosis of hepatoma cells SMMC-7721 induced by Ginkgo biloba seed polysaccharide. World J. Gastroenterol. **8:** 832–836.

25. YANG, J., X. LIU, K. BHALLA, et al. 1997. Prevention of apoptosis by Bcl-2: release of cytochrome c from mitochondria blocked. Science **275:** 1129–1132.
26. JURGENSMEIER, J.M., Z. XIE, Q. DEVERAUX, et al. 1998. Bax directly induces release of cytochrome c from isolated mitochondria. Proc. Natl. Acad. Sci. **95:** 4997–5002.
27. LI, L.H., L.J. WU, S.I. TASHIRO, et al. 2006. The roles of Akt and MAPK family members in silymarin's protection against UV-induced A375-S2 cell apoptosis. Int. Immunopharmacol. **6:** 190–197.
28. NUUTINEN, U., V. POSTILA, M. MATTO, et al. 2006. Inhibition of PI3-kinase-Akt pathway enhances dexamethasone-induced apoptosis in a human follicular lymphoma cell line. Exp. Cell Res. **312:** 322–330.

Ceramide Modulation of Antigen-Triggered Ca^{2+} Signals and Cell Fate

Diversity in the Responses of Various Immunocytes

ENDRE KISS, GABRIELLA SÁRMAY, AND JÁNOS MATKÓ

Department of Immunology, Eotvos Lorand University, Institute of Biology, Pazmany Peter Setany 1/C, 1117, Budapest, Hungary

ABSTRACT: Ceramide is a widely accepted mediator of T cell apoptosis and is released upon receiving various death or stress signals. Recently we have shown that the fate of T cells, life or death, depends strictly on the strength and duration of the ceramide-generating stimulus. Subapoptotic ceramide signals were shown to negatively regulate the antigen-specific activation signaling in T cells. Here we show that these subapoptotic ceramide signals also inhibit the antigen-triggered Ca^{2+} signals in B lymphocytes or the FcεRI-mediated response of mast cells to antigen, but in a differential manner. Burkitt B lymphoma cells, frequently used models of mature B cells, and marginal zone B cells were largely resistant to the inhibitory action of ceramide. The response to cell death–inducing (strong/long duration) ceramide stimuli, resulting in massive apoptosis in T cells, was also differential among the various immunocytes in terms of both the death mechanism and the sensitivity. Our data suggest that ceramide's effects on life and death signaling in immunocytes are cell type-/stage-specific.

KEYWORDS: stress and death signals; sphingolipids; ceramides; rafts; Ca^{2+} signals, antigen-triggered immunocyte activation

INTRODUCTION

Ceramides are widely considered as essential mediators and regulators of cellular apoptosis triggered by various death (FasL, tumor necrosis factor [TNF]-α), inflammatory (IL-1-β), or stress (irradiation, chemotherapeutic drugs, oxidative) signals.[1,2] Ceramides are released upon these stimuli from plasma membrane sphingomyelin by activated sphingomyelinases (SMase).[3]

Address for correspondence: Dr. Janos Matko, Department of Immunology, Eotvos Lorand University, Institute of Biology, Pazmany Peter Setany 1/C, 1117, Budapest, Hungary. Voice: 36-1-381-2175; fax: 36-1-381-2176.
e-mail: matko@cerberus.elte.hu

Recently we have shown that ceramide accumulation in the plasma membrane has a bimodulatory effect on T cells. Induction of T cell apoptosis requires a certain threshold of ceramide stimulus (strength/duration), while below this threshold, ceramide generation does not induce detectable apoptosis.[4]

The subapoptotic ceramide stimuli inhibited the antigen-specific T cell activation signaling without causing significant apoptosis.[4] Several plasma membrane ion channels (Kv1.3, KCa, CRAC) were implicated as potential targets of ceramides,[4-6] but the molecular background of this inhibitory effect still remained largely unexplored. It is also an important question whether this modulatory effect of ceramides is general over all immunocytes, whose calcium signaling is initiated by engagement of multichain immune recognition receptors (MIRRs), such as BCR, TCR or FcεRI, with antigen/allergen.

Here we show that the antigen-triggered calcium signals of immunocytes (T and B lymphocytes or mast cells) of various maturation/differentiation stages are differentially affected (inhibited) by subapoptotic ceramide signals. Interestingly, the sensitivity of the Ca^{2+} signaling to CER inhibition correlated well with the susceptibility of the cell type to cell death (apoptosis) induced by strong (high dose/long duration) ceramide accumulation, and also with their GM_1 ganglioside (raft) expression.

MATERIALS AND METHODS

The murine cell lines IP12-7 T hybridoma (helper T phenotype), X16C B lymphoma (marginal zone B cell phenotype), 38C-13 B lymphoma (immature B cell phenotype), A20 B lymphoma, and 2PK3 B lymphoma were cultured in RPMI-1640 supplemented with 2 mM L-glutamine, 1 mM Na-pyruvate, 50 μM 2-mercaptoethanol, antibiotics, and 5% FCS. The human Jurkat (ATCC TIB152) T leukemia cell line, and the Burkitt's lymphoma cell lines BL41 were cultured in RPMI-1640 supplemented with 2 mM L-glutamine, 1 mM Na-pyruvate, antibiotics, and 10% FCS. ST486 human Burkitt's lymphoma cell line was cultured in the same medium supplemented with 20% FCS. The rat RBL-2H3 cell line of basophilic leukemia origin and displaying mucosal-type mast cell properties was maintained in DMEM supplemented by 5% FCS, 2 mM glutamine, and antibiotics in a humidified atmosphere with 5% CO_2 at 37°C. For the experiments, RBL-2H3 cells were harvested following detachment by 15-min incubation with 10-mM EDTA in DMEM. Splenocytes were isolated from spleens removed from mice killed by cervical dislocation, by collecting and washing the cells in RPMI 1640 culture medium containing 5% fetal calf serum, 1 mM Na-pyruvate, 2 mM L-glutamine, 100 U/mL penicillin, 0.1 mg/mL streptomycin, and 5×10^{-5} M mercaptoethanol, after removal of red blood cells by lysis.

C2-ceramide (*N*-acetyl-D-sphingosine) and its inactive dihydro-ceramide analogue were from Biomol Research Laboratories (Plymouth Meeting, PA) and applied as described earlier.[4] IP12-7 murine Th cells were activated by cross-linking the TCR with anti-CD3 mAb (145-2C11) or by antigen-presenting cell (APC: 2PK3 B lymphoma), while the human Jurkat cells with the anti-human CD3 mAb, MEM-57. Anti-human IgM antibody (Bu1), kind gift from Roy Jefferies, was used to activate B cells, while a murine 2,4-dinitrobenzene-sulfonic acid (DNP)-specific IgE class mAb, A2, and DNP_{11}-BSA antigen was used to activate RBL-2H3 mast cells.

The antigen-induced activation and the ceramide-induced apoptotic cellular responses were monitored by flow cytometry (Becton-Dickinson FACSCalibur) using CellQuest Pro software. Calcium signals were detected using FLUO-3 fluorescent indicator and a method optimized for lymphocyte activation.[7] The mitochondrial depolarization was measured by detection of the percentage of dim and bright cells using DiOC6(3) potential-sensitive fluorescent probe.[4] The late apoptotic phase of the cells was monitored by measuring the percent of subdipolid cells (proportional to the extent of DNA fragmentation), using hypotonic (citrate) extraction and PI staining. The expression of GM_1 gangliosides (rafts) in the various immunocytes was measured by staining the cells with Alexa488-cholera toxin B subunit by flow cytometry.

RESULTS AND DISCUSSION

Recent experiments have shown that the subapoptotic (short: ≤ 10 min and modest: ≤ 25 μM) ceramide signal received prior to TCR stimulation inhibited both phases of the calcium response in helper T lymphocytes: the Ca^{2+} release was slightly suppressed, while the influx phase was suppressed more substantially.[4] Here we investigated whether subapoptotic C2-ceramide accumulation in the plasma membrane of mast cells and B cells of various maturation stage and origin can cause a similar inhibition in their antigen-triggered Ca^{2+} signals. A concentration-dependent inhibition by ceramide was observed in the calcium response of RBL-2H3 mast cells triggered by cross-linking of FcεRI-bound IgE with DNP_{11}-BSA antigen (FIG. 1 D) and in splenocytes activated by BCR cross-linking with anti-μ antibody (FIG. 1 E). The magnitude of the inhibition was similar in these cells to that observed with T lymphocytes (FIGS. 1A and 1C). The inactive analogue, dihydro-ceramide, did not affect the calcium signals (not shown).

In both immature (38C-13) and mature (A20) mouse B cell lines the anti-μ cross-linking-induced calcium signal was also inhibited by the subapoptotic (25 μM, 10 min) C2-ceramide pretreatment, with a remarkably higher sensitivity of immature B cells (FIG. 1F). In contrast, the activation-induced calcium response in B cells of marginal zone origin (X16C) and two human mature (Burkitt lymphoma) B cells (ST486 and BL41) were found resistant to ceramide inhibition in the concentration range of 5–50 μM (FIGS. 1B and 1F). Interestingly, the membrane raft (GM_1 ganglioside) expression of these three

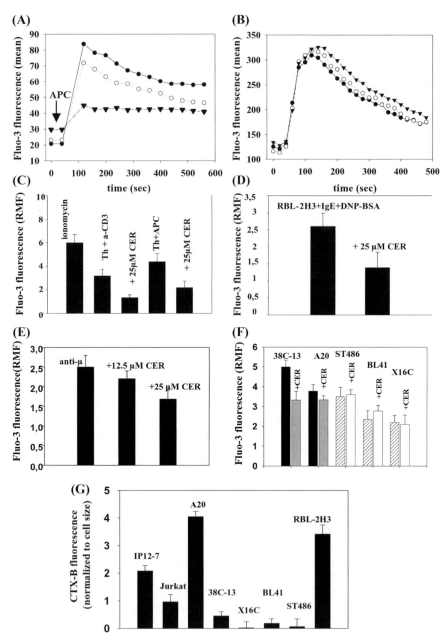

FIGURE 1. Effect of subapoptotic (short, modest) ceramide stimulus on the subsequent antigen-triggered Ca^{2+} response of various immunocytes. (**A**) Inhibition of calcium signals of helper T lymphocytes (IP12-7) induced by antigen-presenting cell (APC: 2PK3 B cell).[4] The calcium signals in the absence (•) and in the presence (10 min) of 12.5 μM (○) and 25 μM (▼) C2-ceramide are shown. (**B**) The calcium response of human ST486 mature

cell lines was extremely low compared to the ceramide-sensitive cell types (FIG. 1G). This might be related to the lack of inhibition because ceramide released in the plasma membrane was reported to restructure the membrane and aggregate lipid rafts, which may contribute to the regulatory effects of ceramides on both life and death signaling.[3,4,6] In addition, several calcium and potassium channels were proposed as direct or indirect targets of ceramide action.[4-6] While Kv1.3 potassium channels were shown to be expressed on both T and B mature, peripheral lymphocytes, expression of voltage-dependent calcium channels (VDCC) was demonstrated only on T cells, so far.[8] Thus, the lack of nifedipine sensitivity in the activation calcium signals of ceramide-resistant cell types (not shown), in contrast to T cells, suggests that VDCCs may also be potential targets of ceramides and the lack of ceramide inhibition may rise partly from their absence in the resistant B cell lines. Further studies are required, however, to confirm this hypothesis and to reveal the molecular background of differential sensitivity of immunocytes to the inhibitory action of plasma membrane ceramide accumulation.

Because it has been shown recently that above a critical signal threshold (high concentration/long duration) ceramide release in the plasma membrane induces massive apoptosis of T cells,[2,4] we also investigated the susceptibility of the other immunocytes to ceramide-mediated cell death. To this end we applied C2-ceramide concentrations of 10–100 μM and an incubation of long duration (several hours). Two critical indicators of apoptosis, the mitochondrial depolarization and DNA-fragmentation, were monitored and simultaneously the propidium iodide–positive necrotic cells were also counted. This analysis also revealed a substantial diversity in the sensitivity and death responses of the various immunocytes investigated. While a substantial fraction of T cells responded to ceramide accumulation by mitochondrion damage–dependent apoptosis accompanied with significant DNA fragmentation, the B cells and the mast cell line responded mostly by necrotic cell death (increased cell volume

---

B cells to BCR cross-linking by anti-μ Ab. The symbols represent the same conditions as in (**A**). (**C**) Maximal amplitudes of calcium responses of IP12-7 Th cells to TCR cross-linking by anti-CD3 Ab (*left*) and to APC (*right*) are shown in the absence and presence of C2 ceramide. For comparison the response to ionomycin Ca^{2+} ionophore is also shown. (**D**) Amplitudes of mast cell calcium response to FcεRI-IgE cross-linking by DNP-BSA antigen are shown in the absence and presence of C2-ceramide. (**E**) Amplitudes of ceramide concentration-dependent calcium responses in mouse splenocytes. (**F**) Amplitudes of calcium responses to BCR stimulation in immature (38C-13) and mature (A20) mouse B cells in the absence (*black*) and presence (*gray*) of 25 μM C2 ceramide. Responses of human mature Burkitt B lymphoma cells (ST486, BL41) and marginal zone B cells (X16-C) to the same stimulation are also shown in the absence (*hatched*) and presence (*white*) of 25 μM C2-ceramide. (**G**) Expression levels of GM_1 gangliosides (*rafts*) monitored by Alexa488-CTX binding are shown for all immunocytes. All the bars represent means of 3–4 independent measurements with the SEM.

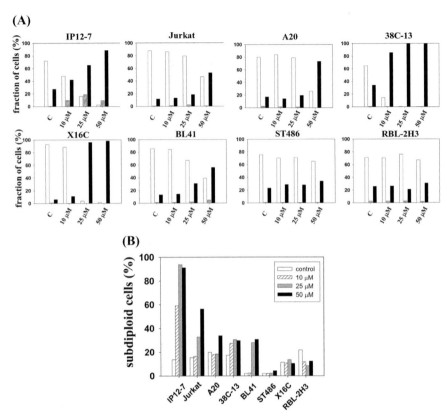

FIGURE 2. Cell death responses of various immunocytes to long-duration ceramide stimuli. (**A**) Fraction of cells displaying normal (*white bars*) and decreased (*striped bar*) mitochondrial membrane potential are shown together with fraction of propidium iodide positive (necrotic) cells (*black bar*). The bars represent means of 2×10^4 cells measured 6 h after addition of C2-ceramide at concentrations indicated. (**B**) Percentage of subdiploid cells measured after 6 h incubation with the indicated concentrations of C2-ceramide by PI staining of hypotonically extracted cells.

followed by significant spontaneous uptake of propidium iodide (PI). Among them, the immature B cells (38C-13) and the marginal zone B cells proved to be most sensitive to the ceramide-mediated necrotic cell death. Interestingly, the two human Burkitt lymphoma B cell lines (ST486, BL41), the mature mouse B cell line (A20), and especially the mast cell line (RBL-2H3) showed a remarkably low sensitivity to ceramide (FIGS. 2A and 2B). The observed differences suggest that the death signaling machinery or the initiating plasma membrane effects of ceramide may be largely different in T cells and various B cells. To understand the fine details of the differential sensitivity to ceramide-mediated cell death requires further studies. The remarkable resistance of the Burkitt lymphomas and mucosal mast cells might reflect either their largely different membrane/microdomain structure or the lack of yet unidentified mediators necessary to execute ceramide-induced death signals.

In conclusion, our data suggest that subapoptotic ceramide accumulation in the plasma membrane is a general mechanism capable of inhibiting antigen-triggered activation calcium signals not only in T cells but in B lymphocytes and mast cells, as well. In addition, strong/long duration ceramide release may differentially induce apoptotic or necrotic cell death among the various immunocytes. These data, together with the resistance of several cell types to both kinds of ceramide effects, suggests that the response of immunocytes to stress or death signal–induced ceramide release is highly cell type-/differentiation stage-specific.

ACKNOWLEDGMENTS

This work was supported by Grants T049696 and TS044711 from the Hungarian National Science Foundation (OTKA).

REFERENCES

1. MATHIAS, S., L.A. PENA & R.N. KOLESNICK. 1998. Signal transduction of stress via ceramide. Biochem. J. **335:** 465–480.
2. GRASSMÉ, H., A. JEKLE, A. RIEHLE, et al. 2001. CD95 signaling via ceramide-rich membrane rafts. J. Biol. Chem. **276:** 20589–20596.
3. VAN BLITTERSWIJK, W.J., A.H. VAN DER LUIT, R.J. VELDMAN, et al. 2003. Ceramide: second messenger or modulator of membrane structure and dynamics? Biochem. J. **369:** 199–211.
4. DETRE, C., E. KISS, Z. VARGA, et al. 2006. Death or survival: membrane ceramide controls the fate and activation of antigen-specific T-cells depending on signal strength and duration. Cell Signal. **18:** 294–306.
5. LEPPLE-WIENHUES, A., C. BELKA, T. LAUN, et al. 1999. Stimulation of CD95 (Fas) blocks T lymphocyte calcium channels through sphingomyelinase and sphingolipids. Proc. Natl. Acad. Sci. USA **96:** 13795–13800.
6. BOCK, J., I. SZABO, N. GAMPER, et al. 2003. Ceramide inhibits the potassium channel Kv1.3 by the formation of membrane platforms. Biochem. Biophys. Res. Commun. **305:** 890–897.
7. RÉTHI, B., C. DETRE, P. GOGOLÁK, et al. 2002. Flow cytometry used for the analysis of calcium signaling induced by antigen-specific T-cell activation. Cytometry **47:** 207–216.
8. KOTTURI, M., D.A. CARLOW, J.C. LEE, et al. 2003. Identification and functional characterization of voltage-dependent calcium channels in T lymphocytes. J. Biol. Chem. **278:** 46949–46960.

Caspase-3 Activation, Bcl-2 Contents, and Soluble FAS-Ligand Are Not Related to the Inflammatory Marker Profile in Patients with Sepsis and Septic Shock

FABIAN KRIEBEL,[a] SILKE WITTEMANN,[b] HSIN-YUN HSU,[b] THOMAS JOOS,[b] MANFRED WEISS,[c] AND E. MARION SCHNEIDER[a]

[a]*Department of Experimental Anesthesiology, University Clinic Ulm, Steinhoevelstrasse 9, 89075 Ulm, Germany*

[b]*Department of Natural and Medical Sciences Institute at the University of Tuebingen, Reutlingen, Germany*

[c]*Department of Anesthesiology, University Clinic Ulm, Steinhoevelstrasse 9, 89075 Ulm, Germany*

ABSTRACT: The current comparative investigation analyses markers of inflammation and apoptosis in peripheral blood of intensive care unit (ICU) patients with postoperative/posttraumatic SIRS (systemic inflammatory response syndrome), sepsis, severe sepsis, or septic shock. Inflammatory markers (C-reactive protein [CRP], cytokines, metalloproteinases [MMPs]) and soluble FAS-Ligand (sCD178) were determined in plasma, and apoptosis-relevant antigens such as active caspase-3, Bcl-2, and sCD178 were quantified in whole-blood cell lysates. These parameters were analyzed daily in 20 postoperative/posttraumatic patients: 2 patients had SIRS, 5 suffered from sepsis (2 died), and 13 had septic shock (5 died). Active caspase-3, Bcl-2, and sCD178 were determined by ELISA and by fluorescence-activated cell sorting (FACS)-array kits using bead-assisted flow cytometry. Cytokines and MMPs were quantified by Luminex-assisted Beadlyte assays. Active caspase-3 was identified in defined samples of whole-blood lysates covering, for example, 5/7, 8/18, and 6/11 consecutive days during the patients' stay on the ICU. Also, sCD178 was detected on successive days. Peaks of active caspase-3 antigen contents in whole blood occurred independently of CRP and inflammatory cytokines such as tumor necrosis factor (TNF)-α and IL-6. In addition, high MMPs 1–3, 7–10, and 13 concentrations were detected. Interestingly, active caspase-3 and cell-associated sCD178 were either elevated simultaneously or in a close time window. The same was true for Bcl-2. In conclusion, activation of apoptosis can be determined in whole blood

Address for correspondence: E. Marion Schneider, Ph.D., Department of Experimental Anesthesiology, University Clinic Ulm, Steinhoevelstrasse 9, 89075 Ulm, Germany. Voice: +49-731-500-60080; fax: +49-731-500-60082.
e-mail: marion.schneider@uni-ulm.de

of postoperative/posttraumatic patients by active caspase-3 and by Bcl-2. Pro- and antiapoptotic effects during sepsis may occur independently of peaks in inflammatory markers. Apoptosis could explain modeling and remodeling of leukocyte subpopulations.

KEYWORDS: whole-blood lysates; active caspase-3; soluble Fas-ligand (sCD178); Bcl-2; inflammatory cytokines; TNF-α; IL-6; sepsis; septic shock; MMPs

INTRODUCTION

Apoptosis, programmed cell death, is a process by which cells undergo inducible, nonnecrotic, cellular suicide. In mouse models mimicking polymicrobial sepsis, the differential occurrence of apoptosis in neutrophils, monocytes/macrophages, and lymphocytes has been reported from different sites in the body.[1,2] In critically ill patients, immediate autopsy revealed massive apoptosis in lymphocytes as assessed by TUNEL assays, DNA gel electrophoresis, and immunohistochemical staining for active caspase-3.[3] Thus, apoptosis may contribute to the sepsis-associated immune dysfunction.[4] Mechanisms of apoptosis include both receptor-mediated and mitochondrial pathways.[5] Because hyperinflammatory cytokines constitute the hallmark of detrimental developments during sepsis, we asked whether inflammatory cytokines might precede or follow apoptosis in either systemic inflammatory response syndrome (SIRS), sepsis, severe sepsis, or septic shock. Therefore, a longitudinal analysis was performed determining markers of inflammation and apoptosis in plasma and whole-blood lysates of postoperative/posttraumatic ICU patients.

MATERIAL AND METHODS

Patients

Upon approval by the Independent Ethics Committee of the University of Ulm, 20 ICU postoperative/posttraumatic patients were included in this prospective pilot study and analyzed daily: 2 with SIRS, 5 suffering from sepsis (2 died), and 13 with septic shock (5 died).

Methods

EDTA blood was lysed on ice with cell lysis buffer (BD-Germany; 10 mM Tris-Hcl, 10 mM NaH_2PO_4/$NaHPO_4$, 130 mM NaCl, 1% Triton X-100, 10 mM PPi, protease inhibitor cocktail of 800 μg/mL benzamidine-HCL, 500 μg/mL o-phenanthroline, 500 μg/mL aprotinin, 500 μg/mL leupeptin, 500 μg/mL pepstatin A, 50nM PMSF). The lysate was centrifuged at

14,000 rpm for 10 min and stored at –20°C. Active caspase-3 was measured with the R&D Quantikine Active Caspase-3 Immunoassay (R&D Systems, Wiesbaden-Nordenstadt, Germany). Bcl-2 was determined by ELISA (Alexis Corp., Grünberg, Germany). Also, active caspase-3 and Bcl-2 were measured by flow cytometry with the BD fluorescence-activated cell sorter FACScaliber (CBA Human Apoptosis Kit, R&D Systems). Cytokines and active metalloproteinases (MMPs) were quantified by Luminex-assisted Beadlyte assays (Qiagen LiquiChip, Hilden, Germany). Soluble FAS-Ligand (sCD178) was determined by ELISA. Spearman rank association analysis was done by means of GraphPad Prism Software. The sepsis score in FIGURES 1 and 2 reflects the severity of the inflammatory response, is based on the international sepsis definition,[6] and is graded: (1) local infection, (2) bacteremia, (3) SIRS, (4) sepsis, (5) severe sepsis, and (6) septic shock.

RESULTS

Laboratory data obtained by plasma and cell lysate analyses demonstrate that active caspase-3 was identified in defined samples of whole-blood lysates (FIGS. 1 and 2), covering (e.g., 5/7, 8/18, 6/11) consecutive days during the patients' stay on the ICU. Also, sCD178 was detected on successive days. This is exemplified in the courses of two patients (FIGS. 1 and 2).

In Patient 1, resection of esophageal carcinoma had been performed, followed by the occurrence of septic shock until day 6; this was followed by postoperative pneumonia and septic shock from day 8 onward. In the course of Patient 1, two episodes with significant increases of active caspase-3 occurred which were mirrored in the kinetics of sCD178 in whole-blood lysates: the first episode was between day 8 and day 10, and the second elevation started on day 21 and remained at a high level until day 33 of the observation period. The elevations of sCD178 determined in the plasma were more distinct than in whole-blood lysates. In two episodes, the plasma sCD178 peaks coincided the caspase-3 and whole-blood peaks. The plasma sCD178 elevations on days 13–15 coincided with a very high tumor necrosis factor (TNF)-α value of 654 pg/mL at maximum. In the first apoptotic episode, the increase in caspase-3 and sCD178 was associated with high values of TNF-α, C-reactive protein (CRP), IL-6, and leukocyte counts. The second apoptotic episode, with an increase in caspase-3 and sCD178, began on day 21, but lacked coincidence with elevated TNF-α, CRP, or IL-6. In this second episode of apoptosis, leukocyte counts of 20,000/μL declined from day 17 to a level of 6–9,000/μL from day 21 onwards. Remarkably, a peak in Bcl-2 contents (day 5) preceded a leukocyte increase and the first apoptotic episode with caspase-3 and sCD178 increase on day 8. Some infection-associated stimulus might have been responsible for high levels of cytokines leading to a stimulation of myelopoiesis. Patient 1 was extubated on day 5. On day 9, tachyarrhythmia absoluta occurred and he

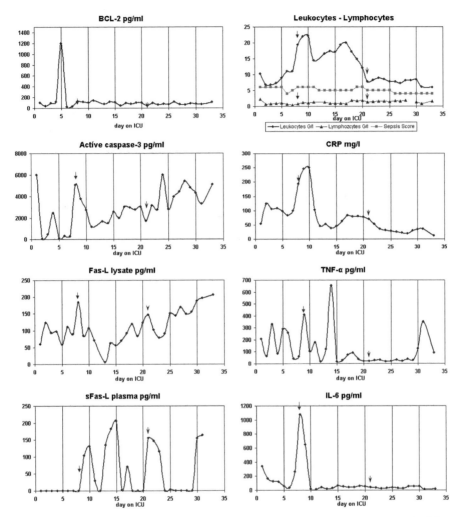

FIGURE 1. Bcl-2, active caspase-3, CD178 in whole-blood lysates, sCD178, CRP, TNF-α in plasma, and leukocyte and lymphocyte counts during the stay on the ICU of Patient 1. The sepsis score is graded: (1) local infection, (2) bacteremia, (3) SIRS, (4) sepsis, (5) severe sepsis, (6) septic shock.

had to be reintubated on account of respiratory insufficiency. Blood cultures were positive for Gram-positive bacteria on days 3, 6, 9, 14, 18, and 28. On day 20, the patient was extubated but had to be reintubated on day 21 because of respiratory insufficiency accompanied by severe sepsis, candidiasis, and wound infection.

Patient 2 suffered from appendicitis and postoperatively developed pneumonia, which caused septic shock until day 4. The courses of sCD178 measured

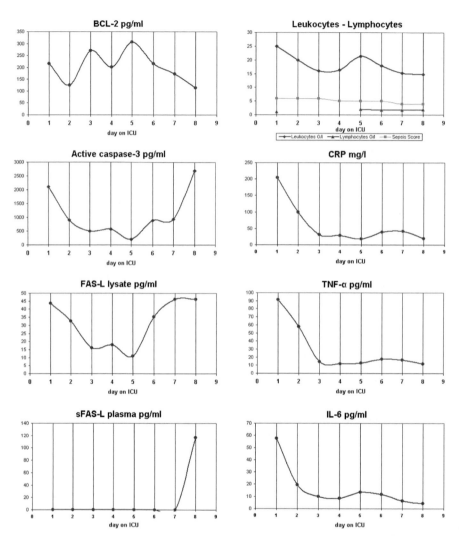

FIGURE 2. Bcl-2, active caspase-3, CD178 in whole-blood lysates, sCD178, CRP, TNF-α in plasma, and leukocyte and lymphocyte counts during the stay on the ICU of Patient 2. The sepsis score is graded: (1) local infection, (2) bacteremia, (3) SIRS, (4) sepsis, (5) severe sepsis, (6) septic shock.

in whole-blood lysates and active caspase-3 are coincident and are inverse to the increase of Bcl-2 between day 3 and day 6. Active caspase-3 was elevated between day 1 and day 3 and started to increase a second time after day 6 until day 8, which was the last observation time point. As with the first patient, we observed a correlation of inflammatory CRP and inflammatory TNF-α and IL-6 only for the first episode of active caspase-3. Again the second apoptosis episode occurred without increased levels of CRP, TNF-α, and IL-6. Fluctuations of total leukocyte counts were according to Bcl-2 contents, being high during leukocyte count increase and low upon declining leukocyte counts. Again, leukocyte counts were reciprocal to caspase-3 and sCD178 in whole-blood lysates. The second episode of caspase-3 elevation was followed by high amounts of sCD178 in the plasma and thus followed high sCD178 in whole-blood lysates. No infectious complications were observed during the second apoptotic episode. However, similar to the observations in the first patient, we found the second apoptotic episode to occur concomitantly with extubation on day 5. Clearly, the two examples demonstrate that apoptotic episodes may or may not coincide with inflammation and infection in patients during sepsis or septic shock. In both patients, the second apoptotic episode was found close to clinical recovery and discharge from intensive care unit (ICU). Both patients survived.

For more patients studied, association of inflammatory markers and apoptosis markers were correlated by Spearman rank association analysis. Elevated levels of active caspase-3 antigen in whole blood were not associated with peaks of the inflammatory or anti-inflammatory molecules. Specifically, Spearman rank associations resulted in the following r-values for active caspase-3: TNF-α: $r = 0.06$, IL-1ß: $r = 0.15$, Eotaxin: $r = 0.15$; IL-13 : $r = 0.10$; IL-1a: $r = 0.04$; IL-10: $r = 0.21$; GM-CSF: $r = 0.14$; IFN-γ: $r = 0.13$; MIP-1α: $r = 0.05$; for the MMPs: MMP1: $r = 0.12$; MMP2: $r = 0.09$; MMP3: $r = -0.10$; MMP7: $r = 0.05$; MMP8: $r = 0.18$; MMP9: $r = 0.21$; MMP10: $r = 0.09$; MMP13: $r = 0.08$, and for CRP: $r = 0.13$.

Also, Bcl-2 contents were not associated with inflammatory markers. In five patients, sFasL did not correlate with IL-6 ($r = 0.04$) nor with TNF-α ($r = 0.00$). In six patients, Bcl-2 did not correlate with IL-6 ($r = 0.01$) nor TNF-α ($r = 0.08$).

However, when comparing the kinetics of active caspase-3 and Bcl-2 protein quantified in whole-blood lysates with plasma or lysate concentrations of sCD178, all three parameters were elevated either simultaneously or in close time window as exemplarily shown in the two patients in FIGURES 1 and 2.

DISCUSSION

The current study analyses the kinetics of apoptosis and inflammation in patients during sepsis, septic shock and state of recovery. Whole-blood lysates were tested for active caspase-3, for Bcl-2 and sCD178. Results were compared with plasma levels of inflammatory cytokines, CRP, the clinical course, and the status of infection. The increase in whole-blood caspase-3 was associated with lysate CD178 indicating that CD95-CD178 apoptosis receptors were involved. Moreover, these apoptosis episodes paralleled declining leukocyte counts. Thus, the extrinsic apoptotic pathway appears to be the major mechanism leading to sepsis-associated leukopenia. Indeed sC178 has been described previously to cause neutropenia in septic complications.[7] Also, leukemias express high levels of CD178.[8] The mechanisms involved clearly centralize sCD178 as being more highly expressed in its soluble form in monocytes and in plasma.[9] However, a number of other reports demonstrate evidence for sCD178 as playing a role in lymphopenia.[10–13] Only recently, the origin of sCD178 as being either vesicle-associated and secreted or being nonspecifically cleaved off the cell membrane has been addressed.[14] Kinetic analysis of individual patients as demonstrated in FIGURES 1 and 2 suggest that neither cell-associated nor plasma levels of sCD178 clearly correlate with declining leukocyte counts. According to Suda and colleagues, membrane CD178 and soluble CD178 even display opposing effects in induction or protection against CD95-induced cell death.[15] Moreover, it has to be taken into account that bacterial infections may cause activation of MMPs such as MMP9 which potently induce the release of sCD178[16] and therefore explain apoptosis independent increases of sCD178. This context may be relevant to occur during days 12–15 in Patient 1, when plasma sCD178 rose significantly, but cell lysate–associated CD178 and active caspase-3 did not to a comparable extent. Going ahead with a blood culture positive for staphylococci, this individualized sCD178 peak coincided with the highest TNF-α peak on day 14. In summary, our results suggest that active caspase-3 and cell-bound CD178 correlate with leukocyte counts in patients with sepsis or septic shock. Active caspase-3, cell-associated CD178, and plasma sCD178 may or may not correlate with elevated inflammatory cytokines (TNF-α, IL-6, CRP).

A fairly clear effect by elevated Bcl-2 contents in whole blood can be shown. Elevated Bcl-2 corresponds to increasing leukocyte counts and less pronounced also, increasing lymphocyte counts. In addition, elevated Bcl-2 behaves reciprocally to lysate caspase-3 and CD178 (FIG. 2). Reasons for not identifying Bcl-2 as an anti-apoptotic and anti-inflammatory molecule by Spearman rank analysis may be due to the highly variable levels of Bcl-2 detected in whole blood lysates of individual patients. Whether Bcl-2 inhibits only the mitochondrial but not the receptor-mediated apoptosis pathway is still controverted.[17] In an extended study performed with Western blots of white blood cells from septic patients, evidence for both mechanisms, receptor and mitochondrial pathways, during sepsis has been provided.[5]

In summary, our analysis provides another hint for the intriguing relevance of apoptosis related to CD95 and CD178 in patients with sepsis, even though only part of these events appear to be directly related to inflammation and/or an infection-associated episode in the course of severe sepsis and septic shock. Increased apoptosis may contribute to immune suppression secondary to apoptotic losses in T cell, B cell, and NK cell populations. These alterations may contribute significantly to the risk of secondary opportunistic infections and to the risk of development of complications of sepsis, such as multiple organ dysfunction syndrome (MODS) and failure (MOF).[18,19] A clear correlation between inflammatory cytokines, apoptosis, and a beneficial effect by blocking caspase-9,[18] as well as a clear association with severity,[20] has been observed. In the present study, high levels of active caspase-3 were detected during the recovery episode. Thus, this late phase of apoptosis may hint at immunological remodeling and reconstitution, which need further investigation.

CONCLUSIONS

Regulation of apoptosis can be determined in whole blood of postoperative/posttraumatic patients by active caspase-3 and by Bcl-2. Interestingly, the high content of caspase-3 may coincide with Bcl-2 and plasma or lysate, sFAS-L in patients with successful reconstitution after septic shock. Peaks of active caspase-3, sFAS-L, and Bcl-2, respectively, may appear simultaneously or follow each other in different combinations. Pro- and antiapoptotic effects during sepsis may occur independently of peaks in inflammatory markers. Apoptosis could explain modeling and remodeling of leukocyte subpopulations.

REFERENCES

1. AYALA, A., C.D. HERDON, D.L. LEHMAN, et al. 1996. Differential induction of apoptosis in lymphoid tissues during sepsis: variation in onset, frequency, and the nature of the mediators. Blood **87:** 4261–4275.
2. AYALA, A., S.M. KARR, T.A. EVANS, et al. 1997. Factors responsible for peritoneal granulocyte apoptosis during sepsis. J. Surg. Res. **69:** 67–75.
3. HOTCHKISS, R.S., P.E. SWANSON, B.D. FREEMAN, et al. 1999. Apoptotic cell death in patients with sepsis, shock, and multiple organ dysfunction. Crit. Care Med. **27:** 1230–1251.
4. HOTCHKISS, R.S., K.W. TINSLEY & I.E. KARL. 2003. Role of apoptotic cell death in sepsis. Scand. J. Infect. Dis. **35:** 585–592.
5. HOTCHKISS, R.S., S.B. OSMON, K.C. CHANG, et al. 2005. Accelerated lymphocyte death in sepsis occurs by both the death receptor and mitochondrial pathways. J. Immunol. **174:** 5110–5118.
6. LEVY, M.M., M.P. FINK, J.C. MARSHALL, et al. 2003. 2001 SCCM/ESICM/ACCP/ATS/SIS International Sepsis Definitions Conference. Crit. Care Med. **31:** 1250–1256.
7. AREF, S., M.E. REFAEI, T. GODA, et al. 2004. Accelerated neutrophil apoptosis in neutropenic patients with hepatosplenic schistosomiasis is induced by serum Fas ligand. Hematol. J. **5:** 434–439.
8. LIU, J.H., S. WEI, T. LAMY, et al. 2000. Chronic neutropenia mediated by fas ligand. Blood **95:** 3219–3222.
9. NWAKOBY, I.E., K. REDDY, P. PATEL, et al. 2001. Fas-mediated apoptosis of neutrophils in sera of patients with infection. Infect. Immun. **69:** 3343–3349.
10. FELMET, K.A., M.W. HALL, R.S. CLARK, et al. 2005. Prolonged lymphopenia, lymphoid depletion, and hypoprolactinemia in children with nosocomial sepsis and multiple organ failure. J. Immunol. **174:** 3765–3772.

11. HOTCHKISS, R.S., K.W. TINSLEY, P.E. SWANSON, et al. 2001. Sepsis-induced apoptosis causes progressive profound depletion of B and CD4+ T lymphocytes in humans. J. Immunol. **166:** 6952–6963.
12. IWAGAKI, H., Y. MORIMOTO, M. KODERA, et al. 2000. Surgical stress and CARS: involvement of T cell loss due to apoptosis. Rinsho Byori **48:** 505–509.
13. ROTH, G., B. MOSER, C. KRENN, et al. 2003. Susceptibility to programmed cell death in T-lymphocytes from septic patients: a mechanism for lymphopenia and Th2 predominance. Biochem. Biophys. Res. Commun. **308:** 840–846.
14. SHUDO, K., K. KINOSHITA, R. IMAMURA, et al. 2001. The membrane-bound but not the soluble form of human Fas ligand is responsible for its inflammatory activity. Eur. J. Immunol. **31:** 2504–2511.
15. SUDA, T., H. HASHIMOTO, M. TANAKA, et al. 1997. Membrane Fas ligand kills human peripheral blood T lymphocytes, and soluble Fas ligand blocks the killing. J. Exp. Med. **186:** 2045–2050.
16. TAMURA, F., R. NAKAGAWA, T. AKUTA, et al. 2004. Proapoptotic effect of proteolytic activation of matrix metalloproteinases by *Streptococcus pyogenes* thiol proteinase (*Streptococcus* pyrogenic exotoxin B). Infect. Immun. **72:** 4836–4847.
17. ROY, S. & D.W. NICHOLSON. 2000. Cross-talk in cell death signaling. J. Exp. Med. **192:** F21–F25.
18. OBERHOLZER, A., L. HARTER, A. FEILNER, et al. 2000. Differential effect of caspase inhibition on proinflammatory cytokine release in septic patients. Shock **14:** 253–257; discussion 257–258.
19. VOLL, R.E., M. HERRMANN, E.A. ROTH, et al. 1997. Immunosuppressive effects of apoptotic cells. Nature **390:** 350–351.
20. DE FREITAS, I., M. FERNANDEZ-SOMOZA, E. ESSENFELD-SEKLER, et al. 2004. Serum levels of the apoptosis-associated molecules, tumor necrosis factor-alpha/tumor necrosis factor type-I receptor and Fas/FasL, in sepsis. Chest **125:** 2238–2246.

Two Forms of the Nuclear Matrix–Bound p53 Protein in HEK293 Cells

MARIA A. LAPSHINA, IGOR I. PARKHOMENKO, AND ALEXEI A. TERENTIEV

Molecular Biology Laboratory, Institute of Problems of Chemical Physics RAS, Chernogolovka, Moscow Region, 142432, Russia

ABSTRACT: Like many other transcription factors, the tumor suppressor protein p53 is bound to the nuclear matrix (NM). To study the interaction of p53 with the NM in more detail, we used alkaline and acidic extractions of NM proteins. It was found that there are two forms of p53, alkali- and acid-soluble, in NM of HEK293 cells and only one alkali-soluble form in NM of actinomycin D–treated MCF-7 cells. We suggest that distinct forms of p53 differ either in interactions with NM proteins or in their charges.

KEYWORDS: nuclear matrix; p53

INTRODUCTION

The nuclear matrix (NM) is suggested to be involved in formation of the higher level of chromatin organization, replication, transcription, and processing and transport of RNA.[1] The ubiquitous transcription factor, tumor suppressor protein p53, is associated with the NM.[3,4] To see whether the interaction of p53 with the NM is related to functional state(s) of p53, we used different extractions of NM proteins. The matrix-bound p53 of HEK293 cells is found to exist in at least two distinct forms with different solubility in alkaline and acidic solutions. In the NM of actinomycin D–treated MCF-7 cells, p53 is present mostly as the alkaline-soluble form.

MATERIALS AND METHODS

For isolation of nuclei, HEK293 and MCF-7 cells grown in DMEM + 10% newborn calf serum (ICN Biomedicals Irvine, CA, USA) were trypsinized and

Address for correspondence: Maria A. Lapshina, Molecular Biology Laboratory, Institute of Problems of Chemical Physics RAS, Chernogolovka, Moscow Region, 142432, Russia. Voice: +7(49652)27–779; fax: +7(096)515-54-20.
  e-mail: lapshina@icp.ac.ru

suspended in 0.25 STC (sucrose-Tris-calcium) buffer (0.25 M sucrose, 50 mM Tris-HCl [pH 9.0], 5 mM $CaCl_2$, 20 mM NH_4Cl, 2 mM PMSF, 1 mM DTT) containing 0.2% (v/v) Triton X-100. The nuclei were purified through a layer of 1 STC (the same as 0.25 STC except for use of 1 M sucrose), suspended in 0.25 STC, purified through cushion of 1 STC again, and washed twice in 0.25 STC.

To prepare nuclear matrixes, nuclei were suspended in TM buffer (10 mM Tris-HCl [pH 7.5], 0.2 mM $MgCl_2$, 1 mM PMSF, 1 mM DTT) containing 1.5 M NaCl and centrifuged. The pellet was briefly washed with TM and DNase I (SERVA) digestion was performed in the digestion buffer (0.25 M sucrose, 50 mM Tris-HCl [pH 7.5], 5 mM $MgCl_2$, 1 mM $CaCl_2$, 1 mM PMSF, 1 mM DTT) at $+4°C$ for 30 min. The suspension was centrifuged, and the pellet was collected as the NM.

For alkaline and acidic extraction of proteins, the NM was suspended in (i) TM containing 0.02 M NaOH for alkaline extraction, (ii) TM containing 0.02 M HCl for acidic extraction, and (iii) TM buffer. After incubation at $+4°C$ for 30 min and centrifugation, supernatants were collected as (i) the "alkaline extract," (ii) the "acidic extract," and (iii) the "TM extract." After brief wash with TM buffer, insoluble pellets of nuclear matrices were collected as (i) "alkali-insoluble NM," (ii) "acid-insoluble NM," and (iii) "TM-insoluble" NM.

Immunoblotting was performed after 10% SDS-PAGE with the primary rabbit polyclonal anti-human p53 antibody and the secondary goat HRP-conjugated anti-rabbit IgG antibody (Santa Cruz Biotechnology, Santa Cruz, CA, USA).

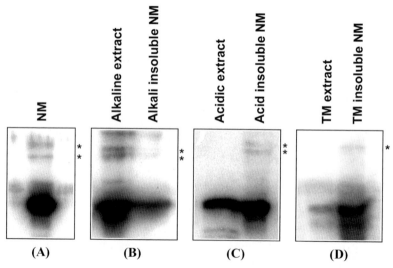

FIGURE 1. Alkaline and acidic extraction of the p53 protein from the NM of HEK293 cells. *Asterisks* mark the higher molecular weight bands of p53.

RESULTS AND DISCUSSION

The p53 protein is associated with the NM of HEK293 cells. The matrix-bound p53 is modified, presumably via ubiquitination, because it is observed as the higher molecular weight bands (FIG.1A). In alkaline solution, p53 is partially extracted from the NM, and most of higher molecular weight bands are detected in the soluble fraction (FIG.1B). During acidic extraction, p53 is also partially extracted to the soluble fraction, but all higher molecular weight bands remained in the acid-insoluble NM (FIG.1C). Extraction in the TM buffer without alkali or acid gave almost no release of p53 to the soluble fraction (FIG.1D).

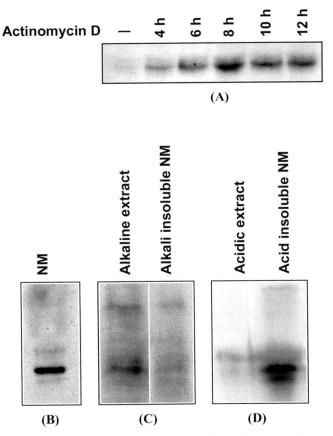

FIGURE 2. Alkaline and acidic extraction of the p53 protein from the NM of actinomycin-D-treated MCF-7 cells. (**A**) accumulation of p53 in the presence of actinomycin D. (**B**) the NM of actinomycin-D-treated MCF-7 cells. (**C**) and (**D**) extraction of actinomycin-D-activated, matrix-bound p53 by alkaline and acidic solutions, correspondingly.

In the MCF-7 cells, p53 is expressed at low levels. In the presence of 5 nM actinomycin D, the expression of p53 increases and reaches maximum after 8 h (FIG.2 A). In MCF-7 cells treated with actinomycin D for 8 h, p53 is associated with the NM (FIG. 2B). The major part of matrix-bound p53 is efficiently extracted from the NM of these cells in the alkaline buffer (FIG. 2C). Under acidic conditions, p53 is not extracted from the NM (FIG. 2D).

Thus, p53 in the NM can be represented by more than one protein pool. The fact that the higher molecular bands of p53 are well extracted by alkali but stable to acid (FIG. 1) demonstrates that applied methods of extraction of matrix-bound proteins allow us to separate at least two forms of matrix-bound p53. Because distinct forms of matrix-bound p53 are differentially sensitive to

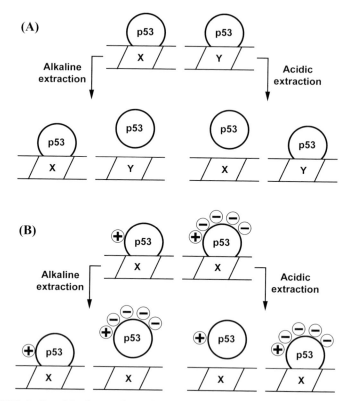

FIGURE 3. Possible forms of p53 in the NM. (**A**) different forms of p53 can be anchored in the NM by distinct proteins ("X" and "Y"). Bonds p53-X and p53-Y can dissociate by alkaline and acidic solution, correspondingly. As an alternative, p53 can be extracted as a complex with "X" protein by alkaline solution or as a complex with "Y" protein by acidic solution. (**B**) different forms of matrix-bound p53 can interact with the same anchoring protein "X," but have different charges. More-acidic forms of p53 are extracted by alkaline solution, whereas more-basic forms are extracted by acidic solution.

alkaline and acidic conditions of extraction, they can differ in their functional properties. It is known that actinomycin D causes activation of p53, thus, we suppose that the alkali-soluble form of p53 represents its active matrix-bound form.

Two forms of p53 can interact with different anchor proteins of the NM, and these interactions can be differentially sensitive to alkaline or acidic conditions (FIG. 3A). On the other hand, two forms of p53 can differ in their charge, as it was shown that p53 is present in multiple forms with different pI in normal liver and HepG2 cells.[5] More-acidic (phosphorylated and/or acetylated) forms of p53 can be more soluble in alkaline solutions, whereas more-basic ones can be extracted in acidic solutions (FIG. 3B).

REFERENCES

1. JACKSON, D.A. 2003. The principles of nuclear structure. Chromosome Res. **11:** 387–401.
2. DEPPERT, W. & M. HAUG. 1986. Evidence for free and metabolically stable p53 protein in nuclear subfractions of simian virus 40-transformed cells. Mol. Cell. Biol. **6:** 2233–2240.
3. JIANG, M., T. AXE, R. HOLGATE, *et al.* 2001. p53 binds the nuclear matrix in normal cells: binding involves the proline-rich domain of p53 and increases following genotoxic stress. Oncogene **20:** 5449–5458.
4. OKOROKOV, A.L., C.P. RUBBI, S. METCALFE & J. MILNER. 2002. The interaction of p53 with the nuclear matrix is mediated by F-actin and modulated by DNA damage. Oncogene **21:** 356–367.
5. SANCHEZ, J.C., P. WIRTH, S. JACCOUD, *et al.* 1997. Simultaneous analysis of cyclin and oncogene expression using multiple monoclonal antibody immunoblots. Electrophoresis **18:** 638–641.

Jpk, a Novel Cell Death Inducer, Regulates the Expression of Hoxa7 in F9 Teratocarcinoma Cells, but not during Apoptosis

EUN YOUNG LEE AND MYOUNG HEE KIM

Department of Anatomy, Embryology Laboratory, Brain Korea 21 Project for Medical Science, Yonsei University College of Medicine, Sodaemoongu Shinchondong 134, Seoul 120-752, Korea

ABSTRACT: Hox proteins play critical role in animal pattern formation during embryogenesis. During the study on the regulation of *Hox* gene expression, a novel gene, *Jpk*, has been isolated as a putative regulatory factor associating with the upstream regulatory sequence of murine *Hoxa7*. Since overexpression of *Jpk* caused cell death in bacteria as well as in eukaryotic cells and Hox has been reported to participate in apoptosis, we tried to analyze the relationship between *Jpk* and *Hoxa7* during apoptosis after confirming the regulatory effect of *Jpk* on the expression of *Hoxa7* in F9 teratocarcinoma cells. For that purpose, an effector (pEGFP-Jpk) and reporter (pGL2-NM307) plasmid containing a luciferase gene under the 307 bp (NM307) of *Hoxa7* upstream regulatory sequence was constructed. In the presence of Jpk (effector), luciferase activity was increased and this enhancement was decreased by siRNA against *Jpk*, suggesting that *Jpk* is a regulatory factor of *Hoxa7*. In order to see whether *Jpk* still regulates the expression of *Hoxa7* during apoptosis, F9 cells were transiently transfected with pcDNA-Jpk, and the expression of *Jpk*, *Hoxa7*, and *CHOP-10* was analyzed using RT-PCR. Hoxa7 and CHOP-10 were not upregulated in the presence of Jpk although Jpk seemed to cause apoptosis, indicating that the regulatory mechanism of *Jpk* on the expression of *Hoxa7* might be different depending on the cell status, that is, an apoptotic or proliferative condition.

KEYWORDS: Hoxa7; Jpk; regulatory factor; apoptosis; proliferation

---

Address for correspondence: Myoung Hee Kim, Department of Anatomy, Embryology Laboratory, Brain Korea 21 Project for Medical Science, Yonsei University College of Medicine, C.P.O. Box 8044, Seoul 120-752, Korea. Voice: +82-2-2228-1647; fax:+82-2-365-0700.
 e-mail: mhkim1@yumc.yonsei.ac.kr

INTRODUCTION

Hox proteins have been known to function as master regulators in pattern formation during embryogenesis and have conserved developmental mechanisms all over the animal species.[1–3] During the analysis of upstream regulatory factors for *Hox* expression, *Jpk* has been isolated as a putative regulatory factor associating with the position-specific regulatory element (PSRE) of murine *Hoxa7* gene by yeast one-hybrid method.[4] Unexpectedly though, Jpk caused cell death through reactive oxygen species (ROS) generation when it was overexpressed in bacteria as well as in eukaryotic cells.[5–7] Interestingly, it has also been reported that Hox proteins participate in apoptosis. Mutation of the Hoxa13 reduced apoptosis in mouse interdigital regions and Antennapedia-induced apoptosis when it was misexpressed in the eye imaginal disc of *Drosophila*.[8,9] In several instances of acute myeloid leukemia, increased expression of both *Meis1* and *Hox* genes, such as *Hoxa7* and *–a9*, has been characterized. Although the overexpression of *Meis1* alone was revealed to induce massive apoptosis, coexpression of *Meis1* and *Hox* was reported to suppress this apoptosis.[10,11] These data altogether suggest that Hox protein might participate in the apoptotic process. Therefore, we tried to investigate the relationship between *Jpk* and *Hoxa7* during apoptosis after confirming the regulatory effect of *Jpk* as a putative *trans*-acting factor of *Hoxa7* in F9 teratocarcinoma cells.

MATERIALS AND METHODS

F9, the murine embryonic teratocarcinoma cells, were cultured in DMEM (Gibco BRL, Carlsbad, CA, USA) containing 10% fetal bovine serum (FBS: JBI, Seoul, Korea), 100 μg/mL penicillin, and 100 μg/mL streptomycin (Gibco) at 37 °C with 5% CO_2. To analyze the regulatory effects of Jpk on the expression of Hoxa7, F9 cells were transfected with effector (pEGFP-Jpk) and reporter (pGL2-NM307, -NR271, -NS218, -OM213, and -BM112) plasmids[6] containing various deletions along the PSRE, together with the pRL-SV40 vectors as an internal control using Lipofectamine™ (Invitrogen, Carlsbad, CA, USA) following the manufacturer's manuals. The luciferase activity was analyzed 24 h after transfection following the Dual-Luciferase® Reporter Assay System (Promega, Madison, WI, USA) and the activity was measured using the TD-20/20 Luminometer (Turner Designs, Sunnyvale, CA, USA). The siRNAs (1, sense 5′-GAA UGC AGG AAG AUC UAA ATT; 2, sense 5′-CUG UUC UAA ACU ACU UGU ATT) for Jpk gene were purchased from Samchully Pharm Co. Ltd. (Siheung, GyeongGi, Korea). To analyze the relationship between *Jpk* and *Hoxa7*, F9 cells were transfected with either pcDNA3 or pcDNA-Jpk when cells reached 70–80% confluency, and then incubated for 24 or 48 h further. Cell viability was measured using 0.4% Trypan blue (Gibco), and RT-PCR was performed using specific primers for Jpk, Hoxa7, CHOP-10, and β-actin as an internal control, after isolating total RNAs using RNAzolB (Tel-Test Inc., Friendswood, TX, USA).

RESULTS AND DISCUSSION

To analyze whether Jpk actually regulates the expression of Hoxa7, a reporter (pGL2-NM307) harboring a luciferase gene under the 307-bp PSRE of *Hoxa7* was transfected into the F9 cells together with an effector (pEGFP:C1-Jpk) expressing Jpk protein as an EGFP-fused form. As expected, reporter activity

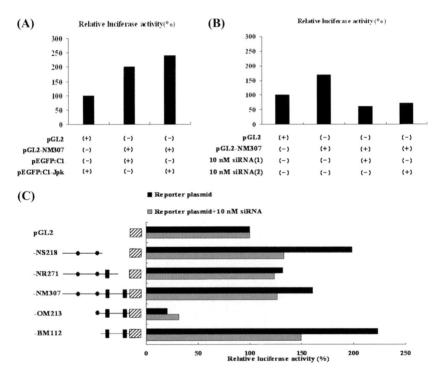

FIGURE 1. (**A**) Analysis of Jpk reactivity on PSRE of *Hoxa7* in F9 cells. (**B**) Knockdown effect of Jpk siRNA in F9 cells without pEGFP-Jpk transfection. (**C**) Analysis of Jpk-responsive regions along the 307-bp PSRE of *Hoxa7*. Reporter constructs containing several deletion forms of PSRE of Hoxa7 are presented. GAGA-binding sites (•), RARE (■), and luciferase gene (▨) are marked. At 24 h after transfection, luciferase activity was measured. Transfection efficiency for each assay was normalized by cotransfection with the *Renilla* luciferase gene. The values of data were calculated by comparison with the normalized luciferase activity of the control group. The normalized luciferase activity from the control group cells was set as 100%. The results are mean value of three independent experiments.

was upregulated in the presence of Jpk (FIG. 1A). However, the empty vector (pEGFP:C1) also upregulated the reporter expression, indicating that F9 cell itself expressed Jpk endogenously. After detecting the endogeneous expression of Jpk in F9 cells through RT-PCR (data not shown), siRNAs (1 and 2) against Jpk were prepared and confirmed the knockdown activity on Jpk expression by detecting the reduction of green fluorescence in the cells expressing EGFP-Jpk fusion protein (data not shown). And then, the reporter plasmid pGL2-NM307 was transfected again into the F9 cells without an effector. As shown in FIGURE 1B, the reporter luciferase activity was upregulated in the absence of pEGFP:C1-Jpk, and this enhancement was downregulated by siRNAs against *Jpk,* demonstrating that the upregulation of reporter expression was indeed induced by the endogenous Jpk. To further map the Jpk-responsive region along the PSRE, several deletion constructs (pGL2-NR271, -NS218, -OM213, and -BM112) as well as pGL2-NM307 were transfected into the F9 cells in the presence or absence of siRNA against Jpk. Reporter activities were increased in all cases and decreased by *Jpk* siRNA except pGL2-OM213 (FIG. 1C). Especially, -NS218 containing two GAGA-binding sites and -BM112 harboring two RAREs were significantly upregulated by the function of Jpk. Since the region containing the GAGA binding site turned out to be important for the anterior boundary of Hoxa7 expression and the region harboring the RARE has shown to be critical for the expression rate,[13] *Jpk* seemed to be required to decide the anterior boundary formation as well as the expression rate of *Hoxa7* during embryogenesis. Interestingly, Jpk significantly transactivated -BM112 harboring the region of two RAREs, whereas -OM213 harboring those of BM112 and one additional GAGA-binding site as well was not, indicating that there seems to be a certain negative regulatory element located in the fragment OM213 where BM112 is not overlapping (FIG. 1C).

Since Jpk regulated the expression of *Hoxa7* through *PSRE* and has been known to cause apoptosis through ROS generation,[7,12] the expression of Hoxa7 was analyzed in the apoptotic condition induced by Jpk overexpression. After constructing pcDNA3-Jpk, it was transfected into the F9 cells and the viability was analyzed using trypan blue exclusion technique. The cells expressing Jpk showed the reduced viability, especially 48 h after transfection (FIG. 2A). The expression of Jpk was highly induced in the pcDNA3-Jpk-transfected cells but not in other cells either transfected with pcDNA3 or not (FIG. 2B). To see whether Jpk has any effect on the expression of Hoxa, the expression level of Hoxa7 was tested through RT-PCR. The expression of *Hoxa7*, however, was not upregulated in cells overexpressing Jpk, 24 and 48 h following transfection of pcDNA3-Jpk (FIG. 2B), suggesting that the regulatory mechanism of Hoxa7 expression by Jpk might be different depending on the cell status, such as proliferation and apoptosis. Whereas the expression of *CHOP-10*, an ER stress-related gene,[14] was slightly increased in all cells following 24-h incubation, and further upregulation was detected in both empty vector- and Jpk-transfected cells 48 h after transfection (FIG. 2B), which led the cells to death. Since

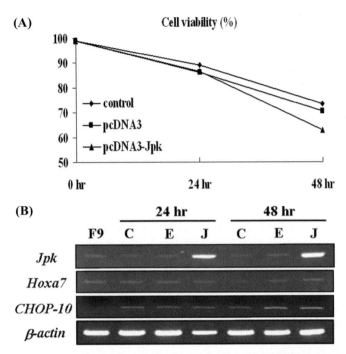

FIGURE 2. Cell viability following transfection with either pcDNA3 or pcDNA3-Jpk (**A**) and the expression profile of *Jpk*, *Hoxa7*, and *CHOP-10* in F9 teratocarcinoma cell (**B**). Either pcDNA3- or pcDNA3-Jpk-transfected F9 cell was incubated for the indicated time and the cell viability was measured using Trypan blue dye exclusion. RT-PCR reaction was carried out for 25 cycles using specific primers and RNAs isolated from the control F9 cells (C), pcDNA3-transfected F9 (E), and pcDNA-Jpk-transfected F9 (J) cells 24 and 48 h after transfection.

the expression level of Hoxa7 was not changed in all cells tested, whether expressing Jpk and/or CHOP-10 or not (FIG. 2B), Jpk does not seem to regulate the expression of Hoxa7 in this particular case, that is, F9 teratocarcinoma cells under apoptosis. These results altogether suggest that the Jpk-mediated expression of Hox and the Hox-related proliferation/apoptosis together with Jpk might be dependent on the cell type as well as the cell context, that is, cancer/apoptosis/normal development and differentiation.

ACKNOWLEDGMENTS

This work was supported by the Korea Research Foundation Grant KRF-2004-E00012 and -2005-204-C00074, and partly by the YBSTI program 2005 and the Research Project on the Production of Bio-Organs (No. 2005 08030901), Ministry of Agriculture and Forestry, Republic of Korea.

REFERENCES

1. ALONSO, C.R. 2002. Hox proteins: sculpting body parts by activating localized cell death. Curr. Biol. **12:** 776–778.
2. MACK, J.A., L. LI, N. SATO, et al. 2005. Hoxb13 up-regulates transglutaminase activity and drives terminal differentiation in an epidermal organotypic model. J. Biol. Chem. **280:** 29904–29911.
3. AKAM, M. 1998. Hox genes in arthropod development and evolution. Biol. Bull. **195:** 373–374.
4. CHO, M., C. SHIN, W. MIN & M.H. KIM. 1997. Rapid analysis for the isolation of novel genes encoding putative effectors to the position-specific regulatory element of murine Hoxa-7. Mol. Cells **7:** 220–225.
5. PARK, S., H.W. PARK & M.H. KIM. 2002. A novel factor associating with the upstream regulatory element of murine Hoxa-7 induces bacterial cell death. Mol. Biol. Rep. **29:** 363–368.
6. KIM, H., S. PARK, H.W. PARK & M.H. KIM. 2002. Isolation and characterization of a novel gene, Jpk, from murine embryonic cDNA library. Kor. J. Genet. **24:** 197–203.
7. KONG, K.A., S.D. PARK, H.W. PARK & M.H. KIM. 2003. A novel gene, Jpk induces apoptosis in F9 murine teratocarcinoma cell through ROS generation. Ann. N.Y. Acad. Sci. **1010:** 433–436.
8. STADLER, H.S., K.M. HIGGINS & M.R. CAPECCHI. 2001. Loss of Eph-receptor expression correlates with loss of cell adhesion and chondrogenic capacity in Hoxa13 mutant limbs. Development **128:** 4177–4188.
9. PLAZA, S., F. PRINCE, J. JAEGER, et al. 2001. Molecular basis for the inhibition of *Drosophila* eye development by Antennapedia. EMBO J. **20:** 802–811.
10. WERMUTH, P.J. & A.M. BUCHBERG. 2005. Meis1-mediated apoptosis is caspase dependent and can be suppressed by coexpression of HoxA9 in murine and human cell lines. Blood **105:** 1222–1230.
11. AFONJA, O., J.E. SMITH, JR., D.M. CHENG, et al. 2000. MEIS1 and HOXA7 genes in human acute myeloid leukemia. Leuk. Res. **24:** 849–855.
12. KONG, K.A., H.S. KIM, H.W. PARK & M.H. KIM. 2005. Membraneous localization of Jpk is not essential to exert cytotoxicity in F9 teratocarcinoma cells. J. Exp. Zoolog. **303A:** 422–429.
13. KIM, M.H., J.S. SHIN, S. PARK, et al. 2002. Retinoic acid response element in HOXA-7 regulatory region affects the rate, not the formation of anterior boundary expression. Int. J. Dev. Biol. **46:** 325–328.
14. WATANABE, Y., O. SUZUKI, T. HARUYAMA & T. AKAIKE. 2003. Interferon-gamma induces reactive oxygen species and endoplasmic reticulum stress at the hepatic apoptosis. J. Cell Biochem. **89:** 244–253.

HGF/SF Regulates Expression of Apoptotic Genes in MCF-10A Human Mammary Epithelial Cells

CATHERINE LEROY,[a] JULIEN DEHEUNINCK,[a]
SYLVIE REVENEAU,[a] BÉNÉDICTE FOVEAU,[a]
ZONGLING JI,[a] CÉLINE VILLENET,[b] SABINE QUIEF,[b]
DAVID TULASNE,[a] JEAN-PIERRE KERCKAERT,[b]
AND VÉRONIQUE FAFEUR[a]

[a]*CNRS UMR 8161, Institut de Biologie de Lille, CNRS/Institut Pasteur de Lille/Université de Lille I/Université de Lille II, B.P. 447, 59021 Lille Cedex, France*

[b]*Plate-forme Biopuces, Université de Lille 2, Faculté de Médecine, Place de Verdun, 59045 Lille, France*

ABSTRACT: Hepatocyte growth factor/scatter factor (HGF/SF) induces scattering, morphogenesis, and survival of epithelial cells through activation of the MET tyrosine kinase receptor. HGF/SF and MET are involved in normal development and tumor progression of many tissues and organs, including the mammary gland. In order to find target genes of HGF/SF involved in its survival function, we used an oligonucleotide microarray representing 1,920 genes known to be involved in apoptosis, transcriptional regulation, and signal transduction. MCF-10A human mammary epithelial cells were grown in the absence of serum and treated or not with HGF/SF for 2 h. Total RNA was reverse-transcribed to cDNA in the presence of fluorescent Cy3-dUTP or Cy5-dUTP to generate fluorescently labeled cDNA probes. Microarrays were performed and the ratios of Cy5/Cy3 fluorescence were determined. The expression of three apoptotic genes was modified by HGF/SF, with A20 being upregulated, and DAXX and SMAC being downregulated. These changes of expression were confirmed by real-time quantitative PCR. According to current-knowledge, A20 is antiapoptotic and SMAC is proapoptotic, while a pro- or antiapoptotic function of DAXX is controversial. The fact that HGF/SF upregulates an antiapoptotic gene (A20) and downregulates a proapoptotic gene (SMAC) is in agreement with its survival effect in MCF-10A cells. This study identified novel apoptotic genes regulated by HGF/SF, which can contribute to its survival effect.

Address for correspondence: Véronique Fafeur, CNRS UMR 8161 Institut de Biologie de Lille, Institut Pasteur de Lille, B.P. 447, 59021 Lille Cedex, France. Voice: 33-3-20-87-10-91; fax: 33-3-20-87-11-11.
  e-mail: veronique.fafeur@ibl.fr

KEYWORDS: HGF/SF; MET; apoptosis; target genes; breast; array

INTRODUCTION

Hepatocyte growth factor/scatter factor (HGF/SF) acts through the MET tyrosine kinase receptor to induce cell survival, growth, scattering, and morphogenesis in various cell types. Both HGF/SF[1,2] and MET[3] are essential for embryonic development. In the normal breast, HGF/SF is expressed primarily by stromal cells, while epithelial cells express MET but not HGF/SF, thus creating a paracrine mechanism in which localized expression of HGF/SF regulates mammary ductal growth and differentiation.[4,5] In contrast to what occurs in normal epithelium, HGF/SF and MET are frequently overexpressed in invasive human breast carcinomas[6,7] as well as in many other cancer types.[8] This high level of HGF/SF expression has been described as an independent predictor of poor overall survival in patients with breast cancer.[9]

HGF/SF and MET are commonly described as antiapoptotic molecules. HGF/SF or MET null mice display a severe reduction in the liver and show massive apoptosis.[1–3] In cell culture, HGF/SF protects a number of cell types against cell toxicity and apoptosis caused by various stimuli, including DNA-damaging agents, serum withdrawal, and activation of death receptors.[10–13] Nonetheless, proapoptotic effects of HGF/SF were also reported in some sarcoma and carcinoma cell lines[14,15] and MET itself was found to be converted to a proapoptotic factor following its cleavage by caspases.[16]

The mechanisms for cell survival by HGF/SF have been explored. HGF/SF was found to inhibit several steps of the apoptotic mitochondrial pathway, including mitochondrial release of cytochrome c and of AIF (apoptosis-inducing factor), and mitochondrial membrane depolarization and activation of several caspases in MDA-MB-453 mammary cells.[17] Accordingly, HGF/SF was found to regulate mitochondrial Bcl family members, both at transcriptional and post-transcriptional levels. This includes inhibition by HGF/SF of the translocation of proapoptotic Bax from the cytosol to the mitochondrial membrane,[18] induction of phosphorylation of Bad, inactivating its proapoptotic function,[19] as well as induction of expression of antiapoptotic molecules, such as Bcl-2[20] or Bcl-xL.[21] In addition, HGF/SF is known to protect cells from apoptosis through activation of both the PI3K-AKT and RAS-ERK pathways.[13] Their downstream actions may include phosphorylation of Bad for the PI3K-AKT pathway [21] and dephosphorylation of the c-Jun N-terminal kinases (JNK) for the RAS-ERK pathway.[12]

Overall, apoptotic genes regulated by HGF/SF are poorly known. In this study, we used an oligonucleotide array to identify novel apoptotic genes regulated by HGF/SF in human mammary cells.

MATERIALS AND METHODS

Cytokines, Drugs, and Cell Cultures

Human recombinant HGF/SF was purchased from Peprotech (Rocky Hill, NJ, USA) and anisomycin from Calbiochem (San Diego, CA, USA). MDA-MB-231 human mammary adenocarcinoma cells were cultured in Dulbecco's modified Eagle's medium (DMEM, Invitrogen, Carlsbad, CA, USA) supplemented with 10% fetal bovine serum (FBS, Invitrogen). MCF-7 human mammary adenocarcinoma cells were cultured in DMEM supplemented with 10% FBS and 10 nM estradiol (Sigma). MCF-10A human mammary epithelial cells, spontaneously immortalized, were cultured in DMEM and HAM's F12 (Invitrogen) (v/v) supplemented with 5% horse serum (HS, Invitrogen), 500 ng/mL hydrocortisone (Calbiochem), 20 ng/mL epidermal growth factor (Peprotech), 10 μg/mL insulin (Sigma-Aldrich, St. Louis, MO, USA), 100 ng/mL cholera toxin (Calbiochem).

Antibodies

Monoclonal antibody to E-cadherin was purchased from Transduction Labs (Lexington, KY, USA), polyclonal antibody to ER-α from Affinity Bioreagents (Golden, CO, USA) and monoclonal antibody to cytokeratin 18 from Oncogene (La Jolla, CA, USA). Monoclonal antibody to MET (DL-21), polyclonal antibody to Phospho MET (Tyr 1234/1235), and polyclonal antibody to GAB1 (CT) were purchased from Upstate Biotechnology (Lake Placid, NY, USA). Monoclonal antibody to actin and polyclonal antibodies to PLCγ1, ERK2 and AKT were purchased from Santa Cruz (Santa Cruz, CA, USA). Polyclonal antibody to Phospho-ERK (pTEpY) was purchased from Promega (Madison, WI, USA) and polyclonal antibody to Phospho-AKT (Ser473) from Cell Signaling Technology (Danvers, MA, USA).

Immunoblot

Cells were suspended in lysis buffer (20 mM Tris-HCl, pH7.8, 50 mM NaCl, 5 mM EGTA, and 1% v/v Triton X-100) containing freshly added protease and phosphatase inhibitors (1 mM phenylmethyl sulfonyl fluoride, 1 μM leupeptin, 2 μM aprotinin, 20 mM β-glycerophosphate, and 1 mM sodium orthovanadate). Lysates were clarified by centrifugation at 4°C, and protein concentration was determined by Bio-Rad protein assay. Western blotting was performed as described previously.[22]

Cell Scattering Assay

Scattering from cell islets was performed as follows. Cells (5,000 cells/24 well-plates) were grown in their respective culture medium. After 48 h, cells were incubated in DMEM-0.5% FBS (MDA-MB-231 and MCF-7) or in DMEM/F12-0.5% HS (MCF-10A) without supplements and in the presence or absence of HGF/SF (10 ng/mL) for 24 h. At the end of the experiment, cells were fixed (methanol-0.02 g/L fast green) and stained using Carazzi/eosin solutions. After examination by light microscopy, photographs were taken.

Survival Assay

MCF-10A cells (200,000 cells/6 well-plates) were grown for 24 h. Cells were then cultured in culture medium deprived of serum and supplements. Cells were then treated or not with anisomycin or HGF/SF. After few hours, when cell detachment was clearly observed, photographs were taken.

Semiquantitative Reverse-Transciption–Polymerase Chain Reaction

Total RNA was isolated using a RNeasy kit (Qiagen, Valencia, CA, USA) and 1 μg of RNA was subjected to single-strand cDNA synthesis using Superscript II reverse transcriptase, oligo dT as a primer (Invitrogen) and a Perkin Elmer DNA thermal cycler. For each polymerase chain reaction (PCR), 2.5 μL of cDNA was used for amplifications. The PCR primer sets and the optimal reaction conditions are shown in TABLE 1. PCR reaction conditions were individually optimized for each gene product. In particular, the cycle number was determined within the linear range of product amplification. The β-actin gene was used as a control for loading. PCR products were analyzed by electrophoresis using 1% agarose gels, and gels were photographed under ultraviolet illumination.

Quantitative RT-PCR

Total RNA was isolated as described above. To quantitatively estimate the mRNA expression of several genes, PCR amplification was performed on a Light-Cycler instrument system (Roche, Mannheim, Germany) using the Light-Cycler-FastStart DNA Master SYBR green I kit (Roche). The PCR primer sets and the optimal reaction conditions are shown in TABLE 2. To confirm the specificity of product, melt curves were generated over a 60°C–95°C range. A negative control, without cDNA, was run with each assay, and

TABLE 1. PCR reaction conditions for semiquantitative RT-PCR assays

Gene name	Primer sequences	Size PCR product (bp)	Annealing temperature (°C)	MgCl$_2$ (mM)	Cycle numbers
A20	sense: CGGCTGCGTATTTTGGGACTC antisense: TCTTCGGGGGCAGGCTCACC	438	60.4	2	33
β actin	sense: CGGTGACGGGGTCACCCACA antisense: AAGCATTTGCGGTGGACGAT	660	55	2.5	30
daxx	sense: CGGCGGCTGCAGGAAAAGGAGT antisense: GCGCCGGGCAACACAGGA	474	62.3	2	36
smac	sense: CACCTGCCCTGTCTCCCCACTCA antisense: CCCTCCCCCTGCCACAACTG	418	57.5	2	33

TABLE 2. PCR reaction conditions for quantitative RT-PCR assays

Gene name	Primer sequences	Size PCR product (bp)	Annealing temperature (°C)
A20	sense: CTTCTCAGTACATGTGGGCGTTCAGG antisense: CCCATTCATCATTCCAGTTCCGAGTATCAT	183	72
β actin	sense: CGGTGACGGGGTCACCCACA antisense: CGCGGTTGGCCTTGGGGTPCAG	260	72
daxx	sense: GGACCCCACAAATGCTGAAAACACTGC antisense: AGGGGATGCGCTGCTCTATGACACG	278	74
smac	sense: AATGATTTAGTAATCGTTCCCTGTTGG antisense: GGCCTGTGTTAAGTCCTGTTGATGT	240	65

the β actin gene was used as an internal control. Relative copy number was calculated using the crossing threshold method, assuming an efficiency of 2 (relative copy number:$2^{-\Delta\Delta CT}$).[23]

Oligonucleotide Array

Total RNA was extracted using a RNeasy Minikit (Qiagen) and their quality was verified by an Agilent Bioanalyzer (Agilent Technologies, Palo Alto, CA, USA). Total RNA was reverse transcribed in the presence of Cy3-dUTP or of Cy5-dUTP. Microarray glass slides were prepared at the Microarray Platform of the Génopole de Lille-Université de Lille 2. Each slide was spotted with oligonucleotide (60 mers, Sigma Genosys, Sigma-Aldrich, St. Louis, MO, USA) sets corresponding to 1,920 genes classified in apoptosis, signal transduction, and transcriptional regulation. The labeled cDNAs were hybridized on microarray slides at 42°C using a Discovery Ventana robot and its optimized conditions of hybridation, washes and reagents. The slides were scanned using a GMS Confocal scanner (Affymetrix, Santa Clara, CA, USA). The scanner output images were localized by overlaying a grid on the fluorescent images, using Jaguar software (Jaguar Software Development, Inc., Findlay, IL, USA). The fluorescent intensities were then calculated and normalized, using TIGR-MIDAS software (The Institute for Genomic Research, Boston, MA, USA). A final reported intensity corresponds to the difference between average probe intensity and average local background intensity. The ratios of the red intensity to the green intensity for all targets were determined and spots with ratio between −2 and +2 were eliminated.

RESULTS

Normal MCF-10A Human Mammary Epithelial Cells Are Responsive to HGF/SF

In order to perform target gene analysis in response to HGF/SF, we selected human mammary epithelial cells from available published data.[24,25] We then chose to characterize responses to HGF/SF in two cell lines expressing the MET receptor: normal MCF-10A human mammary epithelial cells and tumorigenic MDA-MB-231 human mammary adenocarcinoma cells. The tumorigenic MCF-7 human mammary adenocarcinoma cells, which do not express MET, were chosen as a negative control.

Expression of various epithelial markers, including mammary epithelial markers, was checked in these cells (FIG. 1A). Cytokeratin-18, a marker of mammary epithelial cells, was expressed in the three cell lines. E-cadherin, a marker of epithelial junctions, was expressed both by MCF-7 and MCF-10A cells, and, as expected, was not expressed by invasive MDA-MB-231 epithelial cells. The estrogen receptor was expressed only by MCF-7, as expected from estrogen-responsive cells. Expression of the MET receptor was detected both in MDA-MB-231 and MCF-10A cells, and was weakly detected in MCF-7 cells. In addition, GAB1 and PLCγ, two known transducers of MET signaling, were expressed by the three cell lines. This profile of expression confirmed their mammary epithelial characteristic and potential sensitivity to HGF/SF of both MDA-MB-231 and MCF-10A cells.

We then investigated the response of these cells to HGF/SF. In MDA-MB-231 and MCF-10A cells, but not in MCF-7 cells, HGF/SF induced phosphorylation of MET, showing that HGF/SF can activate this receptor in either MDA-MB-231 or MCF-10A cells (FIG. 1B). An efficient cell-scattering response to HGF/SF was only observed in MCF-10A (FIG. 1C). Indeed, the constitutive scattering of MDA-MB-231 cells could not be further induced by HGF/SF, even though HGF/SF was able to induce MET phosphorylation. HGF/SF did not induce cell scattering in MCF-7 cells, in agreement with the low expression (or to the absence of inducible phosphorylation) of MET. We further checked whether MCF-10A cells are sensitive to a survival effect of HGF/SF (FIG. 1D). When cells were treated with anisomycin, a well-known inducer of apoptosis, HGF/SF was found to protect these cells from massive detachment.

These data show that HGF/SF induces MET signaling, scattering, and survival in normal MCF-10A human mammary epithelial cells.

HGF/SF Regulates Expression of Apoptotic Genes

MCF-10A cells were grown for 1 day in the absence of serum and then treated or not with HGF/SF for 2 h. Total RNA was reverse-transcribed in the presence of Cy3-dUTP or Cy5-dUTP. The fluorescently labeled cDNA probes were then hybridized on an oligonucleotide microarray representing 1,920 genes, classified in apoptosis, signal transduction, and transcriptional regulation. The slides were scanned and fluorescent intensities were then calculated and normalized. The ratios of the red intensity to the green intensity for all targets were determined and spots with ratio between -2 and $+2$ were eliminated.

As shown in TABLE 3, 38 genes among the 1,920 genes were differently expressed in response to HGF/SF, with 14 induced and 24 repressed genes. According to their initial classification in the microarray and following analysis of the available literature for these 38 genes, we found that three genes, A20, SMAC and DAXX, were clearly involved in apoptosis. Their differential expressions were confirmed using both semiquantitative and quantitative PCR from an independent experiment (FIG. 2). These results show that HGF/SF induced expression of gene A20 and repressed expression of SMAC and DAXX genes.

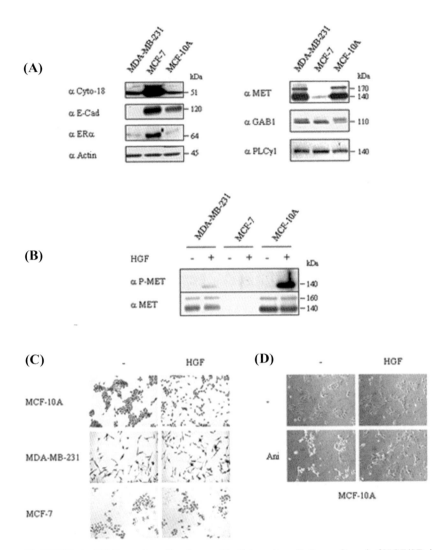

FIGURE 1. (**A**) Detection of various epithelial markers (*left panel*), and of HGF/SF signaling proteins (*right panel*) in MDA-MB-231, MCF-7 and MCF-10A cells. Protein extracts (20 μg) were immunoblotted using specific antibody (antibodies: α: Cadherin E: E-Cad, estrogen receptor α: ERα, cytokeratin 18: Cyto-18, MET, GAB1, actin, phospholipaseC-γ1: PLCγ1). (**B**) Effect of HGF/SF on MET phosphorylation. Cells were treated by HGF/SF (30 ng/mL) for 30 min. Protein extracts (20 μg) were immunoblotted using an anti-phospho-MET antibody. The filter was stripped and reprobed using an anti-MET antibody. (**C**) Cells were treated for 24 h by HGF/SF (10 ng/mL) to induce cell scattering. After fixation and staining, photographs were taken. (**D**) Serum-deprived MCF-10A cells were treated with or without HGF/SF (10 ng/mL) and anisomycin (50 μM) for few hours. When massive cell detachment by anisomycin was observed, photographs were taken.

TABLE 3. Results of the oligonucleotide array analysis

Ratio	Name	Gene name	UniGene ID
−	cd38	CD38 antigen (p45)	Hs.66052
	chl1/call	Cell adhesion molecule L1-like	Hs.210863
	daxx	Death-associated protein 6	Hs.336916
	dfcp1/taff	Double FYVE-containing protein 1	Hs.20047
	dctn2	Dynactin 2	Hs.84153
	etv1/er81	ETS variant gene 1	Hs.89566
	foxh1/fast1	Forkhead box H1	Hs.159251
	gnrh1/lnrh	Gonodotropin-releasing hormone 1/luteinizing-releasing hormone	Hs.82963
	hos1/T-box 5	Holt-Oram syndrome protein	Hs.50947
	iqgap2	IQ motif-containing GTPase-activating protein 2	Hs.78993
	lta/tnfB	Lymphotoxin-alpha	Hs.36
	mknk1/mnk1	MAP kinase-interacting serine/therinine kinase 1	Hs.5591
	map2k3/mek3	Mitogen-activated protein kinase kinase 3	Hs.180533
	myt-1	Myelin transcription factor 1-like	Hs.172619
	nurr/nr4a2	Nuclear receptor-related 1	Hs.82120
	pitpnc1/rdgbb	Phosphtidylinositol transfer protein cytoplasmic 1/retinal degeneration B beta gene	Hs.33212
	pkd212	Polycystic kidney disease 2-like 2 gene	Hs.272418
	ppp1r7	Protein phosphate 1, regulatory submit 7	Hs.36587
	ranbp8/ipo8	Ran-binding protein 8	Hs.119687
	smac/diablo	Second mitochondria-derived activator of caspase	Hs.169611
	sptb	Spectrin beta	Hs.47431
	sycp2	Synaptonemal complex protein 2	Hs.202676
	ttf-1	Thyroid transcription factor-1	Hs.79531
	relB	v-Rel avian reticuloendotheliosis viral oncogene homologue B	Hs.858
+	ebf	Early B-cell factor	Hs.185708
	egr3	Early growth response 3 gene	Hs.74088
	edar	Ectodysplasin 1, anhydrotic receptor	Hs.58346
	flt3lg	Ligand of fms-related tyrosine kinase 3	Hs.428
	il2rg/cd132	Interlukin-2 receptor, gamma	Hs.84
	krt9/k9	Keratin 9	Hs.2783
	nrip1/rip140	Nuclear receptor–interacting protein 1	Hs.155017
	pik3cb/p110-beta	Phosphoinositide 3-kinase, catalytic, beta	Hs.239818
	pmch	Pro-melanin-concentrating hormone	Hs.2182
	rpl6	Ribosomal protein L6	Hs.174131
	smarca2/snf212	SW 1/SNF related, matrix-associated, actin-dependent regulator of chromatin	Hs.198296
	tbr1	T-brain-1 protein	Hs.210862
	tsc22	Transforming growth factor beta–stimulated gene	Hs.114360
	tnfaip3/A20	Tumor necrosis factor-alpha-induced protein 3	Hs.211600

NOTE: List of the 38 genes regulated by HGF/SF in MCF-10A cells (−: repressed, +: induced). Name, gene name, and Unigene ID are indicated.

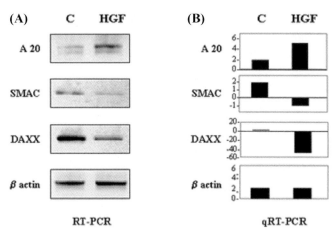

FIGURE 2. Serum-deprived MCF-10A cells were treated or not with HGF/SF (10 ng/mL) for 2 h. mRNA expression of A20, SMAC, and DAXX genes were verified by semi-quantitative RT-PCR (**A**) and by real-time quantitative RT-PCR (**B**).

DISCUSSION

We identified three novel apoptotic genes regulated by HGF/SF in MCF10-A normal human mammary epithelial cells. Indeed, our data demonstrated that HGF/SF induces expression of A20 and represses expression of both SMAC and DAXX. Although these three genes are classified as apoptotic genes, their function in apoptosis is quite distinct.

A20/TNFAIP3 (tumor necrosis factor–induced protein 3) was initially identified as an early response gene of TNFα in endothelial cells[26] and as an inhibitor of TNFα-induced apoptosis.[27] The A20 gene encodes a cytoplasmic zinc finger protein of 90 kDa, which can interact with numerous proteins of TNFα signaling pathways, including TRAF2, TRAF6, ABIN, and IKKγ/NEMO (see Ref. 28). In agreement with these studies, A20-deficient mice are more susceptible to TNFα-stimulated apoptotic cell death, develop severe inflammation and cachexia, and die prematurely.[29] Apart from TNFα-induced cell death, A20 was found to inhibit cell death in a number of other systems, including inhibiting apoptosis induced by serum depletion or by overexpression of p53 (see Ref. 28). A20 is also known to function as a potent inhibitor of NFκB-dependent gene expression. Indeed, A20 expression is regulated by NFκB and can block activation of NFκB through a negative-feedback loop.[30] A similar negative regulation has been shown for IκB, a well-known inhibitor of NFκB.[31] Taken together, these findings support a role of A20 in promoting cell survival against various apoptotic stimuli, as well as in terminating NFκB-dependent gene expression.

SMAC/DIABLO (second mitochondria-derived activator of caspase-3 / direct IAP-binding protein with low pI) was initially purified from subcellular extracts containing high caspase-3 activity.[32] This 25-kDa protein is now known to promote apoptosis by eliminating the functions of IAPs (inhibitor of apoptosis proteins) (see Ref. 33). Indeed, in response to apoptotic stimuli, the mitochondrial SMAC is released into the cytosol, binds to IAPs, and prevents them from sequestering caspases.[32,34,35] Inhibition of IAP functions by SMAC was also found to potentiate TRAIL-induced apoptosis, a death receptor ligand, indicating that SMAC also plays a role in death receptor signaling.[36] It thus appears that Smac/DIABLO can contribute both to the mitochondrial pathway and to the death receptor pathway of apoptosis.

In contrast to A20 and SMAC, the function of DAXX in promoting apoptosis or survival is more controversial (see Ref. 37). DAXX (death-associated protein 6) was originally identified as a cytoplasmic protein that specifically binds to the death domain of the FAS death receptor (also called CD95) and potentiates FAS-induced apoptosis.[38] In the cytoplasm, this protein of about 120 kDa interacts and activates the proapoptotic ASK1 protein[39] and activates JNK, which regulates stress-induced cell death.[40] However, a large proportion of DAXX molecules are nuclear and associate with subnuclear domains. In the nucleus, DAXX represses several transcription factors, including PAX3 and ETS1, and interacts with crucial proteins involved in transcriptional silencing, including histone deacetylase II,[41] providing a potential mechanism by which DAXX can repress transcription. Nonetheless, it is still unclear whether the proapoptotic function of DAXX is dependent on its transcriptional functions. Despite these reports advocating a proapoptotic function for DAXX, several studies have suggested a potential antiapoptotic function of DAXX. Disruption of the DAXX gene in mice results in early embryonic lethality, with extensive apoptosis, showing that DAXX is essential for early development.[42] An antiapoptotic role for DAXX was evidenced from DAXX RNAi experiments showing sensitization of cells to apoptosis induced by FAS, UV or TNFα.[43,44]

This overview of the actual knowledge leads to the proposal that A20 is antiapoptotic, SMAC is proapoptotic, while a pro- or antiapoptotic function of DAXX is still controversial. The fact that HGF/SF induces A20 and represses SMAC expression is therefore in agreement with HGF/SF inducing cell survival. A synthetic view of the known survival mechanisms of HGF/SF and of the main implications of A20, SMAC and DAXX in apoptosis pathways is shown in FIGURE 3. Our study did not evaluate the possible contribution of A20, SMAC, and DAXX in survival signaling pathways of HGF/SF. Functional analysis of these genes by overexpression or downregulation could resolve this issue.

ACKNOWLEDGMENTS

This work was supported by CNRS, Pasteur Institute of Lille, Université de Lille I, and Université de Lille II, INSERM, and by grants from Fondation de France, Ligue Régionale contre le Cancer-Comité Nord and Région Nord-Pas de Calais—FEDER. The Microarray Platform of the Génopole de

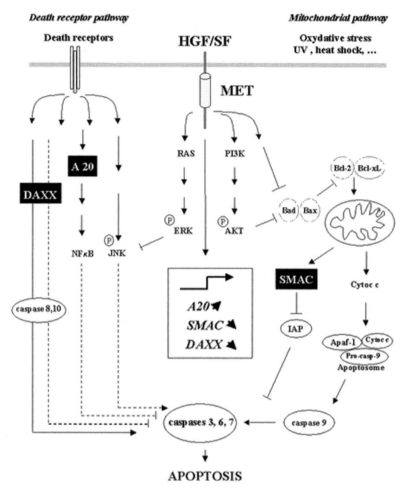

FIGURE 3. Schematic representation of survival mechanisms induced by HGF/SF. The central panel points to the novel target genes (A20, SMAC, and DAXX) of HGF/SF identified in this study and their induced or repressed expression by HGF/SF is indicated by an arrow. The implication of A20, SMAC, DAXX in apoptosis pathways are indicated (see text for details).

Lille-Université de Lille II was supported by "Région Nord-Pas de Calais et Picardie." J.D. was supported by a fellowship from ARC (Association pour la Recherche sur le Cancer) and Z.J. by a fellowship from Fondation de France.

REFERENCES

1. SCHMIDT C., F. BLADT, S. GOEDECKE, *et al.* 1995. Scatter factor/hepatocyte growth factor is essential for liver development. Nature **373:** 699–702.

2. UEHARA, Y., O. MINOWA, C. MORI, et al. 1995. Placental defect and embryonic lethality in mice lacking hepatocyte growth factor/scatter factor. Nature **373**: 702–705.
3. BLADT, F., D. RIETHMACHER, S. ISENMANN, et al. 1995. Essential role for the c-met receptor in the migration of myogenic precursor cells into the limb bud. Nature **376**: 768–771.
4. ANDERMARCHER, E., M.A. SURANI & E. GHERARDI. 1996. Co-expression of the HGF/SF and c-met genes during early mouse embryogenesis precedes reciprocal expression in adjacent tissues during organogenesis. Dev. Genet. **18**: 254–266.
5. NIEMANN, C., V. BRINKMANN, E. SPITZER, et al. 1998. Reconstitution of mammary gland development in vitro: requirement of c-met and c-erbB2 signaling for branching and alveolar morphogenesis. J. Cell Biol. **143**: 533–545.
6. TUCK, A.B., M. PARK & E.E. STERNS. 1996. Coexpression of hepatocyte growth factor and receptor (Met) in human breast carcinoma. Am. J. Pathol. **148**: 225–232.
7. WANG, Y., A.C. SELDEN, N. MORGAN, et al. 1994. Hepatocyte growth factor/scatter factor expression in human mammary epithelium. Am. J. Pathol. **144**: 675–682.
8. JIANG, W.G., T.A. MARTIN, C. PARR, et al. 2005. Hepatocyte growth factor, its receptor, and their potential value in cancer therapies. Crit. Rev. Oncol. Hematol. **53**: 35–69.
9. YAMASHITA, J., M. OGAWA, S. YAMASHITA, et al. 1994. Immunoreactive hepatocyte growth factor is a strong and independent predictor of recurrence and survival in human breast cancer. Cancer Res. **54**: 1630–1633.
10. BOWERS, D.C., S. FAN, K.A. WALTER, et al. 2000. Scatter factor/hepatocyte growth factor protects against cytotoxic death in human glioblastoma via phosphatidylinositol 3-kinase- and AKT-dependent pathways. Cancer Res. **60**: 4277–4283.
11. KOSAI, K., K. MATSUMOTO, S. NAGATA, et al. 1998. Abrogation of Fas-induced fulminant hepatic failure in mice by hepatocyte growth factor. Biochem. Biophys. Res. Commun. **244**: 683–690.
12. REVENEAU, S., R. PAUMELLE, J. DEHEUNINCK, et al. 2003. Inhibition of JNK by HGF/SF prevents apoptosis induced by TNF-alpha. Ann. N. Y. Acad. Sci. **1010**: 100–103.
13. XIAO, G.H., M. JEFFERS, A. BELLACOSA, et al. 2001. Anti-apoptotic signaling by hepatocyte growth factor/Met via the phosphatidylinositol 3-kinase/Akt and mitogen-activated protein kinase pathways. Proc. Natl. Acad. Sci. USA **98**: 247–252.
14. ARAKAKI, N., J.A. KAZI, T. KAZIHARA, et al. 1998. Hepatocyte growth factor/scatter factor activates the apoptosis signaling pathway by increasing caspase-3 activity in sarcoma 180 cells. Biochem. Biophys. Res. Commun. **245**: 211–215.
15. TAJIMA, H., K. MATSUMOTO, T. NAKAMURA, et al. 1991. Hepatocyte growth factor has potent anti-proliferative activity in various tumor cell lines. FEBS Lett. **291**: 229–232.
16. TULASNE, D., J. DEHEUNINCK, F.C. LOURENCO, et al. 2004. Proapoptotic function of the MET tyrosine kinase receptor through caspase cleavage. Mol. Cell Biol. **24**: 10328–10339.
17. GAO, M., S. FAN, I.D. GOLDBERG, et al. 2001. Hepatocyte growth factor/scatter factor blocks the mitochondrial pathway of apoptosis signaling in breast cancer cells. J. Biol. Chem. **276**: 47257–47265.

18. NAKAGAMI, H., R. MORISHITA, K. YAMAMOTO, et al. 2002. Hepatocyte growth factor prevents endothelial cell death through inhibition of bax translocation from cytosol to mitochondrial membrane. Diabetes **51:** 2604–2611.
19. LIU, Y., A.M. SUN & L.D. DWORKIN. 1998. Hepatocyte growth factor protects renal epithelial cells from apoptotic cell death. Biochem. Biophys. Res. Commun. **246:** 821–886.
20. YAMAMOTO, K., R. MORISHITA, S. HAYASHI, et al. 2001. Contribution of Bcl-2, but not Bcl-xL and Bax, to antiapoptotic actions of hepatocyte growth factor in hypoxia-conditioned human endothelial cells. Hypertension **37:** 1341–1348.
21. LIU, Y., E.M. TOLBERT, L. LIN, et al. 1999. Up-regulation of hepatocyte growth factor receptor: an amplification and targeting mechanism for hepatocyte growth factor action in acute renal failure. Kidney Int. **55:** 442–453.
22. PAUMELLE, R., D. TULASNE, Z. KHERROUCHE, et al. 2002. Hepatocyte growth factor/scatter factor activates the ETS1 transcription factor by a RAS-RAF-MEK-ERK signaling pathway. Oncogene **21:** 2309–2319.
23. LIVAK, K.J. & T.D. SCHMITTGEN. 2001. Analysis of relative gene expression data using real-time quantitative PCR and the 2(-Delta Delta C(T)) method. Methods **25:** 402–408.
24. MONTESANO, R., J.V. SORIANO, K.M. MALINDA, et al. 1998. Differential effects of hepatocyte growth factor isoforms on epithelial and endothelial tubulogenesis. Cell Growth Differ. **9:** 355–365.
25. PARR, C. & W.G. JIANG. 2001. Expression of hepatocyte growth factor/scatter factor, its activator, inhibitors and the c-Met receptor in human cancer cells. Int. J. Oncol. **19:** 857–863.
26. OPIPARI, A.W., JR., M.S. BOGUSKI & V.M. DIXIT. 1990. The A20 cDNA induced by tumor necrosis factor alpha encodes a novel type of zinc finger protein. J. Biol. Chem.. **265:** 14705–14708.
27. OPIPARI, A.W., JR., H.M. HU, R. YABKOWITZ, et al. 1992. The A20 zinc finger protein protects cells from tumor necrosis factor cytotoxicity. J. Biol. Chem. **267:** 12424–12427.
28. BEYAERT, R., K. HEYNINCK, & S. VAN HUFFEL. 2000. A20 and A20-binding proteins as cellular inhibitors of nuclear factor-kappa B-dependent gene expression and apoptosis. Biochem. Pharmacol. **60:** 1143–1151.
29. LEE, E.G., D.L. BOONE, S. CHAI, et al. 2000. Failure to regulate TNF-induced NF-kappaB and cell death responses in A20-deficient mice. Science **289:** 2350–2354.
30. KRIKOS, A., C.D. LAHERTY & V.M. DIXIT. 1992. Transcriptional activation of the tumor necrosis factor alpha-inducible zinc finger protein, A20, is mediated by kappa B elements. J. Biol. Chem. **267:** 17971–17976.
31. ARENZANA-SEISDEDOS, F., J. THOMPSON, M.S. RODRIGUEZ, et al. 1995. Inducible nuclear expression of newly synthesized I kappa B alpha negatively regulates DNA-binding and transcriptional activities of NF-kappa B. Mol. Cell Biol. **15:** 2689–2696.
32. DU, C., M. FANG, Y. LI, et al. 2000. Smac, a mitochondrial protein that promotes cytochrome c-dependent caspase activation by eliminating IAP inhibition. Cell **102:** 33–42.
33. VERHAGEN, A.M. & D.L. VAUX. 2002. Cell death regulation by the mammalian IAP antagonist Diablo/Smac. Apoptosis **7:** 163–166.
34. VERHAGEN, A.M., P.G. EKERT, M. PAKUSCH, et al. 2000. Simpson RJ, Vaux DL. Identification of DIABLO, a mammalian protein that promotes apoptosis by binding to and antagonizing IAP proteins. Cell **102:** 43–53.

35. CHAI, J., C. DU, J.W. WU, *et al.* 2000. Structural and biochemical basis of apoptotic activation by Smac/DIABLO. Nature **406:** 855–862.
36. DENG, Y., Y. LIN & X. WU. 2002. TRAIL-induced apoptosis requires Bax-dependent mitochondrial release of Smac/DIABLO. Genes Dev. **16:** 33–45.
37. SALOMONI, P. & A.F. KHELIFI. 2006. Daxx: death or survival protein? Trends Cell Biol. **16:** 97–104.
38. YANG, X., R. KHOSRAVI-FAR, H.Y. CHANG, *et al.* 1997. Daxx, a novel Fas-binding protein that activates JNK and apoptosis. Cell **89:** 1067–1076.
39. CHANG, H.Y., H. NISHITOH, X. YANG, *et al.* 1998. Activation of apoptosis signal-regulating kinase 1 (ASK1) by the adapter protein Daxx. Science **281:** 1860–1863.
40. KHELIFI, A.F., M.S. D'ALCONTRES & P. SALOMONI. 2005. Daxx is required for stress-induced cell death and JNK activation. Cell Death Differ. **12:** 724–733.
41. HOLLENBACH, A.D., C.J. MCPHERSON, E.J. MIENTJES, *et al.* 2002. Daxx and histone deacetylase II associate with chromatin through an interaction with core histones and the chromatin-associated protein Dek. J. Cell Sci. **115:** 3319–3330.
42. MICHAELSON, J.S., D. BADER, F. KUO, *et al.* 1999. Loss of Daxx, a promiscuously interacting protein, results in extensive apoptosis in early mouse development. Genes Dev. **13:** 1918–1923.
43. CHEN, L.Y. & J.D. CHEN. 2003. Daxx silencing sensitizes cells to multiple apoptotic pathways. Mol. Cell Biol. **23:** 7108–7121.
44. MICHAELSON, J.S. & P. LEDER. 2003. RNAi reveals anti-apoptotic and transcriptionally repressive activities of DAXX. J. Cell Sci. **116:** 345–352.

Arsenic Trioxide Represses NF-κB Activation and Increases Apoptosis in ATRA-Treated APL Cells

JULIE MATHIEU AND FRANÇOISE BESANÇON

INSERM Unité 685, Hopital St. Louis, 1 Avenue Claude Vellefaux, 75475 Paris Cedex 10, France

ABSTRACT: Acute promyelocytic leukemia (APL) is characterized by an arrest of granulopoiesis at the promyelocytic stage. The sensitivity of APL cells to all-*trans* retinoic acid (ATRA)-induced differentiation has been successfully exploited for treatment of the disease. We previously reported that ATRA-induced NF-κB activation in APL cells is not essential for granulocytic differentiation, but prolongs the life span of mature cells. This prosurvival effect of NF-κB results from its ability to repress c-jun N terminal kinase (JNK) activation. We here report that arsenic trioxide (As_2O_3) can overcome the antiapoptotic effect of ATRA-induced NF-κB activity. As_2O_3 antagonizes ATRA-induced degradation of the NF-κB inhibitor IκB and consequently decreases NF-κB activation. Also, cotreatment of NB4 cells with ATRA and As_2O_3 results in a higher JNK activation than treatment with ATRA alone. Our results demonstrate a proapoptotic effect of As_2O_3 in ATRA-treated APL cells and suggest that As_2O_3 may be helpful in reducing incidence of side effects linked to accumulation of mature cells, like the ATRA syndrome.

KEYWORDS: NF-κB; apoptosis; JNK; As_2O_3; ATRA; APL

INTRODUCTION

Acute promyelocytic leukemia (APL) is a distinct subtype of acute myelogeneous leukemia characterized by the accumulation, in bone marrow and peripheral blood, of myeloid cells blocked at the promyelocytic stage of differentiation.[1] This pathology is generally associated with the t(15;17) chromosomal translocation that fuses PML and retinoic acid receptor α (RAR-α) genes. The resulting PML-RAR-α translation product is thought to be responsible for blocking expression of genes required for granulocyte differentiation. The specific sensitivity of APL cells to pharmacological doses of all-*trans* retinoic

Address for correspondence: Julie Mathieu, INSERM U685, Centre Hayem, Hôpital St. Louis, 1 Avenue Claude Vellefaux, 75475 Paris Cedex 10. Voice: 01-53-72-21-54; fax: 01- 42-40-95-57.
e-mail: julie.mathieu@stlouis.inserm.fr

acid (ATRA)-induced differentiation has been exploited to achieve high rates of temporary clinical remissions.[1] The major severe complication of ATRA treatment is the retinoic acid syndrome (RAS).[2] This syndrome seems to result from the accumulation of differentiated cells that activate several leukocyte functions allowing myeloid precursors and mature granulocytes to infiltrate various organs. We have recently reported that retinoid-induced NF-κB activation in APL cells is not essential for granulocyte differentiation, but prolongs the life span of mature cells.[3] This signaling could contribute to an accumulation of mature APL cells in patients and participate in the development of RAS. We have shown that the prosurvival effect of NF-κB results from its ability to repress ROS-mediated c-jun N terminal kinase (JNK) activation. We therefore suggest that JNK activators or inhibitors of NF-κB activity used in combination therapy with ATRA may be helpful in reducing RAS incidence.

Arsenic trioxide (As_2O_3), an environmental toxin, was proposed as an alternative therapy to ATRA because of its ability to induce differentiation (at doses < 0.5 μM) and apoptosis (at doses >1 μM) of APL cells.[4,5] Reactive oxygen species (ROS) accumulation and JNK activation were shown to be essential mediators of As_2O_3-induced cell death. The ability of As_2O_3 to inhibit NF-κB activation was also shown to participate in its cytotoxic effect.[5] We have investigated whether As_2O_3 could overcome the antiapoptotic effect of ATRA-induced NF-κB activity and therefore help in reducing the life span of differentiated APL cells.

MATERIALS AND METHODS

Reagents and Cell Culture Conditions

As_2O_3 and ATRA were purchased from Sigma-Aldrich (Saint Quentin Fallavier, France). The NB4 cells[6] were grown in RPMI 1640 medium containing 10% decomplemented fetal calf serum.

Cell Differentiation, Viability, and Apoptosis Assays

Cell viability was determined by the trypan blue dye exclusion method. Granulocytic maturation was evaluated by examination of cellular morphology on cytospin specimen stained with May–Grunwald Giemsa (MGG). Apoptosis was assessed by microscopic observation of morphological changes characteristic of apoptosis (i.e, condensation and fragmentation of nuclei) and DNA fragmentation, which was quantified by flow cytometry using a cell death detection kit (TUNEL assay, Roche, Indianapolis, IN, USA). Ten thousand cells were acquired and analyzed using CellQuest Software (Becton Dickinson, Le Pont de Clay, France). Results presented are those of one experiment of three that gave similar results.

Gel Shift Assays

Nuclear extracts and EMSA were performed as previously described[3] except that the double-stranded NF-κB probe (5'-TAC AAG GGA CTT TCC GCT AT-3'; 5'-ATA GCG GAA AGT CCC TTG TA-3') was end-labeled with biotin on the 5' end. Gels were run at 150 V for 3.5 h and transferred by capillarity on a nylon Hybond-N+ membrane (Amersham, Piscataway, NJ, USA). The probe was fixed to the membrane by UV cross-linking and visualized using the Supershift-LightShift™ Chemiluminescent EMSA Kit (Pierce, Rockford, IL, USA).

Immunoblot Analysis

Cellular extracts were prepared and Western blot analysis was performed as described previously.[3] The following antibodies were used: rabbit polyclonal antibodies against phospho-JNK (Thr183/Tyr185), IκBα (Cell Signaling, Danvers, MA, USA) and JNK (Santa Cruz Biotechnologies, Santa Cruz, CA, USA). Antiactin antibodies (ICN Biomedicals Aurora, OH, USA) were used as a control for protein loading.

RESULTS AND DISCUSSION

We investigated the effect of various doses of As_2O_3 on the viability of NB4 cells, an *in vitro* cellular model for APL,[6] which were induced to differentiate in the presence of a pharmacological concentration of ATRA (1 μM). Morphological features of differentiation and cell viability were examined after 7 days of such treatments. As expected, ATRA treatment of NB4 cells resulted in granulocytic maturation as assessed by the observation of MGG-stained cytosmears (increased proportion of cells with lobed nuclei, reduced nucleus/cytoplasm ratio, and less basophilic cytoplasm) (FIG. 1A). Cotreatment with As_2O_3 decreased the viability of the differentiated cells in a dose-dependent manner (FIG. 1B). Concomitant with the decreased viability, the percentage of cells with typical features of apoptosis, such as fragmentation of nuclei or nucleosomal DNA fragmentation increased (FIG. 1A and C). We conclude that As_2O_3 can overcome the antiapoptotic effect of ATRA-induced NF-κB activity and is efficient in reducing the life span of differentiated APL cells.

We have previously reported that the prosurvival effect of NF-κB in ATRA-treated NB4 cells resulted from its ability to repress JNK activation.[3] To get insight into the mechanism by which As_2O_3 induces apoptosis of such cells, the effect of As_2O_3 on NF-κB and JNK activation was investigated. NB4 cells were treated for 3 days with 1 μM ATRA in the absence or the presence of various concentrations of As_2O_3. Nuclear or total cellular extracts were prepared for

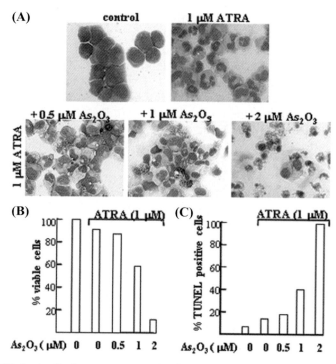

FIGURE 1. As_2O_3 increases apoptosis of ATRA-treated NB4 cells. NB4 cells were cultured with 1μM ATRA without or with 0.5, 1, or 2 μM As_2O_3 for 7 days. (**A**) Morphologic features revealed by MGG staining. (**B**) Viability of cells was assessed by trypan blue exclusion assay. (**C**) DNA fragmentation was quantified by TUNEL staining.

EMSA or Western blot analysis. As shown in FIGURE 2A, As_2O_3 reduced ATRA-induced NF-κB activation in a dose-dependent manner. Efficiency of the various concentrations of As_2O_3 to reduce this activation correlated with their ability to induce cell death (FIGS. 1 and 2A). IκBα is the paradigm of a family of proteins that sequestrate NF-κB in the cytoplasm and therefore impede its nuclear translocation.[7] To investigate the mechanism by which As_2O_3 reduces ATRA-induced NF-κB activation, levels of IκBα expression in NB4 treated with ATRA and/or As_2O_3 were compared. Results are presented in FIGURE 2B. As we previously reported,[3] expression of IκBα was reduced in ATRA-treated cells as compared to control cells. Treatment with As_2O_3 reversed this effect in a dose-dependent manner. In cells treated with 2 μM As_2O_3, IκBα expression was even higher than in control cells (FIG. 2B). These observations are in accordance with previous studies showing that As_2O_3 can inhibit NF-κB activation by inhibiting degradation of IκBα.[8,9] In accordance with our previous report that NF-κB represses proapoptotic JNK activation in ATRA-treated NB4 cells, repression of NF-κB activation by As_2O_3 resulted

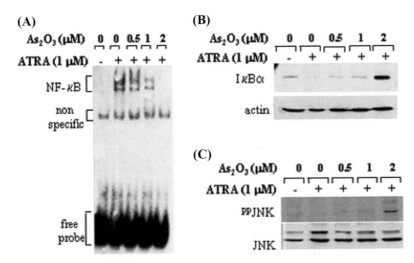

FIGURE 2. As_2O_3 inhibits NF-κB activation and increases JNK activation in ATRA-treated NB4 cells. NB4 cells were cultured with 1 μM ATRA in the absence or presence of indicated concentrations of As_2O_3 for 3 days. (A) Nuclear extracts were prepared and analyzed by EMSA for NF-κB activity. (B) IκBα expression and (C) activation of JNK were determined by Western blot analysis.

in a higher JNK activation in such cells than in cells treated with ATRA alone (FIG. 2C). Altogether, these results strongly suggest that the enhancing effect of As_2O_3 on IκBα expression in ATRA-treated NB4 cells participates in the inhibition of NF-κB activation by this drug and in its proapoptotic effect toward differentiated NB4 cells.

As_2O_3 has proven very effective in treatment of relapsed APL patients who obtained clinical remission with ATRA/chemotherapy. The benefit of ATRA/As_2O_3 combination for remission induction or maintenance in newly diagnosed APL patients is presently under investigation.[10] Our results demonstrate that As_2O_3 when used at concentrations sufficient to inhibit NF-κB activation (>0.5 μM) reduces the life span of mature ATRA-treated APL cells. We suggest that As_2O_3 may be helpful in reducing the incidence of side effects linked to accumulation of mature cells, like the ATRA syndrome.

REFERENCES

1. CASSINAT, B. & C. CHOMIENNE. 2001. Biological features of primary APL blasts: their relevance to the understanding of granulopoiesis, leukemogenesis and patient management. Oncogene **20:** 7154–7160.
2. LARSON, R.S. & M.S. TALLMAN. 2003. Retinoic acid syndrome: manifestations, pathogenesis, and treatment. Best Pract. Res. Clin. Haematol. **16:** 453–461.

3. MATHIEU, J., S. GIRAUDIER, M. LANOTTE & F. BESANCON. 2005. Retinoid-induced activation of NF-κB in APL cells is not essential for granulocytic differentiation, but prolongs the life span of mature cells. Oncogene **24:** 7145–7155.
4. ZHU, J., Z. CHEN, V. LALLEMAND-BREITENBACH & H. DE THE. 2002. How acute promyelocytic leukemia revived arsenic. Nat. Rev. Cancer **2:** 1–9.
5. MILLER, W.H., H.M. SCHIPPER, J.S. LEE, *et al.* 2002. Mechanisms of action of arsenic trioxide. Cancer Res. **62:** 3893–3903.
6. LANOTTE, M., V. MARTIN-THOUVENIN, S. NAJMAN, *et al.* 1991. NB4, a maturation inductible cell line with t(15;17) marker isolated from a human acute promyelocytic leukemia (M3). Blood **77:** 1080–1086.
7. HAYDEN, M.S. & S. GHOSH. 2004. Signaling to NF-kappa B. Genes Dev. **18:** 2195–2224.
8. KAPAHI, P., T. TAKAHASHI, G. NATOLI, *et al.* 2000. Inhibition of NF-kappa B activation by arsenite through reaction with a critical cysteine in the activation loop of Ikappa B kinase. J. Biol. Chem. **275:** 36062–36066.
9. MATHAS, S., A. LIETZ, M. JANZ, *et al.* 2003. Inhibition of NF-kappaB essentially contributes to arsenic-induced apoptosis. Blood **102:**1028–1034.
10. SHEN, Z.X., Z.Z. SHI, J. FANG, *et al.* 2004. All-*trans* retinoic acid/As_2O_3 combination yields a high quality remission and survival in newly diagnosed acute promyelocytic leukemia. Proc. Natl. Acad. Sci. USA **101:** 5328–5335.

Cytotoxicity of TRAIL/Anticancer Drug Combinations in Human Normal Cells

OLIVIER MEURETTE,[a] ANNE FONTAINE,[a] AMELIE REBILLARD,[a] GWENAELLE LE MOIGNE,[a] THIERRY LAMY,[b] DOMINIQUE LAGADIC-GOSSMANN,[a] AND MARIE-THERESE DIMANCHE-BOITREL[a]

[a]*INSERM U620, IFR 140, Université de Rennes 1, 2 avenue du Prof. Léon Bernard, 35043 Rennes cedex, France*

[b]*Département d'Hématologie, Hôpital Pontchaillou, Rennes, France*

> ABSTRACT: TRAIL (TNF-α-Related Apoptosis-Inducing Ligand) is a promising anticancer agent. In fact, it induces apoptosis in cancer cells and not in most normal cells. Nevertheless, certain cancer cells are resistant to TRAIL-induced apoptosis and this could limit TRAIL's efficiency in cancer therapy. To overcome TRAIL resistance, a combination of TRAIL with chemotherapy could be used in cancer treatment. However, sensitivity of human normal cells to such combinations is not well known. We showed in this study that TRAIL/cisplatin, in contrast to TRAIL/5-fluorouracil, was toxic toward human primary hepatocytes and resting lymphocytes. Furthermore, both combinations are toxic toward PHA-IL2-activated lymphocytes. In contrast, freshly isolated neutrophils are resistant to TRAIL in combination or not with anticancer drugs.
>
> KEYWORDS: TRAIL; human primary hepatocytes; lymphocytes; neutrophils

INTRODUCTION

TRAIL (TNF-α-Related Apoptosis-Inducing Ligand) is a member of the tumor necrosis factor (TNF)-α superfamily that is considered as a potential anticancer agent. TRAIL induces apoptosis via activation of the death receptors TRAIL-R1 or TRAIL-R2. TRAIL also possesses two other membrane receptors, TRAIL-R3 and TRAIL-R4, which could function as decoy receptors.[1] Interestingly, TRAIL induces apoptosis in cancer cell but not in most normal cells. Nevertheless, certain cancer cells are resistant to TRAIL-induced

---

Address for correspondence: Marie-Thérèse Dimanche-Boitrel, INSERM U620, IFR140, Faculté de Pharmacie, Université de Rennes 1, 2 avenue Prof. Léon Bernard, 35043 Rennes cedex, France. Voice: 33-0-2-23-23-48-37; fax: 33-0-2-23-23-47-94.

e-mail: marie-therese.boitrel@rennes.inserm.fr

apoptosis and this could limit TRAIL's efficiency in cancer therapy. Combination of TRAIL with anticancer agents restores cell death induction in resistant cancer cells and could thus be used to improve efficiency of TRAIL-based cancer treatment. For example, we have previously described that cisplatin or 5-fluorouracil sensitized human colon cancer cells and hepatocarcinoma cells to TRAIL-induced apoptosis.[2,3] Nevertheless, while the toxicity of TRAIL alone has been intensively studied in normal cells, the toxicity of TRAIL in combination with anticancer drugs remains poorly known. It has been demonstrated that normal keratinocytes, resistant to TRAIL alone, are sensitive to TRAIL/proteasome inhibitor (MG-115) combination.[4] Normal human mesothelial cells, resistant to TRAIL alone, have been shown to become sensitive to TRAIL in combination with cisplatin or doxorubicin.[5] In contrast, human primary hepatocytes are not sensitive to TRAIL/proteasome inhibitor (MG-132 or PS-341) combination,[6] and other studies mentioned the selective effect of TRAIL/anticancer drug combinations in cancer cells.[7,8] Thus, the effect of TRAIL/anticancer agent combinations could depend on the type of normal cells tested and on the anticancer agent used in combination with TRAIL.

Here, we studied the effect of TRAIL/cisplatin and TRAIL/5-fluorouracil combinations in several normal cells: human primary hepatocytes, lymphocytes, and neutrophils. We studied this sensitivity in relation to the TRAIL receptor membrane expression. We demonstrated that TRAIL/cisplatin combination, in contrast to TRAIL/5-flurouracil combination, could be toxic toward human primary hepatocytes and resting lymphocytes. Anticancer agents–associated TRAIL-based cancer therapy could thus have different side effect, depending on the anticancer agent used in combination with TRAIL.

MATERIAL AND METHODS

Human Primary Hepatocyte Isolation and Viability Assay

Human hepatocytes from six adult donors undergoing resection for primary and secondary tumors were obtained by perfusion using a collagenase solution as described previously.[3] Cells were cultured in Williams' E medium (GibcoBRL, Life Technologies, Cergy Pontoise, France) supplemented with 0.2 mg/mL bovine serum albumin (GibcoBRL), 10 µg/mL bovine insulin, 10% fetal calf serum (GibcoBRL), and 2 mM glutamine (GibcoBRL). The medium was discarded 24 h after cell seeding and hepatocytes were thereafter maintained in serum-free medium supplemented with 10^{-7} M hydrocortisone. Cell viability was assessed by a methylene blue colorimetric assay as previously described.[3]

Cell Treatments

The cytotoxic drugs, cisplatin and 5-fluorouracil, were obtained from Merck (Lyon, France). The recombinant human soluble Flag-tagged TRAIL was from Alexis Biochemicals (Coger, Paris, France; http://www.alexis-corp.com) and used at concentration from 12.5–100 ng/mL for 24 h. A total of 2 μg/mL anti-Flag M2 (Sigma-Aldrich, Saint-Quentin Fallavier, France) was added to induce TRAIL oligomerization.

Lymphocyte and Neutrophil Isolation

Fresh peripheral blood mononuclear cells (PBMCs) and neutrophils were isolated from six buffy coat by Ficoll-Hypaque (Amersham Biosciences, Uppsala, Sweden) gradient centrifugation. After 1 h for monocyte adhesion, the supernatant was recovered for lymphocyte culture. Neutrophil isolation was performed as previously described.[9] Lymphocytes and neutrophils were cultured in RPMI medium (GibcoBRL) supplemented with 10% fetal calf serum and 2 mM glutamine (GibcoBRL). Phytohemaglutinin (PHA; 5 μg/mL) (Sigma-Aldrich) was added for 48 h followed by addition of interleukin 2 (IL-2; 100 U/mL) (Sigma-Aldrich) for 5 days.

Quantification of Apoptosis

Microscopic detection of apoptosis was carried out in both floating and adherent cells recovered after treatment using nuclear chromatin staining with 1 μg/mL Hoechst 33342 (Amersham Biosciences, Uppsala, Sweden) for 15 min at 37°C. Cells with apoptotic nuclei (i.e., condensed or fragmented) were counted in comparison with total population ($n = 300$ cells).

Flow Cytometry Analysis of TRAIL Receptor Surface Expression

Flow cytometry analysis of TRAIL receptors was performed as previously described.[2] Mouse anti-human TRAIL receptors (anti-TRAIL receptor-1 to 4 flow cytometry set) were used according to the instructions of Alexis Biochemicals. Secondary fluorescein-isothiocyanate anti-mouse IgG was from Jackson Immunoresearch, Marseille, France.

Statistical Analysis

The statistical analysis was carried out using the unilateral Student's t-test, considering the variances as unequal. The significance is shown as follows: $^*P \leq 0.05$; $^{**}P \leq 0.02$; $^{***}P \leq 0.001$.

RESULTS AND DISCUSSION

Effect of TRAIL/Anticancer Agent Combination in Human Primary Hepatocytes

In contrast to Fas ligand treatment, TRAIL treatment was not toxic toward human primary hepatocytes (FIG. 1A). Sensitivity of human primary hepatocytes to TRAIL-induced apoptosis has been intensively debated. Human primary hepatocytes have been shown to be sensitive to polyhistidine-tagged TRAIL recombinant protein (His-TRAIL), whereas a version of TRAIL that lacks exogenous sequence tags was not toxic.[10] It has recently been shown that general administration of TRAIL in chimeric mice harboring human hepatocytes was not hepatotoxic.[11] Nevertheless, toxicity of TRAIL/anticancer agent combination in human primary hepatocytes is not well known. We tested TRAIL toxicity in combination with cisplatin or 5-fluorouracil. We observed that TRAIL/cisplatin was cytotoxic, whereas TRAIL/5-fluorouracil has no significant cytotoxic effect in human primary hepatocytes (FIG. 1B). Higher doses, up to 400 μg/mL of 5-fluorouracil in combination with concentration up to 200 ng/mL of TRAIL were then tested, but no significant toxicity was observed (data not shown). It has been shown that sensitization to TRAIL-induced apoptosis in some cellular systems could be dependent on upregulation of membrane expression of TRAIL death receptors.[12] We thus measured cell surface expression of TRAIL receptors in human primary hepatocytes before and after a 24-h treatment with cisplatin or 5-fluorouracil. Before treatment, in all human hepatocyte samples, we clearly detected the decoy receptor TRAIL-R4 at the cell membrane, whereas in a few of them, a very weak membrane expression of the death receptor TRAIL-R2 was detected (FIG. 1C). After a 24-h treatment, we observed no significant effect of cisplatin or 5-fluorouracil on TRAIL receptor membrane expression (FIG. 1C). Other modifications of apoptosis signaling factors, like downregulation of c-FLIP or IAP, could account for the abrogation of resistance to TRAIL-induced apoptosis.[1] High-throughput analysis of gene expression modifications following cisplatin treatment in human primary hepatocytes is now under investigation in our laboratory.

Effect of TRAIL/Anticancer Agent Combinations in Lymphocytes and Neutrophils

Haematological cells are primary targets of chemotherapy. We therefore studied toxicity of TRAIL/anticancer agent combinations in freshly isolated resting or PHA+IL-2-activated lymphocytes, and in freshly isolated neutrophils in relation to cell surface expression of TRAIL receptors. We observed that freshly isolated lymphocytes expressed exclusively the decoy receptor TRAIL-R4 (FIG. 2A). PHA+IL-2-activated lymphocytes express a little

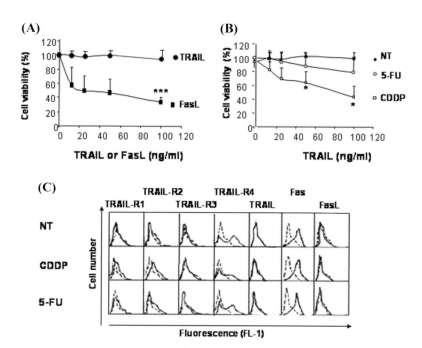

FIGURE 1. TRAIL/cisplatin combination was toxic towards human primary hepatocytes. (**A**) Cell viability was estimated by a methylene blue colorimetric assay after treatment with increased concentrations of TRAIL or Fas ligand (FasL) (12.5, 25, 50, 100 ng/mL cross-linked with 2 μg/mL anti-Flag M2). (**B**) Cell viability was estimated as in **A** after treatment with increased concentrations of TRAIL (12.5, 25, 50, 100 ng/mL cross-linked with 2 μg/mL anti-Flag M2) in combination with cisplatin (CDDP, 5 μg/mL), 5-fluorouracil (5-FU, 100 μg/mL) or not (NT). (**C**) Membrane expression of TRAIL receptors was measured by flow cytometry by using mAbs raised against extracellular domain of TRAIL-R1 (HS201) TRAIL-R2 (HS 201), TRAIL-R3 (HS301), TRAIL-R4 (HS402), Fas (ZB4), or FasL (NOK1) at 1 μg/mL after a 24-h treatment or not (NT) with cisplatin (CDDP; 5 μg/mL) or 5-fluorouracil (5-FU; 100 μg/mL).

amount of TRAIL-R1 and TRAIL-R2 on their cell surface (FIG. 2A). In contrast to the findings of Mirandola et al.[13] we observed no induction of TRAIL-R3 expression following lymphocyte activation by PHA and IL-2, whereas the same antibody was used in both of these studies. The protocol of lymphocyte isolation may account for this discrepancy. Freshly isolated neutrophils expressed exclusively TRAIL-R3 (FIG. 2A) in agreement with other studies.[9]

His-TRAIL has been shown to be toxic toward primary lymphocytes.[14] However, other studies using preparations of His-tagged TRAIL mentioned the absence of TRAIL toxicity toward primary lymphocyte.[12,13] In our study, we observed a weak induction of apoptosis (10%) in human resting or activated lymphocytes following a 24-h Flag-tagged TRAIL treatment. In contrast, PHA+IL-2-activated lymphocytes were sensitive to Fas ligand-induced

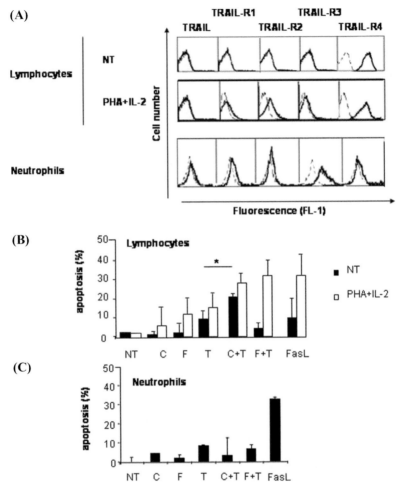

FIGURE 2. Effect of TRAIL/anticancer agent combination in human lymphocytes and neutrophils. (**A**) Membrane expression of TRAIL-receptors was studied as in FIGURE 1 on resting (NT) or activated (PHA+IL-2) lymphocytes and on freshly isolated neutrophils. (**B**) Apoptosis induction in resting (NT) or activated (PHA+IL-2) lymphocytes was measured by Hoechst 33342 staining after a 24-h treatment or not (NT) with cisplatin (C; 5 μg/mL), 5-fluorouracil (5-FU; 100 μg/mL), TRAIL (100 ng/mL cross-linked with 2 μg/mL anti-Flag M2), or combination of TRAIL with cisplatin (C+T) or 5-fluorouracil (F+T), or Fas ligand (FasL; 100 ng/mL cross-linked with 2 μg/mL anti-Flag M2). (**C**) Percentage of apoptosis in neutrophils was measured by Hoechst 33342 staining after a 24-h treatment with cisplatin (C; 5 μg/mL), 5-fluorouracil (5-FU; 100 μg/mL), TRAIL (100 ng/mL cross-linked with 2 μg/mL anti-Flag M2), or a combination of TRAIL with cisplatin (C+T) or 5-fluorouracil (F+T), or Fas ligand (FasL; 100 ng/mL cross-linked with 2 μg/mL anti-Flag M2).

apoptosis (FIG. 2B). Furthermore, treatment with cisplatin or 5-fluorouracil alone induced few apoptotic cells in resting lymphocytes (FIG. 2B). Nevertheless, TRAIL/cisplatin combination became toxic toward human resting lymphocytes (25% of apoptosis), whereas TRAIL/5-fluorouracil was not (FIG. 2B). Regarding PHA+IL-2-activated lymphocytes, both combinations were toxic (30–35% of apoptosis) (FIG. 2B). In order to understand the molecular mechanisms of cell sensitization to TRAIL-induced apoptosis, it would be of great interest to study the effect of anticancer drug treatments on membrane expression of TRAIL-R1 and TRAIL-R2 because it has been shown that fludarabine and chlorambucil increased such an expression in chronic lymphocytic leukemia (CLL) cells but had less effect in normal lymphocytes.[12]

In contrast to Renshaw et al.,[9] we observed no effect of TRAIL treatment in freshly isolated human neutrophils, whereas Fas ligand was toxic (FIG. 2C). Interestingly, the TRAIL used in this study was the Flag-tagged TRAIL, whereas Renshaw et al.[9] used the His-TRAIL version of TRAIL, which has also been shown to be toxic toward human primary hepatocytes.[10] Furthermore, freshly isolated human neutrophils were resistant to both combination of TRAIL with cisplatin or 5-fluorouracil (FIG. 2C).

Altogether, our results show that normal human cells exclusively express the decoy receptors TRAIL-R3 or TRAIL-R4, which could account for their resistance to TRAIL. In fact ectopic expression of TRAIL-R3 has been shown to protect against TRAIL-induced apoptosis.[15] Nevertheless, TRAIL decoy receptor expression is not always correlated with resistance to TRAIL-induced apoptosis.[1] However, sensitivity of normal human cells to TRAIL/anticancer drug combinations depends on the anticancer agent used. Furthermore, TRAIL/cisplatin association could be relatively toxic and compromise the possible use of this combination in clinical cancer therapy.

ACKNOWLEDGMENTS

We thank the Biological Resource Center (BRC) of Rennes for supplying isolated human primary hepatocytes. We also thank the Fondation de France, The Région Bretagne, and Rennes Metropole for their financial support.

REFERENCES

1. ZHANG, L. & B. FANG. 2005. Mechanisms of resistance to TRAIL-induced apoptosis in cancer. Cancer Gene Ther. **12:** 228–237.
2. LACOUR, S. et al. 2001. Anticancer agents sensitize tumor cells to tumor necrosis factor-related apoptosis-inducing ligand-mediated caspase-8 activation and apoptosis. Cancer Res. **61:** 1645–1651.

3. MEURETTE, O. *et al.* 2005. Role of intracellular glutathione in cell sensitivity to the apoptosis induced by tumor necrosis factor {alpha}-related apoptosis-inducing ligand/anticancer drug combinations. Clin Cancer Res. **11:** 3075–3083.
4. LEVERKUS, M. *et al.* 2003. Proteasome inhibition results in TRAIL sensitization of primary keratinocytes by removing the resistance-mediating block of effector caspase maturation. Mol. Cell. Biol. **23:** 777–790.
5. LIU, W. *et al.* 2001. Tumor necrosis factor-related apoptosis-inducing ligand and chemotherapy cooperate to induce apoptosis in mesothelioma cell lines. Am. J. Respir. Cell. Mol. Biol. **25:** 111–118.
6. GANTEN, T.M. *et al.* 2005. Proteasome inhibition sensitizes hepatocellular carcinoma cells, but not human hepatocytes, to TRAIL. Hepatology **42:** 588–597.
7. WU, X.X. *et al.* 2002. Doxorubicin enhances TRAIL-induced apoptosis in prostate cancer. Int. J. Oncol. **20:** 949–954.
8. EVDOKIOU, A. *et al.* 2002. Chemotherapeutic agents sensitize osteogenic sarcoma cells, but not normal human bone cells, to Apo2L/TRAIL-induced apoptosis. Int. J. Cancer **99:** 491–504.
9. RENSHAW, S.A. *et al.* 2003. Acceleration of human neutrophil apoptosis by TRAIL. J. Immunol. **170:** 1027–1033.
10. LAWRENCE, D. *et al.* 2001. Differential hepatocyte toxicity of recombinant Apo2L/TRAIL versions. Nat. Med. **7:** 383–385.
11. HAO, C. *et al.* 2004. TRAIL inhibits tumor growth but is nontoxic to human hepatocytes in chimeric mice. Cancer Res. **64:** 8502–8506.
12. JOHNSTON, J.B. *et al.* 2003. Role of the TRAIL/APO2-L death receptors in chlorambucil- and fludarabine-induced apoptosis in chronic lymphocytic leukemia. Oncogene **22:** 8356–8369.
13. MIRANDOLA, P. *et al.* 2004. Activated human NK and CD8+ T cells express both TNF-related apoptosis-inducing ligand (TRAIL) and TRAIL receptors but are resistant to TRAIL-mediated cytotoxicity. Blood **104:** 2418–2424.
14. MARSTERS, S.A. *et al.* 1996. Activation of apoptosis by Apo-2 ligand is independent of FADD but blocked by CrmA. Curr. Biol. **6:** 750–752.
15. MARSTERS, S.A. *et al.* 1997. A novel receptor for Apo2L/TRAIL contains a truncated death domain. Curr. Biol. **7:** 1003–1006.

Hyperpolarization of Plasma Membrane of Tumor Cells Sensitive to Antiapoptotic Effects of Magnetic Fields

S. NUCCITELLI,[a] C. CERELLA,[a] S. CORDISCO,[a] M.C. ALBERTINI,[c] A. ACCORSI,[c] M. DE NICOLA,[a] M. D'ALESSIO,[a] F. RADOGNA,[a] A. MAGRINI,[b] A. BERGAMASCHI,[b] AND L. GHIBELLI[a]

[a]*Dipartimento di Biologia, Università di Roma Tor Vergata (Rome), Via della Ricerca Scientifica, 1, 00133 Rome, Italy*

[b]*Cattedra Medicina del Lavoro, Università di Roma Tor Vergata (Rome), Via della Ricerca Scientifica, 1, 00133 Rome, Italy*

[c]*Istituto Chimica Biologica, "G. Fornaini," Università di Urbino "Carlo Bo," via Saffi2, 61029, Urbino, Italy*

ABSTRACT: Chemical/physical agents able to prevent apoptosis are receiving much attention for their potential health hazard as tumor promoters. Magnetic fields (MFs), which have been shown to increase the occurrence of some tumors, reduce damage-induced apoptosis by a mechanism involving Ca2+ entry into cells. In order to discover the mechanism of such effect of MFs, we investigated the interference of MFs on cell metabolism and analyzed cell parameters that are involved in apoptotic signaling and regulation of Ca^{2+} fluxes. Here we show that different types (static and extremely low-frequency, ELF pulsating) of MFs of different intensities alter plasma membrane potential. Interestingly, MFs induce plasma membrane hyperpolarization in cells sensitive to the antiapoptotic effect of MFs, whereas cells that are insensitive showed a plasma membrane depolarization. These opposite effects suggest that protection against apoptosis and membrane potential modulation are correlated, plasma membrane hyperpolarization possibly being part of the signal transduction chain determining MFs' antiapoptotic effect.

KEYWORDS: magnetic fields; plasma membrane potential; apoptosis; survival

INTRODUCTION

The influence of magnetic fields (MFs) on life processes and human health is receiving a growing interest. The data reported in the literature suggest a link between MFs and tumorigenicity.[1] MFs have been shown to increase the development of some tumors,[2,3] and to increase tumor cell survival after oncostatic therapies.[4] However, no direct tumorigenic or mutagenic effect has ever been attributed to MFs.[5] Scattered studies showed an interference of MFs on cell metabolism through alteration of specific gene transcription,[6] included c-myc mRNA induction,[7] or through a decrease of spontaneous cell death in culture.[8]

Apoptosis is a cell-intrinsic mechanism that leads healthy cells to programmed cell death and occurs either under physiological conditions or in response to damage. Apoptosis can be considered as a mechanism through which the organism eliminates potentially dangerous mutated or transformed cells. In this view, physical or chemical agents able to prevent apoptosis became dangerous for their ability to increase survival rate of the damaged cells. Indeed, impairment of apoptosis is now considered as the main mechanism of tumor promotion, that is, the mechanism through which agents that are not mutagenic *per se*, promote tumor development by inhibiting removal of tumor cells by apoptosis. We previously reported that MFs reduce apoptosis induced by damaging agents through a mechanism involving Ca^{2+} entry into cells.[9] It has also been reported that MFs induce structural and biophysical changes in cell plasma membrane.[10]

Among the major alterations that may be produced on plasma membrane, changes in transmembrane potential, a parameter controlled by specific ion pumps and responsible for cell homeostasis, are particularly important, because they may alter cell behavior or survival. In some cell types, changes in membrane potential are able to mediate receptor-induced release of intracellular proteins,[11] influence the kinetics of different active transport processes,[12,13] or interfere with ion transport by affecting the transporter affinity.[14,15] Because Ca^{2+} fluxes occur through voltage-dependent membrane channels and alteration of membrane potential can modulate activity of Ca^{2+} channels,[16] we investigated whether MFs are able to induce alteration of membrane potential, and whether they may be related to Ca^{2+} entry into cells.

In the present study we show that MFs do alter plasma membrane potential; in particular, they induce hyperpolarization in cells sensitive to protective effects of MFs and depolarization in insensitive cells. These opposite effects suggest that membrane potential alterations induced by MFs are possibly involved in the antiapoptotic effects of MFs.

MATERIALS AND METHODS

Cells and Culture

U937 and Jurkat cells were kept in RPMI 1640 medium supplemented with 10% inactivated fetal calf serum (FCS), 2mM L-glutamine, 100 IU/mL penicillin, and streptomycin, and kept in a controlled atmosphere (5% CO_2) incubator at 37°C. Experiments were performed at a concentration of 10^6 cells/mL.

MF Application

Static MFs

MFs were produced by metal magnetic disks of known intensities; magnets produce static MFs without producing alternating MFs. Static MFs do not induce any temperature increase. MFs' intensity is given in millitesla (1 T = 10^4 G). Magnets were placed under the culture multiwell plate. Cells were labeled with a fluorescent specific probe for plasma membrane potential (DiBAC) and then MFs were applied for 5 min.

ELF MFs

Cell exposure to ELF MFs was performed as described in previously studies,[17] and cells were exposed only to the magnetic component of ELF inside a Faraday cage. Intensity of MF was measured with a specific sensor placed perpendicularly to the field-force line. Cells were exposed to a range of intensity of MFs, from 0.07 mT to 0.09 mT. Because no metal objects such as thermostatic equipment may be placed in the Faraday cage, experiments were performed with a system of circulating warm water, to guarantee that the waterbath with the cells multiwell plate in the Faraday cage was stably maintained at 37°C (continually monitored by an alcohol thermometer). Control cells were placed in the thermostatic waterbath outside the Faraday cage, in a MF-null area, and considered as the sham-exposed.

Membrane Potential Measurement

DiBAC Staining

Cells (1 × 10^6/mL) were washed in RPMI without phenol red, then loaded with 5 μM anionic oxonol dye bis-(1,3-dibutylbarbituric acid) trimethine oxonol DiBAC (Molecular Probes, Eugene, OR, USA) at 37°C for 15 min in the dark. Cells were stored at RT for 20 min before the measurements. Plasma membrane potential alterations induced by static MFs were measured with fluorescence on conventional flow cytometers (Dako Galaxy and FACScalibur) at 488 nm laser excitation, and DiBAC emission was detected at 535 nm. Increase in fluorescence indicates depolarization, whereas decreased fluorescence indicates hyperpolarization. ELF-induced hyperpolarization was measured with a fluorescence-reversed microscopy equipped with a CoolSNAP digital camera (RS Image Software).

Artificial Depolarization

Artificial depolarization of membrane was induced with depolarizating Hanks' balanced salt solution (Na^+ [100mM], K^+ [42mM]). Control cells were washed and resuspended in normal HBSS, whereas cells to treat were washed and resuspended in depolarizating HBSS. Then, both samples were labelled with DiBAC.

Induction and Detection of Apoptosis

Apoptosis was induced with the protein synthesis inhibitor puromycin (PMC, 10 μg/mL) and was kept throughout the experiments. Apoptosis was measured at 3 h of PMC. Apoptosis was evaluated in terms of apoptotic nuclear morphology detectable by fluorescence microscopy on cells stained with the cell permeant DNA-specific dye Hoechst 33342. The fraction of cells with apoptotic nuclei among the total cell population was calculated by counting 100 per sample in at least three randomly selected microscopic fields; the results are expressed as percentage of apoptotic cells among the total cells counted.

Statistical Analysis

Statistical analysis was performed using Student's *t*-test for impaired data and P values <0.05 were considered significant. Data are presented as mean ± SD.

RESULTS

Static MFs Induce a Membrane Hyperpolarization in U937 Cells and a Membrane Depolarization in Jurkat Cells

Metal magnetic disks producing MFs of known intensities were placed under the culture multiwell plates containing U937 or Jurkat cells in such a way as to get an intensity of 6 mT MF on cells. Cells labelled with the fluorescent-specific probe DiBAC were exposed to magnetic disks for 5 min. Increased DiBAC fluorescence intensity indicates plasma membrane depolarization, whereas decreased fluorescence indicates plasma membrane hyperpolarization. In order

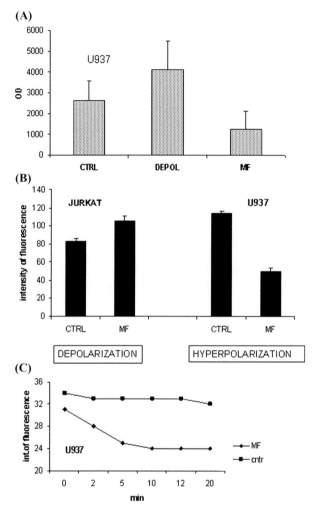

FIGURE 1. MFs alter plasma membrane potential. U937 and Jurkat cells were labelled with DIBAC and fluorescence intensity was measured by flow cytometric analysis in the presence of static MFs. (**A**) Comparison between modulations of membrane potential induced by depolarizating conditions and by MFs in U937 cells. (**B**) Reduction of fluorescence intensity (hyperpolarization) in U937 and increase of fluorescence intensity (depolarization) in Jurkat cells induced by MFs were both statistically significant (Student's t-test: $P < 0.05$). (**C**) Hyperpolarization occurs early in U937, after 2 min of exposure to MFs. The values are the average of three experiments ± SD.

to validate DiBAC staining in our system, we induced depolarizing conditions through unsettling of Na^+/K^+ rate (see the MATERIALS AND METHODS section), as shown in FIGURE 1A. We then measured the effects of MFs. We found that in U937 cells, 5 min of static MFs induce plasma membrane hyperpolarization

(FIG. 1B). Interestingly, we observed an opposite behavior in Jurkat cells (FIG. 1B). A kinetics analysis shows that membrane hyperpolarization in U937 occurs very early, being already detectable after 2 min of exposure to MFs, as shown in FIGURE 1C. We have shown that U937 cells are sensitive to the antiapoptotic effect of MFs,[9,18] whereas Jurkat cells were found to be insensitive to MFs' antiapoptotic action.[18] The intriguing correlation between (i) sensitivity to the antiapoptotic effect of MFs and (ii) effect on plasma membrane polarization suggests that membrane potential alteration induced by static MFs may be involved in the antiapoptotic effects of MFs in tumor cells.

ELF Intensity Producing Antiapoptotic Effects Induces a Membrane Hyperpolarization in U937 Cells

We have applied a range of different intensities of extremely low-frequency pulsating (ELF) MFs to the cells, as described in MATERIALS AND METHODS. We found that ELF produces an antiapoptotic effect similar to that produced by static MFs, though requiring different field intensities. Indeed, whereas static MFs required at least 0.6 mT to achieve a detectable effect, ELF intensities as low as 0.09 mT are sufficient to inhibit apoptosis, whereas 0.07 mT are ineffective (FIG. 2A). After ELF induction, U937 cells were labelled with DiBAC, and membrane potential was measured with a fluorescence-reversed microscopy equipped with a photo-camera (NIH Imaging), as described in MATERIALS AND METHODS. We found that ELF intensity of 0.09 mT induces a hyperpolarization in U937 cells. Instead, the 0.07 mT intensity is not able to alter plasma membrane polarization (FIG. 2B). Also in this case, the correlation between hyperpolarization and protective effects of MFs is maintained, reinforcing the hypothesis of a cause–effect relationship.

DISCUSSION

We previously reported that MFs reduce apoptosis induced by damaging agents through a mechanism involving Ca^{2+} entry into cells.[9] Because Ca^{2+} fluxes occur through voltage-dependent membrane channels and alteration of membrane potential can modulate activity of Ca^{2+} channels,[16] this pushed us to investigate whether MFs were able to induce alteration of membrane potential. Our present finding showed that static MFs do induce a membrane potential modulation. In U937 cells, which we previously demonstrated to be sensitive to antiapoptotic effects of MFs,[9,18] static MFs induce a plasma membrane hyperpolarization, whereas insensitive cells, such as Jurkat cells, undergo a plasma membrane depolarization. These results agree with previous studies showing that MFs induce structural and biophysical changes in cellular plasma membrane.[10]

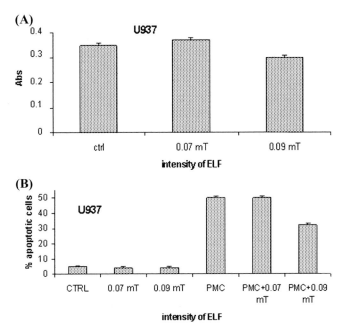

FIGURE 2. ELF intensity producing antiapoptotic effects induces membrane hyperpolarization. U937 cells were exposed to ELF and labelled with DiBAC. The analysis of quantitative fluorescence was performed with a fluorescence-reversed microscopy equipped with a photo-camera (NIH Imaging). The intensity of ELF (0.09 mT) that produces the antiapoptotic effect also induces an hyperpolarization in U937 cells. All values are the average of at least 30 measurements for each treatment. Fluorescence has been converted in absorbance value (OD). The reduction of fluorescence intensity (hyperpolarization) induced by ELF was statistically significant ($P < 0.05$).

Because membrane potential alteration showed an opposite behavior in sensitive and insensitive cells, the findings described here indicate that alteration of membrane potential may be correlated with the antiapoptotic effect of MFs. Furthermore, we showed that, as well as static MFs, a definite intensity of ELF (0.09 mT) induces a hyperpolarization in U937 cells and that the same intensity of ELF is able to produce an antiapoptotic effect in U937 cells. Instead, a lower intensity is ineffective in modulating both parameters. These findings indicate that both static and pulsating MF-induced plasma membrane hyperpolarization may be required for MFs' antiapoptotic effects. It is well known that L-type Ca^{2+} channels are activated by depolarization of plasma membrane and inactivated by hyperpolarization.[19] An involvement of L-type Ca^{2+} channels in the increased Ca^{2+} influx due to MF exposure was suggested[9] because of the sensitivity of this effect to nifedipine, which is a known L-type Ca^{2+} channel inhibitor. However, we show here that MFs increase Ca^{2+} entry in spite of hyperpolarizing plasma membrane. A possible explanation for this apparently paradoxical result may imply that nifedipine may inhibit Ca^{2+} channels additional to the L-type. Indeed, Ca^{2+} channels in non-excitable cells, such as the U937 monocytes, are still poorly investigated. We had shown that MFs increase the capacitative Ca^{2+} influx,[9] which is indeed known to be nifedipine-sensitive in U937 cells.[9] Recent evidence of a role of calmodulin in triggering/modulating capacitative Ca^{2+} influx [20] is particularly interesting in this respect, because we have preliminary evidence that calmodulin is indeed involved in MFs' antiapoptotic effect. As an alternative explanation, hyperpolarization of plasma membrane may attract positive charges from extracellular to intracellular environment, thus non-specifically promoting Ca^{2+} entry into cells.

The mechanism through which MFs antagonize apoptosis is still largely unknown. The importance of the issue is due to the possible health hazard posed by unwanted exposure to MFs, implying a possible role of MFs as tumor promoters. This recommends that great attention be paid to the study of the mechanisms involved in MFs antiapoptotic effects. This study contributes to such interest in two ways: first, it indicates clear alterations on cell parameters directly due to MFs; second, it begins to delineate the biochemical/biophysical differences between cells that are sensitive versus insensitive to the antiapoptotic effect of MFs.

REFERENCES

1. McCann, J., F. Dietrich, C. Rafferty, *et al.* 1993. A critical review of the genotoxic potential of electric and magnetic fields. Mutat. Res. **297:** 61–95
2. Nordenson, I., K.H. Mild, G. Andersson, *et al.* 1994. Chromosomal aberrations in human amniotic cells after intermittent exposure to fifty hertz magnetic fields. Bioelectromagnetics **15:** 293–301
3. Miyakoshi, J., N. Yamagishi, S. Ohtsu, *et al.* 1996. Increase in lipoxanthine-guanine phosphoribosyl transferase gene mutations by exposure to high-density 50 Hz magnetic fields. Mutat. Res. **349:** 109–114
4. Liburdy, R.P., T.R. Sloma, R. Sokolic, *et al.* 1993. ELF magnetic fields, breast cancer, and melatonin: 50 Hz fields block melatonin's oncostatic action on ER+ breast cancer cell proliferation. J. Pineal Res. **14:** 89–97
5. Morandi, M.A., C.M. Pak, R.P. Caren, *et al.* 1996. Lack of an EMF-induced genotoxic effect in the Ames assay. Life Sci. **59:** 263–271
6. Phillips, J.L., W. Haggren, W.J. Thomas, *et al.* 1992. Magnetic field-induced changes in specific gene transcription. Biochim. Biophys. Acta **1132:** 140–144
7. Liburdy, R.P., D.E. Callahan, J. Harland, *et al.* 1993. Experimental evidence for 60 Hz magnetic fields operating through the signal transduction cascade. Effects on calcium influx and c-myc mRNA induction. FEBS Lett. **334:** 301–308
8. Eremenko, T., C. Esposito, A. Pasquarelli, *et al.* 1997. Cell-cycle kinetics of Friend erythroleukemia cells in a magnetically shielded room and in a low-frequency/low-intensity magnetic field. Bioelectromagnetics **18:** 58–66

9. FANELLI, C., S. COPPOLA, R. BARONE, et al. 1999. Magnetic fields increase cell survival by inhibiting apoptosis via modulation of Ca^{2+} influx. FASEB J. **13:** 95–102
10. PARADISI, S., G. DONELLI, M.T. SANTINI, et al. 1993. A 50 Hz magnetic field induces structural and biophysical changes in membranes. Bioelectromagnetics **14:** 247–255
11. FREY, J., M. JANES, W. ENGELHARDT, et al. 1986. Fc gamma-receptor-mediated changes in the plasma membrane potential induce prostaglandin release from human fibroblasts. Eur. J. Biochem. **158:** 85–89
12. ELENO, N., R. DEVES & C.A. BOYD. 1994. Membrane potential dependence of the kinetics of cationic amino acid transport systems in human placenta. J. Physiol. Lond. **479:** 291–300
13. HASLBERGER, A., C. ROMANIN & R. KOERBER. 1992. Membrane potential modulates release of tumor necrosis factor in lipopolysaccharide-stimulated mouse macrophages. Mol. Biol. Cell. **3:** 451–460
14. BLATT, M.R., A. RODRIGUEZ-NAVARRO & C.L. SLAYMAN. 1987. Potassium-proton symport in *Neurospora*: kinetic control by pH and membrane potential. J. Membran. Biol. **98:** 169–189
15. OETTGEN, H.C., C. TERHORST, L.C. CANTLEY, et al. 1985. Stimulation of the T3-T cell receptor complex induces a membrane-potential-sensitive calcium influx. Cell **40:** 583–590
16. PANAGOPOULOS, D.J., N. MESSINI, A. KARABERBOUNIS, et al. 2000. A mechanism for action of oscillating electric fields on cells. Biochem. Biophys. Res. Commun. **272:** 634
17. FIORANI, M., O. CANTONI, P. SESTILI, et al. 1992. Electric and/or magnetic field effect on DNA structure and function in cultured human cells. Mutat. Res. **282:** 25–29
18. GHIBELLI, L., C. CERELLA, S. CORDISCO, et al. 2006. NMR exposure sensitizes tumor cells to apoptosis. Apoptosis **3:** 359–365.
19. TSIEN, R.W., D. LIPSCOMBE, D. MADISON, et al. 1995. Reflections on Ca^{2+}-channels diversity. Trends Neurosci. **18:** 52–54
20. MENE, P., F. PUGLIESE & G.A. CINOTTI. 1996. Regulation of capacitative calcium influx in cultured human mesangial cells: roles of protein kinase C and calmodulin. J. Am. Soc. Nephrol. **7:** 983–990.

Melatonin as an Apoptosis Antagonist

FLAVIA RADOGNA,[a] LAURA PATERNOSTER,[a]
MARIA CRISTINA ALBERTINI,[c] AUGUSTO ACCORSI,[c]
CLAUDIA CERELLA,[a] MARIA D'ALESSIO,[a] MILENA DE NICOLA,[a]
SILVIA NUCCITELLI,[a] ANDREA MAGRINI,[b]
ANTONIO BERGAMASCHI,[b] AND LINA GHIBELLI[a]

[a]*Dipartimento di Biologia, via della Ricerca Scientifica, 1, 00133, Roma, Italy*

[b]*Cattedra Medicina del Lavoro, Università di Roma Tor Vergata, via della Ricerca Scientifica, 1, 00133, Roma, Italy*

[c]*Istituto Chimica Biologica, "G. Fornaini" Università di Urbino Carlo Bo, via Saffi 2, 61029, Urbino, Italy*

ABSTRACT: The pineal hormone melatonin (Mel), in addition to having a well-established role as a regulator of circadian rhythms, modulates nonneural compartments by acting on specific plasma membrane receptors (MT1/MT2) present in many different cell types. Mel plays immunomodulatory roles and is an oncostatic and antiproliferative agent; this led to the widespread belief that Mel may induce or potentiate apoptosis on tumor cells, even though no clear indications have been presented so far. Here we report that Mel is not apoptogenic on U937 human monocytic cells, which are known to possess MT1 receptors at the times (up to 48 h) and doses (up to 1 mM) tested. Mel does not even potentiate apoptosis, but instead, significantly reduces apoptosis induced by both cell-damaging agents (intrinsic pathway) and physiological means (extrinsic pathway). The doses required for the antiapoptotic effect (≥ 100 μM) are apparently not compatible with receptor stimulation (receptor affinity <1 nM). However, receptor involvement cannot be ruled out, because we discovered that the actual Mel concentration active on cells was lower than the nominal one because of sequestration by fetal calf serum (FCS). Accordingly, in FCS-free conditions, Mel doses required for a significant antiapoptotic effect are much lower.

KEYWORDS: melatonin; apoptosis; receptor engagement

INTRODUCTION

It has been historically known that the pineal gland is the major source of melatonin (Mel) in vertebrates. Mel plays a central role in fine-tuning circadian rhythms in vertebrate physiology. Its biosynthesis from L-tryptophan, which is converted to serotonin and then to Mel in a two-step pathway, involves N-acetylation and O-methylation.[1]

Mel, in addition to playing a main role as regulator of circadian rhythms, has recently been shown to modulate immune functions by controlling the behavior of white blood cells (WBCs), which are indeed able to synthesize Mel and possess the specific high affinity (MT1 and MT2, see below). Mel acts as an intracrine, autocrine, or paracrine substance in the immune system.[2]

In the blood stream, Mel binds (sticks) to plasma proteins[3]; it is not clear whether this may serve for Mel targeting, or as sequestration for lowering its blood levels, or for storage, or even for modulating serum protein functions. It is emerging that Mel is involved in many regulatory functions of the cells, possibly through receptor engagement, that is, it modulates the immune response,[2] regulates the apoptotic response on a number of cell types,[4] and regulates signal transduction reactions.[5] As a radical scavenger, Mel may reduce tissue destruction during inflammatory reactions,[6] and Mel also protects both polyunsatured fatty acid from oxidation and nuclear DNA from damage induced by carcinogens and ionizing irradiation.[7] The effects of Mel are mediated by the specific high-affinity receptors localized on the plasma membrane of target cells and coupled to GTP-binding protein. On the basis of the molecular structure, three subtypes of the Mel receptor have been described: Mel_{1A} or mt1 (expressed in mammalian and bird brain), Mel_{1B} or mt2 (expressed mainly in mammalian retina), Mel_{1C} or mt3 (found in amphibian melanophores, brain, and retina, and also in bird and fish brain).[8] The dissociation constant (K_d) of the Mel receptors is in the range of 20–200 pM.[9] Mel receptors regulate several second messengers: cAMP, cGMP, diacylglycerol, inositol trisphosphate, arachidonic acid, and intracellular Ca^{2+} concentration ($[Ca^{2+}]_i$).[5] Nevertheless, some intracellular actions of Mel are independent of any receptor interaction, and may possibly be the consequence of its radical scavenger ability.[7]

It is known that Mel is able to exert a consistent effect on tumor metabolism; in particular, it seems to have an oncostatic action, and several studies have reported that Mel can control tumor growth and inhibit cells proliferation. Mel has been observed exerting strong effects on apoptosis: indeed, Mel has been assumed to exert proapoptotic effects,[10] though no evidence for this has ever been reported. Instead, the recent literature shows a clear antiapoptotic effect on many normal and tumor cells, both at physiological and pharmacological doses.[11] The confusion created by such opposite beliefs is further increased by the contradictory reports on its mechanism of action, which are still being debated as to whether the effects on apoptosis are receptor-mediated or instead depend on its radical scavenger ability. It is a very important target of pharmacologic research to understand Mel's effects at the cellular level in general, and on apoptosis in particular, to be able to exploit Mel's therapeutic potential.

MATERIALS AND METHODS

Cell Culture

For this study, we used the human promonocytic leukemia cell line U937.

Cells were grown in RPMI 1640 medium supplemented with 10% heat-inactivated fetal calf serum (FCS), L-glutamine, and antibiotics (penicillin, streptomycin). The cells were kept in a controlled atmosphere (5% CO_2) incubator at 37°C.

FCS Denaturation

FCS was incubated at 90°C for 1 h and then added to the RPMI 1640 medium.

Treatment with Mel

Mel was used at the final concentration of 1 mM, unless otherwise specified, and added to the culture medium 1 h before apoptosis induction.

Induction of Apoptosis

(1) **By nonoxidative agents:** Apoptosis was induced by the protein synthesis inhibitor puromycin (PMC, 10 µg/mL)
(2) **By physiological agents:** U937 cells were treated with anti-Fas, # 05–201 (FAS, 50 ng/mL).

Identification and Quantification of Apoptotic Cells

Nuclear morphology of control and treated cells was analyzed by fluorescence microscopy after staining with Hoechst 33342; apoptotic cells were characterized by nuclear condensation of chromatin and/or nuclear fragmentation. Apoptosis was evaluated among the Hoechst-stained cells by counting at least 300 cells in at least three randomly selected fields.

Estimation of Mel Concentration in Cell Extract

U937 cells were suspended at a concentration of 7×10^5 cell/mL in the presence of 0, 0.5, 1.0, and 2 mM Mel (final concentrations), in an RPMI medium containing or not FCS. Aliquots of 10×10^6 cells were collected after 4 h of incubation at 37 °C and washed two times with PBS, and suspended in 1 mL of Na phosphate buffer (0.04 M, pH 6.8). Five hundred microliters of Na tetraborate buffer (0.1 M, pH 9.0) and 1 mL of chloroform were added to the pellet and the suspension was vortexed for 3 min and centrifuged at 800 rpm for 5 min. The water phase was discarded, and the organic phase was mixed with 5 mL of borate buffer, vortexed for 3 min, and centrifuged at 800 rpm for 5 min. The water phase was discarded again and the remaining organic phase was dried under nitrogen and finally suspended in chloroform (100 μL). Thirty microliters were loaded onto silica gel 60 thin layer chromatographic (TLC) plates with fluorescent indicator (Merck Sharp & Dohme GMBH, Haar, Germany). A standard Mel solution (85 μM) was prepared and amounts of 70 and 210 μg were loaded onto the TLC plates to localize Mel and quantify its concentration in the cell extracts. This was carried out by UV irradiation of the plates, image acquisition through a CCD camera and by using the Gel Doc 1000/2000 gel documentation system (Quantity One software package for image analysis). The data are expressed in linear intensity units. Quantity One can quantify and analyze the spots through a light/radiation detector that converts signals from TLC samples into digital data. Quantity One displays the digital data on the computer screen in the form of gray-scale images.

RESULTS

Mel Reduces Apoptosis in U937 Cells

U937 human monocytic cells are known to possess MT1 Mel receptors,[12] thus possibly being an ideal system for investigating the effects of Mel on apoptosis.

First of all, we observed that Mel exerts no apoptogenic effects within 48 h of incubation and up to 1 mM doses (not shown). Next, we investigated Mel's effects on stress-induced apoptosis. FIGURE 1A shows that Mel is able to reduce PMC-induced apoptosis in a dose-dependent fashion.

Mel was discovered exerting antiapoptotic effects also on physiologically induced apoptosis, such as that induced by Fas, as shown in FIGURE 1B. This suggests that the Mel antiapoptotic effect involves both the intrinsic and the extrinsic pathways. Thus, from this first set of results, we can conclude that Mel is able to protect U937 cells from apoptosis.

Serum Proteins Sequester Mel in Cellular Culture Medium

We did not find significant protection for doses <200 μM. This seems incompatible with Mel's receptor-engaging function. However, the actual

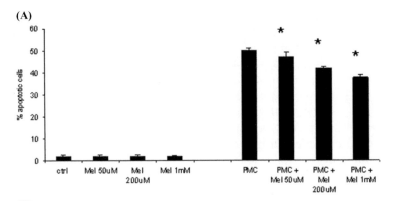
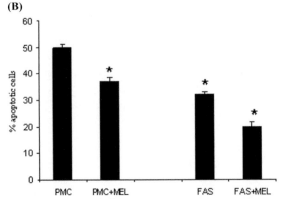

FIGURE 1. Mel antagonizes apoptosis. (**A**) Dose-effect of Mel. U937 cells were incubated in complete medium (presence of 10% FCS) with different doses of Mel for 24 h (*left*); or for 1 h, followed by puromycin (PMC, *right*); apoptosis was evaluated at 24 h of Mel (*left*) or 3 h of PMC (*right*). Results are the average of at least five independent experiments ±SD. Apoptosis reduction at 1 mM Mel is highly significant ($P < 0.05$). (**B**) Mel protects cells from apoptosis induced by both cell-damaging agents (*intrinsic pathway*) and physiological means (*extrinsic pathway*). Apoptosis was estimated at 3 h of PMC, and Fas overnight. Results are the average of at least three independent experiments ±SD. Apoptosis reduction by Fas is significant ($P < 0.05$).

dose able to reach cells may possibly be lower. Indeed, it is known that Mel sticks to protein present in plasma, thus possibly lowering concentration of active Mel. Thus, we performed the same dose–response experiment shown in FIGURE 1A in the absence of FCS: FIGURE 2A shows that indeed in this case, Mel is more effective in its antiapoptotic effect at all the corresponding doses tested.

The TLC analysis of intracellular Mel concentration after 4 h of incubation with the different Mel doses shows that FCS causes a decrease in cellular uptake of about 20-fold (FIG. 2C), thus suggesting that the actual Mel dose in

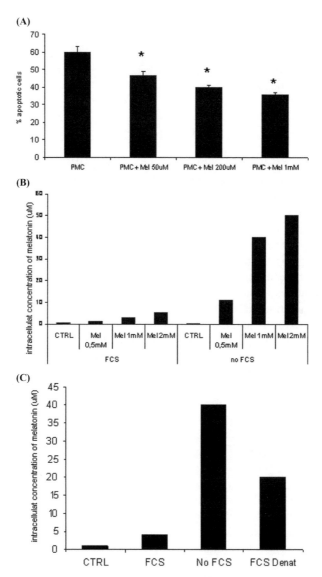

FIGURE 2. Serum proteins sequester Mel in cellular culture medium. (**A**) A concentration of 50 μM protects U937 cells from apoptosis in culture medium lacking in serum. Cells were treated with PMC ± the different doses of Mel in the absence of serum. Results are the average of at three independent experiments ±SD. Apoptosis reduction at all Mel doses is highly significant ($P < 0.05$). (**B**) Serum impairs Mel uptake. Intracellular concentrations of Mel were measured by TLC after 4 h of incubation and are the average of two independent experiments. (**C**) Denatured FCS gave an intracellular uptake that is about five-fold that observed in the presence of regular FCS. Intracellular concentrations of Mel were measured by TLC after 4 h of incubation and are the average of two independent experiments.

the experiments carried out under standard conditions (10% FCS) is much lower than the nominal one. The presence of partially denatured FCS gave intermediate results, with an intracellular uptake that is about five-fold that observed in the presence of regular FCS, suggesting that the ability of sequestering Mel is partially lost by denaturation.

DISCUSSION

With the goal to clarify the effects of Mel on apoptosis, we investigated both the possible proapoptotic or antiapoptotic effect in U937 human monocytic cells, known to possess MT1 Mel receptors,[12] by evaluating the extent of apoptosis on cells treated, respectively, with Mel alone, or Mel plus a direct apoptogenic agent.

Our results confirm that a Mel proapoptotic effect does not exist *per se*, even though such an effect has been reported in several reviews. However, Mel significantly reduces apoptosis induced by two different means: physiological and cell damaging. Thus, the antagonistic action of Mel toward apoptosis may occur via stimulation of two independent pathways, the physiological (or extrinsic) pathway and the stress-induced (or intrinsic) pathway. Thus, this study supports recent studies that have described an antiapoptotic role of Mel[13]; however, the mechanism involved remains to be understood, that is, whether Mel's antiapoptotic effect is generated by receptor stimulation or by Mel's radical scavenger ability.

In malignant cells, such as the human breast cancer cell line MCF-7, micromolar concentrations of Mel have been reported to increase the extent of induced apoptosis.[14]

The doses required for the antiapoptotic effect (≥ 100 μM) in U937 are apparently not compatible with receptor stimulation (receptor affinity <1 nM). We have shown here that this is due, at least in part, to a masking by serum proteins, present in culture medium in large concentration, which stick to Mel,[3] lowering the actual concentration available for cell entry. Thus, in the absence of FCS, Mel is more effective in its antiapoptotic effect at all the doses tested. The TLC analysis of intracellular Mel concentration shows that FCS causes a decrease in cellular uptake of about 20-fold, thus suggesting that the actual Mel dose in the experiments carried on in standard conditions (10% FCS) is much lower than the nominal one.

This is, however, not enough to explain the involvement of receptor engagement; thus, we cannot rule out other possible explanations, such as the role of Mel as an antioxidant in comparing apoptosis. Some groups indeed reported protection against programmed cell death in neuronal cells by the antioxidant activity of Mel.[15] Alternatively, protection may occur through receptor stimulation, but this might require an additional Mel effect in order to achieve the antiapoptotic effect. In conclusion, the antiapoptotic effect of Mel has been demonstrated. However, further investigations are required to elucidate the mechanism of this effect and to what extent receptor engagement and pro-oxidant activity of Mel are involved in its antiapoptotic effect.

REFERENCES

1. REITER, R.J. 1991. Pineal melatonin: cell biology of its synthesis and of its physiological interactions. Endocr. Rev. **12:** 151–180.
2. CARRILLO-VICO, A., J.R. CALVO, P. ABREU, et al. 2004. Evidence of melatonin synthesis by human lymphocytes and its physiological significance: possibile role as intracrine, autocrine, and/or paracrine substance. FASEB J. **18:** 537–539.
3. GILAD, E. & N. ZISAPEL. 1995. High-affinity binding of melatonin to hemoglobin. Biochem. Mol. Med. **56:** 115–120.
4. SAINZ, R.M., J.C. MAYO, C. RODRIGUEZ, et al. 2003. Melatonin and cell death: differential actions on apoptosis in normal and cancer cells. Cell Mol. Life Sci. **60:** 1407–1426.
5. VANEK, J. 1998. Cellular mechanism of melatonin action. Physiol. Rev. **78:** 687–721.
6. REITER, R.J., J.R. CALVO, M. KARBOWNIK, et al. 2000. Melatonin and its relation to the immune system and inflammation. Ann. N. Y. Acad. Sci. **917:** 376–386.
7. REITER, R.J., D.X. TAN, C. OSUNA, et al. 2000. Action of melatonin in the reduction of oxidative stress. J. Biomed. Sci. **7:** 444–458.
8. DUBOCOVICH, M.L. & M. MARKOWSKA. 2005. Functional MT1 and MT2 melatonin receptor in mammals. Endocrine **27:** 101–110.
9. DUBOCOVICH, M.L. 1991. Pharmacological characterization of melatonin binding sites. Adv. Pineal Res. **5:** 167–173.
10. WOLFLER, A., H.C. CALUBA, P.M. ABUJA, et al. 2001. Pro-oxidant activity of melatonin promotes fas-induced cell death in human leukemic Jurkat cells. FEBS Lett. **502:** 127–131.
11. SAINZ, R.M., J.C. MAYO, R.J. REITER, et al. 1999. Melatonin regulates glucocorticoid receptor: an answer to its antiapoptotic action in thymus. FASEB J. **13:** 1547–1556.
12. GARCIA-MAURINO, S., M.G. GONZALEZ-HABA, J.R. CALVO, et al. 1998. Involvement of nuclear binding sites for melatonin in the regulation of IL-2 and IL-6 production by human blood mononuclear cells. J. Immunol. **92:** 76–84.
13. JUKNAT, A.A., M. DEL VALLE, A. MENDEZ, et al. 2005. Melatonin prevents hydrogen peroxide-induced Bax expression in cultured rat astrocytes. J. Pineal Res. **38:** 84–92.
14. ECK, K.M., L. YUAN, L. DUFFY, et al. 1998. A sequential treatment regimen with melatonin and all-trans retinoic acid induces apoptosis in MCF-7 tumour cells. Br. J. Cancer **77:** 2129–2137.
15. POST, A., F. HOLSBOER & C. BEHL. 1998. Induction of NF-kappaB activity during haloperidol-induced oxidative toxicity in clonal hippocampal cells: suppression of NF-kappaB and neuroprotection by antioxidants. J. Neurosci. **18:** 8236–8246.

Prevention of p53 Degradation in Human MCF-7 Cells by Proteasome Inhibitors Does Not Mimic the Action of Roscovitine

CARMEN RANFTLER, MARIETA GUEORGUIEVA, AND JÒZEFA WĘSIERSKA-GĄDEK

Department of Medicine I, Division: Institute of Cancer Research, Medical University of Vienna, Vienna, Austria

ABSTRACT: We have recently observed activation of wild-type (wt) p53 protein in human MCF-7 breast cancer cells upon treatment with roscovitine (ROSC), a potent cyclin-dependent kinase inhibitor. It has been previously suggested that ROSC repressed transcription of Mdm-2, a negative p53 regulator, and that the lack of Mdm-2 contributes to the ROSC-induced upregulation of p53 protein. Therefore, we decided to see whether the prevention of p53 degradation by proteasome inhibitors will mimic the effects generated by ROSC. Exposure of human MCF-7 cells to different proteasome inhibitors resulted in a time-dependent increase of p53. However, unlike ROSC, they failed to modify p53 protein at Ser46 and to induce p53AIP1 protein. Moreover, whereas ROSC arrested MCF-7 cells in the G2-phase of the cell cycle, proteasome inhibitors blocked cells primarily in the S-phase, presumably because of the prevention of cyclin degradation. Our results indicate that prevention of p53 degradation by proteasome inhibitors does not mimic the action of ROSC.

KEYWORDS: p53 tumor suppressor; inhibitors of cyclin-dependent kinases; cell cycle regulation; G2 arrest; apoptosis; depolarization of mitochondria; CDK inhibitors; p53 phosphorylation; p53 stability

INTRODUCTION

Cells are permanently exposed to a variety of stress stimuli. Some of them come from outside the cell whereas the others, such as oxidative stress, are generated inside. Cells defend against the damage of the genome and generate cellular response to repair the DNA lesions and to replace the impaired proteins,

Address for correspondence: Jòzefa Wesierska-Gądek, Cell Cycle Regulation Group, Div. Institute of Cancer Research, Dept. of Medicine I, Medical University of Vienna, Borschkegasse 8 a, A-1090 Vienna, Austria. Voice: +43-1-4277-65247; fax: +43-1-4277-65194.
e-mail: Jozefa.Gadek-Wesierski@meduniwien.ac.at

thereby preventing development of malignancy. The major step in this defense strategy is the transient inhibition of the cell cycle or induction of apoptosis. Cell cycle arrest offers damaged cells a window of time necessary to repair the injury. If the cellular lesions are irreparable, severely damaged cells are eliminated by apoptosis.

The p53 protein, the product of a tumor-suppressor gene, is a key molecule in the signaling and cellular response to different types of stress conditions. The *p53* gene is found in all eukaryotic species except yeast. It consists in humans of approximately 20 kbp (393 codons, 11 exons), the p53 mRNA can be found in all tissues, and the translated protein possesses a molar mass of 53 kDa.[1]

The importance of the p53 gene in the maintenance of genomic stability is reflected by the fact that it is mutated in approximately 50% of human tumors. Moreover, the proper p53 function is inactivated not only by genetic alterations, but also through its nuclear exclusion or by the accelerated degradation mediated by action of virally encoded oncoproteins, such as E6 or HBV pX.[2]

In unstressed normal tissues wild-type (wt) p53 protein is maintained at very low levels on account of the activity of its antagonist, the Mdm-2 protein. The *MDM-2* (human *HDM-2*) oncogene encoding a protein of 491 amino acids[3] acts as a p53-specific E3 ubiquitin ligase, which covalently attaches ubiquitin molecules to the p53 protein, thereby targeting it for degradation in proteasomes. On the other hand, the p53 protein can bind to two different sites on the *MDM-2* gene and negatively regulate the expression of its own negative regulator.[4,5] Susceptibility of p53 for degradation is regulated by posttranslational modifications.[6,7] p53 protein can be acetylated at distinct lysine residues and in this way becomes protected from degradation: acetylation of p53 reduces the ubiquitination rate and its half-life increases.[8] Moreover, the stability and activity of wt p53 protein is additionally regulated by its site-specific phosphorylation. Wt p53 can be phosphorylated at multiple sites. Serine residues at the positions 6, 9, 15, 20, 33, 37, 46, 315, 371, 376, 378, and 392 and threonine residues 18, 55, and 81 have been described to be phosphorylated.[7] The modification of the latter is essential for the regulation of Mdm-2 binding to the amino-terminal domain of p53.[6,7]

After exposure to stress stimuli, p53 protein is induced.[9,10] p53 protein becomes stable, it adopts an active conformation and is localized in the nucleus.[11] Moreover, p53 protein becomes active as the transcription factor,[11] which, depending on the kind of stress stimulus, is able to positively or negatively regulate its target genes.[12,13] Apart from Mdm-2, p53 stimulates the transcription of a variety of target genes like p21WAF1, GADD45, or Bax family members.[12] Interestingly, p53 is sufficient to induce p21^{WAF1} and Mdm-2 without the help of other coactivators.[12] Through the control of the expression of cell cycle regulators, p53 is able to restrict cellular growth by inducing transient cell cycle arrest, terminal senescence, or apoptosis.[12]

Interestingly, there are some additional mechanisms regulating the stability of p53 protein. Another important mechanism to prevent degradation of p53 protein is its interaction with the nuclear enzyme poly(ADP-ribose)polymerase-1 (PARP-1). It has been shown that PARP-1 forms complexes with the carboxy-terminal domain of p53 protein and masks its nuclear export signals, thereby preventing its export and finally degradation in proteasomes.[14] It seems that the phosphorylation of the carboxy-terminal domain of p53 protein is involved in the interaction because dephosphorylated p53 protein failed to bind to PARP-1.[15]

Recently, the upregulation of cellular p53 levels was observed after cell treatment with different cyclin-dependent kinase (CDK) inhibitors: flavopiridol, 5,6-dichloro-1β-D-ribofuranosylbeximidazole (DRB) and roscovitine (ROSC).[16]

The mechanism of CDK inhibitor-mediated p53 stabilization is so far not solved. It has been suggested that they downregulate Mdm-2 levels and that p53 is accumulated on account of the absence of its antagonist. However, the overexpression of Mdm-2 did not prevent the elevation of p53 levels following DRB treatment,[16] which implies that upon ROSC treatment, p53 protein becomes resistant to the Mdm-2-mediated degradation.

For this reason we have decided to explore whether the inhibition of proteasomal activity alone is sufficient for accumulation and site-specific phosphorylation (at serine 46) of p53 in MCF-7 breast cancer cells. Exposure of human MCF-7 cells to different proteasome inhibitors resulted in the elevation of p53. Only two of three tested inhibitors prevented p53 degradation: MG-132 and MG-262. Both agents increased p53 levels in a time-dependent manner. However, unlike ROSC, they failed to modify p53 protein at Ser46. Moreover, whereas ROSC arrested MCF-7 cells in the G2-phase of the cell cycle, proteasome inhibitors blocked cells primarily in the S-phase and after longer treatment in G2. Our results indicate that prevention of p53 degradation by proteasome inhibitors does not mimic the action of ROSC.

MATERIALS AND METHODS

Cells

Human MCF-7 breast carcinoma cells were grown as a monolayer in Dulbecco's medium without phenol red supplemented with 10% FCS at 37°C in an atmosphere of 8% CO_2. Cells were grown up to 60–70% confluence and then treated with ROSC at a final concentration of 20 μM or with proteasome inhibitors MG-132, MG-262, and omuralide at a final concentration of 1–10 μM for the indicated periods of time. Drugs were dissolved as a stock solution in dimethylsulfoxide (DMSO) and stored at −20°C until use.

Antibodies

We used the following antibodies: monoclonal anti-p53 antibody DO-1, a kind gift from Dr. B. Vojtesek (Masaryk Memorial Cancer Institute, Brno) and monoclonal anti-PARP-1 antibodies (C-2-10) from Dr. G. Poirier (Laval University, Quebec). Polyclonal anti-phospho-Ser46-p53 were from New England Biolabs (Beverly, MA, USA). Monoclonal anti-actin (clone C4) antibodies were from ICN Biochemicals (Aurora, OH, USA). Appropriate secondary antibodies linked to horseradish peroxidase (HRP) were from Amersham International (Little Chalfont, Buckinghamshire, UK).

Measurement of DNA of Single Cells by Flow Cytometry

The measurement of DNA content was performed by flow cytometric analysis based on a slightly modified method[17] described previously by Vindelov et al.[18] The cells were detached from substratum by trypsinization; then all cells were harvested by centrifugation and washed in phosphate-buffered saline (PBS). Aliquots of 1×10^6 cells were used for further analysis. Cells were stained with propidium iodide as described previously[17] and then the fluorescence was measured using the Becton Dickinson FACScan after at least 2-h incubation at $+4°C$ in the dark.

Electrophoretic Separation of Proteins and Immunblotting

Total cellular proteins dissolved in SDS sample buffer were separated on 10% or 15% SDS slab gels, transferred electrophoretically onto polyvinylidene difluoride membrane (PVDF) (Amersham International) and immunoblotted as previously described.[19,20] Equal protein loading was confirmed by Ponceau S staining. To determine the phosphorylation status of selected proteins, antibodies recognizing site-specific phosphorylated proteins were diluted to a final concentration of 1:1,000 in 1% BSA in Tris-saline-Tween-20 (TST) buffer.[21] In some cases, blots were used for sequential incubations.

RESULTS

ROSC Inhibits Cell Cycle Progression in MCF-7 Cells at G2

ROSC, a very potent inhibitor of cycle-dependent kinases, blocks very efficiently proliferation of human MCF-7 breast cancer cells. ROSC targeting the activity of CDK2 results in the accumulation of G2-arrested cells in a time- and concentration-dependent manner. As depicted in FIGURE 1, exposure of exponentially growing MCF-7 cells to 20 μM ROSC markedly increased the ratio

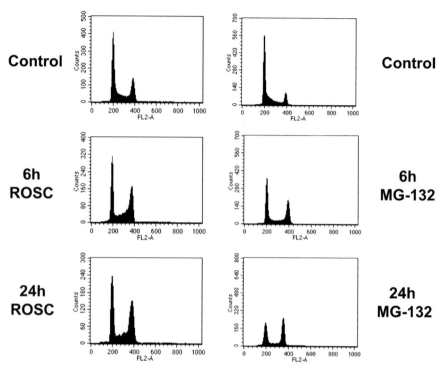

FIGURE 1. ROSC arrests MCF-7 cells in G2 phase of the cell cycle. Exponentially growing MCF-7 cells were exposed to 20 μM ROSC or to 10 μM MG-132 for the indicated periods of time, harvested, and stained with propidium iodide. DNA histograms depicting a representative experiment performed in duplicate were prepared using the ModFIT software.

of the G2 cells. After 24 h 30% of the cells were in G2 phase. Concomitantly, the number of S-phase cells significantly decreased. The G2 arrest coincided with the inhibition of site-specific phosphorylation of CDK2 (not shown).

ROSC Upregulates Wt p53 Protein

Monitoring of the cellular levels of p53 revealed that the exposure of MCF-7 cells to ROSC strongly increases its cellular concentration. As shown in FIGURE 2, p53 protein, barely detectable in control cells, was highly upregulated after ROSC treatment. The marked increase of p53 was observed already after 6 h of ROSC treatment. Interestingly, the extent of p53 response depended on the proliferative status of the cells prior to the onset of treatment. The strongest upregulation was observed after the exposure of exponentially growing cells to the drugs. The confluent cells were less susceptible to ROSC treatment and p53 protein accumulated at lower concentrations (not shown).

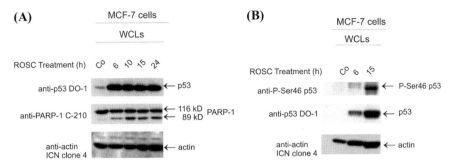

FIGURE 2. Induction of cellular p53 protein in ROSC-treated MCF-7 cells. Whole-cell extracts (WCLs) prepared from human MCF-7 control cells and cells exposed to 20 μM ROSC (30 μg/lane) were separated on a 10% SDS gel and transferred onto a PVDF membrane. The blots were incubated with the indicated primary antibodies in the appropriate concentration. After incubation with secondary antibodies linked to HRP, immune complexes were detected by chemiluminescence using the ECL+ detection system.

ROSC-induced p53 protein was transcriptionally active. As previously shown, it was reflected by the upregulation of a number of its target genes, such as *p21waf1* or *MDM-2*. Moreover, ROSC induced specific phosphorylation of p53 at serine in the position 46 (FIG. 2B). As previously observed, P-Ser-46 already appeared after 4-h ROSC treatment.

Proteasome Inhibitors Differentially Affect the Steady-State of Wt p53 Protein in MCF-7 Cells

Since the intracellular level of wt p53 is primarily regulated by its degradation, we decided to examine the effect of the inhibition of proteasome activity on p53 concentration. We used three distinct proteasome inhibitors MG-132, MG-262, and omuralide. MCF-7 cells were exposed to the inhibitors for increasing periods of time. Surprisingly, only the two inhibitors MG-132 and MG-262 increased cellular levels of p53 protein (FIG. 3). The strongest p53 signal was detected after the inhibition of proteasome activity for 21 h. However, the third agent, omuralide failed to positively affect the steady-state of p53 protein at the same dose. These observations show that the general inhibition of proteasome activity is not sufficient to prevent degradation of p53 protein and indicate that the ability to promote the p53 accumulation depends on the fine specificity of proteasome inhibitors.

Inhibition of Proteasome Activity in MCF-7 Cells Results in Accumulation of p53 That Is Unphosphorylated at Serine 46

In the next step we examined the effect of the proteasome inhibitors on the site-specific phosphorylation of p53 protein. Unlike ROSC, proteasome

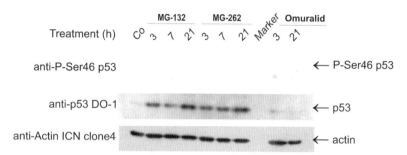

FIGURE 3. Effect of the inhibition of proteasome activity on the cellular levels of p53 tumor-suppressor protein. Human MCF-7 cells were exposed to three proteasome inhibitors for 3, 7, and 21 h. MG-132 and omuralid were added to a final concentration of 5 µM, MG-262 was used at a final concentration of 1 µM. WCLs were loaded on a 10% SDS gel. Conditions of electrophoresis and immunoblotting as described in FIGURE 2.

inhibitors did not induce p53 phosphorylation at serine 46 (FIG. 3). This result substantiates our previous observation that ROSC-induced phosphorylation at serine 46 is a specific process.

Effect of MG-132 on the Cell Cycle Progression

Since MG-132 prevented most efficaciously the degradation of p53 protein in human MCF-7 cells, we used this compound to examine the effect of the inhibition of proteasome activity on the cell cycle. Exposure of MCF-7 cells to MG-132 resulted in the increase of the G2 population (FIG. 4). After treatment for 6 h an accumulation of S-phase cells occurred. The continuous exposure of MCF-7 cells to MG-132 for 24 h additionally led to the accumulation of G2-arrested cells. At 24 h the ratio of G1 cells was essentially reduced. The comparison of the action of MG-132 and ROSC shows a substantial difference.

DISCUSSION

Recently, new anti-cancer drugs inhibiting the activity of CDKs were developed. Some of them target not only the activity of CDKs but additionally affect the expression and functional status of wt p53 protein. The upregulation of cellular p53 levels was observed after cell treatment with flavopiridol, DRB,[16] and ROSC.[22–25] Recently, the new action of ROSC on wt p53 protein was reported.[25] We have observed a strong stabilization of p53 protein in human MCF-7 breast cancer cells exposed to ROSC.[23–25] ROSC phosphorylated p53 protein at serine 46 in a time- and dose-dependent manner, thereby extending its stability.[23–25] The CDK-inhibitor prolonged the half-life of the p53 protein about 40-fold and arrested the cells in G2/M-phase of the cell

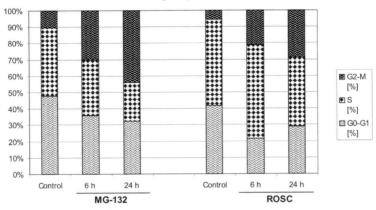

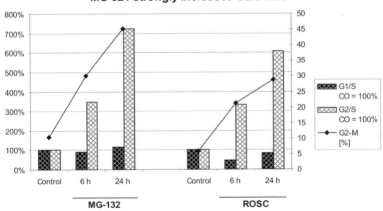

FIGURE 4. MG-132 inhibits cell cycle progression of MCF-7 cells. Cells were exposed to 20 μM ROSC or 10 μM MG-132 for indicated periods of time, harvested, and stained with propidium iodide. Comparison of the drug effects on the distribution of cells in distinct cell cycle phases. Upper panel: values of cells in distinct cell cycle phases. Lower panel: *Bars*: G1/S and G2/S ratios. Values calculated for each sample were normalized to the control; *curve*: the percentage of cells arrested in G2.

cycle.[23–25] Moreover, after longer treatment ROSC induced apoptosis. The induction of apoptosis was closely related to the site-specific phosphorylation of p53 protein. P-Ser-46 p53 protein transcriptionally upregulated its downstream target, p53AIP1 protein.[25] Newly translated p53AIP1 protein entered after a few hours the mitochondrion, its proper compartment. The *de novo* synthesis of mitochondrial p53AIP1 protein preceded the depolarization of mitochondrial membrane and the release of mitochondrial proteins, such as apoptosis-inducing factor (AIF) or cytochrome c.[25]

The mechanism of ROSC-mediated p53 stabilization is so far not solved. It has been suggested that ROSC downregulates Mdm-2 levels and p53 is accumulated because of the absence of its antagonist. However, more recent data indicate that Mdm-2 is not the major player in the ROSC-mediated upregulation of p53 protein. First, we have observed an increase of Mdm-2 protein in cells exposed to ROSC. The increased p53 concentrations were observed despite the Mdm-2 elevation.[25] Secondly, the overexpression of Mdm-2 did not prevent the elevation of p53 following DRB treatment[16] implicating that upon treatment p53 protein becomes resistant to the Mdm-2-mediated degradation. Data shown in this article substantiate this assumption. We observed that the general inhibition of proteasome activity is not sufficient to prevent p53 degradation. Only two of three used inhibitors were able to induce accumulation of cellular p53 protein. Moreover, prevention of the p53 degradation did not induce its site-specific phosphorylation. The fact that only two of three used proteasome inhibitors prevented p53 degradation in MCF-7 cells indicates that they might differ in the specificity and/or capacity to inactivate proteases responsible for p53 destruction. Indeed, it has been previously reported that the efficiency of the action of proteasome inhibitors to prevent p53 degradation strongly varies and additionally depends on the cellular context.[19,20,26] In mouse fibroblasts lacking poly(ADP-ribose) polymerase-1, p53 accumulation was detected after action of PSI, whereas in their normal counterparts the most effective inhibition of proteasome activity was observed after treatment with lactacystin.[20] On the other hand, MG-132 was most efficient in human cervical carcinoma HeLa cells.[19] Unlike lactacystin, PSI or PSII, MG-132 resulted in the accumulation of p53 protein after short treatment despite the presence of human papilloma virus–encoded E6 oncoprotein.[19]

Thus, these results uncoupled the prevention of p53 degradation from the ROSC-mediated p53 induction. The latter is an active process. It seems that ROSC is able to stimulate an unknown cellular protein kinase, which in turn modifies p53 protein.

ACKNOWLEDGMENT

The grant for the study was sponsored by Jubiläumsfonds from Oesterreichische Nationalbank (Grant No. 10364).

REFERENCES

1. LEVINE, A.J. 1992. The p53 tumour suppressor gene and product. Cancer Surv. **12:** 59–79.

2. MOLL, U.M., G. RIOU & A.J. LEVINE. 1992. Two distinct mechanisms alter p53 in breast cancer: mutation and nuclear exclusion. Proc. Natl. Acad. Sci. **89:** 7262–7266.
3. MOMAND, J., G.P. ZAMBETTI, D.C. OLSON, et al. 1992. The mdm-2 oncogene product forms a complex with the p53 protein and inhibits p53-mediated transactivation. Cell **69:** 1237–1245.
4. OREN, M. 2003. Decision making by p53: life, death and cancer. Cell Death Differ. **10:** 431–442.
5. KUBBUTAT, M.H.G., S.N. JONES & K.H. VOUSDEN. 1997. Regulation of p53 stability by Mdm2. Nature **387:** 299–303.
6. JIMENEZ, G.S., S.H. KHAN, J.M. STOMMEL & G.M. WAHL. 1999. p53 regulation by post-translational modification and nuclear retention in response to diverse stresses. Oncogene **18:** 7656–7665.
7. APPELLA, E. & C.W. ANDERSON. 2001. Post-translational modifications and activation of p53 by genotoxic stresses. Eur. J. Biochem. **268:**2764–2772.
8. LI, M., C. LUO, C.L. BROOKS & W. GU. 2002. Acetylation of p53 inhibits ubiquitination by Mdm2. J. Biol. Chem. **277:** 50607–50611.
9. WESIERSKA-GADEK, J., J. WOJCIECHOWSKI, C. RANFTLER & G. SCHMID. 2005. Role of p53 suppressor in ageing: regulation of transient cell cycle arrest and terminal senescence. J. Physiol. Pharmacol. **56:** 15–28.
10. WESIERSKA-GADEK, J. & M. HORKY. 2003. How the nucleolar sequestration of p53 protein or its interplayers contributes to its (re)-activation. Ann. N. Y. Acad. Sci. **1010:** 266–272.
11. LJUNGMAN, M. 2000. Dial 9-1-1 for p53: mechanisms of p53 activation by cellular stress. Neoplasia **3:** 208–225.
12. EL-DEIRY, W.S., T. TOKINO, V.E. VELCULESCU, et al. 1993. WAF1, a potential mediator of p53 suppression. Cell **75:** 817–825.
13. HAUPT, S., M. BERGER, Z. GOLDBERG & Y. HAUPT. 2003. Apoptosis–the p53 network. J. Cell Sci. **116:** 4077–4085
14. WESIERSKA-GADEK, J., J. WOJCIECHOWSKI & G. SCHMID. 2003. Central and carboxy-terminal regions of human p53 protein are essential for interaction and complex formation with PARP-1. J. Cell Biochem. **89:** 220–232.
15. WESIERSKA-GADEK, J., J. WOJCIECHOWSKI & G. SCHMID. 2003. Phosphorylation regulates the interaction and complex formation between wt p53 protein and PARP-1. J. Cell Biochem. **89:** 1260–1284.
16. LJUNGMAN, M., F. ZHANG, F. CHEN, et al. 1999. Inhibition of RNA polymerase II as a trigger for the p53 response. Oncogene **18:** 583–592.
17. WESIERSKA-GADEK, J. & G. SCHMID. 2000. Overexpressed poly(ADP-ribose) polymerase delays the release of rat cells from p53-mediated G1 checkpoint. J. Cell Biochem. **80:** 85–103.
18. VINDELOV, L.L., I.J. CHRISTENSEN & N.J. NISSEN. 1983. A detergent-trypsin method for the preparation of nuclei for flow cytometric DNA analysis. Cytometry **3:** 323–327.
19. WESIERSKA-GADEK, J., D. SCHLOFFER, V. KOTALA & M. HORKY. 2002. Escape of p53 protein from E6-mediated degradation in HeLa cells after cisplatin therapy. Int. J. Cancer **101:** 128–136.
20. WESIERSKA-GADEK, J., E. BOHRN, Z. HERCEG, et al. 2000. Differential susceptibility of normal and PARP knock-out mouse fibroblasts to proteasome inhibitors. J. Cell Biochem. **78:** 681–696.

21. WESIERSKA-GADEK, J., D. SCHLOFFER, M. GUEORGUIEVA, *et al.* 2004. Increased susceptibility of poly(ADP-ribose) polymerase-1 knockout cells to antitumor triazoloacridone C-1305 is associated with permanent G2 cell cycle arrest. Cancer Res. **64:** 4487–4497.
22. WESIERSKA-GADEK, J. & G. SCHMID. 2006. Dual action of the inhibitors of cyclin-dependent kinases: targeting of the cell cycle progression and activation of wild-type p53 protein. Expert Opin. Investi. Drugs **15:** 23–38.
23. WESIERSKA-GADEK, J., M. GUEORGUIEVA, J. WOJCIECHOWSKI & M. HORKY. 2004. Cell cycle arrest induced in human breast cancer cells by cyclin-dependent kinase inhibitors: a comparison of the effects exerted by roscovitine and olomoucine. Pol. J. Pharmacol. **56:** 635–641.
24. WOJCIECHOWSKI, J., M. HORKY, M. GUEORGUIEVA & J. WESIERSKA-GADEK. 2003. Rapid onset of nucleolar disintegration preceding the cell cycle arrest in roscovitine-induced apoptosis of human MCF-7 breast cancer cells. Int. J. Cancer **106:** 486–495.
25. WESIERSKA-GADEK, J., M. GUEORGUIEVA & M. HORKY. 2005. Roscovitine-induced up-regulation of p53AIP1 protein precedes the onset of apoptosis in human MCF-7 breast cancer cells. Mol. Cancer Ther. **4:** 113–124.
26. CHEN, F., D. CHANG, M. GOB, *et al.* 2000. Role of p53 in cell cycle regulation and apoptosis following exposure to proteasome inhibitors. Cell Growth Differ. **11:** 239–246.

Role of ATP in Trauma-Associated Cytokine Release and Apoptosis by P2X7 Ion Channel Stimulation

E. MARION SCHNEIDER,[a] KATRIN VORLAENDER,[a] XUELING MA,[a] WEIDONG DU,[a] AND MANFRED WEISS[b]

[a]*Section of Experimental Anesthesiology, University Clinic Ulm, 89075 Ulm, Germany*

[b]*Department of Clinical Anesthesiology, University Clinic Ulm, 89075 Ulm, Germany*

ABSTRACT: Trauma causes immediate cytokine release and the systemic inflammatory response syndrome (SIRS), often preceding sepsis and septic shock. Mechanisms may involve P2X7 ion channel activation via adenosine 5′-triphosphate (ATP) released from surrounding tissue and platelets. A number of single nucleotide polymorphisms (SNPs) influence the nature and magnitude of P2X7-stimulated cytokine release and apoptosis. In whole blood and isolated mononuclear blood cells (PBMCs) of donors with wild-type and heterozygous mutated genotypes, we found downregulated IL-8 and caspase-3 activation but no reproducible effect on tumor necrosis factor (TNF)-α and IL-1β release. IL-8 and caspase-3 activation were both influenced by paxilline, an inhibitor of calcium-activated potassium channels. Confocal laser scanning microscopy demonstrated that calcium signaling is affected by paxilline as well. We propose that blockade of potassium channels may be relevant to attenuate ATP-induced cytokine responses and apoptosis. The presence of functional SNPs in heterozygous genotypes appears to play a role.

KEYWORDS: ATP; P2X7 ion channel; inflammatory cytokines; caspase-3; calcium-activated potassium channel

INTRODUCTION

Trauma and massive tissue damage contribute to immediate cytokine release and systemic inflammatory response syndrome (SIRS), which may lead to life-threatening sepsis and septic shock. The enormous consequences on innate and adaptive immunity based on the currently accepted danger theory is widely

Address for correspondence: E. Marion Schneider, Ph.D., Sektion Experimentelle Anaesthesiologie, Universitaetsklinikum Ulm, Steinhoevelstrasse 9, 89075 Ulm, Germany. Voice: 0049-0-731-500-60080; fax: 0049-0-731-500-60080.
e-mail: marion.schneider@uni-ulm.de

Ann. N.Y. Acad. Sci. 1090: 245–252 (2006). © 2006 New York Academy of Sciences.
doi: 10.1196/annals.1378.027

accepted.[1] Knowledge about mechanisms explaining hyperinflammation in the absence of infectious complications would improve preventive therapeutic options. We asked whether P2X7 ion channel activation might be involved. This context has been recently discussed for neurological injury.[2]

The P2X7 receptor is ubiquitously expressed and belongs to a family of ligand-gated channels activated by extracellular adenosine 5′-triphosphate (ATP).[3] Binding of extracellular ATP activates multimerization of P2X7 subunits into trimers and hexamers, allowing transmembrane passage of cations and small molecules. Short stimulation with extracellular ATP upregulates cytokine release. Prolonged or repeated exposure to ATP induces the formation of a cytolytic pore and triggers cell death, and a cellular ATPase may accommodate the ATP effects through P2X7 in the local microenvironment.[3] An A to C polymorphism at position 1513 (1513A>C) of the Glu-496 to Ala residue in the intracellular C-terminal tail has been described. Monocytes and lymphocytes from subjects carrying the homozygous substitution 1513A_C showed nonfunctional P2X7, whereas heterozygous individuals had a lower response.[4] H155Y (489C>T) acts as a gain-of-function polymorphism of the P2X7.[5] Significant calcium flux increase was observed in lymphocytes from patients with chronic lymphocytic (CLL) bearing the 489C/T and 489T/T genotypes in association with the 1513A/A genotype. Cells carrying the R307Q at the ATP binding site have reduced or absent P2X7 function due to the failure of ATP binding to the extracellular domain of P2X7.[6] Also a subject who was double heterozygous for R307Q/I568N failed to show any functional P2X7 in all mononuclear cell types including macrophages (loss of function).[6] The functional analysis of the Q460R (1405A>G) conducted both in lymphocytes from CLL patients and in P2X7 transfected HEK293 cells, 1405A>G, indicates a neutral effect on receptor activation.[5] Ile568 to Asn (1729T>A) lies in a trafficking motif of the carboxyl terminus and prevents normal trafficking and surface expression of this receptor. P2X7 is nearly absent in monocytes of heterozygous individuals.[5] Physiologically, the activation of P2X7 receptors via ATP is coupled with the activation of maxi potassium channels, which are responsible for the release of tumor necrosis factor (TNF)-α and IL-1β.[7]

METHODS

P2X7 ion channel stimulation was performed by using Ficoll-separated blood cells. Mononuclear cells were plated with 2×10^5 cells/mL and ATP stimulation was performed with 1 and 5 mM ATP for 6 h and 24 h. Cytokines released were determined by a chemiluminescent assay (Immulite, DPC-Biermann, available at www.dpc-biermann.de). P2X7 single nucleotide polymorphisms (SNPs) were determined by Pyrosequencing (Biotage, Uppsala, Sweden).[8] Ca^{++} signaling was studied by Fluo-4-assisted confocal laser scanning microscopy using a Leica-TCS NT microscope (Leica, Jena, Germany) and argon/krypton laser using semiconfluent HeLa as well as microvascular endothelial cells and adherent macrophages. In detail, cell layers were labeled with 2 μM of Fluo4-AM (Invitrogen, available at www.invitrogen.com) and stimulated with 20 μL of a 1-mM ATP solution. To block ATP-specific channel activation, the cell layer was preincubated with 1.5-μM oxidized ATP before ATP stimulation. To inhibit P2X7-associated maxi potassium channels, cells were preincubated with 10-μM paxilline (Sigma-Aldrich, available at www.Sigma-Aldrich.com) prior to stimulation.

RESULTS

For functional assays, we studied the healthy donors 5163 and 5164, who have been genotyped as shown in TABLE 1.

When we studied the cytokine release in these two donors in whole-blood stimulation assays, we found that the donor 5163 had a lower level of inflammatory IL-8 and that differences in IL-1β and TNF-α were less pronounced (FIG. 1A). Donor 5163 was heterozygous at positions P2X7_1513 and P2X7_1729, both of those SNPs which attenuate P2X7 activation and expression levels. In addition, we found that active caspase-3, indicating apoptosis, was lower in the heterozygous donor than in the wild-type (5164) (FIG. 1B). Bcl-2 contents were antagonistic to activated caspase-3 (FIG. 1B), and Fas-L was apparently independent of ATP stimulation as evaluated in a larger experimental study in addition to the analyses shown here (data not shown).

One prominent effect by P2X7 stimulation via ATP is the activation of protein kinase C and generation of diacyl-glycerol and phosphatidyl-inositol-3-phosphate, which is then translocated to the endoplasmic reticulum and induces the calcium release from intracellular stores. This initial calcium response is responsible for the activation of calcium-dependent potassium channels and additional calcium influx from the extracellular space. Such potassium megachannels may be responsible for a number of physiological effects described for P2X7.[7] We asked whether blockade of the potassium channel by paxilline would influence either cytokine release or apoptosis.

FIGURE 1C shows results on cytokine release by paxilline inhibition of the heterozygous donor 5163: there was no effect on the release of TNF-α, IL-1β, and IL-8. However, paxilline significantly downmodulated caspase-3 activation with 1mM ATP stimulation, but caspase-3 activation was not blocked

TABLE 1. Genotypes of healthy donors 5163 and 5164

SNP DNA Amino acid	489C>T H155Y	946G>A R307Q	1405A>G Q460R	1513A>C E496A	1729T>A I568N
5163	C/T	G/G	A/A	A/C	T/A
5164	C/T	G/G	A/A	A/A	T/T

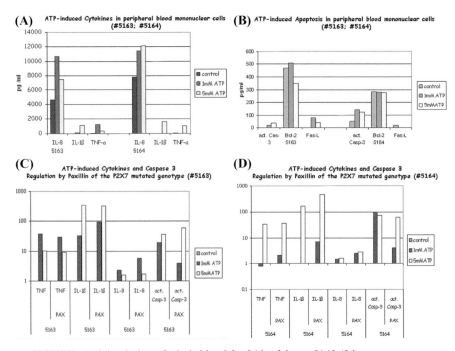

FIGURE 1. Stimulation of whole blood for 24 h of donor 5163 (2 heterozygous mutations in P2X7_1513 and P2X7_1729) and donor 5164 (P2X7 wild type) resulted in increased release of inflammatory cytokines (**A**) as well as caspase-3 (**B**). In isolated PBMC of the same donors, stimulated for 24 h in suspension culture, potassium megachannel inhibitor paxilline (10 μM) caused no alterations in ATP-induced cytokine release (gray columns 1 mM ATP; light columns 5 mM ATP) but significantly inhibited in caspase-3 activation with 1 mM ATP stimulation. No effect occurred at 5 mM ATP (**C**). Identical results were found when stimulating the P2X7 wild-type donor PBMCs, (donor 5164) (**D**).

at higher concentrations of ATP (5 mM). FIGURE 1D shows corresponding data obtained with the wild-type donor 5164. Again, TNF-α, IL-1β and IL-8 are not affected by paxilline. Downmodulation of caspase-3 activation is apparent at 1 mM ATP stimulation but no effect has been observed at 5 mM ATP.

The magnitude of stimulating P2X7 can be measured by quantifying the cytoplasmic calcium release. Calcium-sensitive dyes such as Fluo-4 are excellent tools to be used in microscopic techniques. An experiment of this kind is shown in FIGURE 2. Following stimulation with 5-mM ATP, Fluo-4-labeled HeLa monolayers respond by a calcium release signal within the first 5–15 min after stimulation. Blockade of the ATP signaling occurs by preincubation with oxidized ATP, which abrogates calcium release. Apoptosis induced by ATP stimulation is typically associated with high amounts of calcium translocated to the mitochondrial compartment. Interestingly, preincubation of

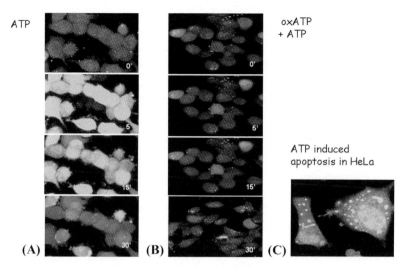

FIGURE 2. HeLa monolayers were labeled with the fluorescent calcium sensor 2 mM Fluo-4AM for 20 min and washed, and calcium release was induced by exogenous ATP (1 mM) (**A**) in HeLa cells and blocked by preincubation with oxidized ATP (1.5 mM oxATP) (**B**); ATP-induced apoptosis occurs with high amounts of calcium accumulating in mitochondria (**C**). Microscopic examination was done by confocal laser scanning microscopy using an Argon/Krypton laser and the Leica TCS-NT software.

Fluo-4-labeled cells with paxilline impairs the calcium release signal. The results were similar to the effect by oxidized ATP, but with a higher background fluorescent staining (data not shown).

DISCUSSION

The current investigation was aimed at the functional analysis of P2X7 receptors to understand the potential effect of ATP released from injured tissue during major trauma.[9,10] Ion channel activation leads to an immediate cytokine release and might thus explain a cytokine-triggered SIRS. Very high ATP ligand concentrations also lead to pore formation and apoptosis.[11,12] Known polymorphisms of the P2X7 ion channel may affect the ATP stimulating process. A SNP (489C>T), leading to an amino acid exchange H155Y, acts as a gain-of-function polymorphism of the P2X7. Significant calcium flux increase was observed in lymphocytes from CLL patients bearing the 489C/T (heterozygous) and 489T/T (homozygous mutated) genotypes in association with another function inactivating SNPs in the 1513A/A genotype (see below). Cells carrying the R307Q corresponding to the 946G>A genotype are affected because this polymorphism influences the ATP binding to the extracellular domain of P2X7. The A1513C polymorphism encodes an E496A change, resulting in defective pore formation,[13] and the T1729A variant causes an I568N substitution that disrupts normal receptor trafficking by preventing surface expression of the receptor.[14] Subjects who are double heterozygous for R307Q/I568N, fail to show any functional P2X7 in all mononuclear cell types including macrophages. This polymorphism constitutes a loss-of-function genotype. The current study presents results of whole blood and isolated blood cells of an individual with two heterozygous P2X7 mutations at positions 1513 and 1729, as compared with blood and blood cells of a wild-type genotype. PBMCs of the two donors exemplified here produce cytokines upon ATP stimulation in a range similar to that of more than 20 donors tested independently of genotyping results (data not shown). Activation of apoptosis by P2X7 has been described previously in the context of macrophages and intracellular killing of infectious agents.[15] Here, apoptosis was measured by activated caspase-3 quantity. The heterozygous responder 5163 activated apoptosis less than the wild-type responder 5164 (FIG. 1C). Bcl-2 levels were unaltered upon ATP stimulation. The modulation of cytokine release and caspase-3 stimulation was further tested in the presence and absence of paxilline, known to affect calcium-dependent potassium channel activation. In contrast to whole blood, donor 5163, being heterozygous for the A1513 (E496A) and the T1729A (I568N) SNPs, displayed a remarkable release of TNF-α and IL-1β with the lower ATP concentration (1 mM), whereas this was not observed in the wild-type. We therefore hypothesize that other polymorphisms or kinetics may affect TNF-α and IL-1 release following ATP stimulation. Specifically, the release of IL-1β is due to the activation of interleukin-1 converting enzyme (ICE)[7] and IL-1β further upregulates TNF-α. In contrast to the cytokine pattern, which is fairly inconsistent regarding P2X7 functional SNPs, the induction of apoptosis is clearly attenuated in the heterozygous individual. Interestingly, the activation of caspase-3 by ATP is susceptible to inhibition of potassium megachannels. The effect occurs by interference with the secondary calcium signal, as demonstrated by calcium flux analysis and confocal laser scanning technology. The primary P2X7 signal involves protein kinase C activation[16] and calcium release from intracellular stores. In addition to activation of the p38 and c-Jun N-terminal kinase (JNK) pathway, which may be responsible for a secondary increase of TNF-α (see above),[17] the calcium-dependent activation of potassium channels may be most important for the induction of apoptosis. The heterozygous P2X7 genotype at position 1513 and 1729 appears to affect apoptosis induction more than cytokine release. The involvement of potassium channels in the regulation of apoptosis has been demonstrated in a number of other cell systems, such as β-amyloid induction of apoptosis,[18] as well as cis-platin-induced apoptosis of tumor cells. A more detailed analysis of P2X7 polymorphisms to guide inflammation, cytokine release, and apoptosis may extend therapeutic options to influence SIRS as well as apoptosis in personalized medicine.

CONCLUSIONS

ATP stimulation of the P2X7 receptor induces the release of inflammatory cytokines in whole blood and in isolated PBMCs. In addition to calcium signaling induced by P2X7 we proved the involvement of calcium-dependent potassium channels to mediate effects induced by ATP stimulation. When testing the blockade of potassium channels by paxilline, we found that the ATP-induced activation of caspase-3 is inhibited, but that cytokine release is not influenced.

ATP stimulation remains an important mechanism for the manifestation of SIRS, which conditions hyperinflammation and immune deviation eventually leading to sepsis and septic shock. We provide evidence that the magnitude of cytokines released is not solely guided by the yet-known polymorphisms of the P2X7 receptor when tested at a given time point and when using peripheral blood leukocytes as responders. However, important evidence is provided that ATP-induced apoptosis is guided by calcium-activated potassium channels rather than by P2X7 directly. This observation implies the option to interfere with ATP/P2X7-induced apoptosis by selective potassium channel interference *in vivo*.

REFERENCES

1. MATZINGER, P. 2002. The danger model: a renewed sense of self. Science **296:** 301–305.
2. ROTHWELL, N. 2003. Interleukin-1 and neuronal injury: mechanisms, modification, and therapeutic potential. Brain Behav. Immun. **17:** 152–157.
3. DI GIROLAMO, M., N. DANI, A. STILLA, *et al.* 2005. Physiological relevance of the endogenous mono(ADP-ribosyl)ation of cellular proteins. FEBS J. **272:** 4565–4575.
4. SLUYTER, R., A.N. SHEMON & J.S. WILEY. 2004. Glu496 to Ala polymorphism in the P2X7 receptor impairs ATP-induced IL-1 beta release from human monocytes. J. Immunol. **172:** 3399–3405.
5. CABRINI, G., S. FALZONI, S.L. FORCHAP, *et al.* 2005. A His-155 to Tyr polymorphism confers gain-of-function to the human P2X7 receptor of human leukemic lymphocytes. J. Immunol. **175:** 82–89.
6. GU, B.J., R. SLUYTER, K.K. SKARRATT, *et al.* 2004. An Arg307 to Gln polymorphism within the ATP-binding site causes loss of function of the human P2X7 receptor. J. Biol. Chem. **279:** 31287–31295.
7. COLOMAR, A., V. MARTY, C. MEDINA, *et al.* 2003. Maturation and release of interleukin-1beta by lipopolysaccharide-primed mouse Schwann cells require the stimulation of P2X7 receptors. J. Biol. Chem. **278:** 30732–30740.
8. AHMADIAN, A., M. EHN & S. HOBER. 2006. Pyrosequencing: history, biochemistry and future. Clin. Chim. Acta **363:** 83–94.
9. NEARY, J.T., Y. KANG, K.A. WILLOUGHBY, *et al.* 2003. Activation of extracellular signal-regulated kinase by stretch-induced injury in astrocytes involves extracellular ATP and P2 purinergic receptors. J. Neurosci. **23:** 2348–2356.

10. PANENKA, W., H. JIJON, L.M. HERX, et al. 2001. P2X7-like receptor activation in astrocytes increases chemokine monocyte chemoattractant protein-1 expression via mitogen-activated protein kinase. J. Neurosci. **21:** 7135–7142.
11. LIANG, L. & E.M. SCHWIEBERT. 2005. Large pore formation uniquely associated with P2X7 purinergic receptor channels. Focus on "Are second messengers crucial for opening the pore associated with P2X7 receptor?" Am. J. Physiol. Cell Physiol. **288:** C240–C242.
12. PFEIFFER, Z.A., M. AGA, U. PRABHU, et al. 2004. The nucleotide receptor P2X7 mediates actin reorganization and membrane blebbing in RAW 264.7 macrophages via p38 MAP kinase and Rho. J. Leukoc. Biol. **75:** 1173–1182.
13. GU, B.J., W. ZHANG, R.A. WORTHINGTON, et al. 2001. A Glu-496 to Ala polymorphism leads to loss of function of the human P2X7 receptor. J. Biol. Chem. **276:** 11135–11142.
14. WILEY, J.S., L.P. DAO-UNG, C. LI, et al. 2003. An Ile-568 to Asn polymorphism prevents normal trafficking and function of the human P2X7 receptor. J. Biol. Chem. **278:** 17108–17113.
15. FERNANDO, S.L., B.M. SAUNDERS, R. SLUYTER, et al. 2005. Gene dosage determines the negative effects of polymorphic alleles of the P2X7 receptor on adenosine triphosphate-mediated killing of mycobacteria by human macrophages. J. Infect. Dis. **192:** 149–155.
16. SHEMON, A.N., R. SLUYTER, A.D. CONIGRAVE, et al. 2004. Chelerythrine and other benzophenanthridine alkaloids block the human P2X7 receptor. Br. J. Pharmacol. **142:** 1015–1019.
17. HIDE, I. 2003. Mechanism of production and release of tumor necrosis factor implicated in inflammatory diseases. Nippon Yakurigaku Zasshi **121:** 163–173.
18. YU, S.P., Z.S. FARHANGRAZI, H.S. YING, et al. 1998. Enhancement of outward potassium current may participate in beta-amyloid peptide-induced cortical neuronal death. Neurobiol. Dis. **5:** 81–88.

Experimental Sepsis

Characteristics of Activated Macrophages and Apoptotic Cells in the Rat Spleen

HELLE EVI SIMOVART,[a] ANDRES AREND,[a] HELLE TAPFER,[a] KERSTI KOKK,[a] MARINA AUNAPUU,[a] ELLE POLDOJA,[a] GUNNAR SELSTAM,[b] AND AADE LIIGANT[a]

[a]*Department of Anatomy, University of Tartu, Estonia*
[b]*Department of Molecular Biology, University of Umeå, Umeå, Sweden*

ABSTRACT: Sepsis, being characterized by massive translocation of bacteria into tissues, induces the suppression of the function of both leukocytes and macrophages. The aim of the study was to count activated macrophages (AMs) and apoptotic (Ao) cells in the rat spleen during the period of experimental sepsis and to clarify the associations of these parameters with each other and with leukocyte migration and bacterial translocation into different organs. The Wistar rats were intraperitoneally inoculated with *Escherichia coli* (*E. coli*) and were sacrificed after 2, 6, 24, 48, and 120 h. Bacteria and leukocytes in tissues were specifically stained. AMs were identified by immunohistological staining and Ao cells by the TUNEL assay. The high counts of *E. coli* at 6h were strongly associated with a low level of the total counts of leukocytes, accompanied by the high translocation of microbes into tissues. In the spleen, lymphocytes, macrophages, and neutrophils with pyknotic nuclei were identified. The count of AMs was highest at 24 h after the inoculation with *E. coli*; at the same time the Ao cell count began to rise and achieved the highest level 24 h later. Our investigation indicates that the molecular peculiarities of macrophages and their responses to the inflammation process are tissue-specific. In the spleen the activation process involving hematopoietic cells and macrophages was remarkable at the late stage of sepsis, characterized by a high count of Ao cells.

KEYWORDS: macrophages; leukocytes; apoptosis; sepsis

INTRODUCTION

One of the central issues in cell biology is the problem of how extracellular signals modulate the cell behavior. Macrophages and leukocytes are of

Address for correspondence: Helle Evi Simovart, Department of Anatomy, University of Tartu, Biomedicum, 19, Ravila Street, Tartu 50411, Estonia. Voice: +372-7-374258; fax: +372-7-374252.
e-mail: helle-evi.simovart@ut.ee

particular interest, as they play a significant role in the host defense against microbes, tumors, and chronic destructive disorders. Gram-negative sepsis is reported to induce massive translocation of bacteria into tissues, which is associated with a decreased macrophage function and a remarkably increased apoptosis of macrophages.[1]

Many activators, bacterial as well as nonbacterial factors, induce local and systemic inflammatory response. At the cellular level sepsis encompasses the upward regulation of both pro- and anti-inflammatory pathways. Inflammation is augmented through the activation of intracellular proinflammatory mediators, including the tumor necrosis factor (TNF)-α and cytokines, interleukins (IL-8, IL-6, IL-1, IL-10), and the granulocyte-macrophage colony-stimulating factor (GM-CSF).[2,3] The activation of inflammatory pathways occurs via the stimulation of host immune effector cells, which subsequently synthesize and release the potent mediators of cell inflammation.

The primary effector cells are activated immunocytes, including polymorphonuclear (PMN) neutrophils, monocyte–macrophages, and lymphocytes. The activation of immune effector cells occurs rapidly. The activation state of circulating immunocytes can be indirectly assessed by measuring the intensity of their cell-surface proteins.

It is still unclear how the molecular interactions between leukocytes, activated macrophages (AMs) and translocated bacteria in different stages of sepsis are related and what is their effect on patient outcome.

The aim of our study was to count the AMs and the apoptotic (Ao) cells in the rat's spleen at different times after inoculation with *Escherichia coli* (*E. coli*) and to detect the associations of these parameters with each other and with leukocyte migration and bacterial translocation.

MATERIAL AND METHODS

Adult male Wistar rats ($n = 48$) with an average weight of 241 g were used. Six rats served as control animals and were injected only with 1.5 mL saline. The other 42 rats were injected intraperitoneally with *E. coli* cells 2×10^7/g of the body weight, suspended in 1.5 mL saline. *E. coli* was isolated from the blood of a septic patient hospitalized in the Hospital of the University of Tartu. The rats had free access to standard laboratory chow and water *ad libitum* during the experiment. The protocol for the research project has been approved by the Ethics Committee of the University of Tartu and it conforms to the provisions of the Declaration of Helsinki in 1995 (as revised in Edinburgh 2000). The rats were anaesthetized with sodium pentobarbital (50 mg/kg of body weight) intraperitoneally and sacrificed by cervical dislocation at different times after inoculation—2 h (Group I), 6 h (Group II), 24 h (Group III), 48 h (Group IV), and 120 h (Group V). A total of 10 rats died at different times after injection and 2 inoculated rats were not included in the experiment, so we formed equal experimental groups consisting of 6 rats. In aseptic conditions, the cardiac blood was taken before the opening of the peritoneal cavity and removal of organs. The same organ specimens were used for light microscopy examinations, immunohistochemical investigation, and the detection of Ao cells.

The total count of leukocytes was measured in a Sysmex K 1000 device (TOA Medical Electronics, Kobe, Japan). Cardiac blood was taken and seeded onto blood agar. The colonies of bacteria were counted after incubation and bacterial counts per milliliter were determined.

The samples for histological investigation were taken from the lung, the liver, the spleen, the ileum, and the colon. The tissue samples were fixed in 10% neutral phosphate-buffered formalin for 24 h, dehydrated, and embedded in paraffin. For the identification of Ao cells, two samples (7 μm and 4 μm) were cut on a sledge microtome to obtain serial sections for light microscopy and for fluorescent microscopy. In order to compare the leukocytes and Ao cells, adjacent sections were compared in light and fluorescent microscopy. Staining of bacteria and nuclei of migrated leukocytes was carried out according to the method of Brown and Hopps.[4] All these detections were performed with the digital optical microscopy system Zeiss AxioPhot 2 (Jena, Germany). The count of bacteria was estimated semiquantitatively and the contamination index (CI) was applied: Grade 0: no bacteria; Grade 1: 1–5 bacteria in any field of view; Grade 2: 6–10 bacteria or an isolated cluster of bacteria in any field of view; Grade 3: 11–20 bacteria or 2 and more clusters of bacteria in any field of view.

For immunohistochemical investigation 3–4 μm-thick paraffin sections were dewaxed and rehydrated. The count of AMs was studied by using the avidin–biotin method. The slides were washed in Tris-buffered saline (TBS), pH 7.4, incubated with proteinase K (Sigma, St. Louis, MO, USA), treated with a solution of bovine serum albumin, and incubated with mouse monoclonal antibody ED[1] (Nordic BioSite AB, Täby, Sweden) as a primary antibody 1:100 in TBS. After that the slides were washed with biotinylated goat anti-mouse IgG (Vector Laboratories, Burlingame, CA, USA) 1:200 in TBS and with streptavidin–AP (Vector Laboratories, Burlingame, CA, USA) 1:1000 in TBS. For visualization the slides were stained with BCIP-NBT (Vector Laboratories) in the dark. The marked macrophages were counted with a light microscope in 168 squares of each viewing field. For each section 10 viewing fields were counted. The area of one microscopic viewing field was 88.872 μm^2.

For the detection of Ao cells, 3–4-μm thick paraffin sections were dewaxed and rehydrated. The *In situ* Cell Death Detection Kit (Roche Molecular Diagnostics GmbH, Penzberg, Germany) was used for the detection of Ao cells according to the manufacturer's protocol. The sections were permeabilized with Proteinase K, PCR Grade (Roche Molecular Diagnostics) and washed with PBS, pH 7.4. Apoptotic nuclei have "nicks" (strand breaks) in their DNA, which are labelled with deoxyuridine triphosphate (dUTP)–fluorescein conjugate via the action of terminal deoxynucleotidyl transferase (TdT) contained in the kit. Apoptotic nuclei appear to be green, whereas nuclei without strand breaks are not labelled. The Ao cells were counted with the microscope Zeiss AxioPhot 2 in UV light. The count of Ao cells in 100 microscopic viewing fields of the tissue sections in each microscopic slide was performed. The area of one microscopic viewing field was 125.660 μm^2.

The Student's t-test as well as correlation analyses were used for the comparison of groups. A P-value less than 0.05 was considered significant. If appropriate, data were tested nonparametrically using Mann-Whitney U test.

RESULTS

Most mortality occurred within the period of 6–48 h after inoculation and the overall mortality was 25%.

Counts of Escherichia coli *in the Blood*

E. coli dissemination in blood rose until 6 h after inoculation, when the counts of microbes were the highest, after which they began to fall and the lowest counts were found in the groups at 48 h and 120 h.

The changes in the total count of leukocytes were opposite to the counts of *E. coli* in blood. The total count of leukocytes decreased sharply at the sixth hour after inoculation, after which it began to rise. A significant negative correlation between the count of *E. coli* in blood and the total count of leukocytes ($r = -0.96$) was found at that time. Statistically significant differences in the count of *E. coli* in blood and in the total count of leukocytes between the groups are shown in FIGURE 1.

Escherichia coli *and Leukocytes in Organs*

E. coli translocation and leukocyte migration into the organs was seen at 6 h after inoculation, whereas at 2 h after injection only single microbes were found. In the liver the bacteria had formed colonies, located mostly in the walls of liver sinusoids, in hepatocytes, in perivascular space, and inside macrophages; in interstitial tissue and in alveolar macrophages in the lungs. At the sixth hour of infection the clusters of microbes and leukocyte infiltration were found in the red pulp and the white pulp of the spleen (FIG. 2). The CI was different between the groups, being the highest at 6 h after injection, after which it began to fall as shown in TABLE 1. No bacteria were found in control animals.

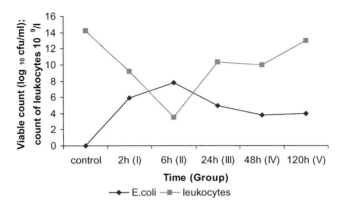

FIGURE 1. Changes in the count of *E. coli* ($\log_{10}$cfu/mL) in the cardiac blood and of the total count of leukocytes (10^9/L) in the blood of infected rats. Statistically significant differences ($P < 0.05$) in the count of *E. coli*: control group versus Group I and II; Group II versus Group IV and V; in the total count of leukocytes: control group versus Group I and II; Group I versus Group II; Group II versus Group III and V.

The leukocytes, which migrated into tissues, were identified by their nuclear characteristics as lymphocytes, neutrophils, and macrophages, in which pyknotic nuclei were detected. In the liver these cells were localized in sinusoids, in perisinusoidal space and perivascularly; in the lungs they were found mostly in the interstitial tissue and in the walls of bronchioles (FIG. 3); in the spleen they were localized in follicles of the white pulp and in the red pulp together with hematopoietic cells.

AM and Ao Cells in the Spleen

The results demonstrated that during the first 6 h the number of AMs did not change. An increase followed after 6 h, achieving the highest level at the twenty-fourth hour after infection (FIG. 4). In the following hours the number of AMs decreased continuously. The changes of the count of AMs and statistically significant differences between the groups are shown in FIGURE 5.

The Ao cell count in the spleen also stayed approximately at a normal level up to 6 h, after which it began to rise quite intensively and reached the maximum level by 48 h after inoculation (FIG. 6). Our investigation in the spleen in this study and in the liver and lungs in our earlier studies with the *In situ* Cell Death Detection Kit demonstrated the green fluorescent nuclei of the positively stained Ao cells in the same areas as mentioned above with migrated leukocytes.[5]

When the counts of AMs and Ao cells were correlated, a negative correlation between them in the 24-h group ($r = -0.44$) and in the 48-h group ($r = -0.85$) was found. Thereafter the Ao cell count decreased sharply and almost reached

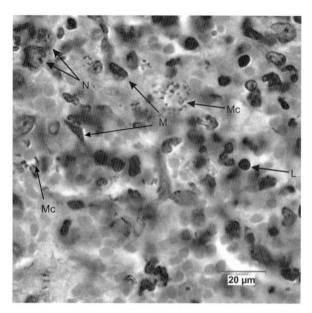

FIGURE 2. Colony of microbes and leukocytes in the spleen (Group II). Mc, microbes; M, macrophages; N, neutrophils; L, lymphocytes.

the normal level at the end of the experiment. The mean values of the Ao cell count in the 120-h group also remained a little higher in comparison with the first hours, but the difference was not statistically significant. The changes of the count of Ao cells and statistically significant differences between the groups are shown in FIGURE 7.

DISCUSSION

In our study, high counts of *E. coli* in blood were found to be strongly associated with a low level of the total count of leukocytes exactly at 6 h after inoculation. At the same time, the highest translocation of bacteria into the organs was found.

Once the inflammatory response is initiated, neutrophils are the first cells (a few minutes after infection) to be recruited to the sites of infection, and after activation they release numerous vasoactive substances. Within the next 24–48 h more neutrophils appear, intermixed with monocytes, and provide the evidence of active phagocytosis. Macrophages play a significant role in the host defense against microbes and tumors and chronic destructive disorders. Mononuclear phagocytes bear around 50 surface receptors. Tissue macrophages, recently arrived from blood, are multipotential and can develop along a variety of disparate routes. Using cytochemical evaluation of sepsis-induced changes in the

TABLE 1. Histological examination of *E. coli* in different tissues

	Group II; 6 h	Group III; 24 h	Group IV; 48 h	Group V; 120 h
Lung	2.00 ± 0.37	1.17 ± 0.48	1.00 ± 0.26	1.00 ± 0.52
Spleen	$2.50 \pm 0.34^{a,b,c}$	1.17 ± 0.17^a	0.83 ± 0.48^b	0.55 ± 0.25^c
Liver	2.33 ± 0.33^d	1.50 ± 0.43	0.83 ± 0.17^d	1.00 ± 0.45
Ileum	1.17 ± 0.17	0.67 ± 0.21	0.83 ± 1.17	0.40 ± 0.25
Cecum	1.00 ± 0.00	0.80 ± 0.20	0.80 ± 0.20	0.80 ± 0.20

The count of bacteria was estimated semiquantitatively and the CI was applied as follows: Grade 0: no bacteria; Grade 1: 1–5 bacteria in any field of view; Grade 2: 6–10 bacteria or an isolated cluster of bacteria in any field of view; Grade 3: 11–20 bacteria or 2 and more clusters of bacteria in any field of view. In control group, bacteria were not seen. Mean ± SEM for each group consisting of six animals was calculated. Statistically significant CI differences ($P < 0.05$; Mann-Whitney U test) in spleen between the groups is indicated as follows: [a]Group II vs. Group III; [b] Group II vs. Group IV; [c] Group II vs. Group V and in liver; [d] Group II vs. Group IV.

rat pulmonary vasculature, investigators have found a large number of activated mononuclear phagocytes in the microvasculature within 48 h of treatment with *E. coli* and found that some pathogens kill macrophages by inducing apoptosis.[6,7]

Bacterial or viral infections induce cell damage, which causes the exit of cell contents into the adjacent tissues. The inflammatory cells including monocytes, lymphocytes, and neutrophils, are accumulated in such lesions where they become activated and release multiple cytotoxic substances (IL-8, IL-6, TNF-α,

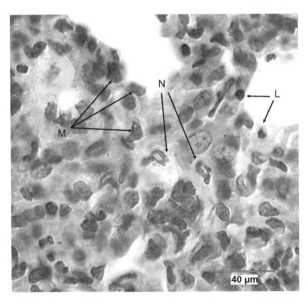

FIGURE 3. Leukocytes in the lung tissue (Group III). M, macrophages; N, neutrophils; L, lymphocytes.

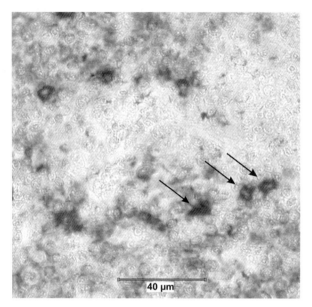

FIGURE 4. AMs in the spleen follicle (Group III).

IL-1β)[8,9] and along with bacterial toxins from *E. coli* induce apoptosis in surrounding cells and leukocytes (T and B cells, macrophages).[10–12]

The coagulation cascade and the complement system become activated, and IL-1 and IL-6 activate T Lymphocytes to produce IFN-γ, IL-2, IL-4, and

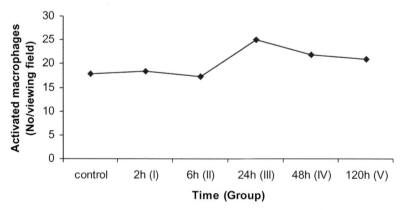

FIGURE 5. Changes of the count of AMs in the spleen in infected rats. Average count per microscopic viewing field. Statistically significant differences ($P < 0.05$) between the groups were as follows: control group versus Group III; Group I versus III and IV; Group II versus III and IV.

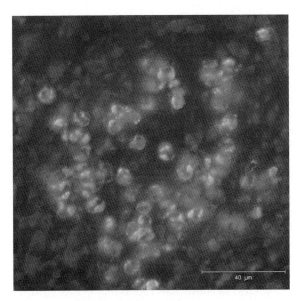

FIGURE 6. Apoptotic cells in the spleen follicle (Group IV).

GM-CSF. Almost all of these agents have direct effects on the vascular endothelium and often the tethered cells activate each other—a process which can be named cross-cellular activation.[13,14]

Our observations of the rat spleen, experimentally infected with *E. coli*, showed that the AM count in the spleen increased after 6 h, reached the

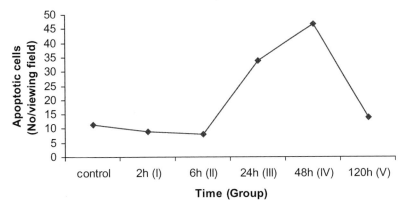

FIGURE 7. Changes of the count of Ao cells in the spleen in infected rats. Average count per microscopic viewing field. Statistically significant differences ($P < 0.05$) between the groups were as follows: control group versus Group III and IV; Group I versus III and IV; Group II versus III and IV.

maximum level at 24 h after inoculation and then slowly decreased toward the end of the experiment. The changes of the AM count in the spleen were similar to the changes in the AM count in the lungs, but different from that in the liver, where the increase in the count of AMs occurred much earlier (2 h after injection) and was very sharp as shown in our previous study.[5] After that, when macrophages were activated and their count began to fall, the count of Ao cells in the spleen reached the maximum level 24 h later (at 48 h after inoculation with *E. coli*). A similar tendency was seen in the lungs; in the liver the rise of the Ao cell count took place 42 h earlier (at 6 h), but still after the AM count had already diminished, as was also found in our earlier study.[5] The mean count of Ao cells was the highest in the spleen compared with the counts in the lungs and the liver, but the AM count was the highest in the liver. The results can be explained with the presence of the Kupffer cells in the liver, which very quickly react to the toxins of microbes. There are plenty of hematopoietic cells in the spleen, which are involved in inflammation and in specific microenvironmental conditions become autoreactive. Unwanted cells are then eliminated by apoptosis together with macrophages. Similar to our experiment, Kurosaka *et al.* have also found that the macrophage response to polymicrobial sepsis appears to be tissue-specific.[15]

The growth factors such as colony-stimulating factors, cytokines and interferon-γ play an important role in apoptosis as well as in the proliferation and differentiation of these hematopoietic cells. Several studies confirm that cytokines (IL-1, IL-6) and TNF released by activated monocytes/macrophages *in vitro* and *in vivo* induce lymphocyte apoptosis and activate lymphoid cells.[16,17] The inflammation process is associated with the functional incapacity of macrophages to release proinflammatory cytokines IL-1β, IL-6, which decreased caspase 1, but increased caspase 3, 8, and 9 activity, which are responsible for the endonuclease activation causing apoptosis.[18]

Many studies indicate that the response of the immune system appears to exhibit a biphasic character. This is thought to take place as a result of the immune cell (neutrophil/monocyte/macrophage) stimulation by the microbes and their products, which were released following the onset of sepsis, and that the induction of apoptosis upon infection results from a complex interaction of bacterial proteins with cellular proteins, finally causing apoptosis.[2,19,20]

The early proinflammatory phase gives way to what appears to be a state of generalized immune cell/organ hyporesponsiveness, a state typified by a loss of phagocytic function and diminished cell-mediated immune responses. These defects are evident in rat/mouse macrophages of the peritoneum, the spleen and the liver.[2]

One of the central components in the control of the immune and inflammatory responses is the ability to sustain or shut down the cells contributing to these responses by the regulation of their cellular suicidal program. Apoptosis is a process by which the selected cell population(s) can be actively deleted from specific tissues, whereas some authors reported that the increased lymphoid Ao seen in tissues might be associated with immune hyporesponsiveness (the decreased proliferative response and the diminished IL-2/IFN-γ release capacity).[21,22] The findings of Ayala *et al.* indicated that the mitogenic stimulation of splenocytes, isolated from mice 24 h after cecal ligation and puncture, causes a significant increase in the rate of AICD (activation-induced cell death), and that splenic lymphoid immune suppression appears to be mediated by an IL-10-induced increase in AICD resulting from the increase of Fas receptors, which is driven in part by Th2-cell action on Th1/Th0 cells, and that the loss of lymphokines may be caused by the increase in lymphocyte Ao seen in polymicrobial sepsis.[22,23]

In conclusion, our investigation reveals that (i) during sepsis the highest level of *E. coli* in the blood as well as the translocation of microbes into tissues were found at 6 h after inoculation; (ii) the AM count was highest at 24 h of sepsis, while the Ao cells count increased at the same time but peaked at 48 h; (iii) the molecular peculiarities of macrophages and their response to the inflammation process were tissue-specific; and (iv) apoptosis of hematopoietic cells gives a reason for a low level of immunocytes in circulation, which refers to immunosuppression. Thus, our investigation indicates that in the spleen the macrophage activation process was remarkable at the late stage of sepsis, and was characterized by a high count of Ao cells; the process, which involved the hematopoietic cells, localized in the spleen as macrophages themselves.

ACKNOWLEDGMENTS

This study was supported by the Estonian Science Foundation Grant No. 4380, the Swedish Institute, and by the targeted financed project No. 0182688505 from the Ministry of Education and Research of Estonia.

REFERENCES

1. AYALA, A. *et al.* 1966. Is sepsis induced apoptosis associated with macrophage dysfunction? Trauma **40:** 568–574.
2. AYALA, A. & I.H. CHAUDRY. 1996. Immune dysfunction in murine polymicrobial sepsis: mediators, macrophages, lymphocytes and apoptosis. Shock **6S:** 27–37.
3. ADRIE, C.H. & M.R. PINSKY. 1999. The inflammatory balance in human sepsis. Intensive Care Med. **36:** 419–428.
4. BROWN, R.C. & H.C. HOPPS. 1973. Staining of bacteria in tissue sections: a reliable gram stain method. Am. J. Clin. Path. **60:** 234–240.
5. SIMOVART, H.E. *et al.* 2003. Changes of activated macrophages and apoptotic cell counts in the organs of rats during experimental sepsis. Medicina **39:** 932–939.
6. SING, B., K.J. DOANE & G.D. NIEHAUS. 1998. Ultrastructural and cytochemical evaluation of sepsis-induced changes in rat pulmonary intravascular mononuclear phagocytes. J. Anat. **192:** 13–23.

7. HILBI, H., A. ZYCHLINSKY & P.J. SANOSETTI. 1997. Macrophage apoptosis in microbial infections. Parasitology **115:** 79–87.
8. CHEN, Y. & A. ZYCHLINSKY. 1994. Apoptosis induced by bacterial pathogens. Microb. Pathog. **17:** 203–212.
9. ROSENBLOOM, A.J. *et al.* 1995. Leukocyte activation in the peripheral blood of patients with cirrhosis of the liver and SIRS. JAMA **274:** 58–65.
10. ELIAS, J.A. *et al.* 1990. Cytokine networks in the regulation of inflammation and fibrosis in the lung. Chest **97:** 1439–1445.
11. MARKS, J.D., C.B. MARKS & J.M. LUCE. 1990. Plasma tumor necrosis factor in patients with septic shock. Am. Rev. Respir. Dis. **141:** 94–97.
12. CALANDRA, T. *et al.* 1990. Prognostic value of tumor necrosis factor/cachectin, interleukin-1, interferon-alpha and inerferon-gamma in the serum of patients with septic shock. J. Infect. Dis. **161:** 982–987.
13. MCGILL, S.N., N.A. AHMED & N.C. CHRISTOU. 1998. Endothelial cells: role in infection and inflammation. World J. Surg. **22:** 171–178.
14. PATEL, K.D. *et al.* 1993. Juxtacrine interactions of endothelial cells with leukocytes: tethering and signaling molecules. Behring Inst. Mitt. **92:** 144–148.
15. KUROSAKA, K., N. WATANABE & I. KOBAYASHI. 2001. Production of proinflammatory cytokines by resident tissue macrophages after phagocytosis of apoptotic cells. Cell Immunol. **211:** 1–7.
16. ALLEN, P.D., A.A. PUSTIN & A.C. NEWLAND. 1993. The role of apoptosis (programmed cell death) in haematopoiesis and immune system. Blood **81:** 63–73.
17. CHUNG, C.S. *et al.* 2000. Neither Fas ligand nor endotoxin is responsible for inducible phagocyte apoptosis during sepsis/peritonitis. J. Surg. Res. **91:** 147–153.
18. HERNANDEZ-CASELLES, T. & O. STUTMAN. 1993. Immune functions of tumor necrosis factor I. Tumor necrosis factor induces apoptosis of mouse thymocytes and can also stimulate or inhibit IL-6 induced proliferation depending on the concentration of mitogenic costimulation. J. Immunol. **151:** 3999–4012.
19. GRASSME, H., V. JENDROSSEK & E. GULBINS. 2001. Molecular mechanisms of bacteria induced apoptosis. Apoptosis **6:** 441–445.
20. OBERHOLZER, C. *et al.* 2001. Apoptosis in sepsis: a new target for therapeutic exploration. FASEB J. **15:** 879–892.
21. HENGARTNER, M.O. 2000. The biochemistry of apoptosis. Nature **407:** 770–776.
22. AYALA, A. *et al.* 2001. IL-10 mediation of activation induced Th2-cell apoptosis during polymicrobial sepsis. Cytokine **14:** 37–48.
23. AYALA, A. *et al.* 2003. Pathological aspects of apoptosis in severe sepsis and shock? Int. J. Biochem. Cell. Biol. **35:** 7–15.

Insulin-Like Growth Factor-I Receptor Correlates with Connexin 26 and Bcl-xL Expression in Human Colorectal Cancer

STANISLAW SULKOWSKI, LUIZA KANCZUGA-KODA, MARIUSZ KODA, ANDRZEJ WINCEWICZ, AND MARIOLA SULKOWSKA

Departments of General and Clinical Pathomorphology, Medical University of Bialystok, Waszyngtona 13, 15-269 Bialystok, Poland

ABSTRACT: Insulin-like growth factor (IGF) and its receptor (IGF-IR) play an important role in mitogenesis, apoptosis, growth, and proliferation of several types of cancers. Overexpression of IGF-IR in colorectal cancer is associated with increase of cancer cell proliferation and migration as well as inhibition of apoptosis. In our previous reports we demonstrated correlations between IGF-IR and apoptosis. Moreover, we observed relationships between connexin26 (Cx26) expression and apoptotic markers in human colorectal cancer. Recently, it has been shown that expression of connexins and gap junction (GJ) functions are also regulated by growth factors, including IGF-I. Therefore, in this study we have focused on the relationships between IGF-IR and Cx26 as well as Bcl-xL expression. A total number of 115 cases of colorectal cancer were examined by immunohistochemistry, using the avidin-biotin-peroxidase method. Associations among the above proteins were assessed in the entire group of colorectal cancer patients and its subgroups, depending on lymph node involvement (N0 and N1), histological grade (G2 and G3), extent of tumor growth (pT1 + pT2 and pT3 + pT4), histopathologic type (adenocarcinoma and mucinous carcinoma), sex, age ($\leq$60 and >60), and tumor site (colon and rectum). The expression of IGF-IR, Cx26, and Bcl-xL was noted in 47%, 56.5%, and 75.6% of the tumors, respectively. In the entire group of patients we found a positive correlation between IGF-IR and Cx26 ($P < 0.0001$, $r = 0.374$) as well as between IGF-IR and Bcl-xL ($P < 0.0001$, $r = 0.344$). Our results may suggest that the insulin-like growth system is involved in regulation of apoptosis and probably connexin expression in colorectal cancer cells.

KEYWORDS: IGF-IR; Cx26; Bcl-xL; connexin; apoptosis; colorectal cancer

---

Address for correspondence: Stanislaw Sulkowski, M.D., Ph.D., Department of Pathology, Medical University of Bialystok, Waszyngtona St 13, 15-269 Bialystok, Poland. Voice: +48-85-7485945; fax: +48-85-7485944.
 e-mail: sulek@zeus.amb.edu.pl

INTRODUCTION

Colorectal cancer is one of the most common kinds of cancer worldwide. Dysregulation of growth factors and their receptors play an important role in colorectal carcinogenesis.[1-3] The insulin-like growth factors (IGF-I and IGF-II) are peptides related to insulin, which play an essential role in regulation of proliferation, cellular growth, and apoptosis of several cells, including epithelial cells.[3,4] Accumulating data suggest that increased IGFs may be associated with development of colon cancer.[5] The actions of IGFs are mediated via the IGF receptors: type I and type II (IGF-IR and IGF-IIR). IGF-IR plays a crucial role in the normal growth and development of several body organs.[4] It has been shown that the IGF-IR gene is overexpressed in most types of cancer, including colorectal tumors. In spite of the fact that functions of IGF-I, IGF-IR, and IRS-1 are well-known and appreciated in regards to their antiapoptotic role, some contrary reports have been published lately that seem to uncover proapototic properties of these agents.[6-8] It was suggested that IGF-I, IGF-IR, and IRS-1 probably cause dual but opposing effects on apoptotic cell death. Therefore, they might affect tumor cell turnover, increase final amount of cancer cells and favor cancer progression.

The resistance to apoptosis is a common phenomenon in cancer cells.[9] It has been well documented that cancerous cells use different strategies to prevent apoptotic processes such as downregulation or mutation of proapoptotic genes as well as overexpression of antiapoptotic genes.[9,10] The main group of genes controlling apoptosis is the Bcl-2 family, which includes both promotors (Bax, Bak, Bad, Bcl-xS) and inhibitors (Bcl-2, Bcl-xL, Mcl-1).[11] Bcl-xL is able to form heterodimers with Bak and Bax. The elevated expression of this protein seems to be an early event in colorectal carcinogenesis.[12] Some studies have demonstrated correlation between intercellular communication and apoptosis.[13,14] In our previous article we found the positive correlation between connexin 26 (Cx26) and Bax expression as well as Cx26 and Bcl-xL in colorectal cancer.[15]

Gap junctional intercellular communication (GJIC) is mediated by gap junctions (GJs), which are formed from transmembrane proteins called connexins. Connexins are encoded by different genes and until now more that 20 connexins were recognized in humans.[16] GJ channels allow small molecules (<1kDa) and ions to flow between adjacent cells. It has been shown that Cx32 and Cx43 are the major GJ proteins that exist in colorectal mucosa. We have recently shown by immunohistochemistry that in normal colon epithelium Cx26 also is present.[17] Normal GJ function has been associated with cellular growth, differentiation, tissue development, and regulation of apoptosis.[18] Changes in the expression and localization of Cxs have been correlated with carcinogenesis. Connexins localized in cytoplasm could regulate expression of different genes and thus might modulate signaling pathways. Because it was demonstrated *in vitro* that IGF-I may increase gap junctional communication (GJC) and connexin protein expression,[19] we evaluated on the basis of immunohistochemistry correlations between Cx26 and IGF-IR expression in colorectal cancer. Furthermore, we focused on the relationships between IGF-IR and antiapoptotic protein Bcl-xL.

MATERIAL AND METHODS

Tissue samples were obtained from 115 patients (61 men and 54 women) who underwent surgical resection because of colon (61 cases) and rectal (54) carcinomas. Our study included 98 colorectal cancers classified histopathologically as adenocarcinoma and 17 as mucinous adenocarcinoma: 83 cases were in G2 grade and 32 cases in G3 grade. There were 12 tumors in pT1 + pT2 stage and 103 in pT3 + pT4 stage. A total of 59/115 (51.3%) patients had involved lymph nodes at the time of diagnosis. The age of patients ranged from 35 to 92 years (mean 65.4 years). Tumor samples with adjacent normal colon mucosa were collected immediately after tumor removal, fixed in 10% buffered formaldehyde solution for 48 h, and then embedded in paraffin blocks at 56°C according to standard procedures. The resected tumors were histopathologically examined using standard hematoxylin-eosin staining.

Immunohistochemistry

Paraffin-embedded tissue sections were subjected to immunostaining using antibodies: rabbit polyclonal antibody (Ab) H-60 recognizing β subunit of IGF-IR (Santa Cruz Biotechnology [SCBt], Santa Cruz, CA, USA), goat polyclonal Bcl-xL Ab (SCBt), and goat polyclonal Cx26 Ab (SCBT) in dilution rate: 1:200, 1:300, and 1:400, respectively. All primary Abs were diluted in phosphate-buffered saline solution (PBS) with 1.5% normal blocking serum. The studies were performed with avidin-biotin-peroxidase complex (ABC Staining System, SCBt). Immunohistochemistry was performed as described previously.[20] Slides were counterstained with hematoxylin. We performed the following immunohistochemical controls: positive controls of colorectal cancers that were reported positive for studied antigens; negative controls included omission of primary Abs. The evaluation of immunostaining for all studied proteins was analyzed in 10 different tumor fields and the mean percentage of tumor cells with positive staining was evaluated. Furthermore, for statistical analysis, colorectal cancers were grouped into IGF-IR, Bcl-xL, and Cx26–positive or negative. The sections were classified as positive when at least 10% of cancer cells expressed the studied antigens.

Statistical Analysis

The significance of the associations was determined using Spearman correlation analysis and the χ^2 (Chi-square) test. Probabilities of $P < 0.05$ were assumed as statistically significant.

RESULTS

To study the relationships between IGF-IR, Cx26, and Bcl-xL, a total number of 115 cases of colorectal cancer were examined. Associations among the above proteins were assessed in the entire group of colorectal cancer (CRC) patients and its subgroups depending on lymph node involvement (N0 and N1), histologic grade (G2 and G3), extent of tumor growth (pT1 + pT2 and pT3 + pT4), histopathologic type (adenocarcinoma and mucinous carcinoma), sex, age (≤ 60 and >60), and tumor site (colon and rectum).

Comparison of IGF-IR to Bcl-xL and Cx26 Expression in Entire CRC Group

The expression of IGF-IR, Cx26, and Bcl-xL was noted in 47%, 56.5%, and 75.6% of the tumors, respectively. In the entire group of patients we found a positive correlation between IGF-IR and Cx26 ($P < 0.0001$, $r = 0.374$) as well as between IGF-IR and Bcl-xL ($P < 0.0001$, $r = 0.344$) (TABLES 1 and 2).

Comparison of IGF-IR to Cx26 Expression in CRC Groups with Different Clinicopathologic Features

IGF-IR correlated with Cx26 more extensively in the group of patients without metastases to the lymph nodes (N[−]) ($P < 0.0001$, $r = 0.539$) compared to the group of CRC patients with involvement of lymph nodes (N[+]) ($P < 0.046$, $r = 0.257$). Similarly, significant relationship was observed in G2 tumors more strongly ($P < 0.001$, $r = 0.349$) than in G3 tumors ($P < 0.019$, $r = 0.412$). Deeper extent of tumor growth (pT3 and pT4) was associated with an undoubtedly indicated highly significant positive correlation between IGF-IR and Cx26 expression ($P < 0.0001$, $r = 0.377$), with no statistical significance in pT1 or pT2 tumors. In parallel, depending on histopathologic type of tumor in colorectal adenocarcinoma, highly significant positive correlation was found ($P < 0.0001$, $r = 0.376$), whereas in mucinous adenocarcinoma there was no significant relationship between studied proteins. Positive correlation between IGF-IR and Cx26 was significantly higher in female CRC group ($P < 0.002$,

TABLE 1. Correlations between Cx26 and IGF-IR in CRC patients by Spearman's correlation test

Groups of patients		Comparison markers Cx26–IGF-IR	
		P	r
All of CRC patients		<0.0001	0.374
N	(−)	<0.0001	0.539
	(+)	0.046	0.257
G	2	0.001	0.349
	3	0.019	0.412
pT	pT1 + pT2	0.236	0.370
	pT3 + pT4	<0.0001	0.377
HP type	Adenocarcinoma	<0.0001	0.376
	Mucinous adenocarcinoma	0.315	0.251
Sex	Males	0.011	0.319
	Females	0.002	0.411
Age	≤60 years	0.027	0.336
	>60 years	<0.0001	0.401
Localization	Rectum	0.001	0.417
	Colon	0.010	0.326

NOTE: HP type: histopathologic type; pT: tumor size; N: lymph node involvement; G: grading of cell differentiation.

$r = 0.411$), older individuals ($P < 0.0001$, r = 0.401) or in rectal locations of tumors ($P < 0.001$, $r = 0.417$) (TABLE 1).

Linkages of IGF-IR and Bcl-xL in Groups with Different Clinicopathologic Features

Expressions of IGF-IR and Bcl-xL matched each other in statistically significant manner and correlated positively in patients without metastases to the lymph nodes ($P < 0.0001$, $r = 0.485$). There was no statistical significance in patients with metastatic lymph nodes ($P > 0.05$, $r = 0.243$). Degree of cancer histologic differentiation of G2 was associated with an undoubtedly indicated trend toward highly significant positive correlation between expression of IGF-IR and Bcl-xL ($P = 0.001$, $r = 0.356$), with no statistical significance in G3 tumors. Similarly positive correlation between these proteins was observed in the group of CRC patients with adenocarcinomas ($P = 0.001$, $r = 0.332$) and in CRC subjects at the age of more than 60 years ($P = 0.001$, $r = 0.384$) as well as in female patients ($P < 0.0001$, $r = 0.464$), while subjects with mucinous carcinoma and at the age ≤60 years as well as in male CRC group did not show any linkage of this kind. There was an indication of high positive correlation with statistical significance in tumors of pT3 or pT4 ($P = 0.001$, $r = 0.314$) more extensively than in tumors of pT1 or pT2. In contrast to the foregoing

TABLE 2. Correlations between IGF-IR and Bcl-xL in CRC patients by Spearman's correlation test

Groups of patients		Comparison markers IGF-IR–Bcl-xL	
		P	r
All of CRC patients		<0.0001	0.344
N	(−)	<0.0001	0.485
	(+)	0.061	0.243
G	2	0.001	0.356
	3	0.080	0.319
pT	pT1 + pT2	0.039	0.626
	pT3 + pT4	0.001	0.314
HP type	Adenocarcinoma	0.001	0.332
	Mucinous adenocarcinoma	0.153	0.351
Sex	Males	0.102	0.213
	Females	<0.0001	0.464
Age	≤60years	0.073	0.283
	>60 years	0.001	0.384
Localization	Rectum	0.012	0.341
	Colon	0.005	0.347

NOTE: HP type: histopathologic type; pT: tumor size; N: lymph node involvement; G: grading of cell differentiation.

presented results of correlation between IGF-IR and Cx26 expression, correlation between IGF-IR and Bcl-xL was significantly higher in individuals with colonic location of tumor ($P = 0.005$, $r = 0.347$) (FIG. 1A–D; TABLE 2).

DISCUSSION

The IGF-I receptor (IGF-IR), transmembrane tyrosine kinase receptor, composed of two extracellular α and two intracellular β chains, is the major receptor for IGF-I and IGF-II.[21] The IGF-IR mediates mainly growth and differentiation and increases cell survival.[22] The IGF receptors are found in both mucosal and muscular layers of intestine.[23] In colorectal cancer cells, overexpression of IGF receptors was observed.[24] When receptors are stimulated by IGF-I, they inhibit apoptosis and allow progression through the cell cycle. In spite of the fact that functions of IGF-I and IGF-IR are well-known in regard to its antiapoptotic role, some contrary reports have been published lately that seem to reveal proapotoctic properties of these agents.[7,8] It was suggested that IGF-I and IGF-IR probably cause dual but opposing effects on apoptotic cell death, thereby possibly affecting tumor cell turnover and increasing the final amount of cancer cells. Consequently, IGFs by promotion of mitogenesis and cell survival, as well as by enhancement of apoptosis, could contribute to acceleration of both cancer cell

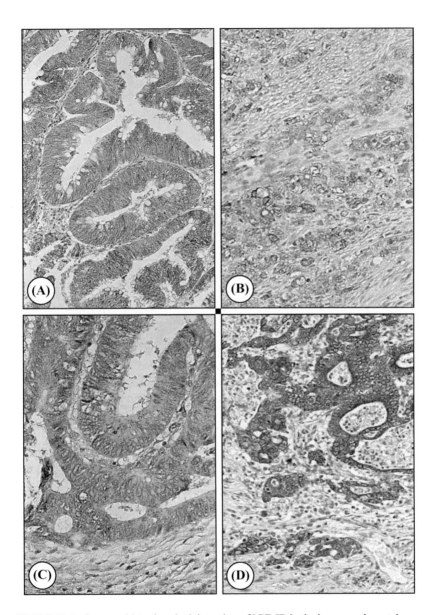

FIGURE 1. Immunohistochemical detection of IGF-IR in the human colorectal cancer. **(A)** Strong cytoplasmic immunostaining in G2 grade colorectal cancer (original magnification ×100). **(B)** Cytoplasmic localization of immunostaining in colorectal cancer classified as G3 (original magnification ×200). **(C)** Immunohistochemical detection of Cx26 in the human colorectal cancer. Granular staining of Cx26 localized mainly in cytoplasm of colorectal cancer cells in the tumor classified as G2 (original magnification × 200). **(D)** Immunohistochemical detection of Bcl-xL. Strong cytoplasmic immunostaining of Bcl-xL in the majority of colorectal cancer cells (original magnification ×200).

turnover and cancer progression. For example, it was shown *in vitro* that activation of IGF-IR upregulated expression of antiapoptotic proteins Bcl-2 and Bcl-xL.[25,26] On the other hand, it has been demonstrated lately that IGF-I and IGF-IR can also stimulate the process of apoptosis.[7,8] Raile *et al.*[7] found that IGF-I in spite of promotion of proliferation can also enhance caspase-3 activation. Moreover, proapoptotic properties of a truncated mutant IGF-IR were described. We have recently noted the positive correlation between IGF-IR and proapoptotic proteins Bax and Bak in human colorectal cancer.[27] Positive correlation between IGF-IR and Bcl-xL in the present study is another evidence that the IGF system could influence apoptosis in colorectal carcinogenesis and the ratio between pro- and antiapoptotic proteins. We suppose that IGF-IR could be associated with enhanced turnover of different cancer cells and increased risk of distant metastases.

Some studies have demonstrated association between connexins and programmed cell death.[13,14] It is probable that regulation of apoptosis by connexins could be, among others, a result of a control of apoptotic markers such as Bcl-2 family proteins. Recently, in our immunohistohemical study of Cx26 and apoptotic marker expression in colorectal cancer, we have revealed the positive correlation between Cx26 and proapoptotic Bax as well as between Cx26 and antiapoptotic Bcl-xL.[15] Consequently, our conclusion was that cytoplasmic Cx26 could perform additional functions in malignant cell (for example, it might be involved in the control of apoptotic process). We suppose that Cxs, despite its well-known role in stimulation of apoptosis, could also interact with antiapoptotic proteins. Consequently, Cxs localized in the cytoplasmic compartment of malignat cells, similar to IGF-IR, might increase tumor cell turnover and colorectal cancer progression.

Gap junctional communication (GJC) is a subject to regulation by numerous substances, which may be also involved in synthesis and recruitment of connexins. It has been shown that several substances affect this process, such as transforming growth factor-β1 (TGF-β1),[28] fibroblast growth factors (FGFs),[29–31] or vascular endothelial growth factor (VEGF).[32] Furthermore, it was revealed that growth hormone (GH) increased amount of Cx43 and may influence gap junction (GJ) formation in the brain.[33] Moreover, Aberg *et al.*[19] showed that IGF-I may directly stimulate astrocyte primary cultures to synthesize Cx43 and interact with IGF binding proteins (IGFBPs), possibly inducing functional GJC. Interestingly, Lin *et al.*[34] revealed that IGF-I induced a decrease in GJC in lens epithelial cells through the binding of Cx43 to PKCγ and phosphorylation of Cx43 by PKCγ, resulting in disassembly of GJ plaques and decreased GJ activity. On the other hand, the reverse relationship has been previously reported: Bradshaw *et al.*[35] revealed that C6 glioma cells transfected with a connexin43 cDNA and expressing abundant connexin43 have shown decreased synthesis of the growth factor IGF-I together with decreased levels of a positive modulator IGFBP-3, and the increased levels of a negative modulator IGFBP-4 in the extracellular milieu. They supposed that it is one of potential mechanisms that may be responsible for the reduced proliferative capability of malignant cells. In our previous papers we described cytoplasmic localization of connexins in breast and colorectal cancer.[36,37] We suggested that connexin proteins accumulate in the cytoplasm of cancer cells, and it is probable that in this localization they could play a distinct role from physiological functions. It is very likely that connexins may control tumor progression by their influence on the expression of genes which are responsible for regulation of cell growth and differentiation and other functions of cancerous cells. Association between Cx26 and IGF-IR expression in colorectal cancer cells might partly explain this theory, but additional functional studies on the role of connexins in signaling pathways are required.

REFERENCES

1. GRYFE, R., C. SWALLOW, B. BAPAT, et al. 1997. Molecular biology of colorectal cancer. Curr. Probl. Cancer **21:** 233–300.
2. WATSON, D.S., I. BROTHERIC, B.K. SHENTON, et al. 1999. Growth dysregulation and p53 accumulation in human primary colorectal cancer. Br. J. Cancer **80:** 1062–1068.
3. MOSCHOS, S.J. & C.S. MANTZOROS. 2002. The role of the IGF system in cancer: from basic to clinical studies and clinical applications. Oncology **63:** 317–332.
4. LEROITH, D. & C.T. ROBERTS. JR. 2003. The insulin-like growth factor system and cancer. Cancer Lett. **195:** 127–137.
5. DURAI, R., W. YANG, S. GUPTA, et al. 2005. The role of the insulin-like growth factor system in colorectal cancer: review of current knowledge. Int. J. Colorectal Dis. **20:** 203–220.
6. KAWAKAMI, A., T. NAKASHIMA, M. TSUBOI, et al. 1998. Insulin-like growth factor I stimulates proliferation and Fas-mediated apoptosis of human osteoblasts. Biochem. Biophys. Res. Commun. **247:** 46–51.
7. RAILE, K., R. HILLE, S. LAUE, et al. 2003. Insulin-like growth factor I (IGF-I) stimulates proliferation but also increases caspase-3 activity, Annexin-V binding, and DNA-fragmentation in human MG63 osteosarcoma cells: co-activation of pro- and anti-apoptotic pathways by IGF-I. Horm. Metab. Res. **35:** 786–793.
8. WU, J., K. HAUGK & S.R. PLYMATE. 2003. Activation of pro-apoptotic p38-MAPK pathway in the prostate cancer cell line M12 expressing a truncated IGF-IR. Horm. Metab. Res. **35:** 751–757.
9. IGNEY, F.H. & P.H. KRAMMER. 2002. Death and anti-death: tumour resistance to apoptosis. Nat. Rev. Cancer **2:** 277–288.
10. THOMPSON, C.B. 1995. Apoptosis in the pathogenesis and treatment of disease. Science **267:** 1456–1462.
11. REED, J.C. 1998. Bcl-2 family proteins. Oncogene **17:** 3225–3236.
12. KRAJEWSKA, M., S.F. MOSS, S. KRAJEWSKI, et al. 1996. Elevated expression of Bcl-X and reduced Bak in primary colorectal adenocarcinomas. Cancer Res. **56:** 2422–2427.
13. HUANG, R.P., M.Z. HOSSAIN, R. HUANG, et al. 2001. Connexin 43 (cx43) enhances chemotherapy-induced apoptosis in human glioblastoma cells. Int. J. Cancer **92:** 130–138.

14. TANAKA, M. & H.B. GROSSMAN. 2004. Connexin 26 induces growth suppression, apoptosis and increased efficacy of doxorubicin in prostate cancer cells. Oncol. Rep. **11:** 537–541.
15. KANCZUGA-Koda, L., S. SULKOWSKI, M. KODA, et al. 2005. Connexin 26 correlates with Bcl-xL and Bax proteins expression in colorectal cancer. World J. Gastroenterol. **11:** 1544–1548.
16. SOHL, G. & K. WILLECKE. 2004. Gap junctions and the connexin protein family. Cardiovasc. Res. **62:** 228–232.
17. Kanczuga-KODA, L., S. SULKOWSKI, M. KODA, et al. 2004. Expression of connexins 26, 32 and 43 in the human colon: an immunohistochemical study. Folia Histochem. Cytobiol. **42:** 203–207.
18. TROSKO, J.E. & R.J. RUCH. 2002. Gap junctions as targets for cancer chemoprevention and chemotherapy. Curr. Drug Targets **3:** 465–482.
19. ABERG, N.D., F. BLOMSTRAND, M.A. ABERG, et al. 2003. Insulin-like growth factor-I increases astrocyte intercellular gap junctional communication and connexin43 expression in vitro. J. Neurosci. Res. **74:** 12–22.
20. KODA, M., S. SULKOWSKI, C. GAROFALO, et al. 2003. Expression of the insulin-like growth factor-I receptor in primary breast cancer and lymph node metastases: correlations with estrogen receptors α and β. Horm. Metab. Res. **35:** 794–801.
21. GRIMBERG A. & P. COHEN. 2000. Role of insulin-like growth factors and their binding proteins in growth control and carcinogenesis. J. Cell. Physiol. **183:** 1–9.
22. REINMUTH, N., W. LIU, F. FAN, et al. 2002. Blockade of insulin-like growth factor I receptor function inhibits growth and angiogenesis of colon cancer. Clin. Cancer Res. **8:** 3259–3269.
23. HOWARTH G.S. 2003. Insulin-like growth factor-I and the gastrointestinal system: therapeutic indications and safety implications. J. Nutr. **133:** 2109–2112.
24. OUBAN, A., P. MURACA, T. YEATMAN, et al. 2003. Expression and distribution of insulin-like growth factor-1 receptor in human carcinomas. Hum. Pathol. **34:** 803–808.
25. PARRIZAS, M. & D. LEROITH. 1997. Insulin-like growth factor-1 inhibition of apoptosis is associated with increased expression of the bcl-xL gene product. Endocrinology **138:** 1355–1358.
26. PUGAZHENTHI, S., E. MILLER, C. SABLE, et al. 1999. Insulin-like growth factor-I induces bcl-2 promoter through the transcription factor cAMP-response element-binding protein. J. Biol. Chem. **274:** 27529–27535.
27. KODA, M., J. RESZEC, M. SULKOWSKA, et al. 2004. Expression of the insulin-like growth factor-I receptor and proapoptotic bax and bak proteins in human colorectal cancer. Ann. N.Y. Acad. Sci. **1030:** 377–383.
28. ROBE, P.A., B. ROGISTER, M.P. MERVILLE, et al. 2000. Growth regulation of astrocytes and C6 cells by TGF beta 1: correlation with gap junctions. Neuroreport **11:** 2837–2841.
29. NADARAJAH, B., H. MAKARENKOWA, D.L. BEKER, et al. 1998. Basic FGF increases communication between cells of the developing neocortex. J. Neurosci. **18:** 7881–7890.
30. LE, A.C. & L.S. MUSIL. 2001. A novel role for FGF and extracellular signal-regulated kinase in gap junction-mediated intercellular communication in the lens. J. Cell. Biol. **154:** 197–216.
31. REUSS, B., M. HERTEL, S. WERNER, et al. 2000. Fibroblast growth factors-5 and -9 distinctly regulate expression and function of the gap junction protein connexin43 in cultured astroglial cells from different brain regions. Glia **30:** 231–241.

32. PIMENTEL, R.C., K.A. YAMADA, A.G. KLEBER, et al. 2002. Autocrine regulation of myocyte Cx43 expression by VEGF. Circ. Res. **90:** 671–677.
33. ABERG, N.D., B. CARLSSON, L. ROSENGREN, et al. 2000. Growth hormone increases connexin-43 expression in the cerebral cortex and hypothalamus. Endocrinology **141:** 3879–3886.
34. LIN, D., D.L. BOYLE & D.J. TAKEMOTO. 2003. IGF-I-induced phosphorylation of connexin 43 by PKCgamma: regulation of gap junctions in rabbit lens epithelial cells. Invest. Ophthalmol. Vis. Sci. **44:** 1160–1168.
35. BRADSHAW, S.L., C.C. NAUS, D. ZHU, et al. 1993. Alterations in the synthesis of insulin-like growth factor binding proteins and insulin-like growth factors in rat C6 glioma cells transfected with a gap junction connexin43 cDNA. Regul. Pept. **48:** 99–112.
36. KANCZUGA-KODA, L., S. SULKOWSKI, J. TOMASZEWSKI, et al. 2005. Connexins 26 and 43 correlate with Bak, but not with Bcl-2 protein in breast cancer. Oncol. Rep. **14:** 325–329.
37. KANCZUGA-KODA, L., S. SULKOWSKI, M. KODA, et al. 2005. Alterations in connexin26 expression during colorectal carcinogenesis. Oncology **68:** 217–222.

Characterization of the Proapoptotic Intracellular Mechanisms Induced by a Toxic Conformer of the Recombinant Human Prion Protein Fragment 90–231

VALENTINA VILLA,[a] ALESSANDRO CORSARO,[a]
STEFANO THELLUNG,[a] DOMENICO PALUDI,[b] KATIA CHIOVITTI,[c]
VALENTINA VENEZIA,[a] MARIO NIZZARI,[a] CLAUDIO RUSSO,[d]
GENNARO SCHETTINI,[a] ANTONIO ACETO,[c] AND TULLIO FLORIO[a]

[a]*Department of Oncology, Biology and Genetics, University of Genova, 16132 Genova, Italy*

[b]*Department of Scienze degli Alimenti, Veterinary School, University of Teramo, Teramo, Italy*

[c]*Department of Biomedical Sciences, University G. D'Annunzio, Chieti, Italy*

[d]*Department of Health Sciences, University of Molise, Campobasso, Italy*

ABSTRACT: Prion diseases comprise a group of fatal neurodegenerative disorders that affect both animals and humans. The transition of the prion protein (PrP) from a mainly α-structured isoform (PrP^C) to a prevalent β-sheet-containing protein (PrP^{Sc}) is believed to represent a major pathogenetic mechanism in prion diseases. To investigate the linkage between PrP neurotoxicity and its conformation, we used a recombinant prion protein fragment corresponding to the amino acidic sequence 90–231 of human prion protein (hPrP90–231). Using thermal denaturation, we set up an experimental model to induce the process of conversion from PrP^C to PrP^{Sc}. We report that partial thermal denaturation converts hPrP90–231 into a β-sheet-rich isoform, displaying a temperature- and time-dependent conversion into oligomeric structures that share some physico-chemical characteristics with brain PrP^{Sc}. SH-SY5Y cells were chosen to characterize the potential neurotoxic effect of hPrP90–231 in its different structural conformations. We demonstrated that hPrP90–231 in β-conformation, but not when α-structured, powerfully affected the survival of these cells. hPrP90–231 β-structured caused DNA fragmentation and a significant increase in caspase-3 proteolytic activity (maximal effects + 170%), suggesting the occurrence of apoptotic cell death. Finally, we investigated the involvement of MAP kinases in the regulation of β-hPrP90–231-dependent apoptosis. We observed that the p38 MAP kinase blocker SB203580 prevented the apoptotic cell

Address for correspondence: Prof. Tullio Florio, Department of Oncology, Biology and Genetics, University of Genova, Viale Benedetto XV, 16132 Genova, Italy. Voice and fax: +39-010-3538806.
e-mail: tullio.florio@unige.it

death evoked by hPrP90–231, and Western blot analysis revealed that the exposure of the cells to the peptide induced p38 phosphorylation. In conclusion, we demonstrate that the hPrP90–231 elicits proapoptotic activity when in β-sheet-rich conformation and that this effect is mediated by p38 and caspase-3 activation.

KEYWORDS: prion protein; apoptosis; p38 MAP kinase; caspases

INTRODUCTION

Transmissible spongiform encephalopathies (TSEs) are lethal degenerative disorders of the nervous system caused by transmissible particles that contain a pathogenic isoform of the prion protein (PrP), a normal constituent of cell membranes. Among TSEs, Creutzfeldt-Jakob Disease (CJD), Gerstmann Straussler Scheinker (GSS), and kuru in humans and scrapie and bovine spongiform encephalopathy (BSE) in domestic animals[1] represent the more frequent conditions. The pathogenesis of TSEs involves conformational conversions of normal cellular PrP (PrP^C) to an abnormal isoform named PrP scrapie or PrP^{Sc}.[2] PrP^{Sc} is characterized by a high content of what is considered the likely causative agent of prion disease. Its accumulation in the brain produces a decline of cognitive and motor function with a pathognomonic triad: spongiform vacuolation of the gray matter, neuronal death, and glial proliferation, with amyloid deposition less constantly observed. PrP^{Sc} is characterized by a high content of β-sheet forming insoluble prontese K–resistant amyloid aggregates and it is derived from its normal cellular isoform (PrP^C), which is soluble and rich in α-helical structure.[3] While the primary structure of PrP^C is identical to that of PrP^{Sc}, secondary and tertiary structural changes are responsible for the distinct physico-chemical properties of the two isoforms. In particular, it was proposed that structural changes in the helix-3 are the main determinant of the PrP^C–PrP^{Sc} conversion.[4] It is commonly accepted that PrP^{Sc} represents the infectious form of PrP, but whether it is also the neurotoxic form is still debated. A major gap in our understanding exists about which isoform of PrP actually kills neurons and the involved molecular mechanisms. It was recently proposed that either intermediates formed along the prion replication process or a bioproduct of PrP^C–PrP^{Sc} conversion, rather than PrP^{Sc} itself, may be the neurotoxic entities responsible for the cell death.[5] To study a precise relationship between molecular structure and neurotoxic effect of PrP^{Sc}, we synthesized a recombinant PrP fragment corresponding to the amino acids 90–231 (hPrP90–231) of the human sequence corresponding the protease-resistant core of the PrP^{Sc}, also called PrP27–30. After purification, hPrP90–231 is structured as a soluble protein with a high content of α-helix and PK sensitivity, all of them characteristics of PrP^{C}.[6] We used this recombinant peptide to study the molecular mechanisms of the transition toward the pathological form and the induction of cell death. To obtain hPrP90–231 conversion into a β-sheet-rich, amyloidogenic and partially PK-resistant isoform, all features described in PrP[Sc], we used a simple thermal denaturation of hPrP90–231 (incubation for 1 h at 53°C) and we demonstrated that in this condition hPrP90–231 powerfully induced p38- and caspase-mediated apoptosis of the SH-SY5Y human neuroblastoma cells.

MATERIALS AND METHODS

Peptide Purification

Escherichia coli XL1-blue transformed with pGEX-4T-2:PrP90–231 were grown at 30°C till $OD_{600} = 0.6$. Fusion protein expression was induced with 0.4 mM IPTG (4 h at 30°C), with supplemented aeration. After centrifugation, the bacterial pellet was dissolved by sonication, cell debris removed by centrifugation, and crude lysate added to equilibrated glutathione-Sepharose resin. The binding of glutathione-S-transferase (GST) to glutathione was accomplished overnight at 4°C, with gentle shaking. Beads loaded with GST:PrP^C were washed and equilibrated in thrombin cleavage buffer containing 10 U/mg of bovine thrombin and shaken for 4 h at 4°C. The released protein was recovered; thrombin was inactivated with phenylmethylsulfonyl fluoride (PMSF) and further purified by subsequent gel filtration.[6]

PK Digestion

hPrP90–231 in its different spatial conformation was digested with different concentrations of PK (Sigma Aldrich, Milano, Italy).The reaction was performed in physiological phosphate buffer (pH 7.3) for 1 h at 37°C. The reaction was stopped by boiling the sample in Loemli buffer. hPrP90–231/PK protein concentration ratios (w:w) ranged from 100:1 to 1:1.

Cell Culture and Treatments

SH-SY5Y human neuroblastoma cells were cultured in Dulbecco's modified Eagle's medium (DMEM) (Gibco, Rockville, MD) supplemented with 10% fetal bovine serum (FBS) (Gibco), 2 mM glutamine, 100 U/mL penicillin, and 100 μg/mL streptomycin; cells were maintained at 37°C in a humidified incubator under 95% air and 5% CO2.

Survival Assay

Mitochondrial function, as index of cell viability, was evaluated by measuring the reduction of 3-(4,5-dimethylthiazol-2-yl)-2,5-diphenyltetrazolium bromide (MTT, Sigma Aldrich). The cleavage of MTT to a purple formazan product by mitochondrial dehydrogenase was quantified spectrophotometrically, as previously reported.[7]

Bis-Benzimide Staining

Cells, plated on 25-mm glass coverslips, were treated with hPrP90–231 in 2% FBS containing medium for different times. At the end of the experiment, SH-SY5Y cells were fixed in 4% paraformaldehyde (30 min) and then incubated with 1 μg/mL bis-benzimide at 37°C for 30 min; after extensive washing, coverslips were analyzed for condensed or disrupted nuclei by means of a fluorescent microscope (Leica DM 2500, Leica Microsystems, Wetzlar, Germany). At least 1,000 cells per coverslip were analyzed, and experiments, performed in duplicate, were repeated at least three times.

Cytoplasmic Mono- and Oligonucleosome ELISA

Apoptosis was evaluated using the Cell Death Detection Enzyme Linked Immunoadsorbent Assay Kit (Roche Molecular Biochemicals, Indianapolis, IN) following the manufacturer's instructions.

Caspase-3/7 Activity

To quantify caspase 3/7 activity induced by hPrP90–231, we used the Apo-ONE Homogeneous Caspase-3/7 Assay (Promega, Milano, Italy) following the manufacturer's instructions.

Statistical Analysis

Experiments were performed in quadruplicate and repeated at least three times. Statistical analysis was performed by means of one-way ANOVA. A P-value less than 0.05 was considered statistically significant.

Chemicals

The 3F4 antibody was purchased from Signet (London, UK), anti-GST from Amersham Biosciences (Milano, Italy), and all other chemicals were, unless otherwise stated, from Sigma-Aldrich (Milano, Italy).

RESULTS

hPrP90–231 Transition in a β-Sheet-Rich Isoform

By controlled thermal denaturation (incubation at 53°C for 1h) we obtain the transition of hPrP90–231 from α-helix to a prevalent β-sheet structure, as demonstrated in circular dichroism (CD) experiments. The purified recombinant hPrP90–231 diluted in phosphate buffer (pH 7.2, 0.4 mg/mL) shows, in native condition, the typical α-helix conformation (two minima at 208 and 222 nm and an α-helix content of 42.4% and β-sheet content of 6.8%). After partial thermal denaturation, the CD spectrum of hPrP90–231 showed a single minimum at about 218 nm, indicating that a large number of β-structures are present in the peptide (α-helix content: 8.6%, β-sheet content: 58.8%). We characterized the structural differences between PrP^C and PrP^{Sc}, analyzing the resistance to PK proteolysis in native conditions and after thermal denaturation. Natively configured hPrP90–231 was completely degraded at a PrP:PK (w:w) ratio of 100:1 (FIG. 1a); after thermal denaturation the recombinant protein acquired partial resistance to digestion even when incubated with PK at the ratio of PrP:PK = 50:1 (FIG. 1b). Following prolonged incubation of hPrP90–231

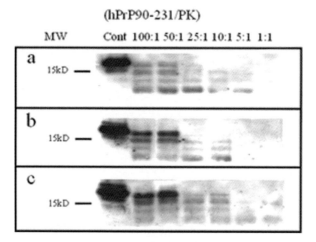

FIGURE 1. Protease K resistance of hPrP90–231. hPrP90–231 in its native form (**a**), after 1 h of incubation at 53°C (**b**), or after prolonged incubation (1 h at 53°C followed by 5 days at 37°C) (**c**) was incubated with protease K (1 h at 37°C). After digestion, protein content was size-fractionated by SDS-PAGE, and hPrP90–231 immunoreactivity detected using the 3F4 antibody. Apparent molecular weights (in kDa) are shown on the *left*, and the PrP/PK ratio 1000:1 to 1:10 (from 10 μg/25 μL hPrP90–231/10 ng/25 μL PK, to 10 μg/25μL hPrP90–231/100 μg/25 μμL PK) is indicated *above each lane*. PK digestion of α-hPrP90–231 causes the reduction of the 16-KDa band along with the appearance of cleaved products whose production was reduced by thermal denaturation.

(5 days at 37°C after 1 h at 53°C), PK resistance of the peptide further increased, being a small amount of the peptide still detectable in Western blot for a PrP:PK ratio of 10:1 (FIG.1c). These results are in accord with the high resistance of PrP^{Sc} in its C-terminal region to digestion with PK, generating a fragment starting near residue 90.[8] Thus, we demonstrate that a temperature- and time-dependent conversion of recombinant hPrP90–231 leads to the acquisition of some physico-chemical characteristics of PrP^{Sc}, including high content of β-structure and partial resistance to PK digestion. In opposition to previous reports, in which wild-type protein conversion was possible only in the presence of denaturing agents of low pH or reducing condition,[9] we obtained the *in vitro* conversion of hPrP90–231 in a PrP^{Sc}-like isoform under physiological buffer conditions.

β-Sheet-Structured hPrP90–231 Induces SH-SY5Y Cell Toxicity

To study the cytotoxicity properties of hPrP90–231 cytotoxicity properties in both conformations (either as α-helix-structured peptide or in a predominant β-rich conformation, obtained by thermal denaturation), we used a human neuroblastoma cell line (SH-SY5Y). This cell line is widely used to investigate neuronal apoptosis and necrosis[7,10,11] and represents a good model to characterize the PrP-dependent cell death *in vitro*.[12] The effects of hPrP90–231 were examined by adding the peptide directly to the culture medium, mimicking the interaction of PrP^{Sc} deposits with the neurons as occurred *in vivo*.[13] Cell viability and induction of apoptosis have been considered to identify and quantify the possible hPrP90–231 cytotoxicity. Treatments with hPrP90–231 have been managed in culture supplemented with 2% fetal bovine serum (FBS). The reduction of the FBS concentration to 2% inhibited cell proliferation, without a significant reduction in cell viability. By MTT reduction, quantified SH-SY5Y cell viability hPrP90–231-dependent cell death occurred in a structure- and concentration-dependent manner (TABLE 1). hPrP90–231 was scarcely toxic for SH-SY5Y cells in its native α-structured conformation (−8% of cell viability after 3 days of treatment). Conversely, β-rich hPrP90–231, in concentration of 100 nM, induced a significant reduction of SH-SY5Y viability after 1 day of treatment ($P < 0.05$ vs. its respective control), with maximal effect occurring after 3 days of treatment (−51%; $P < 0.01$). When higher concentrations were used (1 μM, TABLE 1), the effect of β-structured hPrP90–231 was more pronounced (−74%) after 3 days of treatment. These data strongly support the hypothesis that prion toxicity is associated with its secondary structure conformation. High concentrations of recombinant GST or PrP23–231 purified under the same experimental conditions as hPrP90–231, were used to demonstrate the specificity of hPrP90–231 cytotoxicity. In FIGURE 2, we show that only hPrP90–231, after thermal denaturation, induced a significant toxicity in SH-SY5Y cells. Conversely, treatments

TABLE 1. Dose response and time course of SH-SY5Y cell death induced by hPrP90–231

Concentration	Days of treatment		
	1	2	3
100 nM			
hPrP90–231 α	100.6 ± 5	97 ± 4.6	92 ± 4.6
hPrP90–231 β	83.9 ± 4.2*	82.9 ± 3.3	66.9 ± 3.3**
1 μM			
hPrP90–231 α	95 ± 4.8	83 ± 4.2*	58 ± 2.9**
hPrP90–231 β	67.7 ± 3.4**	55.5 ± 2.8**	26.8 ± 1.3**

Cells were treated with two different concentrations of α- and β-structured hPrP90–231 (100 nM and 1 μM), and cell viability was assessed by MTT test after 1, 2, and 3 days. Data are expressed as a percentage of vehicle-treated samples and each point represents the average of three experiments performed in quadruplicate. *$P < 0.05$ and **$P < 0.01$ versus control values.

with nondenatured hPrP90–231, hPrP23–231 with or without thermal denaturation, or high concentrations of bacterial purified recombinant GST, did not affect cell survival. This control was relevant because it excluded, on one hand, the possibility that bacterial contaminants may trigger the SH-SY5Y cell death and, on the other, that the effect was somehow nonspecifically mediated by denatured proteins. Moreover, these data underline the possibility that truncated PrP90–231, corresponding to species actually accumulating in TSE-affected brains, may behave in a substantially different way compared to hP23–231.

The peptide, corresponding to the amino acids 106–126 of PrP, was reported to retain *in vitro* most of the features of the pathogenic form of PrP.[14] For this reason it was used in many experimental studies on TSE. We decided to compare the toxic effects of this synthetic peptide with those of the recombinant hPrP90–231 fragment in its β-structured toxic isoform. As shown in FIGURE 3, hPrP90–231 reached, for lower concentration (10 or 100 nM), a maximal cell death effect after 5 days. Conversely, PrP 106–126 caused a significant cell death only after 5 days of treatment (−37% of cell viability) with PrP106–126 100 μM. These data indicate that the recombinant peptide is much more toxic than PrP106–126, showing higher potency and efficacy.

β-Sheet-Structured hPrP90–231 Induces SH-SY5Y Cell Apoptosis

In order to investigate the mechanisms involved in hPrP90–231 cell toxicity, we analyzed whether apoptosis or necrosis occurred in SH-SY5Y cells treated with this peptide. Apoptosis after 1 and 3 days of treatment was quantified by bis-benzimide staining (percentage of condensed fragmented nuclei in cells treated with vehicle and hPrP90–231, either α- or β-structured) (FIG. 4) After cell treatment with β-structured hPrP90–231, we observed a

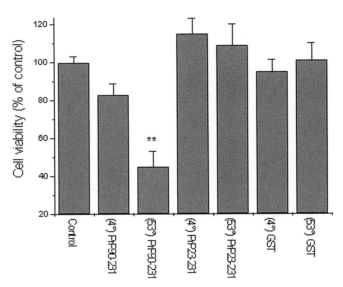

FIGURE 2. Effect of hPrP90–231 (1 μM), hPrP23–231 (1 μM), or GST (10 mM) on SH-SY5Y cell viability. Effect on SH-SY5Y cell viability induced by 4 days' treatment with hPrP90–231 (1 μM), hPrP23–231 (1 μM), or GST (10 mM) used as control unrelated protein, without and with thermal denaturation (4°C and 53°C). We observed that only hPrP90–231, after thermal denaturation, induces a significant toxicity for SH-SY5Y cells. Conversely, the treatment with nondenatured hPrP90–231 or hPrP23–231, with or without thermal denaturation, or with high concentrations of bacterial purified recombinant GST, did not affect cell survival. Data are expressed as a percentage of vehicle-treated samples and each point represents the average of three experiments performed in quadruplicate. **$P < 0.01$ versus control values.

highly significant induction of apoptosis after 1 (+10%) and 3 (+50%) days of treatments. Conversely, hPrP90–231 in α-conformation induced apoptosis only after prolonged treatment (3 days) and to a much lower extent than the hPrP90–231 in β-structure (about 8% vs. 50% of apoptotic nuclei). Apoptosis induced by the peptide was also determined by a specific ELISA, which measures the amount of mono- and oligonucleosomes released in the cytosol by the apoptotic endonucleases. Treatment with hPrP90–231(1 μM) in its α-native conformation showed a smaller, although significant apoptotic effect (+19% over control). The occurrence of apoptosis was conversely significantly detectable after 3 days of treatment with 1μM β-rich peptide (79% over control). To test whether hPrP90–231 may induce necrosis, we evaluated the release of lactate dehydrogenase (LDH) from the cytosol of cells treated with the peptide. We did not observe LDH release after 3 day of hPrP90–231 (1 μM) treatment independently of the thermal denaturation (data not shown), excluding that necrosis was primarily involved in hPrP90–231-induced cell death.

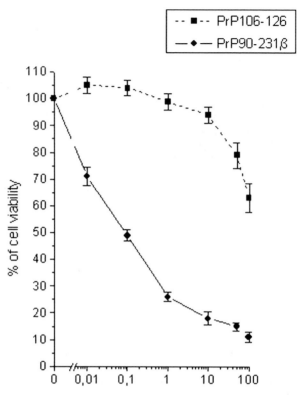

FIGURE 3. Comparison of potency and efficacy of the toxicity induced by PrP106–126 and β-hPrP90–231. Cells have been placed in 2% FBS-containing medium and treated with PrP106–126 (10-50-100 μM) or hPrP 90–231β-structured (10–100 nM and 1 μM). Cell viability has been measured by the MTT test after 5 days of treatment. Data are expressed as percentages of control and each value is the mean of three experiments in quadruplicate.**$P \leq 0.01$, value are express as percentages of untreated controls; $^{oo}P \leq 0.01$, value are express as percentages versus PrP 90–231β-structured.

Caspases and p38 MAP Kinase Involvement in hPrP90–231 Apoptotic Activity

To identify the molecular executors of hPrP90–231-induced apoptosis, we analyzed the involvement of caspases and p38 MAP kinase in the toxic effects of the peptide. Among the number of intracellular factors that contributes to the activation and execution of apoptosis, a central role is played by caspases. Caspases are cysteine proteases that cleave different substrates with high specificity and take part in the apoptotic process at various levels.[15] The possible caspase involvement in the toxic activity of the prion fragment was measured,

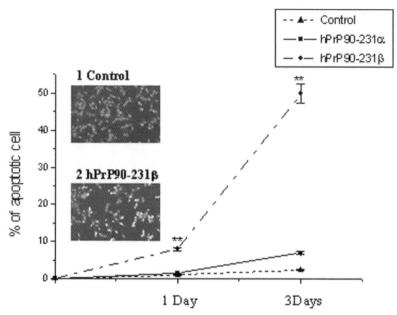

FIGURE 4. β-structured hPrP90–231-induced apoptosis of SH-SY5Y cells. After 1 and 3 days of treatment, with vehicle (1 μM) or hPrP90–231 after thermal denaturation (1 μM), SH-SY5Y cells were fixed in paraformaldehyde and stained with the fluorescent nuclear bis-benzimide and microphotographed. Apoptosis was evidenced by nuclear fragmentation and condensation. Apoptosis was quantified after 1 or 3 days of treatment with hPrP90–231 (native and β-structured) by counting at least 1,000 cells for each experimental point. Data are expressed as percentage of apoptotic cells evidenced by nuclear fragmentation and condensation. $*P < 0.05$, $**P < 0.01$, versus control

evaluating the effects of hPrP90–231 on SH-SY5Y cell viability in the presence of the caspase inhibitor Z-VAD-FMK. As shown in FIGURE 5A, after 3 days of treatment with hPrP90–231 (β-structured, 1 μM), PrP-dependent cell death was significantly inhibited in the presence of the caspase blocker Z-VAD-FMK (1 μM), as assessed by MTT reduction. By quantification of apoptotic production of oligonucleosomes, we showed that hPrP90–231-dependent apoptosis was significantly inhibited in the presence of the caspase blocker Z-VAD-FMK (1 μM) after 3 days of treatment with β-structured hPrP90–231 (1 μM) (FIG. 5B). Finally, we observed, using a caspase 3–7 activity fluorimetric assay, that after 5 h of treatment, hPrP90–231 (1 μM) caused a significant increase in caspase proteolytic activity that was completely abolished by coincubation with Z-VAD-FMK (1 μM) (FIG. 5C). In addition to caspases, the mitogen-activated protein kinases (MAP kinases) are important regulators of several degenerative processes.[16] Therefore, we evaluated the role of p38 MAP kinase in the

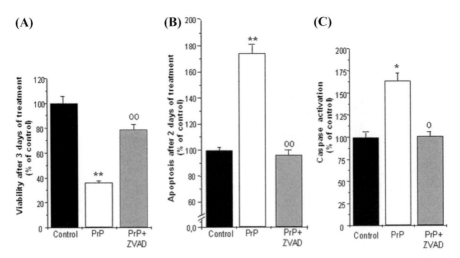

FIGURE 5. Caspase involvement in activation of β-hPrP90–231 apoptotic activity. (**A**) Effect of Z-VAD-FMK (1 μM) on SH-SY5Y cell death induced by hPrP90–231 in its β-structured isoform. Cells were placed in 2% FBS-containing DMEM and treated with peptide (1 μM) in the absence (*white bar*) or presence (1 μM) (*gray bar*) of Z-VAD-FMK. Cell viability was assessed by MTT test after 3 days of treatment. **$P < 0.01$ versus control values. °°$P < 0.01$ versus hPrP90–231 values. (**B**) Apoptosis induction assayed, using an ELISA test for the production of oligonucleosomes, was evaluated after 2 days of treatment with hPrP90–231 (1 μM, β-structured) in the absence (*white bar*) or (*gray bar*) presence of Z-VAD-FMK (1μM). Data are expressed as percentages of the control values. **$P < 0.05$ versus control values. °°$P < 0.05$ versus hPrP90–231 values. (**C**) Effect of hPrP90–231 (1 μM, β-structured, for 5 h) on caspase 3–7 activity (*white bar*) in SH-SY5Y cells measured using the profluorescent substrate Z-DEVD-rhodamine110 (1μM). The coincubation with Z-VAD-FMK completely reversed this effect (*gray bar*). Data are expressed as percentages of the control values. *$P < 0.05$ versus control values. °$P < 0.05$ versus hPrP90–231 values.

proapoptotic effects of β-structured hPrP90–231. As shown in FIGURE 6A, hPrP90–231 (1 μM) used with the selective blocker SB203580 (1μM) totally abolished cell death and DNA cleavage induced by hPrP90–231. Moreover, the activation/phosphorylation of p38 MAP kinase by hPrP90–231 was directly assessed by Western blot. Cells were treated (FIG. 6C) with vehicle, hPrP90–231 (1 μM), or hPrP90–231 + SB203580 (1 μM) for 24 h and p38 phosphorylation was assessed by Western blot using phosphospecific antibodies. The treatment with hPrP90–231 (1 μM) caused a significant activation of p38 (FIG. 6 C), while the addition of SB203580 caused a reduction of phosphorylation of p38. These data indicate that the activation of p38 MAP kinase pathway represents a critical step for the execution of apoptosis induced by hPrP90–231 in our cell model.

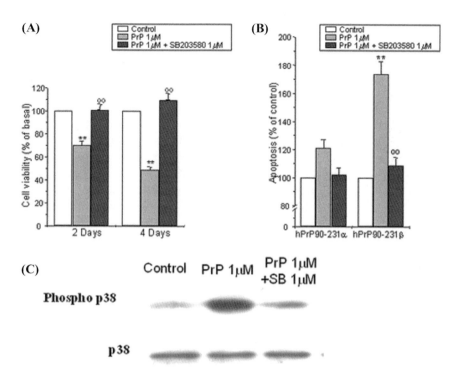

FIGURE 6. p38 MAP kinase involvement in β-hPrP90–231 apoptotic activity. (**A**) Effect of SB203580 (1 μM) on SH-SY5Y cell death induced by hPrP90–231 in its β-structured isoform. Cells were placed in 2% FBS-containing DMEM and treated with peptide (1 μM) in the absence (*light gray bars*) or presence of SB203580 1μM (*dark grey bars*). Cell viability was assessed by MTT test after 2 and 4 days of treatment. **$P < 0.01$ versus control values. $^{\circ\circ}P < 0.01$ versus hPrP90–231 values. (**B**) Apoptosis induction assayed, using an ELISA test for the production of oligonucleosomes, was evaluated after 2 days of treatment with hPrP90–231 (1 μM, β-structured) in the absence (*light gray bars*) or presence of SB203580 (1 μM) (*dark gray bars*). Data are expressed as percentages of the control values. **$P < 0.01$ versus control values. $^{\circ\circ}P < 0.01$ versus hPrP90–231 values. (**C**) hPrP90–231 induced MAP kinase p-38 phosphorylation/activation in SH-SY5Y cell line. Cells were treated with vehicle (**C**), hPrP90–231 (1 μM), or hPrP90–231 + SB203580 (1 μM) for 24 h and p38 phosphorylation was assessed by Western blot using phosphospecific antibodies.

DISCUSSION

The comprehension of the molecular and cellular mechanisms that cause neuronal loss represents a crucial step to clarify the pathophysiology of prion diseases and to develop therapies. PrP misfolding is believed to represent the causative event of the extracellular accumulation of protease-resistant amino terminally truncated peptides, called PrP27–30. PrP27–30 accumulation was hypothesized to play a role in neuronal death.[17]

Treatments of neuronal cell cultures with PrP^{Sc} extracts or prion-derived peptides are considered a suitable approach to mimic PrP^{Sc} intracerebral accumulation and toxicity. Recombinant peptides corresponding to the amino-terminally truncated PrP^{Sc} are now widely used to study PrP^{C}–PrP^{Sc} refolding on account of their capability to adopt either α-helical or β-sheet predominant structure *in vitro*. Most studies employed small peptides encompassing specific PrP domains believed to be crucial for the neurotoxicity. In particular, PrP106–126 was used to highlight possible death pathways activated in TSE-affected brains. However, a major limitation in the use of these peptides is their low efficacy and potency: for instance, toxicity after long time of treatment (up to 7–10 days) and extremely high concentrations of peptides (50–100 μM).[18–22] Moreover, PrP106–126, reproducing only a small portion of PrP and being constitutively β-structured, does not allow a comparison between the three-dimensional structure and the gain in toxicity. Last but not least, its pathophysiological relevance may be questioned since it has never been demonstrated to originate from the partial cleavage of any pathological prion strain. Thus, the developing of more complete experimental models to assess PrP^{Sc} neurotoxicity mechanisms still represents an important research goal. Recombinant PrP-related polypeptides, expressed and purified in *E. coli*, may offer a new and very promising tool to investigate physical, chemical, and biological properties of both PrP^{C} and PrP^{Sc}. We decided to set up a protocol in which it was possible to obtain a recombinant peptide corresponding to the PrP amino acids 90–231 in two different molecular three-dimensional conformations to verify the possible relationship between toxicity and structure in prion protein. Therefore, we expressed and purified in *E. coli* the recombinant hPrP90–231, preserving its native, soluble, and monomeric structure.[6] Using this soluble PrP fragment, we identified a simple protocol (i.e., limited thermal denaturation 1 h at 53°C without chemical denaturation or alteration of pH and ionic strength) to induce its conversion from an α-structured protein (resembling PrP^{C}) to a β-sheet-containing protein (a model of PrP^{Sc}).

In our experiments we demonstrate that the PrP toxicity is highly enhanced after thermal denaturation. These data represent novel and relevant information showing, for the first time, in an *in vitro* model of toxicity, a precise relationship between the spatial structure of a recombinant pathology-related PrP fragment and its biological effects. Indeed, in most of the previously published studies aimed at identifying experimental models to evaluate the conversion from PrP^{C} to PrP^{Sc}, the possible cellular effects of this structural transition are not evaluated.[23,24] In our experiments, we demonstrated that hPrP90–231 toxicity occurred at concentrations and incubation times significantly shorter than previously reported for synthetic peptides or other recombinant PrP-derived recombinant proteins.[14,25,26,27]

One of the aims of this work was to investigate some molecular aspects of the cell death machinery activated by hPrP90–231, focusing, in particular, on the role of caspases and MAP kinases in this process. For this purpose, we used the neuronal-like SH-SY5Y cell line. The SH-SY5Y cells are widely used to study intracellular events linked to neurotoxic insults, such as trophic factor withdrawal, oxidative stress, DNA damage, and neurodegenerative disorders.[28,29] A prolonged exposure of SHSY5Y cells to the peptide induced apoptosis in a time- and concentration-dependent manner. By MTT reduction, we demonstrated that PrP toxicity is associated with its secondary structure conformation; indeed, recombinant hPrP90–231 in its native α-helix structure displays minimal effects on cell survival, while, when β-structured, it exerts a powerful cell death. To investigate whether hPrP90–231 cell death showed hallmarks of apoptosis, we determined whether hPrP90–231 treatment caused a direct activation of caspases and if the inhibition of these enzymes could prevent the cell death induced by the peptide. Our data show that PrP peptide-dependent cell death was significantly inhibited in the presence of the pan-caspase blocker Z-VAD-FMK (1 μM) after 3 days of treatment with β-structured hPrP90–231. Then we focused our study on the role of the MAP kinase enzymatic cascades. The p38 MAP kinase, beyond its role in immune response,[30] represents intracellular effectors that transduce different cell insults, including trophic factor withdrawal, oxidative stress, and glutamate toxicity.[28,31,32] Western blot analysis showed an increase of the immunoreactivity for the phosphorylated/activated p38 MAP kinase when SH-SY5Y cells were exposed to β-structured hPrP90–231 after 24 h of treatment. Moreover, the cotreatment of the SHSY5Y with β-structured hPrP90–231 in the presence of the p38 inhibitor SB203580 completely blocked the apoptosis induced by 2 days of treatment with β-structured hPrP90–231 and prevented the reduction of cell viability caused by 2 and 4 days of exposure to the peptide. These results support the hypothesis of the p38 MAP kinase involvement in the SH-SY5Y death and indicate that the increase of p38 phosphorylation plays a critical role in the cell death triggered by the peptide, rather than being a consequence of cellular stresses induced through other pathways. In conclusion, this is the first report in which the biological *in vitro* effects of a recombinant PrP fragment are completely related to its three-dimensional structure, as it is supposed to occur for PrP during TSE. Moreover, in our opinion the identification of specific intracellular pathways related to hPrP90–231 toxicity represents a relevant indication to monitor the effectiveness of compounds possibly able to prevent or reverse the gain of toxicity of pathogen isoforms of PrP.

ACKNOWLEDGMENTS

The financial support provided by Italian Ministry of University and Research (MIUR) FIRB2001 (RBNE01ARR4˙004) and PRIN2004 to T. Florio and MIUR PRIN 2004 to A. Aceto is gratefully acknowledged.

REFERENCES

1. PRUSINER, S.B. 2001. Shattuck lecture–neurodegenerative diseases and prions. N. Engl. J. Med. **344:** 1516–1526.
2. PRUSINER, S.B., M.R. SCOTT, S.J. DEARMOND & F.E. COHEN. 1998. Prion protein biology. Cell **93:** 337–348.
3. AGUZZI, A. & M. POLYMENIDOU. 2004. Mammalian prion biology: one century of evolving concepts. Cell **116:** 313–327.
4. GALLO, M., D. PALUDI, D.O. CICERO, et al. 2005. Identification of a conserved N-capping box important for the structural autonomy of the prion alpha 3-helix: the disease associated D202N mutation destabilizes the helical conformation. Int. J. Immunopathol. Pharmacol. **18:** 95–112.
5. CHIESA, R. & D.A. HARRIS 2001. Prion diseases: what is the neurotoxic molecule? Neurobiol. Dis. **8:** 743–763.
6. CORSARO, A., S. THELLUNG, C. RUSSO, et al. 2002. Expression in *E. coli* and purification of recombinant fragments of wild type and mutant human prion protein. Neurochem. Int. **41:** 55–63.
7. THELLUNG, S., V. VILLA, A. CORSARO, et al. 2002. p38 MAP kinase mediates the cell death induced by PrP106-126 in the SH-SY5Y neuroblastoma cells. Neurobiol. Dis. **9:** 69–81.
8. PRUSINER, S.B. 1998. Prions. Proc. Natl. Acad. Sci. USA **95:** 13363–13383.
9. MORILLAS, M., D.L. VANIK & W.K. SUREWICZ. 2001. On the mechanism of alpha-helix to beta-sheet transition in the recombinant prion protein. Biochemistry **40:** 6982–6987.
10. FLORIO, T., D. PALUDI, V. VILLA, et al. 2003. Contribution of two conserved glycine residues to fibrillogenesis of the 106–126 prion protein fragment. Evidence that a soluble variant of the 106–126 peptide is neurotoxic. J. Neurochem. **85:** 62–72.
11. CORSARO, A., S. THELLUNG, V. VILLA, et al. 2003. Prion protein fragment 106–126 induces a p38 MAP kinase-dependent apoptosis in SH-SY5Y neuroblastoma cells independently from the amyloid fibril formation. Ann. N. Y. Acad. Sci. **1010:** 610–622.
12. O'DONOVAN, C.N., D. TOBIN & T.G. COTTER. 2001. Prion protein fragment PrP-(106–126) induces apoptosis via mitochondrial disruption in human neuronal SH-SY5Y cells. J. Biol. Chem. **276:** 43516–43523.
13. HETZ, C., M. RUSSELAKIS-CARNEIRO, K. MAUNDRELL, et al. 2003. Caspase-12 and endoplasmic reticulum stress mediate neurotoxicity of pathological prion protein. Embo. J. **22:** 5435–5445.
14. FORLONI, G., N. ANGERETTI, R. CHIESA, et al. 1993. Neurotoxicity of a prion protein fragment. Nature **362:** 543–546.
15. HENGARTNER, M.O. 2000. The biochemistry of apoptosis. Nature **407:** 770–776.
16. MIELKE, K. & T. HERDEGEN. 2000. JNK and p38 stresskinases–degenerative effectors of signal-transduction-cascades in the nervous system. Prog. Neurobiol. **61:** 45–60.
17. ZOU, W.Q., S. CAPELLARI, P. PARCHI, et al. 2003. Identification of novel proteinase K-resistant C-terminal fragments of PrP in Creutzfeldt-Jakob disease. J. Biol. Chem. **278:** 40429–40436.
18. FORLONI, G., R. DEL BO, N. ANGERETTI, et al. 1994. A neurotoxic prion protein fragment induces rat astroglial proliferation and hypertrophy. Eur. J. Neurosci. **6:** 1415–1422.

19. THELLUNG, S., T. FLORIO, A. CORSARO, et al. 2000. Intracellular mechanisms mediating the neuronal death and astrogliosis induced by the prion protein fragment 106–126. Int. J. Dev. Neurosci. **18:** 481–492.
20. THELLUNG, S., T. FLORIO, V. VILLA, et al. 2000. Apoptotic cell death and impairment of L-type voltage-sensitive calcium channel activity in rat cerebellar granule cells treated with the prion protein fragment 106–126. Neurobiol. Dis. **7:** 299–309.
21. BROWN, D.R., B. SCHMIDT & H.A. KRETZSCHMAR. 1996. Role of microglia and host prion protein in neurotoxicity of a prion protein fragment. Nature **380:** 345–347.
22. FLORIO, T., S. THELLUNG, C. AMICO, et al. 1998. Prion protein fragment 106–126 induces apoptotic cell death and impairment of L-type voltage-sensitive calcium channel activity in the GH3 cell line. J. Neurosci. Res. **54:** 341–352.
23. JACKSON, G.S., A.F. HILL, C. JOSEPH, et al. 1999. Multiple folding pathways for heterologously expressed human prion protein. Biochim. Biophys. Acta **1431:** 1–13.
24. SWIETNICKI, W., M. MORILLAS, S.G. CHEN, et al. 2000. Aggregation and fibrillization of the recombinant human prion protein huPrP90-231. Biochemistry **39:** 424–431.
25. BROWN, D.R. 2000. PrPSc-like prion protein peptide inhibits the function of cellular prion protein. Biochem. J. **352:** 511–518.
26. DANIELS, M., G.M. CEREGHETTI & D.R. BROWN. 2001. Toxicity of novel C-terminal prion protein fragments and peptides harbouring disease-related C-terminal mutations. Eur. J. Biochem. **268:** 6155–6164.
27. PILLOT, T., B. DROUET, M. PINCON-RAYMOND, et al. 2000. A nonfibrillar form of the fusogenic prion protein fragment [118–135] induces apoptotic cell death in rat cortical neurons. J. Neurochem. **75:** 2298–2308.
28. GHATAN, S., S. LARNER, Y. KINOSHITA, et al. 2000. p38 MAP kinase mediates bax translocation in nitric oxide-induced apoptosis in neurons. J. Cell Biol. **150:** 335–347.
29. MORIYA, R., T. UEHARA & Y. NOMURA. 2000. Mechanism of nitric oxide-induced apoptosis in human neuroblastoma SH-SY5Y cells. FEBS Lett. **484:** 253–260.
30. HAN, J., J.D. LEE, L. BIBBS & R.J. ULEVITCH. 1994. A MAP kinase targeted by endotoxin and hyperosmolarity in mammalian cells. Science **265:** 808–811.
31. KUMMER, J.L., P.K. RAO & K.A. HEIDENREICH. 1997. Apoptosis induced by withdrawal of trophic factors is mediated by p38 mitogen-activated protein kinase. J. Biol. Chem. **272:** 20490–20494.
32. KAWASAKI, H., T. MOROOKA, S. SHIMOHAMA, et al. 1997. Activation and involvement of p38 mitogen-activated protein kinase in glutamate-induced apoptosis in rat cerebellar granule cells. J. Biol. Chem. **272:** 18518–18521.

Role of NADPH Oxidase and Calcium in Cerulein-Induced Apoptosis

Involvement of Apoptosis-Inducing Factor

JI HOON YU,[a] KYUNG HWAN KIM,[a] AND HYEYOUNG KIM[b]

[a]*Department of Pharmacology and Institute of Gastroenterology, Brain Korea 21 Project for Medical Science, College of Medicine, Yonsei University, Seoul 120-752, Korea*

[b]*Department of Food Science and Nutrition, Brain Korea 21 Project, College of Human Ecology, Biomedical Secretion Research Center, Yonsei University, Seoul 120-749, Korea*

> ABSTRACT: Apoptosis linked to oxidative stress has been implicated in pancreatitis. NADPH oxidase has been considered as a major source of reactive oxygen species (ROS) during inflammation and apoptosis in pancreatic acinar cells. Recently we demonstrated that NADPH oxidase subunits Nox1, $p27^{phox}$, $p47^{phox}$, and $p67^{phox}$ are constitutively expressed in pancreatic acinar cells and may contribute to apoptosis in pancreatic acinar AR42J cells stimulated with cerulein. The present study aims to investigate the apoptotic mechanism of pancreatic acinar cells stimulated with cerulein by determining whether cerulein induces apoptosis-inducing factor (AIF) expression and whether cerulein-induced expression of AIF is inhibited by transfection with antisense oligonucleotide (AS ODN) of $p47^{phox}$ or $p67^{phox}$ or treatment with a Ca^{2+} chelator BAPTA-AM. As a result, cerulein induced the expression of apoptotic gene AIF. Transfection with AS ODN of $p47^{phox}$ or $p67^{phox}$ or treatment with BAPTA-AM inhibited cerulein-induced AIF expression in pancreatic acinar AR42J cells. These results demonstrate that NADPH oxidase and calcium have a role in cerulein-induced apoptosis in pancreatic acinar AR42J cells by inducing the expression of AIF. In conclusion, the increase in intracellular Ca^{2+} and NADPH oxidase activity may be the upstream event of apoptotic gene (AIF) expression, which contributes to cerulein-induced apoptosis in pancreatic acinar AR42 cells.
>
> KEYWORDS: cerulein; NADPH oxidase; AIF; Ca^{2+}; AR42J cells

---

Address for correspondence: Hyeyoung Kim, Department of Food Science and Nutrition, Yonsei University College of Human Ecology, Seoul 120-749, Korea. Voice: 82-2-2228-1734; fax: 82-2-313-1894.
  e-mail: kim626@yumc.yonsei.ac.kr

INTRODUCTION

Reactive oxygen species (ROS) play a role in the pathogenesis of acute pancreatitis.[1] Studies in experimental models of pancreatitis indicate that pancreatic oxidative stress occurs during the early stage of induction.[2] Scavenger therapy for ROS has attained some success in experimental pancreatitis models. High doses of cerulein, a cholecystokinin (CCK) analogue, result in acute pancreatitis, which is characterized by a dysregulation of the production and secretion of digestive enzymes, the inhibition of pancreatic secretion, and an elevation in their serum levels, cytoplasmic vacuolization, the death of acinar cells, and edema formation.

Recent studies revealed that apoptosis is a major mechanism of acinar cell death in various experimental models of acute pancreatitis, suggesting a role for apoptosis in the pathophysiology of the disease.[3] Mitochondria play an important role in the regulation of apoptosis. When the mitochondrial pathway is triggered, pro-apoptotic proteins, such as apoptosis-inducing factor (AIF), translocate from their mitochondrial locations into the cytosol.[4] In the caspase-independent pathway, the cytosolic AIF translocates to the nucleus, resulting in DNA fragmentation.[5] Supraphysiologic concentrations of cerulein induce apoptosis in the rat pancreatic acinar AR42J cells. We previously showed that high concentration of cerulein induced the expression of pro-apoptotic gene bax and p53 and DNA fragmentation in AR42J cells, which was mediated by intracellular Ca^{2+}.[6] However, the mechanism of how cerulein induces apoptosis in pancreatic acinar cells remains to be investigated.

There is evidence that a major source of ROS during inflammation and apoptosis is NADPH oxidase.[7] In phagocytic cells, the NADPH oxidase is composed of the membrane-bound subunits $gp91^{phox}$ and $p22^{phox}$ as well as the cytosolic subunits $p67^{phox}$ and $p47^{phox}$. Upon activation of the enzyme, a complex of the cytosolic subunits translocates to the membrane and facilitates NADPH-dependent formation of superoxide (O_2^-), which in turn gives rise to the production of other secondary ROS such as hydrogen peroxide (H_2O_2). Since apoptosis is linked to oxidative stress in pancreatitis, NADPH oxidase may have a novel role in the apoptotic mechanism in pancreatic acinar cells. There is rapid elevation of cytosolic Ca^{2+}, resulting in the activation of proteolytic enzymes and mitochondrial damage, eventually leading to cell membrane disruption and cell death. Mitochondrial Ca^{2+} accumulation may lead to opening of the mitochondrial permeability transition pore, with release of the pro-apoptotic factor from the intermembrane space followed by apoptosis. Recently, we found the presence of NADPH oxidase subunits Nox1, $p27^{phox}$, $p47^{phox}$, and $p67^{phox}$ in pancreatic acinar cells, which was activated by cerulein to induce IL-6 expression and apoptosis.[8]

This study investigates the apoptotic mechanism of pancreatic acinar cells stimulated with cerulein by determining whether cerulein induces AIF expression. To investigate the possible involvement of NADPH oxidase and intracellular Ca^{2+} on AIF expression, antisense oligonucleotides (AS ODN) or sense oligonucleotides (S ODN) for NADPH oxidase subunit $p27^{phox}$ or $p47^{phox}$ or a Ca^{2+} chelator, BAPTA-AM (1,2-bis [o-aminophenoxy]ethane-N,N,N′,N′-tetraacetic acid tetra[acetoxymethyl] ester) was applied to the cells.

METHODS

AR42J cells (pancreatoma, ATCC CRL 1492) were cultured in Dulbecco's modified Eagle's medium supplemented with 10% fetal bovine serum and antibiotics. mRNA and protein expression of AIF were determined in the cells cultured in the presence or absence of cerulein (10^{-7} M). The cells were either treated with or without 10 μM BAPTA-AM or transfected with AS ODN or S ODN. The cells were treated with BAPTA-AM for 30 min prior to the stimulation with cerulein. The cells were stimulated with cerulein for 8 h (mRNA expression of AIF) and 24 h (protein expression of AIF). For transfection with AS ODN or S ODN, phosphothioate-modified oligonucleotides (ODNs) were produced commercially (Gibco-BRL, Grand Island, NY, USA). The sequences of $p22^{phox}$ antisense (AS) and sense (S) ODNs were GATCTG-CCCCATGGTGAGGACC and GGTCCTCAC-CATGGGGCAGATC. The sequences of the $p47^{phox}$ AS and S ODN were CTGTTGAAGTACTCGGTGAG and CTCAC CGAGTACTTC-AACAG. The cells were treated with ODNs using transfection reagent DOTAP (Boehringer-Mannheim, Mannheim, Germany). Protein expression was determined by Western blotting and standardized by actin as an internal control. mRNA expression was determined by reverse transcriptase-polymerase chain reaction (RT-PCR) and standardized by co-amplification with the housekeeping gene GAPDH as an internal control. The sequences of primers used in the PCR are as follows: AIF forward ACTCCAAGAAGTCTGTCTGC-TATCGA, reverse CTGACTCCAACTGATTGTACAATT-GC; GAPDH forward ACCACAGTCCATGCCATCAC, reverse TCCACCACCCTGTTGCT GTA. After co-amplification with 32 to 35 cycles, the PCR products were separated on 1.5% agarose gels and visualized by UV transillumination.

RESULTS

To determine the relation among intracellular Ca^{2+}, activation of NADPH oxidase, and AIF expression in the cells stimulated with cerulein, the cells were treated with BAPTA-AM or transfected with AS ODNs or S ODN of $p22^{phox}$ or $p47^{phox}$ and cultured in the presence of cerulein (10^{-7} M). Cerulein-induced expression (mRNA and protein) of AIF was inhibited by transfection with AS ODNs of $p22^{phox}$ and $p47^{phox}$, but not by transfection with S ODNs (FIG. 1A, B). Cerulein-induced mRNA and protein expression of AIF were

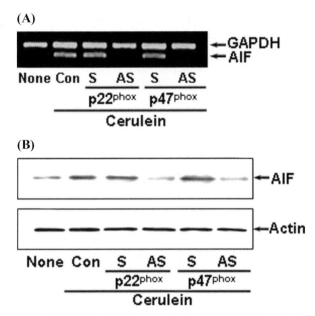

FIGURE 1. Cerulein-induced AIF expression in AR42J cells transfected with or without AS ODNs or S ODN of p22^{phox} or p47^{phox}. The cells were transfected with or without AS ODNs or S ODN of p22^{phox} or p47^{phox} and cultured in the presence or absence of cerulein (10^{-7} M) for 8 h (mRNA) and 24 h (protein) (**A, B**). mRNA expression was determined by RT-PCR and standardized by co-amplification with the housekeeping gene GAPDH as an internal control (**A**). Protein expression was determined by Western blotting and standardized by actin as an internal control (**B**). None, the cells without treatment; Con, the cells with cerulein treatment; S, the cells transfected with S ODN with cerulein treatment; AS, the cells transfected AS ODN with cerulein treatment.

inhibited in the cells treated with BAPTA-AM (FIG. 2A, B). Inhibition of NADPH oxidase activation or Ca^{2+} chelation by transfection with AS ODNs of p22^{phox} or p47^{phox} or treatment with BAPTA-AM significantly suppressed cerulein-induced AIF expression in the cells, as compared to the cells without treatment or those transfected with the corresponding S ODNs.

DISCUSSION

Oxidative stress is regarded as a pathogenic factor in acute pancreatitis. An increase in intracellular Ca^{2+} and ROS is involved in pancreatitis induced by CCK and its analogue cerulein.[8,9] Our previous study showed that NADPH oxidase subunits gp91^{phox}, homologue Nox1, p22^{phox}, p47^{phox}, and p67^{phox} are constitutively expressed in pancreatic acinar AR42J cells. Therefore, cerulein-induced ROS production could be derived from NADPH oxidase activated

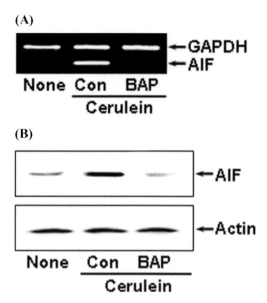

FIGURE 2. Cerulein-induced AIF expression in AR42J cells treated with or without BAPTA-AM. The cells were treated with or without BAPTA-AM (10 μM) and cultured in the presence or absence of cerulein (10^{-7} M) for 8 h (mRNA) and 24 h (protein) (**A, B**). mRNA expression was determined by RT-PCR and standardized by co-amplification with the housekeeping gene GAPDH as an internal control (**A**). Protein expression was determined by Western blotting and standardized by actin as an internal control (**B**). None, the cells without treatment; Con, the cells with cerulein treatment; BAP, the cells with BAPTA-AM treatment.

by inducing the translocation of cytosolic subunits $p47^{phox}$ and $p67^{phox}$ to the membrane in pancreatic acinar AR42J cells.

ROS produced during pancreatitis are mediators of apoptosis, which is involved in the pathogenesis and progression of acute pancreatitis.[10] ROS production and apoptosis were closely related in cerulein-induced pancreatitis.[11] In the present study, we focused on the possible relations among cerulein-induced activation of NADPH oxidase, increase in intracellular Ca^{2+}, and the expression of apoptotic gene AIF. We found that cerulein induced AIF expression, which was suppressed by the inhibition of NADPH oxidase activity or chelation of intracellular Ca^{2+}. The results suggest that ROS produced by activated NADPH oxidase together with increase in intracellular Ca^{2+} may upregulate the expression of apoptotic genes such as AIF and thus induce apoptosis in cerulein-stimulated AR42J cells.

In conclusion, inhibition of NADPH oxidase activation or suppression of the increase in intracellular Ca^{2+} may alleviate cerulein-induced apoptosis in pancreatic acinar cells by inhibiting the expression of the apoptotic gene, AIF.

ACKNOWLEDGMENT

This study was supported by a grant from the Korea Research Foundation (2005-041-E00092) to K.H.K. The study was supported by the Brain Korea 21 Project, Yonsei University College of Human Ecology.

REFERENCES

1. SCHENBERG, M.H., D. BIRK & H.G. BERGER. 1995. Oxidative stress in acute and chronic pancreatitis. Am. J. Clin. Nutr. **62:** 1306S–1314S.
2. GOUGH, D.B., B. BOYLE, W.P. JOYCE, et al. 1990. Free radical inhibition and serial chemiluminescence in evolving experimental pancreatitis. Br. J. Surg. **77:** 1256–1259.
3. KAISER, A.M., A.K. SALUJA, A. SENGUPTA, et al. 1995. Relationship between severity, necrosis, and apoptosis in five models of experimental acute pancreatitis. Am. J. Physiol. **269:** C1295–C1304.
4. WANG, X., C. YANG, J. CHAI, et al. 2002. Mechanisms of AIF-mediated apoptotic DNA degradation in *Caenorhabditis elegans*. Science **298:** 1587–1592.
5. CREGAN, S.P., V.L. DAWSON & R.S. SLACK. 2004. Role of AIF in caspase-dependent and caspase-independent cell death. Oncogene **23:** 2785–2796.
6. YU, J.H., J.W. LIM, K.H. KIM, et al. 2005. NADPH oxidase and apoptosis in cerulein-stimulated pancreatic acinar AR42J cells. Free Radic. Biol. Med. **39:** 590–602.
7. HIRAOKA, W., N. VAZQUEZ, W. NIEVES-NEIRA, et al. 1998. Role of oxygen radicals generated by NADPH oxidase in apoptosis induced in human leukemia cells. J. Clin. Invest. **102:** 1961–1968.
8. YU, J.H., J.W. LIM, H. KIM & K.H. KIM. 2005. NADPH oxidase mediates interleukin-6 expression in cerulein-stimulated pancreatic acinar cells. Int. J. Biochem. Cell Biol. **37:** 1458–1469.
9. RARATY, M., J. WARD, G. ERDEMLI, et al. 2000. Calcium-dependent enzyme activation and vacuole formation in the apical granular region of pancreatic acinar cells. Proc. Natl. Acad. Sci. USA **97:** 13126–13131.
10. SANDOVAL, D., A. GUKOVSKAYA, P. REAVEY, et al. 1996. The role of neutrophils and platelet-activating factor in mediating experimental pancreatitis. Gastroenterology **111:** 1081–1091.
11. GUKOVSKAYA, A.S., I. GUKOVSKY, V. ZANINOVIC, et al. 1997. Pancreatic acinar cells produce, release, and respond to tumor necrosis factor-alpha. Role in regulating cell death and pancreatitis. J. Clin. Invest. **100:** 1853–1862.

Signaling for Integrin α5/β1 Expression in *Helicobacter pylori*–Infected Gastric Epithelial AGS Cells

SOON OK CHO,[a] KYUNG HWAN KIM,[a] JOO-HEON YOON,[b] AND HYEYOUNG KIM[c]

[a]*Department of Pharmacology and Institute of Gastroenterology, Brain Korea 21 Project for Medical Science, Yonsei University College of Medicine, Seoul 120-752, Korea*

[b]*Department of Otorhinolaryngology and Biomolecule Secretion Research Center, Brain Korea 21 Project for Medical Science, Yonsei University College of Medicine, Seoul 120752, Korea*

[c]*Department of Food and Nutrition and Biomolecule Secretion Research Center, Brain Korea 21 Project, Yonsei University College of Human Ecology, Seoul 120-749, Korea*

ABSTRACT: Integrin expression in cancer tissues demonstrates its possible contribution to tumor progression, invasion, and metastasis. *Helicobacter pylori* (*H. pylori*) infection is related to gastric cancer and gastric inflammation. *H. pylori* induced upregulation in expression of integrin in gastric epithelia cells. Reactive oxygen species (ROS) are considered as an important regulator in the pathogenesis of *H. pylori*–induced gastric ulceration and carcinogenesis. Integrin expression may be regulated by oxidant-sensitive transcription factors, nuclear factor-κB (NF-κB) and activator protein-1 (AP-1). The present study aims to investigate whether *H. pylori* in a Korean isolate (HP99) induces the expression of integrin α5 and integrin β1, and whether *H. pylori*–induced expression of integrin α5 and integrin β1 are inhibited in the cells transfected with mutant genes for Ras (ras N-17), c-Jun (TAM-67), and IκBα(MAD-3) or treated with DPI, an inhibitor of NADPH oxidase. As a result, *H. pylori* induced the expression of integrin α5 and integrin β1 in gastric adenocarcinoma (AGS) cells time-dependently. Treatment of DPI or transfection with mutant genes for Ras (ras N-17), c-jun (TAM67), and IκBα(MAD3) inhibited *H. pylori*–induced expression of integrin α5 and integrin β1 in AGS cells. In conclusion, *H. pylori* activates Ras, NF-κB, and AP-1 and thus induces the expression of integrin α5 and integrin β1 in gastric epithelial cells. Inhibition of ROS production by DPI suppressed the expression of integrin α5 and integrin β1 in gastric epithelial cells. The results suggest

Address for correspondence: Hyeyoung Kim, Department of Food and Nutrition, Yonsei University College of Human Ecology, Seoul 120-749, Korea. Voice: 82-2-2123-3125; fax: 82-2-364-5781.
e-mail: kim626@yonsei.ac.kr

the possible involvement of NADPH oxidase for ROS production in *H. pylori*–infected gastric epithelial cells.

KEYWORDS: *Helicobacter pylori*; integrin α5/β1; AGS cells

INTRODUCTION

Helicobacter pylori (*H. pylori*) infection leads to gastroduodenal inflammation, peptic ulceration, and gastric carcinoma. *H. pylori* has been considered as a major etiological agent causing chronic gastritis, along with other features, including lymphoid follicles or lymphoid aggregates, surface epithelial degradation with mucous depletion, and intestinal metaplasia. According to previous reports, active gastritis could lead to gastric mucosal atrophy, which is thought to be a high-risk factor for the development of gastric cancer. Many studies have been undertaken in an attempt to elucidate the pathogenic mechanisms of *H. pylori* infection. One characteristic event in inflammation is the infiltration of inflammatory cells into the subepithelial gastric lamina propria. Inflammatory cells, mainly neutrophils and macrophages, produce large amounts of reactive oxygen species (ROS) in host defense reaction. In addition, *H. pylori* itself produces ROS in gastric epithelial cells even in the absence of inflammatory cells. ROS are supposed to be involved in tumor initiation and enhance the expression of oncogenes and stimulate cell proliferation. Several studies have demonstrated that *H. pylori* stimulated gastric hyperproliferation, an essential step in the preliminary stage for the development of gastric carcinoma. *H. pylori*–induced activation of nuclear factor-κB (NF-κB) and activator protein-1 (AP-1) contributes to the expression of several genes involved in inflammation such as IL-8 and cell adhesion such as ICAM-1.[1]

Synthesis of toxic ROS by human neutrophils is an essential component of innate defense. Conversion of molecular oxygen into superoxide anions is catalyzed by the multicomponent NADPH oxidase.[2] Diphenyleneiodonium (DPI) produces noncompetitive inhibition of NADPH oxidase via binding covalently to FAD when the enzyme is activated. Although DPI is a nonspecific inhibitor of flavoenzymes, decrease of cellular ROS production in the presence of DPI has been increasingly interpreted as resulting from inhibition of an NADPH oxidase by DPI.

NF-κB is a member of the Rel family, including p50 (NF-κB1), p52 (NF-κB2), Rel A (p65), c-Rel, Rel B, and *Drosophila* morphogen dorsal gene product. Its activity is controlled by its cytoplasmic inhibitory protein IκBα. Previously, we found that *H. pylori* increased lipid peroxidation, an indication of oxidative damage, and induced the activation of two species of NF-κB dimers (a p50/p65 heterodimer and a p50 homodimer) in gastric epithelial cells.[3] Activator protein-1 (AP-1) is a redox-sensitive transcription factor and

Ras is reported to mediate upstream signaling for AP-1 activation. MAD-3 is a mutant gene for IκBα which substitute two serine residues to alanine. Transfection with MAD-3 into the cells inhibits NF-κB activation. Ras N-17 is a dominant negative mutant of ras. Transfection with ras N-17 interferes with ras function by the expression of a dominant inhibitory mutation in c-Ha-ras. TAM-67 is a dominant negative mutant lacking the transactivation domain of c-jun. Transfection with TAM-67 inhibits the function of wild-type c-Jun or c-Fos through either a quenching or blocking mechanism.

Integrins mediate cell–cell and cell–ECM adhesion. Integrin expression in cancer tissues demonstrates its possible contribution to tumor progression, invasion and metastasis.[4] It has been reported that the expression of integrin subunits such as integrin α2, α3, α5, and α6 was correlated with cell adhesion, invasion, and metastasis of gastric carcinoma cells.[5] *H. pylori*–induced upregulation in expression of integrin αM and αX in Kato 3 cells[6] and integrin α7 in AGS cells were reported after 24-h culture.[7] Su *et al.* reported that the monoclonal antibody against integrin α5 reduced *H. pylori* adhesion to AGS cells, indicating that *H. pylori* invasion into AGS cells was mediated by integrin α5.[8] Therefore, the increased expression of integrin α5 by *H. pylori* might contribute to increase in invasion of *H. pylori* itself into gastric epithelial cells.

The present study aims to investigate whether *H. pylori* in a Korean isolate (HP99) induces the expression of integrin α5 and integrin β1, and whether *H. pylori*–induced expression of integrin α5 and integrin β1 are inhibited in the cells transfected with mutant genes for Ras (ras N-17), c-Jun (TAM-67), and IκBα(MAD-3) or treated with DPI, an inhibitor of NADPH oxidase.

METHODS

H. pylori strain (HP99) was isolated from gastric antral mucosa obtained from a Korean patient with duodenal ulcer. HP99 was kindly provided by Dr. H.C. Jung (Seoul National University College of Medicine, Seoul, Korea). These bacteria were inoculated onto chocolate agar plates (Becton Dickinson Microbiology Systems, Cockeysville, MD) at 37° under microaerophilic conditions using an anaerobic chamber (BBL Campy Pouches System, Becton Dickinson Microbiology Systems). A human gastric epithelial cell line AGS (gastric adenocarcinoma, ATCC CRL 1739) was purchased from the American Type Culture Collection. The cells were grown in complete medium, consisting of RPMI 1640 medium supplemented with 10% fetal bovine serum, 2 mM glutamine, 100 U/mL penicillin, and 100 μg/mL streptomycin (Sigma, St. Louis, MO). AGS cells were seeded in 10-cm dishes at 5×10^6 cells and cultured to reach 80% confluency. Prior to the stimulation, each dish was washed twice with fresh cell culture medium containing no antibiotics. *H. pylori* was harvested, washed with phosphate-buffered saline (PBS), and then resuspended into antibiotic-free cell culture medium. *H. pylori* was added to AGS cells at bacterium/cell ratio of 300:1 and cultured for 24 h (timecourse, DPI treatment, and transfection for Western blot analysis). The cells were treated with an NADPH oxidase inhibitor DPI (10 μM). A mutated IkBa gene MAD-3 is a double-point mutant gene (substitution of two serine residues at positions 32 and 36 by alanine residues). By transfection of MAD-3 into the cells, IκBα could not be phosphorylated, which inhibits NF-κB activation. A dominant negative mutant of c-Jun, called TAM-67, is a potent inhibitor of AP-1-mediated transactivation. Transfection of a dominant negative mutant of ras, called ras N-17, into the cells interferes with ras function by the expression of a dominant inhibitory mutation in c-Ha-ras. This mutation changes serine-17 to arginine-17 in the gene product and thus inhibits ras activity. The control vector pcDNA3 (Invitrogen Corp., Carlsbad, CA) was transfected to the cells instead of mutant genes for ras (ras N-17), c-Jun (TAM-67), and IκBα (MAD-3). These cells were considered as a relative control and named pcN-3. AGS cells without transfection were cultured in the presence of *H. pylori* (control) or absence of *H. pylori* (none).

RESULTS

Protein expression of integrin α5 and integrin β1 were determined in *H. pylori*–infected AGS cells. As shown in FIGURE 1, *H. pylori* induced the expression of integrin α5 and integrin β1 in a time-dependent manner up to 24 h. Actin was constitutively expressed in AGS cells and not changed with incubation time. DPI was pretreated to the culture medium 2 h before the

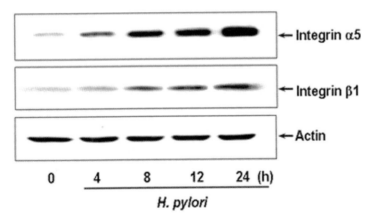

FIGURE 1. Time-dependent induction of integrin α5 and integrin β1 in *H. pylori*–infected AGS cells. The cells were seeded in 10-cm dishes at 5×10^6 cells and cultured to reach 80% confluency. The cells were cultured with various time periods in the presence of *H. pylori* at a bacterium/cell ratio of 300:1. Protein expression of integrin α5 and integrin β1 were determined by Western blotting. Actin was used for protein loading control.

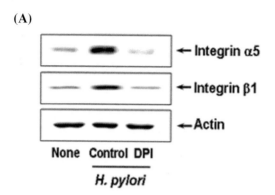

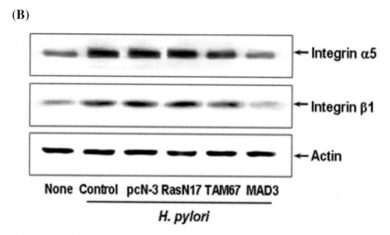

FIGURE 2. (**A**) Expression of integrin α5 and integrin β1 in AGS cells stimulated with *H. pylori* treated with or without DPI. DPI (10 μM) was pretreated to the cells 2 h prior to the stimulation with *H. pylori*. (**B**) Expression of integrin α5 and integrin β1 in AGS cells stimulated with *H. pylori* transfected with or without Ras N-17, TAM-67, or MAD-3. Protein expression of integrin α5 and integrin β1 were determined by Western blotting. Actin was used for protein-loading control. None, the cells without *H. pylori* stimulation; Control, the cells stimulated with *H. pylori*; DPI, the cells stimulated with *H. pylori* and treated with DPI; pcN-3, the cells transfected with control vector (pcDNA) and stimulated with *H. pylori*; RasN17, the cells transfected with ras N-17 and stimulated with *H. pylori*; TAM67, the cells transfected with TAM-67 and stimulated with *H. pylori*; MAD3, the cells transfected with MAD-3 and stimulated with *H. pylori*.

treatment of *H. pylori*. The bacterial cells were added to the cultured cells at a bacterium/cell ratio of 300:1 for 24 h. *H. pylori*–induced expression of integrin α5 and integrin β1 was inhibited by treatment with DPI (FIG. 2A). AGS cells transfected with or without mutant genes for ras (ras N-17), c-Jun (TAM-67), and IκBα (MAD-3) were cultured in the presence of *H. pylori* for 24 h. (FIG. 2B). *H. pylori*–induced expression of integrin α5 and integrin β1 was inhibited in the cells transfected with mutant gene for ras (ras N-17), c-Jun (TAM-67), and IκBα (MAD-3) compared to the cells transfected with pcDNA (pcN-3). The results suggest the involvement of Ras, NF-κB, and AP-1 in *H. pylori*–induced expression of integrin α5 and integrin β1 in gastric epithelial cells.

DISCUSSION

The main finding of this study is that *H. pylori* activates Ras, NF-κB, and AP-1 and thus induces the expression of integrin α5 and integrin β1 in gastric epithelial cells. Inhibition of ROS production by DPI suppressed the expression of integrin α5 and integrin β1 in gastric epithelial cells. The results suggest the possible involvement of NADPH oxidase for ROS production in *H. pylori*–infected gastric epithelial cells. Since integrin expression is related to tumor progression, invasion, and metastasis,[4] inhibition of upstream signaling of Ras, NF-κB, or AP-1 may be beneficial for prevention of *H. pylori*–induced gastric diseases including gastric cancer. It has been reported that the expression of integrin subunits such as integrin α2, α3, α5, and α6 was correlated with cell adhesion, invasion, and metastasis of gastric carcinoma cells.[5] *H. pylori* induced upregulation of integrin αM and αX in Kato 3 cells[6] and integrin α7 in AGS cells.[7] The monoclonal antibody against integrin α5 reduced *H. pylori* adhesion to AGS cells.[8] Therefore, identification of upstream signaling for integrin expression will provide beneficial information for a therapeutic approach to *H. pylori*–related gastric disease.

ACKNOWLEDGMENT

This study was supported by the Brain Korea 21 Project, Yonsei University College of Human Ecology.

REFERENCES

1. LIM, J.W., H. KIM & H.K. KIM. 2003. Cell adhesion-related gene expression by *Helicobacter pylori* in gastric epithelial AGS cells. Int. J. Biochem. Cell Biol. **35:** 1284–1296.
2. DELEO, F.R., & M.T. QUINN. 1996. Assembly of the phagocyte NADPH oxidase: molecular interaction of oxidase proteins. J. Leukocyte Biol. **60:** 677–691.
3. SEO, J.H., J.W. LIM, H. KIM & K.H. KIM. 2004. *Helicobacter pylori* in a Korean isolate activates mitogen-activated protein kinases, AP-1, and NF-kappaB and induces chemokine expression in gastric epithelial AGS cells. Lab. Invest. **84:** 49–62.
4. GIANCOTTI, F.G. & F. MAINIERO. 1994. Integrin-mediated adhesion and signaling in tumorigenesis. Biochim. Biophys. Acta **1198:** 47–64.

5. KOIKE, N., T. TODOROKI, H. KOMANO, *et al.* 1997. Invasive potentials of gastric carcinoma cell lines: role of $\alpha 2$ and $\alpha 6$ integrins in invasion. J. Cancer Res. Clin. Oncol. **123:** 310–316.
6. COX, J.M., C.L. CLAYTON, T. TOMITA, *et al.* 2001. cDNA microarray analysis of *cag* pathogenicity island-associated *Helicobacter pylori* epithelial cell response genes. Infect. Immun. **69:** 6970–6980.
7. SEPULVEDA, A.R., H. TAO, E. CARLONI, *et al.* 2002. Screening of gene expression profiles in gastric epithelial cells induced by *Helicobacter pylori* using microarray analysis. Aliment. Pharmacol. Therap. **16:** 145–157.
8. SU, B., S. JOHANSSON, M. FALLMAN, *et al.* 1999. Signal transduction-mediated adherence and entry of *Helicobacter pylori* into cultured cells. Gastroenterology **117:** 595–604.

Human Recombinant Vasostatin-1 May Interfere with Cell–Extracellular Matrix Interactions

VALENTINA DI FELICE,[a] FRANCESCO CAPPELLO,[a] ANTONELLA MONTALBANO,[a] NELLA ARDIZZONE,[a] CLAUDIA CAMPANELLA,[a] ANGELA DE LUCA,[a] DANIELA AMELIO,[b] BRUNO TOTA,[b] ANGELO CORTI,[c] AND GIOVANNI ZUMMO[a]

[a]*Human Anatomy Section, Di.Me.S., University of Palermo, 90127, Palermo, Italia*

[b]*Sezione di Fisiologia Organismale, Dipartimento di Biologia Cellulare, Università della Calabria, 87036, Arcavacata di Rende, Cosenza, Italia*

[c]*DIBIT, Department of Oncology, San Raffaele H Scientific Institute, 20132, Milano, Italia*

ABSTRACT: Vasostatin-1 (VS-1), the N-terminal fragment derived from the cleavage of chromogranin A (CgA), has been shown to exert several biological activities on several tissues and organs. Recently, it has been reported that human recombinant VS-1 (STA-CGA$_{1-78}$) may alter myocardial contractility in eel, frog, and rat hearts. In this article we have explored if STA-CGA$_{1-78}$ can induce intracellular cascades interacting both with adhesion molecules and/or extracellular matrix (ECM), components, that is, involvement of the heat shock protein 90 (HSP90) and the endothelial NOS (eNOS), known to be implicated in signal transduction mechanisms affecting myocardial contractility. We used 3D cultured adult rat cardiomyocytes cultivated over fibronectin or fibroblasts or embedded in matrigel or collagen type I. Aurion-conjugated VS-1 (Au-STA-CGA$_{1-78}$) has been used to identify possible sites of interaction of this molecule with the cell membrane. We found that in our 3D culture, cell–ECM interactions played a crucial role in the cellular localization of HSP90 as well as in the expression of eNOS. VS-1 appeared to modulate cell–ECM interactions, thereby remarkably leading to a different cellular localization of HSP90. Moreover, Au-STA-CGA$_{1-78}$ was never detected inside the cell nor overlapping the plasma membrane, but nearby the outer side of the cardiomyocyte plasmalemma, at a particular distance, typical of integrins. On the whole, these data suggest that VS-1 does not have a classic receptor on the membrane but that integrins may represent a nonconventional VS-1 receptor modulating eNOS signaling pathway.

Address for correspondence: Valentina Di Felice, M.Sc., Ph.D., Human Anatomy Section, Department of Experimental Medicine, Via del vespro 129, 90127 Palermo, Italia. Voice: 0039-091-6553575; fax: 0039-091-6553580.
 e-mail: valentina.difelice@unipa.it

KEYWORDS: vasostatin-1; ECM–cell interactions; 3D cultures; cardiomyocytes

INTRODUCTION

Chromogranin A (CgA) belongs to the family of chromogranins, which are secretory proteins localized into the electron-dense granules of several endocrine and neuroendocrine cells[1,2] and released into circulation. Peptides deriving from CgA have been detected in several tissues[3] and exhibited unusual biological properties. Among them, they have been identified two human vasostatins (VSs): CgA_{1-76} (vasostatin-1; VS-1) and CgA_{1-113} (vasostatin-2; VS-2). They can act as regulatory peptides in an autocrine, paracrine, or endocrine manner.[4]

Recently, it has been reported that a human recombinant VS-1 (STA-CGA_{1-78}) can decrease important parameters of myocardial inotropism, such as stroke volume (SV) and stroke work (SW), in eel[5] and frog[6] hearts, and left ventricular pressure (LVP) and rate pressure product (RPP) in the Langendoff-perfused rat heart.[7] Both in rat and eel hearts this negative inotropism seemed to be dependent from the presence of the endocardial endothelium (EE) and from the NO-cGMP signal transduction pathway, while in the frog heart it seemed independent.

The enzyme responsible for the production of most of the NO in myocardium is the endothelial nitric oxide synthase (eNOS). eNOS is usually localized in the plasma membrane caveolae of cardiomyocytes. *In vitro* inactive eNOS can associate with caveolin-3 and can be activated, following appropriate stimuli, through phosphorylation by Akt/PKB kinase, coupled to eNOS by heat shock protein 90 (HSP90) binding.[8–10]

In this article we used 3D-cultured adult rat cardiomyocytes to explore possible sites of interactions of STA-CGA_{1-78}, containing the VS-1 CGA_{1-76} sequence, with the cell membrane or with extracellular components of myocardial extracellular matrix (ECM). Two proteins have been considered as markers of STA-CGA_{1-78} action and aurion-conjugated STA-CGA_{1-78} (Au-STA-CGA_{1-78}) has been used to identify putative sites of VS-1 interactions with components of either the cardiac cells or ECM.

The results show that the expression of HSP90 and eNOS proteins in 3D-cultured cardiomyocytes can be affected by STA-CGA_{1-78} and that these phenomenon can be related to the ECM components used as culture substrates. Furthermore, Au-STA-CGA_{1-78}, was always visible extracellularly nearby the plasma membrane.

MATERIALS AND METHODS

Freshly dissociated cardiomyocytes, obtained from rat ventricles treated with 50 U/mL collagenase II were maintained in culture for no more than 10 days in M-199 medium supplemented with 20% fetal calf serum (FCS), with antibiotics and antimycotics.

Four different coating substrates were used in the cultures: rat collagen type I; matrigel basement membrane (GFR); human plasma fibronectin; and a monolayer of cardiac fibroblasts obtained from the same myocardium. After 24 h from plating, cells were treated with STA-CGA$_{1-78}$ 5–10 μg/mL for 30 min. After treatment cells were fixed and used for immunofluorescence and electron microscopy. Recombinant Ser-Thr-Ala-hCGA$_{1-78}$ (STA-CGA$_{1-78}$) was obtained as previously described.[7,11]

For 3D-culture gels, aliquots of isolated cardiomyocytes were suspended in a collagen type I solution or in a matrigel solution. The gels were superfused with M-199 medium-20% FCS and placed in a cell culture incubator.

For transmission electron microscopy 5×10^4 cardiomyocytes were mixed with matrigel and placed into 24-well plate inserts. After STA-CGA$_{1-78}$ treatment, cells were fixed, dehydrated until ethanol 70% (v/v), and embedded into L.R.White resin. For experiments with Au-STA-CGA$_{1-78}$ (3- to 4-nm-diameter beads), one out of two inserts was treated with 5 μg/mL Au-STA-CgA$_{1-78}$ for 1 h, before fixation.

For immunofluorescence experiments methanol fixed cells were blocked and incubated with primary antibodies (1:50, anti-HSP90 α/β and anti-eNOS). Cells were then incubated with fluorescent secondary antibody (FITC-conjugated anti-mouse secondary antibody and FITC-conjugated anti-rabbit secondary antibody). Imaging was done on a Leica Laser Scannin Confocal Microscope (Leica Microsystems GmbH, Wetzlar, Germany).

For Western blotting analysis cardiac fibroblasts were lysated with RIPA lysis buffer containing proteases and phosphatases inhibitors. Total proteins were determined using the DC Protein Assay Kit (Biorad, Hercules, CA). The same amount of proteins was run on SDS-PAGE and transferred to PVDF membranes. Secondary antibody was detected using ECL.

RESULTS

At electron microscopy mitochondria of isolated cardiomyocytes appeared intact and used as a control for cellular viability. In some cardiomyocytes myofibrils lost their organization. Cardiomyocytes cultured over fibronectin or a layer of cardiac fibroblasts, or embedded in matrigel or collagen type I were treated with 5 or 10 μg/mL STA-CGA$_{1-78}$ after 24 h from plating.

HSP90 α/β protein was localized only in small compartments inside the cell and its level did not change after treatment when cells were cultivated over fibronectin or embedded in collagen type I. The same protein was localized beneath the cardiomyocyte plasma membrane when cells were cultivated embedded in matrigel or over a layer of cardiac fibroblasts. After exposure to 10 μg/mL STA-CGA$_{1-78}$, HSP90 α/β moved to small internal compartments near the nucleus. eNOS enzyme expression in matrigel cultures varied after treatment with STA-CGA$_{1-78}$ at the concentrations of 5–10 μg/mL.

To study the possible binding sites of STA-CGA$_{1-78}$ to the cellular membrane, we used Au-STA-CGA$_{1-78}$, with 3- to 4 nm-diameter beads. Cardiomyocytes, plated inside a 50-μm-thick matrigel gel, were treated with 5 μg/mL Au-STA-CGA$_{1-78}$. In samples treated for electron microscopy, Au-STA-CGA$_{1-78}$ was visible outside the cell and nearby the outer side of the cardiomyocyte plasma membrane at a minimum distance in the range between 16 and 25 nm (FIG. 1). No molecules were found inside the cell or overlapping the plasma membrane.

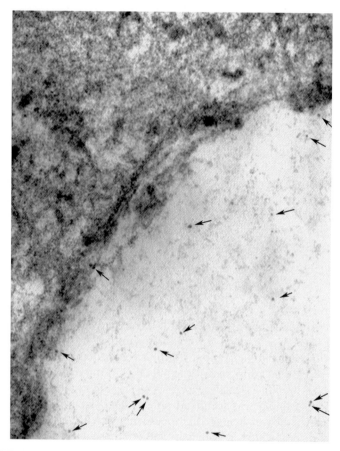

FIGURE 1. Transmission electron microscopy shows the position of 3- to 4-nm-aurion beads, representing Au-STA-CGA$_{1-78}$ molecules in 3D cultures of cardiomyocytes (*black arrows*).

DISCUSSION

These results demonstrate for the first time that STA-CGA$_{1-78}$ affects HSP90 protein cellular localization and eNOS expression in cardiomyocytes cultured in matrigel three dimensionally. They also show that Au-STA-CGA$_{1-78}$ exclusively localizes outside the plasma membrane of cardiomyocytes.

Mitochondria integrity, HSP90 basal expression, and eNOS induction, apart from their cellular function, can be considered markers of cell viability. We used different substrates in the cultures: fibronectin, collagen type I, matrigel, and cardiac fibroblasts. An important finding of the present work is that HSP90 protein changes localization depending on the substrate used. In particular HSP90 was detected beneath the plasma membrane when cardiomyocytes were cultured on cardiac fibroblasts or in matrigel, suggesting the importance of cell–ECM molecular interactions in the subcellular compartmentalization of this protein.

Cardiomyocytes cultured in matrigel appeared to be the best system to study VS-1 effects also on eNOS expression. After exposure to 5–10 μg/mL STA-CGA$_{1-78}$ for 30 min, eNOS protein level increases perhaps for new protein production.

Even if HSP90 and eNOS did not show the same cellular localization, we think that their expressions are correlated, because it is already known in literature that eNOS activation is dependent on HSP90 caveolar localization.[12]

An interference with cell–ECM interactions has been postulated as a possible mechanism of action of STA-CGA$_{1-78}$ on cultured cardiomyocytes. To study this kind of interactions 3D cultures were treated with Au-STA-CGA$_{1-78}$. In our experiments, the aurion conjugated molecule was always found at a particular distance from the plasma membrane (between 16 and 25 nm). This distance is typical of interactions between cells and ECM proteins.

Our findings may lead to the working hypothesis that natural VS-1 may act *in vivo* interfering with cell–ECM interactions. However, neither a conventional receptor nor the action sites of VS-1 are known. Two putative domains of the human recombinant VS-1 could be taken into consideration for its binding either to the cell membrane or to the ECM components: an RGD sequence at residues 43-45[13] and a net positively charged domain at residues 47–70.[14] Anyway, the RGD site of CgA is not conserved among different species and its involvement in the regulation of cell–ECM interactions remains to be proved.

Another possibility is that VS-1-induced HSP90 localization and eNOS expression could result from VS-1 interaction with membrane phospholipids and consequent enhancement of membrane fluidity. However, a combination of the two mechanisms can be postulated.

REFERENCES

1. WINKLER, H. & R. FISCHER-COLBRIE. 1992. The chromogranins A and B: the first 25 years and future perspectives. Neuroscience **49:** 497–528.
2. DAY, R. & S.U. GORR. 2003. Secretory granule biogenesis and chromogranin A: master gene, on/off switch or assembly factor? Trends Endocrinol. Metab. **14:** 10–13.
3. HELLE, K.B. 2004. The granin family of uniquely acidic proteins of the diffuse neuroendocrine system: comparative and functional aspects. Biol. Rev. Camb. Philos. Soc. **79:** 769–794.
4. HELLE, K.B., M.-H. METZ-BOUTIGUE & D. AUNIS. 2001. Chromogranin A as a calcium-binding precursor for a multitude of regulatory peptides for the immune, endocrine and metabolic systems. Curr. Med. Chem. Immun. Endoc. Metab. Agents **1:** 119–140.
5. IMBROGNO, S., M.C. CERRA & B. TOTA. 2003. Angiotensin II-induced inotropism requires an endocardial endothelium-nitric oxide mechanism in the in-vitro heart of Anguilla anguilla. J. Exp. Biol. **206:** 2675–2684.
6. CORTI, A., C. MANNARINO, R. MAZZA, et al. 2004. Chromogranin A N-terminal fragments vasostatin-1 and the synthetic CGA 7–57 peptide act as cardiostatins on the isolated working frog heart. Gen. Comp. Endocrinol. **136:** 217–224.
7. CERRA, M.C., L. DE IURI, T. ANGELONE, et al. 2005. Recombinant N-terminal fragments of chromogranin-A modulate cardiac function of the Langendorff-perfused rat heart. Basic Res. Cardiol. **100:** 1–10.
8. TAKAHASHI, S. & M.E. MENDELSOHN. 2003. Synergistic activation of endothelial nitric-oxide synthase (eNOS) by HSP90 and Akt: calcium-independent eNOS activation involves formation of an HSP90-Akt-CaM-bound eNOS complex. J. Biol. Chem. **278:** 30821–30827.
9. CHEN, J.X. & B. MEYRICK. 2004. Hypoxia increases Hsp90 binding to eNOS via PI3K-Akt in porcine coronary artery endothelium. Lab. Invest. **84:** 182–190.
10. FONTANA, J., D. FULTON, Y. CHEN, et al. 2002. Domain mapping studies reveal that the M domain of hsp90 serves as a molecular scaffold to regulate Akt-dependent phosphorylation of endothelial nitric oxide synthase and NO release. Circ. Res. **90:** 866–873.
11. RATTI, S., F. CURNIS, R. LONGHI, et al. 2000. Structure-activity relationships of chromogranin A in cell adhesion. Identification of an adhesion site for fibroblasts and smooth muscle cells. J. Biol. Chem. **275:** 29257–29263.
12. GRATTON, J.P., J. FONTANA, D.S. O'CONNOR, et al. 2000. Reconstitution of an endothelial nitric-oxide synthase (eNOS), hsp90, and caveolin-1 complex in vitro. Evidence that hsp90 facilitates calmodulin stimulated displacement of eNOS from caveolin-1. J. Biol. Chem. **275:** 22268–22272.
13. GASPARRI, A., A. SIDOLI, L.P. SANCHEZ, et al. 1997. Chromogranin A fragments modulate cell adhesion. Identification and characterization of a pro-adhesive domain. J. Biol. Chem. **272:** 20835–20843.
14. MANDALA, M., J.F. BREKKE, G. SERCK-HANSSEN, et al. 2005. Chromogranin A-derived peptides: interaction with the rat posterior cerebral artery. Regul. Pept. **124:** 73–80.

Microgravity Signal Ensnarls Cell Adhesion, Cytoskeleton, and Matrix Proteins of Rat Osteoblasts

Osteopontin, CD44, Osteonectin, and α-Tubulin

YASUHIRO KUMEI,[a] SADAO MORITA,[a] HISAKO KATANO,[b] HIDEO AKIYAMA,[c] MASAHIKO HIRANO,[c] KEI'ICHI OYHA,[a] AND HITOYATA SHIMOKAWA[a]

[a]*Graduate School of Tokyo Medical and Dental University, Tokyo 113-8549, Japan*

[b]*Institute of Medical Science, University of Tokyo, Tokyo 108-8639, Japan*

[c]*Toray Research Center, Kamakura 248-8555, Japan*

ABSTRACT: Rat osteoblasts were cultured for 4 or 5 days aboard the Space Shuttle and solubilized during spaceflight. Post-flight analyses by quantitative reverse transcriptase-polymerase chain reaction (RT-PCR) determined the relative mRNA levels of matrix proteins, adhesion molecules, and cytoskeletal proteins including osteopontin (OP), osteonectin (ON), CD44, α-tubulin, actin, vimentin, fibronectin (FN), and β1-integrin. The mRNA levels of OP and α-tubulin in the flight cultures were decreased by 30% and 50% on day 4 and day 5 of flight, as compared to the ground controls. In contrast, the CD44 mRNA levels in the flight cultures increased by 280% and 570% of the ground controls on day 4 and day 5. The mRNA levels of ON and FN in the flight cultures were slightly increased as compared to ground controls. The mRNA levels of actin, vimentin, or β1-integrin did not change in spaceflight conditions. The matrix proteins, adhesion molecules, and cytoskeletal proteins may form dynamic network complexity with signaling molecules as an adaptive response to perturbation of mechanical stress under microgravity.

KEYWORDS: spaceflight; gravity; osteoblast; osteopontin; α-tubulin; CD44; osteonectin; fibronectin

Address for correspondence: Dr. Yasuhiro Kumei, Department of Hard Tissue Engineering, Section of Biochemistry, Graduate School of Tokyo Medical and Dental University, Tokyo 113-8549, Japan. Voice/fax: +81-3-5803-4555.
 e-mail: kumei.bch@tmd.ac.jp

Ann. N.Y. Acad. Sci. 1090: 311–317 (2006). © 2006 New York Academy of Sciences.
doi: 10.1196/annals.1378.034

INTRODUCTION

It is well documented that bone mass loss is induced by spaceflight. However, the mechanisms for this are unknown. The external stimuli are transduced into intracellular signals through the extracellular matrix (ECM), adhesion molecules, and the cytoskeletal structure. These signals regulate cellular motility, proliferation, differentiation, and apoptosis/survival. Tubulin and actin constitute the microtubules and microfilaments, while vimentin is a major element of the intermediate filaments.

A matrix protein osteopontin (OP) plays a pivotal role in bone metabolism: hydroxyapatite crystal formation and cell attachment to the bone surface. Mechanical unloading did not induce bone loss in the OP-deficient mice.[1] OP binding to the hyaluronic acid receptor CD44 of osteoblasts is important for bone metabolism.[2] Another matrix protein, osteonectin (ON), is widely expressed in bone. We examined microgravity effects on the gene expression for ECM, adhesion molecules, and cytoskeletal proteins, including OP, ON, CD44, α-tubulin, actin, vimentin, β1-integrin, and fibronectin (FN) in primary rat osteoblasts.

MATERIALS AND METHODS

Osteoblasts prepared from rat femur marrows were cultured aboard the Space Shuttle.[3] On day 4 of the flight, half of the cultures were treated with 1α,25 dihydroxyvitamin D3 for 24 h and solubilized on board by guanidine solution (Experiment I). On day 5 of the flight, the entire sequence of this procedure was repeated for the remaining half of the cultures (Experiment II). The ground (1 × g) control experiments were conducted synchronously with flight cultures under identical conditions except for microgravity. After return to the ground, the relative mRNA levels were analyzed by quantitative reverse transcriptase-polymerase chain reaction (RT-PCR), using gene-specific primers: OP sense 5'-CCTCTGAAGAAACGGATGACT-3', antisense 5'-CTGGGCAACTGGGATGACCTT-3'; CD44 sense 5'-ACCG ACCTTCCCACTTCACAG-3', antisense 5'-GCACTACACCCCAATCTTC AT-3'; ON sense 5'-CTCTTCCTGCCACTTCTTTGC-3'; antisense 5'-CTTG TTGATGTCCTGCTCCTT-3'; α-tubulin sense 5'-TGTCTTCCATCACTGC TTCC-3' antisense 5'-CACGCTTGGCATACATCAGA-3'; FN sense 5'-ACCAGTGGGATAAGCAGCAT-3', antisense 5'-CCTTCCAGCGACCCGT AGAG-3'. Data were normalized against a housekeeping gene glyceraldehyde 3-phosphate dehydrogenase.[3]

RESULTS AND DISCUSSION

We conducted a set of osteoblast cultures for 4 or 5 days aboard the Space Shuttle. The mRNA levels of a matrix protein OP in the flight cultures were

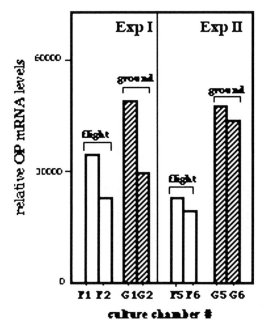

FIGURE 1. The relative mRNA levels of osteopontin (OP) of rat osteoblasts that were cultured during spaceflight. Data from duplicate experiments on day 4 (*Exp I*) and day 5 (*Exp II*) of flight.

decreased by 30% and 50 % relative to the ground controls on day 4 and day 5, respectively (FIG. 1). OP binds to the hyaluronic acid receptor CD44, which also makes the ligand for another matrix protein FN. The CD44 mRNA levels in the flight cultures were increased and were 280% and 570% higher than those of the ground controls on day 4 and day 5 (FIG. 2). The FN mRNA levels in the flight cultures were slightly increased, approximately by 150% of the ground control levels on day 4 (FIG. 3). The upregulation of FN might be caused by increased PGE_2 release in the flight cultures,[3] since PGE_2 is known to stimulate FN expression in rat osteoblasts.[4]

OP plays a pivotal role in bone formation and resorption under mechanical stress.[5] OP deficiency inhibited cell adhesion to matrix and induced apoptosis.[6] CD44 binding to OP protected cells from apoptosis by activating the antiapoptotic genes.[7] Previously we reported that gravity unloading (similar to spaceflight conditions) elevated the mitochondrial apoptotic index (ratio of Bax/Bcl-2) in human osteoblasts, but simultaneously induced the antiapoptotic gene IAP (inhibitor of apoptotic protein) to prevent apoptosis.[8] The CD44 upregulation in the flight cultures may be involved in the compensative regulation to counteract the apoptotic trend that is induced by OP reduction under microgravity.

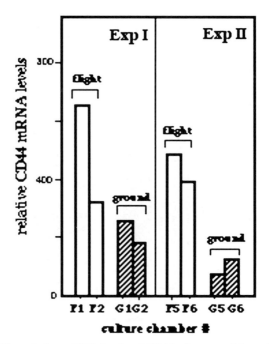

FIGURE 2. The relative mRNA levels of CD44 of rat osteoblasts that were cultured during spaceflight. Data from duplicate experiments on day 4 (*Exp I*) and day 5 (*Exp II*) of flight.

Interaction between OP and CD44 is mediated by β1-integrin.[9] CD44 and integrins are transmembrane adhesion molecules that are associated with actin filaments inside the cell and matrix proteins outside. The mRNA levels of β1-integrin, actin, or vimentin were not affected by spaceflight (data not shown). However, the α-tubulin mRNA levels in the flight cultures were decreased by approximately 30% and 50% of the ground control levels on day 4 and day 5, respectively (FIG. 4). The mRNA levels of another matrix protein ON in the flight cultures were increased by approximately 150% of the ground controls (data not shown). Although ON are involved in bone formation,[10] the precise role of ON in bone metabolism remains unknown. Association of ON with tubulin suggests potential involvement of ON in the rearrangement of cytoskeletal structure under microgravity.

Mechanical perturbation resulted in 60% decrease of OP and 75% decrease of tubulin expression, while increasing FN by 150% without affecting β1-integrin or actin production in primary osteoblast cultures.[11] It is well documented that PGE_2 mediates the response to mechanical stress and change of the cell shape of osteoblasts. Parabolic flight (up to a 20-sec exposure to microgravity in an airplane) conditions induced the morphological change and higher PGE_2 production relative to ground controls in rat osteoblastic cells.[12] Hypergravity induced a 30% decrease of the microtubule network height concomitant

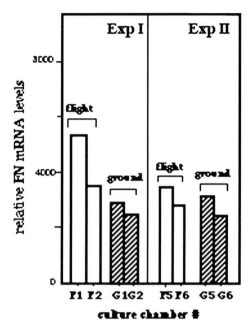

FIGURE 3. The relative mRNA levels of fibronectin (FN) of rat osteoblasts that were cultured during spaceflight. Data from duplicate experiments on day 4 (*Exp I*) and day 5 (*Exp II*) of flight.

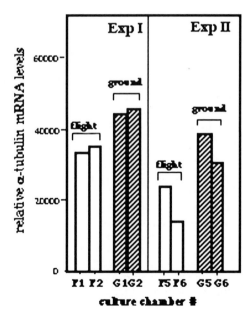

FIGURE 4. The relative mRNA levels of α-tubulin of rat osteoblasts that were cultured during spaceflight. Data from duplicate experiments on day 4 (*Exp I*) and day 5 (*Exp II*) of flight.

with a fivefold increase of PGE_2 release in rat osteoblasts.[13] All of these data were consistent with our previous findings that spaceflight conditions increased PGE_2 release in rat osteoblasts by 136-fold of the ground controls.[3]

Since the amounts of RNA obtained from the spaceflight experiments were extremely limited, we could not evaluate the results statistically. However, results of similar tendency were reproduced in the duplicate experiments in the same spaceflight. In conclusion, spaceflight conditions modulated the expression of some of ECM, adhesion molecules, and cytoskeletal proteins. Concerted participation of these molecules may act as a dynamic network toward an adaptive response to the perturbation of mechanical stress under microgravity. Data may provide some hints for understanding the mechanism underlying bone mass loss that is induced by spaceflight.

REFERENCES

1. ISHIJIMA, M., K. TSUJI, S.R. RITTLING, et al. 2002. Resistance to unlading-induced three-dimensional bone loss in osteopontin-deficient mice. J. Bone Miner. Res. **17:** 661–667.
2. SODEK, J., B. ZHU, M.-H. HUYNH, et al. 2002. Novel functions of the matricellular proteins osteopontin and osteonectin/SPARC. Conn. Tissue Res. **43:** 308–319.
3. KUMEI, Y., H. SHIMOKAWA, H. KATANO, et al. 1996. Microgravity induces prostaglandin E2 and interleukin-6 production in normal rat osteoblasts: role in bone demineralization. J. Biotechnol. **47:** 313–324.
4. TANG, C.-H., R.-S. YANG & W.-M. FU. 2005. Prostaglandin E2 stimulates fibronectin expression through EP1 receptor, phospholipase C, protein kinase Cα, and c-Src pathway in primary cultured rat osteoblasts. J. Biol. Chem. **280:** 22907–22916.
5. MORINOBU, N., M. ISHIJIMA, S.R. RITTLING, et al. 2003. Osteopontin expression in osteoblasts and osteocytes during bone formation under mechanical stress in the calvarial suture *in vivo*. J. Bone Miner. Res. **18:** 1706–1715.
6. WEINTRAUB, A.S., L.M. SCHNAPP, X. LIN & M.B. TAUBMAN. 2000. Osteopontin deficiency in rat vascular smooth muscle cells is associated with an inability to adhere to collagen and increased apoptosis. Lab. Invets. **80:** 1603–1615.
7. KHAN, S.A., C.A. LOPEZ-CHUA, J. ZHANG, et al. 2002. Soluble osteopontin inhibits apoptosis of adherent endothelial cells deprived of growth factors. J. Cell Biocehm. **85:** 728–736.
8. NAKAMURA, H., Y. KUMEI, S. MORITA, et al. 2003. Suppression of osteoblastic phenotypes and modulation of pro- and anti-apoptotic features in normal human osteoblastic cells under a vector-averaged gravity condition. J. Med. Dent. Sci. **50:** 167–176.
9. KATAGIRI, Y.U., J. SLEEMAN, H. FUJII, et al. 2000. CD44 variants but not CD44s cooperate with beta1-containing integrins to permit cells to bind to osteopontin independently of arginine-glycine-aspartic acid, thereby stimulating cell motility and chemotaxis. Cancer Res. **59:** 219–226.
10. DELANY, A.M., M. AMLING, M. PRIEMEL, et al. 2000. Osteopenia and decreased bone formation in osteonectin-deficient mice. J. Clin. Invest. **105:** 915–923.

11. MEAZZINI, M.C., C.D. TOMA, J.L. SCHAFFER, *et al*. 1998. Osteoblast cytoskeletal modulation in response to mechanical strain in vitro. J. Orthop. Res. **16:** 170–180.
12. GUIGNANDON, A., L. VICO, C. ALEXANDRE & M.H. LAFAGE-PROUST. 1995. Shape changes of osteoblastic cells under gravitational variations during parabolic flight–relationship with PGE2 synthesis. Cell Struct. Funct. **20:** 369–375.
13. SEARBY, N.D., C.R. STEELE & R.K. GLOBUS. 2005. Influence of increased mechanical loading by hypergravity on the microtubule cytoskeleton and prostaglandin E2 release in primary osteoblasts. Am. J. Physiol. Cell Physiol. **289:** C148–C158.

14-3-3 Proteins Bind Both Filamin and $\alpha_L\beta_2$ Integrin in Activated T Cells

SUSANNA M. NURMI, CARL G. GAHMBERG,
AND SUSANNA C. FAGERHOLM

Division of Biochemistry, Faculty of Biosciences, PB 56 (Viikinkaari 5D), 00014 University of Helsinki, Helsinki, Finland

ABSTRACT: Engagement of the T cell receptor (TCR) initiates intracellular signaling cascades that result in T cell activation, differentiation, acquisition of effector functions, or apoptosis. The signals from the TCR are coupled to distal signaling pathways by adapter proteins leading to dramatic changes in the cytoskeleton, transcription, and activation of integrins, which mediate adhesion. LFA-1 (leukocyte function-associated antigen-1) integrin ($\alpha L\beta 2$ or CD11a/CD18) plays an important role in adhesion, for example, by linking extracellular ligands to the actin cytoskeleton. The intracellular tails of integrins contain several phosphorylation sites, making them candidate-binding partners for 14-3-3 proteins, which are adaptor proteins that bind to phosphorylated ligands. In a screen for 14-3-3 binding partners in T cells, we identified both β_2 integrins and filamin. The integrin β_2 chain binds to 14-3-3 proteins through phosphorylated Thr758 after TCR ligation and this association regulates integrin-mediated cell spreading, which is necessary for adhesion. Here, we show that filamin associates with 14-3-3 proteins in activated T cells. 14-3-3 association with T cell membrane and cytoskeleton proteins after cell stimulation may mediate numerous T cell functions.

KEYWORDS: integrin; 14-3-3 protein; phosphorylation; LFA-1; T cell

INTRODUCTION

14-3-3 proteins comprise a family of ubiquitously expressed 28–35-kDa acidic polypeptides that mediate several important cellular regulatory processes including cell adhesion.[1] They are highly conserved between different species and they spontaneously form hetero- and homodimers.[2] Both

Address for correspondence: Susanna C. Fagerholm, Division of Biochemistry, Faculty of Biosciences, PB 56 (Viikinkaari 5D), 00014 University of Helsinki, Helsinki, Finland. Voice: +358-9-19159683; fax: +358-9-19159068.
e-mail: Susanna.Fagerholm@helsinki.fi

monomers can independently bind one ligand, which enables this adaptor protein to express much diversity in regulatory processes. Probably the most intriguing fact about these proteins is that they bind almost exclusively to phosphorylated ligands and usually the binding sequence is conserved.[2] After ligand binding, 14-3-3 proteins can regulate the functions of its binding partner in several different ways, for example, acting as adaptor proteins between two interacting proteins or sequestering ligands in specific locations.[3]

A common model for cell activation that involves numerous phosphorylation events is the T lymphocyte. T cells are activated through T cell receptor recognition of peptidic fragments of antigen bound to major histocompatibility complex (MHC) molecules on antigen-presenting cells (APCs). In addition, T cells require a costimulatory signal delivered through contact with other ligands on the APCs, especially B7 and intercellular adhesion molecule-1 (ICAM-1), which bind to CD28 and LFA-1, respectively. As a result of T cell activation through the TCR, the avidity of LFA-1 for ICAM-1 is rapidly increased, leading to the formation of an adhesion complex between the T cell and the APCs; this complex is known as the immunological synapse and it facilitates antigen recognition.[4,5]

LFA-1 belongs to the integrin superfamily of proteins These type I transmembrane glycoproteins are expressed in plasma membranes as heterodimers (α and β subunits) and link extracellular ligands to the actin cytoskeleton through proteins such as filamin, talin, and α-actinin.[6] Each integrin polypeptide consists of a large extracellular domain, a single transmembrane segment, and a relatively short cytoplasmic tail. The adhesiveness of the integrins must be carefully regulated, so that adhesive reactions occur only after cell activation. Integrin activation has been postulated to happen in two different ways; affinity and valency regulation. Both of these mechanisms have been proposed to be important for the regulation of integrin activity and it is believed that these mechanisms are not mutually exclusive.[7,8] One possible mechanism for regulating the functions of integrins is by phosphorylation of integrin cytoplasmic tails and both the α- and β-chain contain several possible phosphorylation sites.[9] Through T cell receptor ligation and chemokine or phorbol-ester stimulation site-specific phosphorylation of the cytoplasmic tail takes place and the LFA-1 becomes activated. The phosphorylation sites of the β_2 integrin polypeptide have been mapped.[9-13] The threonine triplet Thr758-Thr760 in the LFA-1 integrin cytoplasmic domain is important for integrin function[14] by regulating the interactions with the actin cytoskeleton and modulation of cell spreading and adhesion.[10,14] The large cytoskeletal protein filamin binds directly to the cytoplasmic tail of LFA-1, but the detailed functional implications of this association have remained unclear. Filamin A is known to play important roles in the mechanical stability of plasma membrane and cortex, formation of cell shape, mechanical responses of cells, and cell locomotion. It localizes to the cortical actin cytoskeleton and along the length of stress fibers, but is also found in some focal adhesions.[6]

MATERIALS AND METHODS

Reagents and Antibodies

Phorbol 12,13-dibutyrate (PDBu) was obtained from Sigma-Aldrich (Zwijndrecht, the Netherlands). Recombinant ICAM-1 was purchased from R&D Systems (Minneapolis, MN, USA). The monoclonal activating antibody against CD3, OKT3, was purified from ascites fluid produced by hybridoma cells (clone CRL 8001; ATCC). The mAb R2E7B against the human β_2 subunit of the CD11/CD18 leukocyte integrins has been described previously.[15] The 14-3-3 antibody K-19 was obtained from Santa Cruz Biotechnology, Inc. (Santa Cruz, CA, USA) and the filamin antibody MAB1678 was from Chemicon International Inc. (Temecula, CA, USA) BMH1 and BMH2 constructs were a kind gift from C. MacKintosh, Dundee.

Cell Fractionation

Human T cells were either left untreated or were treated with an OKT3 antibody (10 μg/mL) against the TCR, or 200 nM PDBu and lysed as described previously[16] and sheared by mechanical force. After discarding the nuclei by low-speed centrifugation, the samples were ultracentrifuged at 100,000 g for 45 min and the soluble and membrane fractions were separated. Samples were analyzed by the 14-3-3 overlay method as described previously[16] and by Western blotting with the β_2 antibody R2E7B.

14-3-3 Affinity Chromatography

Purified BMH1 and BMH2 or BSA was coupled to Sepharose.[17] Human T cells were either left untreated or were stimulated with OKT3 or PDBu. The cells were lysed as described previously.[11] The lysates were mixed with the affinity matrix for 1 h and washed extensively with 500 mM NaCl. Bound proteins were eluted with SDS and analyzed by Western blotting with the 14-3-3 antibody (K-19, Santa Cruz Biotechnology) or the filamin antibody (MAB1678, Chemicon International Inc).

Co-Immunoprecipitation

Human T cells were activated with OKT3 or PDBu or left untreated. Cells were lysed as described previously,[11] and lysates were precleared with protein G-Sepharose. Immunoprecipitations were made with the MAB1678 filamin antibody coupled to protein G-Sepharose, and the immunoprecipitates were washed four times with lysis buffer. The bound proteins were eluted with

SDS and analyzed by Western blotting with the 14-3-3 and filamin antibodies. Immunofluorescence staining was performed as described.[1]

RESULTS

Activation of T Cells Leads to Increased 14-3-3–Target Protein Association

T cell activation leads to the phosphorylation of numerous proteins in the cell. To characterize 14-3-3 protein associations in T cells, we fractionated cells into soluble and membrane fractions and used a 14-3-3 overlay method[16] (FIG. 1A). There was a large increase in binding to 14-3-3 proteins when the T cells were activated through the TCR or with phorbol esters. We were interested in identifying some of these binding partners and since Fagerholm et al.[13] had shown that Thr758 phosphorylated β2 integrin peptides bound 14-3-3 proteins, the cytoplasmic domain of the β2 integrin polypeptide was one of the possible targets (FIG. 1B). Using 14-3-3 affinity chromatography and immunoblotting, we were able to identify β2 integrin among the numerous 14-3-3 binding partners.[1] Activation of cells with phorbol ester or through CD3-ligation increased the binding of LFA-1 to 14-3-3 proteins, which was clearly seen in the membrane fractions[1] (FIG. 1A).

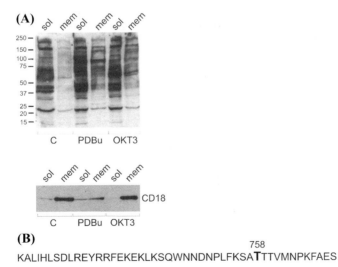

FIGURE 1. (**A**) Binding of soluble/membrane fractionated T cell lysates (resting C, PDBu-activated or OKT3-activated) to 14-3-3 proteins was assessed by an 14-3-3 overlay method (*upper panel*).[16] The success of fractionation was verified by detecting CD18 from the same nitrocellulose membrane by Western blotting using blotting β$_2$ antibody R2E7B (*lower panel*). (**B**) Sequence of β$_2$ integrin cytoplasmic domain with the 14-3-3 protein binding site in boldface and numbered.

Filamin Binds to 14-3-3 Proteins in Vitro *and* in Vivo

To examine whether any other proteins would bind to 14-3-3s more avidly after cell stimulation, affinity chromatography and immunoblotting were used. 14-3-3 proteins and filamin indeed interacted with each other *in vitro* and the binding was markedly increased after cell stimulation (FIG. 2A). To determine whether this interaction takes place *in vivo*, we performed co-immunoprecipitation experiments and detected 14-3-3 and filamin using immunoblotting. These proteins do indeed interact in cell and the binding of 14-3-3 proteins to filamin was markedly increased after T cell activation (FIG. 2B). Co-localization of these proteins was also observed by immunofluorescence, but only after cells were treated with phorbol ester (data not shown).

DISCUSSION

This study was initiated with the aim of examining the role of 14-3-3 proteins in T cell activation. T cells were fractionated into membrane and soluble fractions and a 14-3-3 overlay method was used to investigate the binding of

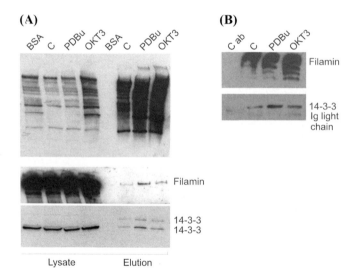

FIGURE 2. (**A**) Binding of resting (C), phorbol-ester activated (PDBu), or OKT3-activated T cell lysates to 14-3-3 or BSA (BSA) affinity columns. The binding of lysate samples and eluted samples was assessed by the 14-3-3 overlay method (*upper panel*). Filamin and 14-3-3 proteins were detected using Western blot (*lower panel*). (**B**) Increased binding of 14-3-3 and filamin in TRC-ligated and PDBu-activated T cells *in vivo*. Endogenous filamin was immunoprecipitated from human T cell lysates with filamin mAb (MAB1678) and co-immunoprecipitating 14-3-3 proteins were detected by Western blotting with the 14-3-3 antibody K-19. Immunoprecipitated filamin was detected to demonstrate equal loading.

different proteins to the 14-3-3 proteins. The activation of T cells through the TCR or with phorbol esters markedly increased the binding to the proteins in the membrane fraction, but T cell activation had very little or no effect on the binding of soluble proteins to the 14-3-3 proteins. Fagerholm et al.[13] had shown that phosphorylated β_2 integrin peptides bind to 14-3-3 proteins, and thus the β_2 integrin chain was a possible candidate for binding to 14-3-3 proteins. The interaction was indeed shown to exist *in vitro* and *in vivo* and the binding was discovered to be site-specific as well as phosphorylation-dependent.[1] This association mediates integrin-dependent cell spreading necessary for adhesion.

A potential 14-3-3 binding protein was filamin, which had previously been shown to bind 14-3-3 proteins in HEK293-cells.[18] The binding of this prospective partner was studied by 14-3-3 affinity chromatography and the protein was identified by immunoblotting. Filamin indeed bound to 14-3-3 proteins already in resting T cells, but there was a distinct increase in binding when T cells were activated. Co-immunoprecipitation and co-immunofluorescence studies confirmed that the interaction also occurs *in vivo*. We also identified protein kinase C (PKC) in the complex obtained by 14-3-3 affinity chromatography, but the binding of PKC[19] was not increased after activation (results not shown). Thus, since the interaction of 14-3-3 proteins with β_2 chain and filamin was observed both *in vivo* and *in vitro* and the PKC interaction was confirmed, the 14-3-3 affinity chromatography proved to be a valuable tool for identifying 14-3-3 binding partners in activated T cells.

A direct interaction between the β_2 integrin cytoplasmic tail and filamin has been shown previously.[20] Thus, we were interested in finding out whether 14-3-3 proteins acted as adaptor proteins between β_2 integrins and filamin, and this way creating a link between the cell membrane and the actin cytoskeleton. However, in co-immunoprecipitation studies we were not able to identify β_2 integrins, 14-3-3 and filamin in the same complex. Filamin binds directly to actin,[6] but we could not see any effect of ARAApSAPA-penetratin (a cell-permeable peptide sequence that blocks the interaction of 14-3-3 protein with its partners) on actin binding of filamin (data not shown). Filamin also binds to numerous other intracellular signaling proteins, and it is possible that the 14-3-3 protein binding modulates some of these binding interactions. The results of this study, however, establish a novel interaction between 14-3-3 proteins and filamin in activated T cells. In conclusion, the activation of T cells recruits 14-3-3 proteins to the cell membrane and cytoskeleton to function in different T cell processes.

ACKNOWLEDGMENTS

This study was supported by the Academy of Finland, the Sigrid Juselius Foundation, the Finnish Cancer Society, the Magnus Ehrnrooth Foundation, and the Liv och Hälsa Foundation.

REFERENCES

1. FAGERHOLM, S.C., T.J. HILDEN, S.M. NURMI & C.G. GAHMBERG. 2005. Specific integrin {alpha} and {beta} chain phosphorylations regulate LFA-1 activation through affinity-dependent and -independent mechanisms. J. Cell Biol. **171**: 705–715.
2. TZIVION, G. & J. AVRUCH. 2002. 14-3-3 proteins: active cofactors in cellular regulation by serine/threonine phosphorylation. J. Biol. Chem. **277**: 3061–3064.
3. HERMEKING, H. 2003. 14-3-3 cancer connection. Nat. Rev. Cancer **3**: 931–934.
4. SPRINGER, T.A. 1990. Adhesion receptors of the immune system. Nature **346**: 425–434.
5. GAHMBERG, C.G. 1997. Leukocyte adhesion: CD11/CD18 integrins and intercellular adhesion molecules. Curr. Opin. Cell Biol. **9**: 643–650.
6. LIU, S., D.A. CALDERWOOD & M.H. GINSBERG. 2000. Integrin cytoplasmic domain-binding proteins. J. Cell. Sci. **113**: 3563–3571.
7. CALDERWOOD, D.A. 2004. Integrin activation. J. Cell. Sci. **117**: 657–666.
8. VAN KOOYK, Y. & C.G. FIGDOR. 2000. Avidity regulation of integrins: the driving force in leukocyte adhesion. Curr. Opin. Cell Biol. **12**: 542–547.
9. FAGERHOLM, S.C., T.J. HILDEN & C.G. GAHMBERG. 2004. P marks the spot: site-specific integrin phosphorylation regulates molecular interactions. Trends Biochem. Sci. **29**: 504–512.
10. HIBBS, M.L., S. JAKES, S.A. STACKER, *et al.* 1991. The cytoplasmic domain of the integrin lymphocyte function-associated antigen 1 beta subunit: sites required for binding to intercellular adhesion molecule 1 and the phorbol ester-stimulated phosphorylation site. J. Exp. Med. **174**: 1227–1238.
11. VALMU, L., T.J. HILDEN, G. VAN WILLIGEN & C.G. GAHMBERG. 1999. Characterization of beta2 (CD18) integrin phosphorylation in phorbol ester-activated T lymphocytes. Biochem. J. **339**: 119–125.
12. HILDEN, T.J., L. VALMU, S. KARKKAINEN & C.G. GAHMBERG. 2003. Threonine phosphorylation sites in the beta 2 and beta 7 leukocyte integrin polypeptides. J. Immunol. **170**: 4170–4177.
13. FAGERHOLM, S., N. MORRICE, C.G. GAHMBERG & P. COHEN. 2002. Phosphorylation of the cytoplasmic domain of the integrin CD18 chain by protein kinase C isoforms in leukocytes. J. Biol. Chem. **277**: 1728–1738.
14. PETER, K. & T.E. O'TOOLE. 1995. Modulation of cell adhesion by changes in alpha L beta 2 (LFA-1, CD11a/CD18) cytoplasmic domain/cytoskeleton interaction. J. Exp. Med. **181**:215–326.
15. NORTAMO, P., M. PATARROYO, C. KANTOR, *et al.* 1988. Immunological mapping of the human leucocyte adhesion glycoprotein gp90 (CD18) by monoclonal antibodies. Scand. J. Immunol. **28**: 527–546.
16. POZUELO, M., C. MACKINTOSH, A. GALVAN & E. FERNANDEZ. 2001. Cytosolic glutamine synthetase and not nitrate reductase from the green alga *Chlamydomonas reinhardtii* is phosphorylated and binds 14-3-3 proteins. Planta **212**: 264–269.
17. MOORHEAD, G., P. DOUGLAS, V. COTELLE, *et al.* 1999. Phosphorylation-dependent interactions between enzymes of plant metabolism and 14-3-3 proteins. Plant J. **18**: 1–12.
18. JIN, J. *et al.* 2004. Proteomic, functional, and domain-based analysis of *in vivo* 14-3-3 binding proteins involved in cytoskeletal regulation and cellular organization. Curr. Biol. **14**: 1436–1450.

19. VAN DER HOEVEN, P.C., J.C. VAN DER WAL, P. RUURS, et al. 2000. 14-3-3 isotypes facilitate coupling of protein kinase C-zeta to Raf-1: negative regulation by 14-3-3 phosphorylation. Biochem. J. **345:** 297–306.
20. SHARMA, C.P., R.M. EZZELL & M.A. ARNAOUT. 1995. Direct interaction of filamin (ABP-280) with the beta 2-integrin subunit CD18. J. Immunol. **154:** 3461–3470.

Grb2-Associated Binder 1 (Gab1) Adaptor/Scaffolding Protein Regulates Erk Signal in Human B Cells

ADRIENN ANGYAL,[a] DAVID MEDGYESI,[b] AND GABRIELLA SARMAY[a]

[a]*Department of Immunology, Eotvos Lorand University, Budapest 1117, Hungary*

[b]*Research Group of the Hungarian Academy of Sciences at the Department of Immunology, Eotvos Lorand University, Budapest 1117, Hungary*

ABSTRACT: **RNA silencing experiments showed that knocking down Gab1 adaptor protein in BL41 human Burkitt lymphoma cells significantly reduced B cell receptor (BCR)–induced Erk phosphorylation, indicating that Gab1 plays a pivotal role in regulating Erk activity in B cells.**

KEYWORDS: Gab; B cell; RNA silencing

INTRODUCTION

Gab1, a member of the Gab/dos family of adaptor/scaffolding proteins, was originally isolated as a 110-kDa binding protein for growth factor receptor–bound protein 2 (Grb2) by screening an expression cDNA library constructed from a glial tumor.[1] This protein is involved in the signal transduction pathway of a variety of growth factors, cytokines, and antigen receptors (e.g., epidermal growth factor receptor [EGFR] hepatocyte growth factor receptor [HGF], IL-3, gp130, B cell receptor [BCR], and T cell receptor [TCR]). Gab1 has been shown to possess a highly conserved N-terminal pleckstrin homology domain, a central proline-rich domain including a c-Met binding domain, and multiple tyrosine residues, which undergo phosphorylation upon EGF and insulin stimulation. It consequently binds signaling molecules with SH2 domains such as SHP-2 tyrosine phosphatase and phosphatidyl inositol 3-kinase (PI3-K), Crk/CrkL, src homology 2 domain containing adaptor (Shc), and phospholipase C-γ (PLC-γ). Grb2 constitutively binds to proline-rich sequences of Gab1, while through its pleckstrin homologue (PH) domain, Gab1 is able to bind to PIP3 in the cell membrane.[2,3]

Address for correspondence: Gabriella Sarmay, Eotvos Lorand University, Pazmany Peter Setany 1/c, 1117 Budapest, Hungary. Voice: 36-1-2090-555; ex. 8662, fax: 36-1-3812-176.
e-mail: sarmayg@cerberus.elte.hu

Gab1 was shown to participate in activating ras/MAPK pathway upon EGFR signaling, which is mediated by the SHP-2 tyrosine phosphatase and the ras regulating protein, rasGAP.[4] SHP-2 activated by the C terminal tyrosine phosphorylated motif(s) of Gab1 may dephosphorylate the rasGAP binding phosphotyrosine on Gab1, resulting in a sustained activation of ras/MAPK pathway. Furthermore, via positive feedback regulation, Gab1 enhances PI3-K activity, resulting in a high level of phosphatidyl inositol 3,4,5 trisphosphate (PIP3) in the plasma membrane.[4,5] These data suggest that Gab1 is a positive regulator of EGFR-mediated signaling.

Knocking out the Gab1 gene is lethal for the embryos, and thus studying the role of Gab1 on the immune response is difficult. The *in vivo* role of Gab1 was studied in irradiated mice reconstituted by Gab1−/− fetal liver cells. In this model Gab1 has been shown to be a specific negative regulator of T-independent 2 response of marginal zone B cells in the spleen.[6]

Gab1 may also have a critical role in B cell activation because it becomes phosphorylated on several tyrosine residues upon BCR cross-linking and thus may serve as a platform to translocate cytoplasmic proteins to the cell membrane, where they can act on their membrane-bound targets.[7] Phosphorylation of the appropriate tyrosine residues on Gab1 allows it to bind the SH2 domains of Shc, PI3K, and SHP2.[8] Gab1 overexpression increased the magnitude of BCR-induced Akt phosphorylation, which was shown to depend on its activity to bind PI3-K. Binding to Gab1 promoted tyrosine phosophorylation of SHP2 and its binding to Grb2, potentially influencing downstream signaling.[9]

We have shown previously that co-clustering BCR and FcγRIIb induces dephosphorylation of Gab1 in BL41 Burkitt lymphoma cells, resulting in a decreased PI3-K activity.[10] Here we further revealed the role of Gab1 in B cell signaling. Our aim was to clarify the function of Gab1 in BCR-mediated signaling.

We explored this question by temporarily knocking down Gab1 in BL41 cells using small interfering RNA (siRNA). We found that Erk activation was almost completely blocked, while Akt phosphorylation was partially reduced in the Gab1 knocked-down samples. These results suggest that in human B cells Gab1 is a major regulator of Erk, and a less potent regulator of Akt/PKB.

MATERIALS AND METHODS

BL41 Burkitt lymphoma cells were grown in RPMI 1640 medium supplemented with 5% heat-inactivated FCS. A total of 5×10^6 cells were transfected with 2 μg siCONTROL nontargeting siRNA #2 (Dharmacon D-001210-02-05), fluorescence-labeled siGLO cyclophilin B siRNA (human/mouse) (Dharmacon D-001610-01-05), and siRNA specific for GAB1 (Quiagene High Performance 2-For-Silencing siRNA Duplexes), with the help of the Amaxa nucleofecting system, following the guidelines. The cells were cultured for 24,

48, and 72 h after this treatment, and then the cells were washed and stimulated for 2 min by 0.5-μg biotin-SP-conjugated affinity pure F(ab')$_2$ fragment goat anti-human IgG+IgM (H+L) (Jackson 109-066-127) or left untreated. The cell pellets were lysed with 20 mM Tris (pH 7.5) buffer containing 150 mM NaCl, 1 mM EDTA, 1 mM EGTA, 1% Triton X-100 in the presence of protease- and phosphatase inhibitors (2.5 mM sodium pyrophosphate, 1mM β-glycerolphosphate, 1 mM Na$_3$VO$_4$ and 1 μg/mL leupeptin). Lysate samples were exposed to SDS-PAGE and analyzed by Western blotting. Silencing effect on Gab1 was detected with Gab1-specific antibody (Cell Signaling #3232) and activation of PKB/Akt and MAPK signaling pathways was detected by antibodies specific for the phosphorylated forms of Akt and Erk. The Western blots were developed by horseradish peroxidase–conjugated species-specific second antibodies followed by enhanced chemiluminescence substrate (Pierce Biotechnology, Rockford, IL, USA) and analyzed by densitometry.

RESULTS AND DISCUSSION

To unravel the function of Gab1 docking protein in BCR-induced signaling we applied small interfering RNA (siRNA) to temporarily downregulate Gab1 expression. The efficacy of the siRNA transfection in BL41 cells appeared to be about 35% as determined by fluorescence-labelled siGLO cyclophilin B siRNA. We could not reach a higher degree of transfection in this cell line with any technology used. To optimize Gab1 silencing, the expression of the protein was monitored by Western blot in cells cultured for 24, 48, and 72 h following the transfection. As compared to the samples treated with irrelevant siRNA, transfection with Gab1-specific siRNA resulted in a downregulation of the protein level after 24 h, which gradually reappeared later. After 48 h the silencing RNA was probably degraded and protein expression levels reached the control values (FIG. 1A).

It was shown previously that downregulation of either Gab1 or Gab2 by siRNAs in human breast carcinoma cells effectively inhibited the EGF-stimulated ERK activation pathway[11]; thus, as a possible functional consequence of knocking down Gab1 in B cells, the phosphorylation of Erk due to BCR clustering was tested. Probing the membranes by antibodies specifically recognizing the phosphorylated form of Erk (pErk) has shown that BCR-induced Erk activation is considerably diminished in Gab1 downregulated samples cultured for 24 h, and in parallel with the Gab1 expression, it gradually returned to the control value after 48 h and 72 h (FIG. 1A).

Gab1 was shown to link BCR to PI 3-K /Akt signaling pathway in WEHI-231 mouse B cells.[9] Our previous results showed that PI3-K, Grb2, and SHP-2 associate with phosphorylated Gab1 in BL41 Burkitt lymphoma cells.[10] To assess the relative importance of Gab1 in BCR-induced activation of the mitogen-activated protein kinase (MAPK) and PI 3-K, we compared the two main

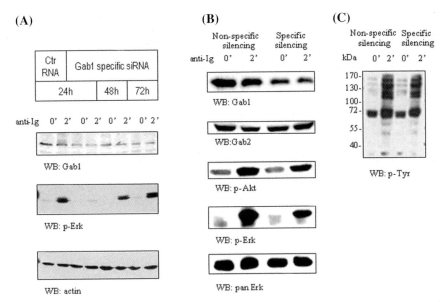

FIGURE 1. The effect of Gab1-specific siRNA in BL41 cells. The cells were treated with nonspecific or Gab1 specific siRNA and then cultured for 24, 48, and 72 h, respectively. After cell culture was terminated, the cells were washed and stimulated by anti-Ig for 2 min or left untreated, and then peletted and lysed. Detergent extracts of cell samples were exposed to SDS-PAGE, and analyzed by Western blotting (**A**). BL41 cells were treated with control or Gab1-specific siRNA for 24 h, and then treated as above (**B**) and (**C**). The membranes were consecutively probed with the specific antibodies.

signaling pathways by monitoring Erk and Akt phosphorylation in Gab1-specific siRNA-treated samples. Western blot analysis of untreated and anti-Ig stimulated cell samples has shown that downregulation of Gab1 significantly diminished BCR-stimulated Erk phosphorylation, while it was less efficient in reducing Akt phosphorylation. Expression level of Gab2, another member of the Gab adaptor family, did not show any alteration in the same samples (FIG. 1B). Because Gab 1 and Gab2 are homologous proteins and have similar structural features, they may functionally overlap, and Gab2 may substitute some of Gab1's function in BL41 cells.

The overall tyrosine phosphorylation of proteins was similar in control and Gab1 knocked-down, anti-Ig triggered cell samples, confirming the selectivity of the effect (FIG. 1C).

Statistical evaluation of the data has shown that although the downregulation of Gab1 was only partial, it reduced phosphorylation of Erk by 67% (ranging between 85% and 55%), even in the presence of Gab2, suggesting that in human B cells Gab1 is the primary regulator of Erk. At the same time phosphorylation of Akt was lowered only by approximately 30%, indicating that Gab1 is a less potent regulator of Akt/PKB (FIG. 2).

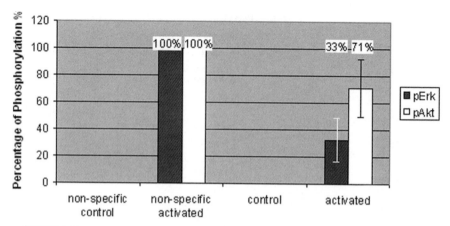

FIGURE 2. Densitometric analysis of BCR-induced Akt and Erk phosphorylation in BL41 cells transfected with Gab1-specific siRNA. After 24 h the cells were harvested, washed, stimulated by anti-Ig, and then tested as above. The membranes were consecutively probed by antibodies specific for the phosphorylated form of Erk and Akt, respectively. Equal loading was controlled by probing the membranes with pan-Erk-specific antibodies. The data were normalized by comparing the intensity of each specific band with the intensity of band detected by pan-Erk-specific antibody for the same sample. Values of the nonspecific siRNA-treated, BCR-activated samples were taken as 100%, and the decrease of phosphorylation of Erk and Akt in the Gab1 downregulated samples was calculated and illustrated. Average of at least three independent experiments.

Activation of Akt depends on PI-3-K activity because its product, PIP3, interacts with the PH domain of PDK1 and PDK2, which phosphorylate Akt. Recruitment of PI3-K to the cell membrane is a prerequisite of its activation. Membrane docking of PI3-K is also possible through the phosphorylated CD19 molecule, a member of the BCR coreceptor complex, so we investigated to what extent CD19 could bind PI3-K. We found that the tyrosine phosphorylation of CD19 was not significant after BCR cross-linking on BL41 cells, and thus it may not bind PI3-K sufficiently. Alternatively, Gab2, the other member of the adaptor/scaffolding protein family, having consensus binding motifs for PI3-K p85 SH2 domains, may recruit PI3-K to the cell membrane, contributing to its activation. Because silencing Gab1 resulted in different degrees of inhibition of Erk and Akt activation, we suppose that Gab1 and Gab 2, although playing overlapping role in B cells, are not equal functionally; Gab1 preferentially regulates Erk, while perhaps Gab2 regulates Akt in BL41 cells.

CONCLUSION

Downregulation of Gab1 by siRNAs effectively inhibited the BCR-stimulated ERK activation pathway, and thus Gab1 represent an important link between BCR and downstream Erk signaling.

ACKNOWLEDGMENTS

This work was supported by the Hungarian Scientific Research Fund (OTKA TS 044711) and by the National Research and Development Programs (NKFP 1A040-04).

REFERENCES

1. HOLGADO-MADRUGA, M., D.R. EMLET, D.K. MOSCATELLO, et al. 1996. A Grb2-associated docking protein in EGF- and insulin-receptor signalling. Nature **379**: 560–564.
2. HIBI, M. & T. HIRANO 2000. Gab-family adapter molecules in signal transduction of cytokine and growth factor receptors, and T and B cell antigen receptors. Leuk. Lymphoma. **37**: 299–307.
3. GU, H., B.G. NEEL. 2003. The "Gab" in signal transduction. Trends Cell Biol. **13**: 122–130.
4. MONTAGNER, A., A. YART, M. DANCE, et al. 2005. A novel role for Gab1 and SHP2 in epidermal growth factor-induced Ras activation. J. Biol. Chem. **280**: 5350–5360.
5. MATTOON, D.R., B. LAMOTHE, I. LAX & J. SCHLESSINGER. 2004. The docking protein Gab1 is the primary mediator of EGF-stimulated activation of the PI-3K/Akt cell survival pathway. BMC Biol. **2**: 24.
6. ITOH, S., M. ITOH, K. NISHIDA, et al. 2002. Adapter molecule Grb2-associated binder 1 is specifically expressed in marginal zone B cells and negatively regulates thymus-independent antigen-2 responses. J. Immunol. **168**: 5110–5116.
7. NISHIDA, K., Y. YOSHIDA, M. ITOH, et al. 1999. Gab-family adapter proteins act downstream of cytokine and growth factor receptors and T- and B-cell antigen receptors. Blood **93**: 1809–1816.
8. GOLD M.R., R.J. INGHAM, S.J. MCLEOD, et al. 2000. Targets of B-cell antigen receptor signaling: the phosphatidylinositol 3-kinase/Akt/glycogen synthase kinase-3 signaling pathway and the Rap1 GTPase. Immunol. Rev. **176**: 47–68.
9. INGHAM, R.J., L. SANTOS, M. DANG-LAWSON, et al. 2001. The Gab1 docking protein links the b cell antigen receptor to the phosphatidylinositol 3-kinase/Akt signaling pathway and to the SHP2 tyrosine phosphatase. J. Biol. Chem. **276**: 12257–12265.
10. KONCZ, G., G.K. TOTH, G. BOKONYI, et al. 2001. Co-clustering of Fcgamma and B cell receptors induces dephosphorylation of the Grb2-associated binder 1 docking protein. Eur. J. Biochem. **268**: 3898–3906.
11. MENG, S., Z. CHEN, T. MUNOZ-ANTONIA & J. WU. 2005. Participation of both Gab1 and Gab2 in the activation of the ERK/MAPK pathway by epidermal growth. Biochem. J. **391**: 143–151.

CXC Receptor and Chemokine Expression in Human Meningioma

SDF1/CXCR4 Signaling Activates ERK1/2 and Stimulates Meningioma Cell Proliferation

FEDERICA BARBIERI,[a] ADRIANA BAJETTO,[a] CAROLA PORCILE,[a] ALESSANDRA PATTAROZZI,[a] ALESSANDRO MASSA,[a] GIANLUIGI LUNARDI,[a] GIANLUIGI ZONA,[b] ALESSANDRA DORCARATTO,[c] JEAN LOUIS RAVETTI,[c] RENATO SPAZIANTE,[b] GENNARO SCHETTINI,[a] AND TULLIO FLORIO[a]

[a]Section of Pharmacology, Department of Biology, Oncology and Genetics, University of Genova, Genova 16132, Italy

[b]Section of Neurosurgery, Department of Neuroscience, Ophthalmology and Genetics, University of Genova, Genova 16132, Italy

[c]Division of Pathology, San Martino Hospital, Genova 16132, Italy

ABSTRACT: Recent evidence indicates that cancer cells express chemokine (CK) receptors and that their signaling is crucial for tumor proliferation, migration, and angiogenesis. The profiles of expression of CXC CK receptors (CXCR1-5) and their main ligands (growth-related oncogene, GRO1-2-3/CXCL1-2-3; interleukin 8, IL-8/CXCL8; monokine-induced γ-interferon MIG/CXCL9; γ-interferon-inducible-protein-10, IP-10/CXCL10; stromal cell-derived factor-1, SDF1/CXCL12; B-cell activating CK-1, BCA-1/CXCL13) were analyzed by reverse transcription polymerase chain reaction (RT-PCR) in surgical samples of human meningiomas. All the five receptors displayed high percentages of positive cases: 92% CXCR1, 89% CXCR2, 83% CXCR3, 78% CXCR4, and 94% CXCR5. Conversely, their ligands showed a lower pattern of expression: 40% IL-8, 42% GRO1-3, 42% IP-10, 28% MIG, 53% SDF1, and 3% BCA-1. SDF1/CXCR4 interaction plays a pivotal role in cancer proliferation. Thus, the signaling mechanisms activated by the exclusive binding between SDF1 and CXCR4 was investigated in 12 primary cultures from meningioma tissues. CXCR4 was functionally coupled as demonstrated by the significant increase of DNA synthesis in meningioma cells in response to SDF1, measured by [^{3}H]-thymidine uptake. In three primary cultures, the SDF1-dependent mitogenic activity

Address for correspondence: Tullio Florio, Section of Pharmacology, Department of Biology, Oncology and Genetics, University of Genova, V. le Benedetto XV 2, 16132 Genova, Italy. Voice and fax: +39-010-3538806.
 e-mail: tullio.florio@unige.it

Ann. N.Y. Acad. Sci. 1090: 332–343 (2006). © 2006 New York Academy of Sciences.
doi: 10.1196/annals.1378.037

was associated with a marked phosphorylation of extracellular signal-regulated kinase (ERK1/2) as evaluated by Western blots. PD98059 (a MEK inhibitor) significantly reduced ERK1/2 activation, thus linking the SDF1/CXCR4 pathway to meningioma cell proliferation via ERK1/2 signal transduction. We demonstrate, for the first time in human meningiomas, the simultaneous expression of CXCR1-5 and their CKs and the mitogenic activity of SDF1/CXCR4, suggesting a pivotal role of these receptor–ligand pairs in meningeal tumors.

KEYWORDS: chemokine; receptor; meningioma; CXCR4; SDF1; primary culture; cell proliferation; ERK1/2

INTRODUCTION

Chemokines (CKs) play an essential role in cellular migration and intercellular communications during inflammatory response,[1] but emerging evidence suggests that CKs also regulate tumor proliferation, angiogenesis, metastasis, and multiple signaling pathways. CKs exert their activity by binding to G-protein-coupled receptors: to date six receptors (CXCR1-6) for the CXC cytokine group (CXCL1-16) have been described in humans. Several types of cancer cells show different CXC receptor profiles,[2] but CXCR4 is most frequently detected, being present in a variety of cancer cell lines and tissues. CXCR4 expression is often associated with other CXC receptors: B cell lymphomas express both CXCR3 and CXCR5,[3] and human melanoma cells coexpress CXCR3 and CXCR4.[4] In addition, the stromal cell–derived factor 1 (SDF1/CXCL12), the unique ligand of CXCR4, has been found in glioma,[5–7] medulloblastoma,[8] lymphoma,[9] and ovarian[10,11] and pancreatic cancer.[12] Similarly, the concomitant expression of IL-8 and its receptors CXCR1 and CXCR2 is involved in prostate cancer cell proliferation.[13] Besides SDF1, other CKs such as GRO-1[14] and IL-8[15,16] act as mediators of tumorigenesis. Moreover, SDF1 stimulates the growth of glioblastoma cell lines and normal astrocytes,[7,17] medulloblastoma cell lines, and xenografted tumors.[8]

Meningiomas account for approximately 20% of all intracranial brain tumors diagnosed in adults. They are graded as benign (WHO I), atypical (WHO II), or malignant (WHO III).[18] Histological grading remains one of the most useful parameters, together with the extent of resection and the proliferative index, in predicting tumor recurrence and/or progression. Benign meningiomas are the most common type, although all the tumor types are associated with significant morbidity and mortality. The primary treatment approach for meningiomas is surgical resection and radiation for unresectable, high-grade or recurrent tumors; however, no effective chemotherapy exists once surgery and radiation have failed.[19] To date, while improved diagnostic modalities are available, several clinical and molecular aspects of meningiomas are understudied. In addition, there is a wide heterogeneity in clinical outcomes, even within benign tumors, and thus the study of gene expression profiling, such as of CXCRs and CKs, may help deepening these processes. Despite the important role of SDF1/CXCR4 interaction in tumor cell growth, no data are available about the signaling pathways that mediate these effects in meningioma proliferation and malignancy. In many cell types the transduction of proliferative signals involves the extracellular signal-regulated kinases 1/2 (ERK1/2) whose enzymatic activity increases in responses to mitogenic stimuli.[20] Recently, we demonstrated that SDF1 induces astrocytes and glioblastoma cell proliferation through CXCR4 activation by ERK1/2 and Akt.[7] In this study, we have investigated, for the first time, the pattern of mRNA expression of CXC receptors 1–5 and their main ligands in 55 human surgical meningioma samples. In addition, taking into account the critical role played by CXCR4 and SDF1 in the promotion of growth of central nervous system tumor cells, the study was focused on identifying the molecules involved in signal transduction, particularly the kinase pathway activated in most human tumors. *In vitro* experiments on primary meningioma cultures have been carried out to evaluate the ability of SDF1 to induce cell proliferation and regulate ERK1/2 phosphorylation.

MATERIALS AND METHODS

Tissue Samples

A total of 55 surgical tumor samples were collected between 1997 and 2005 from the Section of Neurosurgery, University of Genova (Genova, Italy). Patients entered in the study were 30 females and 25 males (median age 65 years, range 35–80 years) including 47 WHO grade I (benign: 28 meningothelial, 11 fibrous, 8 transitional), 5 WHO grade II (atypical), and 3 grade III (anaplastic) tumors. After histological examination according to the World Health Organization (WHO) classification,[21] part of the tumor was immediately frozen and stored at $-80°C$ until reverse transcription polymerase chain reaction (RT-PCR) analysis. In 12 cases, a fresh fragment of the sample was dissociated for short-term cultures.

Reverse-Trancription (RT)-PCR Analysis

Total RNA from surgical specimens was isolated using the acidic-phenol technique and subjected to RT-PCR as previously reported.[11] Primer sequences (TIB MolBiol, Genova, Italy) are reported in TABLE 1. Amplification profile was $94°C$ for 5 min, 40 cycles at $94°C$ for 1 min, $60°C$ for 1 min, and $72°C$ for 1 min, followed by 7 min at $72°C$. Resulting PCR products were electrophoresed through 2% agarose gels stained with ethidium bromide. The levels of the housekeeping β-actin transcript were used as a PCR reaction efficiency control.

TABLE 1. Primer sequences of CXCRs and CK

Probe	Sense (5'→3')	Antisense (5'→3')	Accession No.
CXCR1	GGGGCCACACCAACCTTC	AGTGCCTGCCTCAATGTCTCC	L19591
CXCR2	CCGGGCGTGGTGGTGAG	TCTGCCTTTTGGGTCTTGTGAATA	X73969
CXCR3	TCTGCCTTTTGGGTCTTGTGAATA	AGGAAGATGAAGTCTGGGAG	X95876
CXCR4	GGCCCTCAAGACCACAGTCA	TTAGCTGGAGTGAAAACTTGAAG	NM_003467
CXCR5	ATCTTCTTCCTCTGCTGGTC	GTTCCTCTAGCTACCCCAAA	X68149
DF1	ATGAACGCCAAGGTCGTGGTC	CTTGTTTAAAGCTTTCTCCAGGTACT	NM_000609
IP-10	ATGAATCAAACTGCGATTCTGATT	TTAAGGAGATCTTTTAGACATTTC	NM_001565
MIG	ATGAAGAAAAGTGGTGTTCTTTC	TTATGTAGTCTTCTTTTGACGAGA	X72755
IL-8	TGGGTGCAGAGGGTTGTG	CAGACTAGGGTTGCCAGATTTA	M28130
BCA-1	ATGAAGTTCATCTCGACATCTCTG	TACTTCCATCATTCTTTGTATCCA	AF044197
GRO 1-	CTCTCCGCCGCCCCAGCAATCCC	TTCAGCATCTTTTCGATGATTTCT	NM_001511
2-			NM_002089
3-			NM_002090
β-actin	ACGGGGTCACCCACACTG	TGCTTGCTGATCCACATCTGC	M10277

Primary Cultures

Cultures of meningioma cells of WHO grade I were obtained by mechanical disruption of 12 fresh resected tumors, cultured in Dulbecco's modified essential medium (D-MEM) supplemented with 10% fetal calf serum (FCS), 2 mM glutamine, 100 U/mL penicillin/streptomycin, all purchased from Euroclone (Milano, Italy). Cells were subcultured at least twice before the experiments were performed, after removal of debris and nonadhering cells. The cultures were characterized by immunocytochemical staining with specific cellular markers (epithelial membrane antigen expression, EMA).[22]

[^{3}H]-Thymidine Proliferation Assay

Cells (8×10^5/well) from primary cultures were serum-starved for 24 h before being treated with human SDF1α (Pepro Tech EC Ltd., London, United Kingdom) for 24 h and then pulsed with [^{3}H]-methyl-thymidine (1 μCi/mL, Amersham Biosciences, Milano, Italy) in the last 4 h of treatment and harvested onto glass fiber filters (Millipore, Bedford, MA); the incorporated radioactivity was measured by scintillation counting. When indicated, then PD98059 (10 μM, Calbiochem, San Diego, CA) was added to the cells for 10 min, before SDF1.

Western Blot

Cells were serum-starved for 24 h before the exposure to SDF1 (25 nM) for 2 min and lysed for 10 min at 4°C in 1% NP-40, 20 mM Tris HCl pH 8, 137 mM NaCl, 10% glycerol, 2 mM EDTA, 1 mM PMSF, 1μg/mL leupeptin, 1 mM sodium orthovanadate, 10 mM NaF (all from Sigma-Aldrich, Milano, Italy). A total of 10 μg was resolved by SDS-PAGE and transferred onto PVDF membrane (Bio-Rad Laboratories, Hercules, CA). Blots were probed with anti-phospho-ERK1/2 (Cell Signaling New England BioLabs, Beverly, MA, USA) and reprobed with anti-ERK1/ (Sigma Aldrich) to ensure equal loading. Proteins were detected using chemioluminescence (ECL, Amersham Biosciences).

Statistical Analysis

Results are expressed as the means ± SE of replicate determinations and statistical significance (P value ≤ 0.05) was assessed by the Student's t-test.

TABLE 2. CXCR and ligand pairs mRNA expression evaluated by RT-PCR in human meningioma samples

	WHO grade (positive/total)			
	I	II	III	Positive/total
CXCR1	28/31	5/5	1/1	34/37 (92%)
CXCR2	25/29	5/5	1/1	31/35 (89%)
IL-8	11/29	3/5	0/1	14/35 (40%)
GRO1-2-3	13/30	2/5	0/1	15/36 (42%)
CXCR3	24/30	5/5	1/1	30/36 (83%)
IP-10	12/30	3/5	0/1	15/36 (42%)
MIG	8/30	2/5	0/1	10/36 (28%)
CXCR4	37/47	4/5	2/3	43/55 (78%)
SDF1	25/47	3/5	1/3	29/55 (53%)
CXCR5	28/30	5/5	1/1	34/36 (94%)
BCA-1	1/29	0/5	0/1	1/35 (3%)

RESULTS

CXC Receptor and Ligand Expression in Human Meningiomas

We screened the meningioma samples for the expression of CXCR1-5 and CKs by RT-PCR. TABLE 2 reports the patterns of mRNA expression of CXC receptors and ligands among the different WHO grades. All the five receptors displayed high percentages of positive cases ranging from 78% (CXCR4) to 94% (CXCR5): CXCR1 was expressed in 34/37 samples, CXCR2 in 31/35, CXCR3 in 30/36, CXCR4 in 43/55, and CXCR5 in 34/36 cases. Conversely, their ligands show a lower pattern of expression, from 3% (BCA) to 53% (SDF1): IL-8 and GRO1-3 were detected in 14/35 and 15/36 cases, respectively; IP-10 in 15/36, MIG in 10/36, SDF1 in 29/55 and BCA-1 in 1/35. Focusing on CXCR4 and SDF1 expression, they resulted coexpressed in 27/55 (49%) cases. CXCR4 mRNA was present in 78% of WHO I, 80% of WHO II, and 66% of WHO III, whereas the SDF1 mRNA was identified in 53% of WHO I, 60% of WHO II, and 33% of WHO III; in addition only 10/55 (18%) were negative for the expression of both proteins.

SDF1 Effects on Primary Meningioma Cultures

We evaluated the ability of SDF1 to induce cell proliferation in primary cultures derived from 12 meningioma specimens by the [^{3}H]-thymidine incorporation assay. In these cultures, all belonging to WHO grade I, immunostaining with EMA and CXCR4 antibodies was used to confirm the presence of meningeal tumor cells and the expression of this receptor at the time of the *in vitro* experiments (data not shown). Cells were starved for 24 h before stimulation with increasing concentrations (6.25, 12.5, 25.0, 37.5, 50.0 nM) of SDF1 for 24 h. SDF1 induced a statistically significant increase in [^{3}H]-thymidine uptake in 8 of 12 of the meninigoma cultures (FIG. 1). In particular, Cases 1 and 2 reached a maximum increase in DNA synthesis at 25 nM SDF1 (4.2- and 3.5-fold over the basal, respectively) while Cases 3, 4, 5, and 7 showed the peak of stimulation after exposure to 12.5 nM SDF1 (2.2-, 1.9-, 2.0-, and 1.6-fold over the basal, respectively). Case 8, despite a marked increase of cell growth (+1.6-fold) at 25 nM, did not reach the statistical significance ($P = 0.06$). In Case 10, the cell proliferation was significantly stimulated at 37.5 nM SDF1 (+1.3-fold). Three cases that did not significantly respond to SDF1 (Cases 6, 11, and 12) even if slight mitogenic effects were observed (+1.3-, +1.2-, and +1.2-fold, respectively).

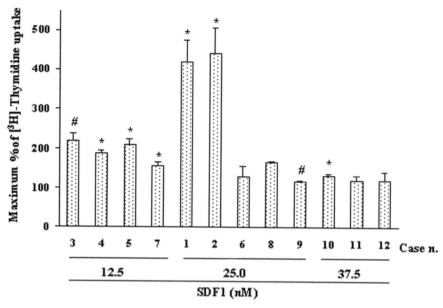

FIGURE 1. Maximum SDF1-induced DNA synthesis in 12 primary meningioma cultures by [^{3}H]-thymidine uptake assay. Values are expressed as percentage of the basal incorporation. Statistical significance versus basal values: *$P < 0.01$, #$P < 0.05$, NS = not significant.

SDF1/CXCR4 Interaction Activates ERK1/2 in Meningioma Cells

To study the intracellular signaling mediated by CXCR4/SDF1 interaction, we investigated whether SDF1 could activate ERK1/2. Serum-starved cultures derived from 3 meningothelial tumors (Cases 1, 4, and 5) were stimulated with 25 nM SDF1 and cell lysates were analyzed by Western blot. SDF1 treatment of meningioma cells induced a rapid increase of phospho-ERK1/2 levels from 1.7- to 2-fold over basal, as quantified by densitometric analysis (FIG. 2A). To further investigate the relationship between SDF1-induced proliferation and ERK1/2 activation, we analyzed the SDF1-stimulated DNA synthesis in the presence of PD98059 (10 μM), an inhibitor of MEK (MAP kinase kinase), by [^{3}H]-thymidine incorporation assay (FIG. 2B). When cultures were treated with PD98059, before addition of SDF1, the uptake of [^{3}H]-thymidine induced by SDF1 was strongly inhibited (-72%, -73%, -48% for Cases 1, 4, and 5) and in two cases (4 and 5) the compound also reduced the basal DNA synthesis (-56% and -40%, respectively). Overall, these results show that ERK1/2-mediated intracellular signaling pathway is involved in meningioma cell proliferation induced by SDF1.

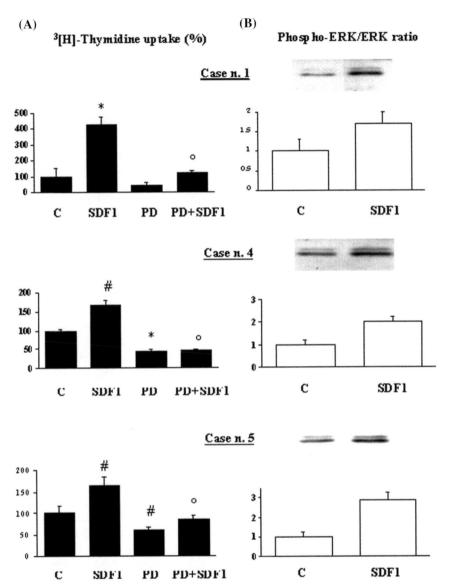

FIGURE 2. (A) Effects of PD98059 on the SDF1-induced meningioma cell proliferation. Serum-starved cells were treated with SDF1 (25 nM) in the presence or absence of PD98059 (10 μM) and DNA synthesis was measured by [^{3}H]-thymidine uptake assay. Values are expressed as percentage of the basal incorporation. Statistical significance versus basal values: $^*P < 0.01$, $^\#P < 0.05$. (B) SDF-1-induced ERK1/2 phosphorylation in three primary meningioma cells (Cases 1, 4, and 5). Cell lysates from untreated (C) or treated with 25.0 nM SDF1 were analyzed by Western blot with anti-phspho-ERK1/2. Densitometry was performed on the blots and the ratio of phospho-ERK1/2/total ERK1/2 was calculated.

DISCUSSION

Most meningiomas are slowly growing benign tumors; however, many of them, as well as atypical and anaplastic meningiomas, show an aggressive biological and clinical behavior associated with high rates of recurrence and long-term unfavorable prognosis. The molecular mechanisms involved in meningioma initiation, proliferation and progression are not yet fully understood. In particular, the role of growth factors and their receptors in downstream signaling pathways in meningiomas is often unclear, despite the overexpression of PDGF,[23] EGF,[24] TGF-alpha,[25] and insulin-like growth factor II (IGF-II),[26] and contributes to tumor growth.

Besides their role in cell migration and inflammation processes, CKs and their receptors may also signal cell proliferation and survival in several human cancer types including brain tumors. To date, little is known about the role of CKs in meningioma cells, while data are available about the role of meningeal cells as an important source for the expression of SDF1 during brain development, where it acts as a chemotactic factor.[27] In addition meningeal cells may be a local and early source of CKs induced by pathogenic bacteria,[28] particularly the CXC chemokine IL-8.[29]

In the present study, we investigated the presence of mRNA for CXC CKs and their receptors in 55 meningiomas by RT-PCR. CXCR1-5 appear to be constitutively expressed with high frequency, while the corresponding ligands show a lower pattern of expression. The large number of meningiomas expressing CKs and their receptors clearly suggest a significant role for such molecules in the biology of these tumors. CKs and their receptors are involved in several physiological and pathological effects in the brain. In particular, SDF1 and CXCR4, through their unique interaction, exert a critical role during brain development[27] and tumor progression.[30,5,8] Previous studies by our group addressed the ability of SDF1 to induce astrocyte proliferation through ERK1/2 activation[31] and, more recently, to exert a mitogenic activity also in glioma and ovarian cell lines, likely through an autocrine/paracrine mechanism.[7,11] The pathogenesis of meningioma had not yet been fully understood, although alterations of signaling processes and genetic modifications might have a pivotal role in pathways leading to its development. CKs and CXCRs mediate multiple signal tranduction pathways and several cellular processes, mainly investigated in immune cells. Among them, the CK SDF1 and its cognate receptor CXCR4 have recently taken on substantial interest because of their role in growth several human neoplasms.[32] However, information linking cellular functions and signaling pathways mediated by SDF1/CXCR4 interaction within cancer cells, particularly meningioma, have been scantly investigated. It has been reported that the growth factor–mediated activation of MAPK intracellular signaling cascade contributes to the meningioma proliferation.[24,33,34] Indeed, some reports have suggested that the MAPK pathway may be implicated in the meningioma cells' growth: recently, it was demonstrated that MAPK activation is required for both proliferation and apoptosis of malignant meningioma and that activation of MAPK was associated with malignant meningioma proliferation.[34] In this study the ability of SDF1 to induce cell proliferation was tested in primary cell culture obtained from 12 meningioma tissues in which SDF1 leads to a significant increase of DNA synthesis, thus confirming that the interaction of this CK with CXCR4 activates a proliferative signal. The SDF1-induced proliferation is reduced by PD98059, a MEK inhibitor, indicating that ERK1/2 is involved in the proliferative signal of SDF1. These data demonstrated that primary tumor cells expressed a functional CXCR4 in agreement with other tumor types and suggest that activation of ERK1/2 has growth-promoting effects in benign meningiomas. Taken together these results showed, for the first time, that CXCR4 and SDF1 are involved in the proliferation of meningeal tumor cells, contributing to the biological features of this neoplasm, such as the ability to survive and to grow autonomously.

The functional role of SDF1-induced ERK1/2 activation in cell proliferation may represent a therapeutic target; at present novel pharmacological modulators SDF1/CXCR4-induced responses are currently under investigation in preclinical studies.[8,35] In addition, the broad expression of CXCR1-5 in meningiomas suggests that these receptors could play important functions in tumorigenesis similarly to those identified for CXCR4. As relations between histopathology, molecular features and biological behavior of meningioma become clarified, the approach to its management will become more specifically addressed and CKs and their receptors may represent relevant targets.

REFERENCES

1. MURPHY, P.M. 2001. Chemokines and the molecular basis of cancer metastasis. N. Engl. J. Med. **345:** 833–835.
2. BALKWILL, F. 2004. Cancer and the chemokine network. Nat. Rev. Cancer **4:** 540–550.
3. JONES, D., R.J. BENJAMIN, A. SHAHSAFAEI, et al. 2000. The chemokine receptor CXCR3 is expressed in a subset of B-cell lymphomas and is a marker of B-cell chronic lymphocytic leukemia. Blood **95:** 627–632.
4. ROBLEDO, M.M., R.A. BARTOLOME, N. LONGO, et al. 2001. Expression of functional chemokine receptors CXCR3 and CXCR4 on human melanoma cells. J. Biol. Chem. **276:** 45098–50105.
5. REMPEL, S.A., S. DUDAS, S. GE, et al. 2000. Identification and localization of the cytokine SDF1 and its receptor, CXC chemokine receptor 4, to regions of necrosis and angiogenesis in human glioblastoma. Clin. Cancer Res. **6:** 102–111.
6. ZHOU, Y., P.H. LARSEN, C. HAO, et al. 2002. CXCR4 is a major chemokine receptor on glioma cells and mediates their survival. J. Biol. Chem. **277:** 49481–49487.
7. BARBERO, S., R. BONAVIA, A. BAJETTO, et al. 2003. Stromal cell-derived factor 1alpha stimulates human glioblastoma cell growth through the activation of both extracellular signal-regulated kinases 1/2 and Akt. Cancer Res. **63:** 1969–1974.

8. RUBIN, J.B., A.L. KUNG, R.S. KLEIN, et al. 2003. A small-molecule antagonist of CXCR4 inhibits intracranial growth of primary brain tumors. Proc. Natl. Acad. Sci. USA **100:** 13513–13518.
9. CORCIONE, A., L. OTTONELLO, G. TORTOLINA, et al. 2000. Stromal cell-derived factor-1 as a chemoattractant for follicular center lymphoma B cells. J. Natl. Cancer Inst. **92:** 628–635.
10. SCOTTON, C.J., J.L. WILSON, K. SCOTT, et al. 2002. Multiple actions of the chemokine CXCL12 on epithelial tumor cells in human ovarian cancer. Cancer Res. **62:** 5930–5938.
11. PORCILE, C., A. BAJETTO, F. BARBIERI, et al. 2005. Stromal cell-derived factor-1alpha (SDF-1alpha/CXCL12) stimulates ovarian cancer cell growth through the EGF receptor transactivation. Exp. Cell Res. **308:** 241–253.
12. KOSHIBA, T., R. HOSOTANI, Y. MIYAMOTO, et al. 2000. Expression of stromal cell-derived factor 1 and CXCR4 ligand receptor system in pancreatic cancer: a possible role for tumor progression. Clin. Cancer Res. **6:** 3530–3535.
13. MURPHY, C., M. MCGURK, J. PETTIGREW, et al. 2005. Nonapical and cytoplasmic expression of interleukin-8, CXCR1, and CXCR2 correlates with cell proliferation and microvessel density in prostate cancer. Clin. Cancer Res. **11:** 4117–4127.
14. ZHOU, Y., J. ZHANG, Q. LIU, et al. 2005. The chemokine GRO-alpha (CXCL1) confers increased tumorigenicity to glioma cells. Carcinogenesis **26:** 2058–2068.
15. DESBAILLETS, I., A.C. DISERENS, N. TRIBOLET, et al. 1997. Upregulation of interleukin 8 by oxygen-deprived cells in glioblastoma suggests a role in leukocyte activation, chemotaxis, and angiogenesis. J. Exp. Med. **186:** 1201–1212.
16. BRAT, D.J., A.C. BELLAIL & E.G. VAN MEIR. 2005. The role of interleukin-8 and its receptors in gliomagenesis and tumoral angiogenesis. Neuro-oncol. **7:** 122–133.
17. BAJETTO, A., R. BONAVIA, S. BARBERO, et al. 2001. Chemokines and their receptors in the central nervous system. Front. Neuroendocrinol. **22:** 147–184.
18. CLAUS, E.B., M.L. BONDY, J.M. SCHILDKRAUT, et al. 2005. Epidemiology of intracranial meningioma. Neurosurgery **57:** 1088–1095; discussion 1088–1095.
19. WHITTLE, I.R., C. SMITH, P. NAVOO, et al. 2004. Meningiomas. Lancet **363:** 1535–1543.
20. GUTKIND, J.S. 1998. Cell growth control by G protein-coupled receptors: from signal transduction to signal integration. Oncogene **17:** 1331–1342.
21. KLEIHUES, P., D.N. LOUIS, B.W. SCHEITHAUER, et al. 2002. The WHO classification of tumors of the nervous system. J. Neuropathol. Exp. Neurol. **61:** 215–225; discussion 226–229.
22. ARENA, S., F. BARBIERI, S. THELLUNG, et al. 2004. Expression of somatostatin receptor mRNA in human meningiomas and their implication in in vitro antiproliferative activity. J. Neurooncol. **66:** 155–166.
23. JOHNSON, M.D., A. WOODARD, P. KIM, et al. 2001. Evidence for mitogen-associated protein kinase activation and transduction of mitogenic signals by platelet-derived growth factor in human meningioma cells. J. Neurosurg. **94:** 293–300.
24. CARROLL, R.S., P.M. BLACK, J. ZHANG, et al. 1997. Expression and activation of epidermal growth factor receptors in meningiomas. J. Neurosurg. **87:** 315–323.
25. HALPER, J., C. JUNG, A. PERRY, et al. 1999. Expression of TGFalpha in meningiomas. J. Neurooncol. **45:** 127–134.
26. NORDQVIST, A.C., M. PEYRARD, H. PETTERSSON, et al. 1997. A high ratio of insulin-like growth factor II/insulin-like growth factor binding protein 2 messenger RNA as a marker for anaplasia in meningiomas. Cancer Res. **57:** 2611–2614.

27. REISS, K., R. MENTLEIN, J. SIEVERS, et al. 2002. Stromal cell-derived factor 1 is secreted by meningeal cells and acts as chemotactic factor on neuronal stem cells of the cerebellar external granular layer. Neuroscience **115:** 295–305.
28. WELLS, D.B., P.J. TIGHE, K.G. WOOLDRIDGE, et al. 2001. Differential gene expression during meningeal-meningococcal interaction: evidence for self-defense and early release of cytokines and chemokines. Infect. Immun. **69:** 2718–2722.
29. CHRISTODOULIDES, M., B.L. MAKEPEACE, K.A. PARTRIDGE, et al. 2002. Interaction of *Neisseria meningitidis* with human meningeal cells induces the secretion of a distinct group of chemotactic, proinflammatory, and growth-factor cytokines. Infect. Immun. **70:** 4035–4044.
30. ZLOTNIK, A. 2004. Chemokines in neoplastic progression. Semin. Cancer Biol. **14:** 181–185.
31. BAJETTO, A., S. BARBERO, R. BONAVIA, et al. 2001. Stromal cell-derived factor-1alpha induces astrocyte proliferation through the activation of extracellular signal-regulated kinases 1/2 pathway. J. Neurochem. **77:** 1226–1236.
32. BALKWILL, F. 2003. Chemokine biology in cancer. Semin. Immunol. **15:** 49–55.
33. JOHNSON, M.D., E. OKEDLI, A. WOODARD, et al. 2002. Evidence for phosphatidylinositol 3-kinase-Akt-p7S6K pathway activation and transduction of mitogenic signals by platelet-derived growth factor in meningioma cells. J. Neurosurg. **97:** 668–675.
34. MAWRIN, C., T. SASSE, E. KIRCHES, et al. 2005. Different activation of mitogen-activated protein kinase and Akt signaling is associated with aggressive phenotype of human meningiomas. Clin. Cancer Res. **11:** 4074–4082.
35. COSCIA, M. & A. BIRAGYN. 2004. Cancer immunotherapy with chemoattractant peptides. Semin. Cancer Biol. **14:** 209–218.

Reduction of Bcr-Abl Function Leads to Erythroid Differentiation of K562 Cells via Downregulation of ERK

A. BRÓZIK,[a] N.P. CASEY,[b] Cs. HEGEDŰS,[a] A. BORS,[c] A. KOZMA,[d] H. ANDRIKOVICS,[c] M. GEISZT,[d] K. NÉMET,[e] AND M. MAGÓCSI[a]

[a]*National Medical Center, Department of Molecular Cell Biology, Budapest, Hungary*

[b]*University of Tasmania, Division of Medicine, Hobart, Australia*

[c]*National Medical Center, Department of Molecular Diagnostics, Budapest, Hungary*

[d]*National Medical Center, Department of Hematology and Transplantation, Budapest, Hungary*

[d]*Semmelweis University, Department of Physiology, Budapest, Hungary*

[e]*National Medical Center, Department of Experimental Gene Therapy, Budapest, Hungary*

ABSTRACT: The chimeric *bcr-abl* gene encodes a constitutively active tyrosine kinase that leads to abnormal transduction of growth and survival signals leading to chronic myeloid leukemia (CML). According to our previous observations, *in vitro* differentiation of several erythroid cell lines is accompanied by the downregulation of extracellular signal-regulated kinases (ERK)1/2 mitogen-activated protein kinase (MAPK) activities. In this work we investigated whether ERKs have a decisive role in either the erythroid differentiation process or apoptosis of *bcr-abl*[+] K562 cells by means of direct (MEK1/2 inhibitor UO126) and indirect (reduced Bcr-Abl function) inhibition of their activities. We found that both Gleevec and UO126 induced hemoglobin expression. Gleevec treatment reduced the phosphorylation of Bcr-Abl, ERK and STAT-5 for up to 24 h, decreased Bcl-XL levels, and induced caspase-3-dependent apoptosis. In contrast, UO126 treatment resulted in only a transient decrease of ERK activity and did not induce cell death. For studying the effect of reduced Bcr-Abl function on erythroid differentiation at the level of the *bcr-abl* transcript, we applied the siRNA approach. Stable degradation of *bcr-abl* mRNA was achieved by using a retroviral vector with enhanced green

Address for correspondence: Maria Magocsi, National Medical Center, Department of Molecular Cell Biology, H-1113 Budapest, Dioszegi 64, Hungary. Voice: 36-1-372-4353; fax: 36-1-372-4353.
e-mail: magocsi@biomembrane.hu

fluorescent protein (EGFP) reporter. Despite a high (>90%) transduction efficiency we detected only a transient decrease in Bcr-Abl protein and in phosphorylated ERK1/2 levels. This transient change in Bcr-Abl signaling was sufficient to induce hemoglobin expression without significant cell death. These results suggest that by transiently reducing Bcr-Abl function it is possible to overcome the differentiation blockade without evoking apoptosis in CML cells and that reduced ERK activity may have a crucial role in this process.

KEYWORDS: erythroid differentiation; apoptosis; CML; Bcr-Abl; MEK/ERK MAPK pathway; stable RNA interference

INTRODUCTION

The extracellular signal-regulated kinases 1 and 2 (ERK1 and ERK2) play a well-recognized role in mitogenic signal transduction and they are constitutively active in many tumor cell lines. The MEK/ERK signaling pathway has also been implicated to be involved in cellular differentiation such as the monocytic[1] and megakaryocytic differentiation of hematopoietic cells.[2,3] The cells of K562 pluripotent human chronic myeloid leukemia (CML) cell line can be differentiated along the erythroid lineage by a variety of chemical compounds, including butyrate,[4] hemin, and different antitumor agents, like cisplatin or ara-C. Changes in ERK1/2 activities have been reported to show distinct characteristics during erythroid differentiation of K562 cells depending on the inducing agent.[5] At the same time, an increasing body of evidence suggests that downregulation of ERK1/2 activities is preferred for erythroid differentiation.[6–9] In K562 cells, the chimeric *bcr-abl* gene encodes a constitutively active tyrosine kinase that leads to abnormal transduction of growth and survival signals via multiple signal transduction pathways, including the ERK mitogen-activated protein kinase (MAPK) cascade.[10] Selective inhibition of the Bcr-Abl fusion protein with Gleevec (imatinib mesylate, STI-571) leads to mitochondria-dependent apoptosis and differentiation toward the erythroid lineage.[11,12] Signaling mechanisms regulating these different cellular responses have not been clearly distinguished yet. Apoptosis and erythroid differentiation have recently been reported as simultaneous consequences of Gleevec treatment in two CML-derived cell lines, K562 and JURL-MK1.[13] In the present work we investigated the role of ERK1/2 kinases in cell fate decisions of K562 cells, such as erythroid differentiation or apoptosis, by means of indirect (with reduced Bcr-Abl function) and direct (with the MEK1/2 inhibitor UO126) inhibition of their activities. We also studied the kinetics of ERK1/2 activities in order to characterize their influence on cell death and/or differentiation.

MATERIALS AND METHODS

Cell Line, Culture Conditions, and Stimulation

Bcr-Abl+ K562 cells were grown in RPMI medium without nucleosides, supplemented with 10% (v/v) fetal bovine serum and with 2 mM glutamine, 50 units/mL penicillin, and 50 μg/mL streptomycin (Invitrogen-Gibco, Carlsbad, CA, USA) at 37°C in humidified air/CO^2 (19:1) atmosphere. Cells (2×10^5/mL) were treated with various drugs for 3 days. Viability was determined by trypan blue exclusion and MTT incorporation.[14] UO126 (Sigma, St. Louis, MO, USA) and Gleevec (Vichem, Budapest, Hungary) were dissolved in dimethylsulfoxide (DMSO).

Preparation of Cell Extracts and Analysis by Western Blotting

Cells were lysed as described previously.[15] 50 μg of lysed protein was separated by SDS-PAGE and transferred to PVDF membrane using the Mini-Protean III system (Bio-Rad, Hercules, CA, USA). The following antibodies were used: anti-(cAbl), anti-(ERK1), anti-(STAT-5), and anti-(Bcl-XL) (Santa Cruz, Biotechnology, Santa Cruz, CA, USA) polyclonal and anti-(phospho-STAT-5), anti-(phospho-ERK1/2) (Santa Cruz, Upstate), anti-(phospho-Tyr, clone 4G10) (Upstate, Dundee, Scotland, UK) and anti-PARP (Upstate) monoclonal IgGs. To develop immuno-staining we used HRP-conjugated secondary antibodies (Jackson Immunoresearch Europe, New market, UK) and the ECL detection system (Amersham, GE Healthcare Bio-Sciences AB, Uppsala, Sweden). For quantitative analysis of Western blots, the Bio Rad ChemDoc (Bio-Rad) system was used with the Quantity One 4.4 program.

Preparation of Total RNA and Quantitative Measurement of Globin mRNA Levels by the LightCycler System

Total RNA was prepared using the manufacturer's protocol recommended for TRIZOL reagent (Invitrogen, Carlsbad, CA, USA). Total cDNA was prepared by reverse transcription with random hexamers using 1 μg RNA. A total of 1 μL of total cellular cDNA was amplified using the LightCycler FastStart DNA Master SYBR Green I kit (Roche, Mannheim, Germany). The different mRNA quantities, relative to β2 microglobulin, were calculated by comparing the actual values to standard curves. Primer-sequences used for reverse transciption polymerase chain reaction (RT-PCR) experiments: B2Mex1F: 5′GAGTATGCCTGCCGTGTG3′B2Mex3R:5′ATCCAAATGCGGCATCT3′, HsglobinF:5′TACCCTTGGACCCAGAGGTTCTTT3′, HsglobinR:5′CCCAG

GAGCCTGAAGTTCTC3'. Primers were designed to detect the elements of the beta-globin gene cluster, (b,d,e,g1,g2) but not the elements of the alpha-globin cluster.

Silencing Bcr-Abl Using the siRNA Method

DNA template for anti-*bcr-abl* short double-stranded RNA was the sequence used by Li et al.[16] and was synthesized chemically. DNA template for the nonsense siRNA was the original noncoding siRNA sequence provided by the Ambion p*Silencer* 3.1-H1 vector (Ambion, Austin, TX, USA) with human H1 polymerase promoter. For the construction of siRNA expression cassettes, H1-siRNA coding sequences were amplified by PCR, using primers recommended for sequencing extended with AgeI restriction sites at both ends. The sequenced H1-siRNA expression cassettes were than inserted into the pLPCX retroviral vector at a location where the risk of promoter interference is minimal (between the extended viral packaging signal and the puromycin resistance gene). Enhanced green fluorescent protein was chosen as a reporter. The enhanced green fluorescent protein (EGFP) gene was taken from the pEGFP-N1 plasmid (BD Biosciences, Erembodegem, Belgium) and cloned into the multiple cloning site of the pLPCX vector. Stable transduction of K562 cells was achieved by using a double-packaging cell system Phoenix/PG13.[17] PG13 cells were selected with 1.5 (μg/mL puromycin before collecting the virus supernatants. K562 cells were further selected with the same amount of puromycin from the second day after transduction. Flow-cytometric analyses were performed to detect EGFP expression and propidium iodide incorporation (FACSCalibur System, BD Biosciences). FISH analysis was made as described.[18]

RESULTS

The effects of pharmacological inhibition of the Bcr-Abl tyrosine kinase (by Gleevec) and the MEK 1/2 MAPK kinase (by its inhibitor UO126) were examined on major signaling events that play a decisive inhibitory role in differentiation or apoptosis of K562 cells (FIGS. 1A–1F).

In order to assess the effects of Gleevec and UO126 treatments on cell fate decisions such as differentiation or cell death, we measured the hemoglobin content of K562 cells (as a marker of erythroid differentiation) and followed cell viability for 3 days post treatment. Gleevec was used at a concentration of 0.5 μM, which corresponds to its IC_{50} value in K562 cells,[19] while UO126 was applied at 5 μM, which reached the maximal effect on hemoglobin production. Quantitative RT-PCR data revealed that hemoglobin expression showed a significant increase 24 h after Gleevec or UO126 treatments (FIG. 1A), which corresponds to earlier findings.[4,11] Changes in cell number due to Bcr-Abl or MEK1/2 inhibition were followed using the trypan-blue exclusion

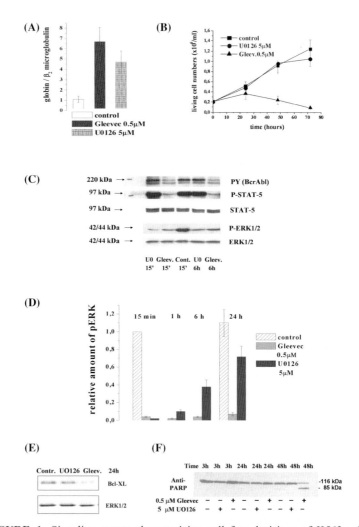

FIGURE 1. Signaling events characterizing cell fate decisions of K562 cells after Gleevec treatment or inhibition of ERK1/2 activity by the MEK1/2 inhibitor UO126. (**A**) Hemoglobin mRNA level determined by quantitative RT-PCR after 24 h of 0.5 μM Gleevec or 5 μM UO126 treatments, using *β2 microglobulin* as a reference gene. Data were normalized to the nontreated control. (**B**) Cells were treated with Gleevec or UO126 and changes in the number of living cells were detrmined by trypan-blue exclusion at time points indicated. (**C**) A typical Western blot developed by a set of different antibodies (as indicated) from 50 μg of total cell lysates of UO126 or Gleevec-treated K562 cells. STAT-5 and ERK1/2 stainings served as loading controls. (**D**) Quantitative evaluation of Western blots developed by anti-P-ERK1/2 and anti-ERK1 antibodies. Samples were prepared at time points indicated. P-ERK1/2 protein levels were determined relative to optical densities of ERK1 stainings by using the Bio Rad ChemDoc^{TM} system. Data were normalized to the nontreated controls. (**E, F**) Typical Western blots of Gleevec- or UO126-treated cells developed by anti-Bcl-XL (**E**) or anti-PARP (**F**) antibodies.

test (FIG. 1B). Significant decrease in the number of living cells was detected after 48 h of 0.5 μM of Gleevec treatment, compared to the nontreated control. A total of 5 μM UO126, which is able to induce erythroid differentiation, did not alter significantly the cell number even up to 72 h after the treatment. Samples were prepared for the quantitative RT-PCR experiments at 24 h, when the apoptotic effect of Gleevec was not apparent.

Signaling mechanisms leading to the observed cellular responses have been studied by the Western blot technique, using phospho-specific antibodies that recognize only the active forms of signal-transducing molecules (Bcr-Abl, STAT-5, ERK1/2). Antibodies recognizing all forms of a given protein (phosphorylated and nonphosphorylated) were used as loading controls and as references for the quantitative analyses, because their quantities did not change during the experiments. Samples were prepared under strong phosphatase inhibition conditions at different time points after treatments as indicated on figures. FIGURE 1C shows a representative Western blot, developed by a set of different antibodies. Inhibition of Bcr-Abl tyrosine phosphorylation by Gleevec results in a decrease of STAT-5 and ERK1/2 activities. Conversely, pharmacological inhibition of MEK1/2 by UO126 causes a decrease only in ERK1/2 activities, suggesting that ERK1/2 inhibition *per se* is sufficient to induce hemoglobin expression in K562 cells.

ERK activity changes were followed in time to compare the long-term effects of Gleevec and UO126. FIGURE 1D shows that inhibition of MEK1/2 by UO126 causes a transient decrease in ERK1/2 activities. Minimal phosphorylated ERK level is detectable within 1 h, after which ERK activity starts to reappear, returning near to the level of the nontreated control in 24 h. In the case of Gleevec treatment, ERK1/2 phosphorylation is strongly suppressed even after 24 h.

In order to elucidate the processes leading to different cell viability after Gleevec or UO126 treatments, well-known markers of apoptosis were examined (FIGS. 1E and 1F). Protein amount of the antiapoptotic protein Bcl-XL was estimated by the Western blot technique, using anti-ERK1 immunostaining as a loading control (FIG. 1E). Gleevec, but not UO126 was able to decrease the protein level of Bcl-XL compared to the nontreated control. The poly-ADP-ribosylase (PARP) protein is a substrate of the caspase-3 protease, an enzyme that is active only in apoptotic cells. Cleavage products of the 116-kDa PARP protein are detectable by Western blot as an evidence of programmed cell death. In our case, apoptosis appears 48 h after 0.5 (μM Gleevec treatment, which is in a good correlation with the cell numbers assessed after 72 h. A total of 5 (μM UO126 treatment does not induce caspase-3-dependent apoptosis in K562 cells.

To monitor the effect of reduced Bcr-Abl function on the erythroid differentiation capability of K562 cells at the level of the *bcr-abl* transcript, we applied stable RNA interference (RNAi). The DNA template sequence for short double-stranded RNA was as described[16] and cloned into the pLPCX retroviral vector together with an H1 promoter. EGFP was used as a reporter gene. Under the same conditions, a DNA-template for a noncoding hairpin (nonsense) RNA was used as control, due to interferon-like cellular responses induced by short double-stranded RNAs.[20] Changes in cell number, viability, apoptosis, erythroid differentiation as well as alterations in cellular signaling were followed in transduced cells from day 3 up to day 21.

Evidence of successful reduction of the Bcr-Abl protein amount by our retrovirus-based siRNA expression system is presented in FIGURE 2A. Changes in Bcr-Abl protein levels were followed by the Western blot technique. The relative amount of the fusion protein was calculated from Western blots using ERK1 as a reference protein (see the MATERIALS AND METHODS). Optical density values were normalized to the nontransduced control. The protein amount of Bcr-Abl was significantly lower in cells transduced by the anti-*bcr-abl* siRNA construct than those transduced by the nonsense siRNA or in the nontransduced controls from the third day after transduction. Later the protein level of Bcr-Abl started to recover and the effect of siRNA disappeared in 6–10 days. The same Western blots (used for FIG. 2A) were developed by anti-phospho-ERK1/2 antibody to estimate the activity of the MEK/ERK pathway downstream of Bcr-Abl (FIG. 2B). Arbitrary units obtained from optical densities were calculated as mentioned before. The obtained effects are in good correlation with the changes in Bcr-Abl protein levels, representing that anti-*bcr-abl* siRNA expression results in a significant decrease of ERK1/2 activities. Results of a typical experiment of the three independent transductions are represented on FIGURE 2C.

Flow cytometric measurements of transduced K562 cells (grown in 1.5 μg/mL puromycin) were made every second day to estimate the proportion and the fluorescent intensity of EGFP-positive cells as a marker of transduction efficiency. FACS analyses showed an increase in the number of EGFP-positive cells in time due to puromycin selection, reaching a maximum at about 90% of living cells in the case of both constructs (data not shown). In cells transduced with the nonsense construct, mean fluorescence values rose remarkably on account of puromycin selection, but this phenomenon was missing in the case of the sense construct. Cell death, measured by simultaneous propidium-iodide staining, did not increase significantly in cells expressing the anti-*bcr-abl* or the nonsense siRNA constructs with similar fluorescence intensity values of the reporter, while the presence of apoptosis was detectable in cells expressing the nonsense construct with the highest intensity. This latter effect may be attributed to the toxic effect of EGFP expressed in a great amount. Cells with lowered Bcr-Abl protein level (3–6 days after transduction) did not show significant growth disadvantages, as compared to those expressing the nonsense construct or to the nontransduced control (data not shown).

In order to assess erythroid differentiation in K562 cells with reduced Bcr-Abl function, hemoglobin content was examined in cells transduced with

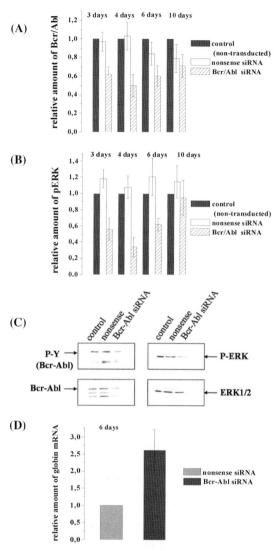

FIGURE 2. Investigation of stable anti-*bcr-abl* siRNA on changes in cellular signaling and erythroid differentiation of K562 cells. (**A, B**) K562 cells were transduced with retroviral vectors containing DNA template for anti-*bcr-abl* (Bcr/Abl siRNA) or for noncoding (nonsense siRNA) siRNAs. Cell lysates were prepared at time points indicated. Western blots were developed by anti-Bcr-Abl, anti-ERK1/2 and anti-ERK1 antibodies. Optical densities were evaluated (as described before) using ERK1/2 as loading control and were normalized to the protein levels of the nontransduced cells. **A** shows changes in Bcr-Abl; **B** shows changes in phosphorylated-ERK1/2 levels in time. **C** represents the result of a typical Western blot (used for FIGS. 2A and 2B) of the three independent transductions. **D** shows hemoglobin mRNA level in transduced K562 cells measured by quantitative RT-PCR at 6 days.

retroviral vectors containing nonsense or anti-*bcr-abl* siRNA templates by quantitative RT-PCR (FIG. 2D). β2-microglobulin was used as a reference gene. Data were normalized to the nonsense control on account of interferon-like cellular responses induced by short double-stranded RNAs. Taken together, our results show for the first time that transiently lowered function of Bcr-Abl achieved by the siRNA technique is able to induce a significant increase in hemoglobin expression without causing significant apoptosis or even growth disadvantage in K562 cells.

DISCUSSION

Tyrosine phosphorylation of Bcr-Abl substrates activates multiple signal transduction cascades shared by cytokines known to influence growth and differentiation of hemopoietic cells. Bcr-Abl interacts with various adaptor signaling molecules, which in turn leads to the activation of Ras, Raf-1, and subsequently the MEK/ERK pathway, resulting in promitotic transcriptional reorganization. Bcr-Abl also activates the PIP3K (phosphoinositide 3-kinase), which leads to an antiapoptotic signal, mediated by Bcl-XL.[21]

When examining the selective inhibition of Bcr-Abl autophosphorylation, we found that Gleevec induces erythroid differentiation in K562 cells together with downregulation of ERK 1/2 activities. Furthermore, the same treatment induces apoptosis via decreasing Bcl-XL levels and activating the caspase-3 cascade. Pharmacological inhibition of the MEK/ERK signaling pathway by UO126 results in a considerable increase in hemoglobin level without inducing a significant decrease in cell number or causing any sign of apoptosis. In contrast with Gleevec, UO126 does not reduce STAT-5 phosphorylation, demonstrating that downregulation of STAT-5 activity is not essential for the induction of erythroid differentiation in K562 cells. On the other hand, STAT5 downregulation might contribute to the apoptotic effect of Gleevec, similar to another Bcr-Abl inhibitor, PD180970.[22]

Comparison of long-term effects of UO126 and Gleevec on the MEK/ERK MAPK pathway reveals that even a transient decrease in ERK1/2 activities is sufficient to induce globin expression *per se*, and that erythroid differentiation and apoptosis are two clearly distinguishable cellular events in K562 cells. Our results suggest that the ERK1/2 MAP kinases might have an important negative regulatory role in the erythroid differentiation process of K562 cells, with a probable role in maintaining the transformed phenotype. However, prolonged inhibition of ERK1/2 activities may contribute to apoptotic cell death observed in the case of Gleevec.

Because it is not possible to eradicate some cases of CML by Gleevec treatment, alternative approaches targeting primary or secondary resistance are of great interest. Killing of Bcr-Abl$^+$ leukemic cells by RNA interference was reported by several research groups.[16,23–26] In our experiments, retroviral transduction of DNA template for short interfering RNA that eliminates *bcr-abl* mRNA resulted in partial downregulation of the Bcr-Abl protein expression from days 3 to 6 post transduction. We show here for the first time that this transient change in Bcr-Abl signaling is sufficient to evoke a detectable decrease in ERK1/2 activities and a significant increase in globin mRNA levels. On the basis of our FISH experiments (not shown), parental K562 cells carry more than 10 copies of *bcr-abl*, which might explain the partial (and transient) success in eliminating the protein. This might also be the reason for the lack of apoptosis. In spite of the fact that puromycin selection increased the number of EGFP-positive cells, it was unable to induce cell death in the case of the sense construct. In correspondence with our findings, Scherr *et al.*[26] have recently demonstrated that stable but not transient anti-*bcr-abl* RNA interference can deplete efficiently K562 cells. More than three lentiviral integrations were required for effective depletion of K562 cells, while transduced cells with lower copy number of the vector recovered after prolonged cell culture.

Taken together, our results suggest that among the multiple signal transduction pathways operating with lowered intensity at reduced Bcr-Abl function, decreased ERK1/2 activities might be the key determinant to overcome the differentiation blockade typical of myeloproliferative disorders and to induce erythroid differentiation in CML-derived K562 cells.

ACKNOWLEDGMENTS

We are grateful to Monika Batkai and Zoltan Ondrejo for technical assistance and to Gergely Szakacs and Balazs Sarkadi for critical reading of the manuscript. This work was supported by Hungarian Research Founds: OTKA T046965 and NKFP 1A-060/04.

REFERENCES

1. HE, H., X. WANG, M. GOROSPE, *et al*. 2006. Life Sci. **78:** 1217–1224.
2. MELEMED, A.S., J.W. RYDER & T.A. VIK. 1997. Blood **90:** 3462–3470.
3. SEVINSKY, J.R., A.M. WHALEN & N. AHN. 2004. Mol. Cell Biol. **24:** 4534–4545.
4. WITT, O., K. SAND & A. PEKRUN. 2000. Blood **95:** 2391–2396.
5. WOESSMANN, W. *et al*. 2004. Cell Biol. Int. **28:** 403–410.
6. KANG, C.D., I.R. DO, K.W. KIM, *et al*. 1999. Exp. Mol. Med. **31:** 76–82.
7. WOESSMANN, W. & N.F. MIVECHI. 2001. Exp. Cell Res. **264:** 193–200.
8. KAWANO, T. *et al*. 2004. Mol. Cell Biochem. **258:** 25–33.
9. KUCUKKAYA, B., D.O. ARSLAN & B. KAN. 2006. Life Sci. **78:** 1217–1224.
10. ZOU, X. & K. CALAME. 1999. J. Biol. Chem. **274:** 18141–18144.
11. FANG, G., C.N. KIM & C.L. PERKINS, *et al*. 2000. Blood **96:** 2246–2253.
12. JACQUEL, A. *et al*. 2003. FASEB J. **17:** 2160–2162.
13. KUZELOVA, K., D. GREBENOVA, I. MARINOV & Z. HRKAL. 2005. J. Cell Biochem. **95:** 268–280.

14. MOSMANN, T. *et al.* 1983. J. Immunol. Methods **65:** 55–63.
15. KOLONICS, A. *et al.* 2001. Cell. Signal. **13:** 743–754.
16. LI, M.J. *et al.* 2003. Oligonucleotides **13:** 401–409.
17. UJHELLY, O. *et al.* 2003. Hum. Gene. Ther. **14:** 403–412.
18. PARDUE, L.M. & J.G. GALL. 1969. Proc. Natl. Acad. Sci. USA **168:** 1356–1358.
19. BOREN, J., M. CASCANTE, S. MARIN, *et al.* 2001. J. Biol. Chem. **276:** 37747–37753.
20. BRIDGE, A.J., S. PEBERNARD, A. DUCRAUX, *et al.* 2003. Nat. Genet. **34:** 263–264.
21. SMITH, D.L. & J. BURTHEM, WHETTON. 2003. Expert Rev. Mol. Med. **2003:** 1–27.
22. HUANG, M. *et al.* 2002. Oncogene **21:** 8804–8816.
23. WILDA, M., U. FUCHS, W. WOSSMANN & A. BORKHARDT. 2002. Oncogene **21:** 5716–5724.
24. WOHLBOLD, L., H. VAN DER KUIP, C. MIETHING, *et al.* 2003. Blood **102:** 2236–2239.
25. SCHERR, M., K. BATTMER, T. WINKLER, O. HEIDENREICH, *et al.* 2003. Blood **101:** 1566–1569.
26. SCHERR, M., K. BATTMER, B. SCHULTHEIS, *et al.* 2005. Gene Ther. **12:** 12–21.

The MAPK Pathway and HIF-1 Are Involved in the Induction of the Human PAI-1 Gene Expression by Insulin in the Human Hepatoma Cell Line HepG2

ELITSA Y. DIMOVA AND THOMAS KIETZMANN

Department of Biochemistry, Faculty of Chemistry, University of Kaiserslautern, D-67663 Kaiserslautern, Germany

ABSTRACT: Enhanced levels of plasminogen activator inhibitor-1 (PAI-1) are considered to be a risk factor for pathological conditions associated with hypoxia or hyperinsulinemia. The expression of the PAI-1 gene is increased by insulin in different cells, although, the molecular mechanisms behind insulin-induced PAI-1 expression are not fully known yet. Here, we show that insulin upregulates human PAI-1 gene expression and promoter activity in HepG2 cells and that mutation of the hypoxia-responsive element (HRE)–binding hypoxia-inducible factor-1 (HIF-1) abolished the insulin effects. Mutation of E-boxes E4 and E5 abolished the insulin-dependent activation of the PAI-1 promoter only under normoxia, but did not affect it under hypoxia. Furthermore, the insulin effect was associated with activation of HIF-1α via mitogen-activated protein kinases (MAPKs) but not PDK1 and PKB in HepG2 cells. Furthermore, mutation of a putative FoxO1 binding site which was supposed to be involved in insulin-dependent PAI-1 gene expression influenced the insulin-dependent activation only under normoxia. Thus, insulin-dependent PAI-1 gene expression might be regulated by the action of both HIF-1 and FoxO1 transcription factors.

KEYWORDS: insulin; human plasminogen-activator inhibitor-1; hypoxia-inducible factor-1; MAPK

Address for correspondence: E.Y. Dimova, University of Kaiserslautern, Faculty of Chemistry/Department of Biochemistry, Erwin-Schroedinger Strasse 54, 67663 Kaiserslautern, Germany. Voice: +49-631-205-3032; fax: +49-631-205-3419.
e-mail: edimova@gwdg.de

INTRODUCTION

Plasminogen activator inhibitor-1 (PAI-1) controls the regulation of the fibrinolytic system in blood by inhibiting both urokinase-type and tissue-type plasminogen activators. PAI-1 can be secreted by platelets, vascular endothelial cells,[1] vascular smooth muscle cells,[2] and several nonvascular cell types.[3,4] Abnormal PAI-1 expression is associated with hyperinsulinemia, obesity, hypertension, diabetes type 2, and cancer. Furthermore, PAI-1 has been correlated with an increased risk of primary and secondary cardiovascular events as well as a poor prognosis in these syndromes (reviewed in Ref. 5).

Insulin can regulate the transcription of a large number of genes either in a positive or in a negative manner.[6] It has long been suggested that the effects of insulin are mediated through a common insulin-responsive element (IRE) and a transcription factor that binds to an IRE.[7–9] However, up to now at least eight distinct IREs have been characterized[10] suggesting that there is no single consensus IRE.

Various transcription factors have been suggested to mediate the insulin response, including cAMP-response element binding protein (CREB),[11] members of the FoxO subfamily[12–15] as well as GA binding protein (GABP),[16] Fra-2/Jun D,[17,18] early growth response gene-1 (Egr-1), known also as nerve growth factor I-A, zif 268, or Krox 24,[19] hypoxia-inducible factor-1α (HIF-1α),[20] insulin-response element-A binding protein (IRE-ABP),[7] nuclear factor-1 (NF-1),[21] serum-response factor (SRF),[22] and upstream stimulatory factors (USFs).[23,24]

Insulin has been shown to increase the endogenous PAI-1 gene expression in HepG2 cells[25] and the transcription of a human PAI-1 promoter Luc construct in human umbilical vein endothelial cells.[26,27] Several groups searched for IRE within the human PAI-1 promoter using different model systems, but the results obtained so far are conflicting. The IRE in the human PAI-1 promoter was first suggested to be in the regions $-93/-62$, $-157/-128$, and $-777/-741$ when HepG2 cells were used.[28] In addition, a mutation at $-52/-45$ also abolished the insulin responsiveness of the human PAI-1 promoter in GH4 cells.[13] By contrast, it was shown with the rat PAI-1 gene that the transcription factor HIF-1 mediated the induction by insulin in primary rat hepatocytes.[20] Although this factor normally induces expression of genes under hypoxic conditions by binding hypoxia-responsive elements (HREs), this activation can also occur in the presence of insulin and other growth factors.[29–34] Because the human PAI-1 promoter contains a HRE binding HIF-1[35] and two classical E-boxes that bind HIF-1 and USFs,[36,37] it seems likely that the insulin effects on the human PAI-1 gene are mediated via HIF-1 or USFs. Thus, it was the aim of our study to investigate the transcriptional mechanisms, the signaling pathway, and the transcription factor(s) that mediate the insulin-dependent induction of the human PAI-1 gene under normoxia and hypoxia using human HepG2 hepatoma cells as a model system.

MATERIALS AND METHODS

Cell Culture, Cell Transfection, and Luciferase Assay

HepG2 cells (4×10^5 per 60-mm dish) cultured in MEM supplemented with 10% fetal bovine serum (Invitrogen, Karlsruhe, Germany), 1% nonessential amino acids (Invitrogen) and 0.5% antibiotics were transfected by using the calcium phosphate method as described.[38] A total of 24 h before harvesting, insulin (dissolved in 0.9% NaCl containing 0.1% bovine serum albumin) was applied to 500 nM final concentration to the serum-free medium, controls were treated with solvent, and cells were cultured under normoxic and hypoxic conditions. For Western blot experiments, cells were transfected with 10-μg expression vectors for dominant-negative PDK1, TRB3, dominant negative Raf-1, or an empty control vector. A total of 4 h or 15 min before harvesting, the cells were treated with insulin or/and hypoxia and in controls with solvent. For Northern blot analyses, cells were treated with insulin and/or hypoxia and solvent in controls for only about 4 h unless otherwise indicated.

Plasmid Constructs

pGl3-hPAI-806, pGl3-hPAI-806-HREm, pGl3-hPAI-806-M4, pGl3-hPAI-806-M5, and hPAI-806-HREm45 plasmids have been already described.[36] The pGl3PAI-FoxO1m and pGl3PAI-HREmFoxO1m plasmids were constructed using a QuickChange Site-Directed Mutagenesis kit (Stratagene, La Jolla, CA) and the following mutant primer 5'-gggtggaacatgagtt catct**ggta**cctgcccacatctgg-3'.

RNA Preparation and Northern Analysis

Isolation of total RNA, Northern analysis, hybridizations, and detections were performed as described before.[39] Blots were quantified with a videodensitometer (Biotech Fischer, Reiskirchen, Germany).

Western Blot Analysis

PAI-1 Western blot analysis was carried out as described.[40,41] The primary mouse antibodies against human PAI-1 (American Diagnostics, Pfungstadt, Germany) and HIF-1α (Novus, Littleton, CO) were used in a 1:100 and 1:1,000 dilution, respectively. The phospho-PKB (Ser437) or the phosho-Erk1/2 (Thr202/Tyr204) antibodies (Cell Signaling, Frankfurt/M, Germany) were used in a dilution of 1:1,000. ANTI-FLAG (Sigma, St. Louis, MO) and the α-Myc-tag 9B11 (Cell Signaling) antibodies were used in a dilution of 1:1,000, while the human β-actin antibody (Sigma) was used in a dilution of 1:10,000. The secondary antibody was a goat anti-mouse or anti-rabbit IgG (Santa Cruz Biotechnology, Santa Cruz, CA) and used in a dilution of 1:5,000. The ECL Western blotting system (Amersham, Freiburg, Germany) was used for detection.

RESULTS

Human PAI-1 mRNA and Protein Levels Are Upregulated by Insulin in HepG2 Cells

To determine how insulin modulates PAI-1 gene expression in the human hepatoma cell line HepG2, we analyzed PAI-1 mRNA levels after a 4-h treatment with hypoxia, insulin, or both stimuli together. In accordance with previous studies,[36,40,42] we observed that hypoxia upregulated PAI-1 mRNA by about twofold (FIG. 1). The admission of insulin caused an increase in PAI-1 mRNA by about threefold under normoxia and by about fourfold under hypoxia. The increase in PAI-1 protein levels by insulin was about 1.7-fold under normoxia and about threefold under hypoxia (FIG. 1).

Insulin-Enhanced Expression of Human PAI-1 Promoter Luc Constructs in HepG2 Cells

Because PAI-1 is induced under hypoxia and on account of the findings that insulin can enhance the levels of the transcription factor HIF-1α,[20,29,32] we hypothesized that insulin enhanced PAI-1 via the HIF binding sites, HRE, E4, and E5, within the human PAI-1 promoter. To explore this, we performed transient transfection experiments using wild-type PAI-1 promoter Luc constructs as well as different mutants (FIG. 2). The wild-type human PAI-1 promoter construct could be activated not only by hypoxia, but also by insulin, as detected by the reporter luciferase assay. Mutation of the HRE in the construct pGl3hPAI-HREm abolished insulin- and hypoxia-dependent activation of the hPAI-1 promoter (FIG. 2). Mutation of E-box 4 and 5 in the human PAI-1 promoter constructs pGl3hPAI-M4 and pGl3hPAI-M5 abolished the induction by insulin under normoxia, whereas the insulin response under hypoxia was still present. The triple mutant construct pGl3hPAI-HREmM45 responded neither to insulin nor to hypoxia, thus indicating again the major involvement of the HRE in the stimulation of human PAI-1 by insulin and hypoxia (FIG. 2). These data suggest that the effect of insulin on the human PAI-1 promoter is mediated mainly by the HRE.

It was proposed that at least a part of the IRE could be located between positions −52/−45 containing a putative forkhead-related response element

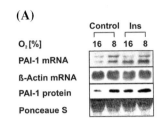
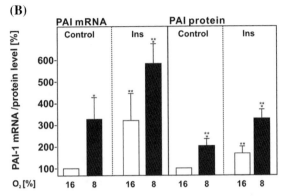

FIGURE 1. Insulin-dependent increase of human PAI-1 mRNA and protein levels in HepG2 cells. (**A**) The PAI-1 mRNA and protein levels were measured by Northern and Western blots, respectively. Representative Northern and Western blots are seen. Twenty μg of total RNA isolated from cultured HepG2 cells were subjected to Northern blot analyses and hybridized with DIG-labeled PAI-1 and β-actin antisense RNA probes. A total of 100 μg of protein from the HepG2 cell culture medium was subjected to Western blotting with an antibody against hPAI-1. Autoradiographic signals were detected by chemiluminescence and quantified by videodensitometry. (**B**) The hPAI-1 mRNA expression and protein levels under normoxia (16% O_2) were set to 100%. Values represent means ± SEM of three independent experiments. Statistics: Student t-test for paired values: *significant difference 16% O_2 versus 8% O_2, **significant difference 16% O_2 versus 16% O_2 + insulin and 8% O_2 versus 8% O_2 + insulin; $P \leq 0.05$.

(TATTT).[13] To prove that, we created two constructs: the first was mutated in the putative binding site for FoxO1 (−52/−45), and the second contained mutations in both the HRE and the FoxO1 site. Like the construct containing E4 and E5 mutations, the mutation at the position –52/−45 in the pGl3hPAI-FoxO1m construct abolished the insulin effect under normoxia, whereas this effect was preserved under mild hypoxia. The pGl3PAI-HREmFoxO1m construct was induced neither by insulin under normoxia and hypoxia nor by hypoxia alone (FIG. 2). These results indicate that the −52/−45 region of the human PAI-1 promoter in addition to E4 and E5 might contribute to the insulin-dependent induction, but only under normoxia.

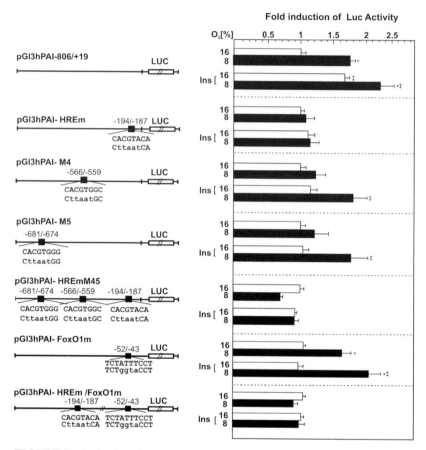

FIGURE 2. Induction of human PAI-1 promoter constructs by insulin in HepG2 cells. HepG2 cells were transfected with Luc gene constructs driven by the wild-type hPAI promoter (pGL3-hPAI-806) or the hPAI-1 promoter, mutated at HRE, E-box 4, E-box 5 and putative FoxO1 binding site. The wild-type sequences are shown on the *upper strand*; mutations are indicated on the *lower strand* by lower case letters. The Luc activity was expressed as fold induction compared to the Luc activity, measured in the respective controls. Values represent means ± SEM of five independent experiments, each performed in duplicate. Statistics: Student *t*-test for paired values: *significant difference 16% O_2 versus 8% O_2, **significant difference 16% O_2 versus 16% O_2 + insulin and 8% O_2 versus 8% O_2 + insulin; $P \leq 0.05$.

Involvement of the MAPK Pathway in the Enhancement of HIF-1α Levels by Insulin

Two major pathways, PI(3)K and mitogen-activated protein kinase (MAPK), are involved in insulin receptor signaling. To elucidate the signaling pathway implicated in the enhancement of HIF-1α levels by insulin, these

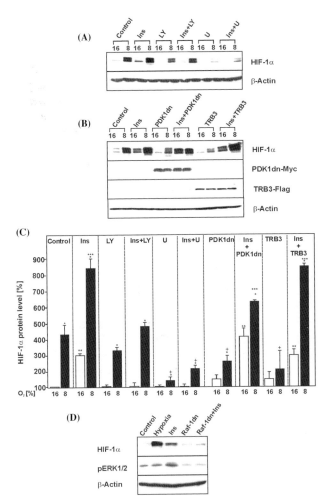

FIGURE 3. Modulation of the insulin-mediated HIF-1α induction by the PI(3)-kinase inhibitor LY294002, the MEK-1 inhibitor U0126, dominant-negative PDK1, the PKB inhibitor TRB3 and dominant-negative Raf-1. HepG2 cells treated with 10 μM LY294002, 10 μM U0126 or cells transfected with either expression vectors for dominant-negative PDK1, the PKB inhibitor TRB3, dominant-negative Raf-1 or empty control vector were incubated with or without 500 nM insulin under normoxia or hypoxia for 4 h. (**A, B, D**) Representative HIF-1α Western blot and expression levels of dominant-negative PDK1 and TRB3. A total of 100 μg of protein from HepG2 cells were analyzed by Western blotting with an antibody against HIF-1α and Myc-Tag, Flag-Tag or pERK, respectively. (**C**) Statistical analyses of the HIF-1α levels. The HIF-1α protein levels under normoxia (16% O_2) were set to 100%. Values represent means ± SEM of at least three independent experiments. Statistics: Student t-test for paired values: *significant difference 16% O_2 versus 8% O_2; ** significant difference respective controls at 16% O_2 versus treatment with insulin; ***significant difference control at 8% O_2 versus treatment with insulin; + significant difference 8% O_2 versus 8% O_2 with inhibitors or with dominant negative constructs; $P \leq 0.05$.

phosphorylation cascades were blocked by LY294002, which inhibits PI(3)K, and by the MEK-1 inhibitor U0126, which blocks the MAPK pathway. Hypoxia enhanced HIF-1α levels by about fourfold; stimulation with insulin increased HIF-1α protein levels by about threefold under normoxia and eightfold under hypoxia. Consistent with previous findings[20] treatment of HepG2 cells with the PI(3)-kinase inhibitor, LY294002, decreased HIF-1α under mild hypoxia. Furthermore, LY294002 blocked the induction of HIF-1α by insulin under normoxia, but did not affect it as strongly under hypoxia (FIGS. 3A and 3C). After pretreatment with LY294002 and subsequent stimulation with insulin, HIF-1α protein levels were increased to the same extent as under hypoxia alone. Pretreatment with the MEK-1 inhibitor U0126 almost abolished both hypoxia-induced and insulin-induced HIF-1α protein levels (FIGS. 3A and 3C). Together, these results suggest involvement of the MAPK pathway in the activation of HIF-1α by insulin in HepG2 cells. To confirm these results we overexpressed dominant-negative PDK1 and observed a decrease in HIF-1α protein levels under normoxia and hypoxia. Insulin treatment of HepG2 cells transfected with dominant-negative PDK1 did not result in inhibition of HIF-1α protein levels. Overexpression of the PKB inhibitor TRB3 decreased basal HIF-1α protein levels under normoxia and hypoxia. However, in the presence of insulin, HIF-1α protein levels were still increased (FIGS. 3B and 3C). In addition, the overexpression of dominant negative Raf-1 prevented the enhancement of HIF-1α protein levels by hypoxia and insulin (FIG. 3D). Thus, these data suggest that in HepG2 cells the MAPK pathway, but not PDK1 and PKB mediated the insulin-dependent increase of HIF-1α.

DISCUSSION

In this study we investigated the human PAI-1 expression in response to insulin. Our data demonstrated several new findings with respect to human PAI-1 and HIF-1α regulation. First, it was found that, in contrast to an earlier proposal, insulin acted mainly via the HRE within the promoter. Second, the low-affinity HIF-1 but high-affinity USF binding E-boxes, E4 and E5, affected the insulin-dependent induction only under normoxia. Third, the −52/−45 element was also involved in the normoxic response to insulin. Fourth, PI(3)K and ERK, but not PDK1 and PKB, mediated the insulin effects via HIF-1α to PAI-1 in HepG2 cells.

The findings that insulin acted mainly via the HRE can be explained by the low affinity of the E-boxes E4 and E5 for HIF-1 and their high affinity for USFs.[36,37] However, mutation of E4 and E5 diminished the normoxia-mediated response, indicating that the insulin effects on PAI-1 might be partially mediated via USFs, which would be in line with findings from the fatty acid synthase promoter.[23,24] Further, these data are in line with the results available for the rat PAI-1 promoter, where insulin effects were also mediated through the HRE in primary rat hepatocytes.[20] However, our data appeared to be contradictory to those of another study, where the IRE within the human PAI-1 promoter was first suggested to be in the regions −93/−62, −157/−128, and −777/−741 when HepG2 cells were used.[28] In addition, as in the present study, a mutation at −52/−45 also abolished the insulin responsiveness of the human PAI-1 promoter in GH4 cells.[13] However, mutation of this site abolished only the insulin-dependent induction under normoxia (FIG. 2). Because this element resembles a forkhead-like binding sequence that is similar to those found in the PEPCK[10] and IGFBP-1[43,44] gene, it might be a FoxO1 binding site. The members of the forkhead transcription factor family are thought to play an important role in development and cell differentiation and appeared to be regulated by insulin in several cell lines of hepatic origin.[45–47] In mammalian cells, Foxo1 (previously known as FKHR), Foxo3 (previously FKHRL1), and Foxo4 (previously AFX) represent the FoxO subfamily. These proteins have been found to be phosphorylated by PKB in response to insulin at three conserved residues, namely, Thr24, Ser256, and Ser319. However, in the absence of insulin, the FoxO transcription factors reside in the nonphosphorylated form in the nucleus and induce gene expression. Upon insulin treatment, the FoxO proteins become phosphorylated, which results in their nuclear export and cytoplasmic retention and subsequently in inhibition of target gene expression.[14,15,48–50] Thus, a mutation of the FoxO1-like binding site in the human PAI-1 promoter should reduce basal promoter activity and should lead to a loss of the inhibitory effect of insulin. However, our study indicated that mutation of the FoxO1-like binding site within the human PAI-1 promoter did not reduce basal promoter activity. Further, if FoxO proteins would bind to that site, insulin should have mediated a reduction in promoter activity, but this was not the case. Instead, insulin induced promoter activity, and mutation of this site abolished the insulin effect only under normoxia. Thus, it seems likely that this element binds an other or atypical FoxO protein which has not yet been identified. In addition, the insulin effect under hypoxia was not abolished and thus we consider that this factor is not the major regulator of insulin-dependent PAI-1 expression in HepG2, cells although the possible interrelation between HIF-1 and FoxO1 proteins needs to be further elucidated.

In our study, the insulin-induced enhancement of HIF-1α was inhibited by the specific PI(3)K inhibitor LY294002 under normoxia, whereas only slight effects were visible under hypoxia. Thus, the role of PI(3)K in that process remained uncertain. In addition, the insulin-dependent increase of HIF-1α was completely abolished by the MEK-1 inhibitor U0126 (FIG. 3). These results implicated a major involvement of the MAPK pathway, whereas studies performed in prostate carcinoma-derived cell lines,[30] the arising retinal pigment epithelia cell line-19 (ARPE-19)[32] and primary rat hepatocytes[20] showed the involvement of a PI(3)K-dependent pathway. In addition, the overexpression of dominant negative PDK1 as well as the tribbles homologue TRB3, which inhibits Akt/PKB activation,[51] reduced the basal HIF-1α protein levels but did not influence the insulin-induced HIF-1α protein levels, thus indicating again the possible involvement of the MAPK pathway.

Together, we have found that the insulin effect on PAI-1 and HIF-1α expression is dependent on the MAPK pathway in HepG2 cells. To our knowledge, these studies provide the first report that insulin can activate HIF-1α via the MAPK pathway in HepG2 cells. Interestingly, another study showed transcriptional activation of PAI-1 by insulin due to PI(3)K activation, followed by protein kinase C and ERK2 phosphorylation in HepG2 cells.[28] Thus, further investigations are necessary to elucidate cross-talks between the signaling pathways mediating the different response to insulin under normoxia and hypoxia in HepG2 cells.

ACKNOWLEDGMENTS

We thank Dr T.D. Gelehrter (Department of Human Genetics, University of Michigan Medical School, Ann Arbor) for the kind gift of the PAI-1 cDNA.

This study was supported by the Deutsche Forschungsgemeinschaft SFB 402 Teilprojekt A1, GRK 335 Deutsche Krebshilfe 106429, and Fonds der Chemischen Industrie.

REFERENCES

1. ERICKSON, L.A., C.M. HEKMAN & D.J. LOSKUTOFF. 1985. The primary plasminogen-activator inhibitors in endothelial cells, platelets, serum, and plasma are immunologically related. Proc. Natl. Acad. Sci. USA **82:** 8710–8714.
2. REILLY, C.F. & R.C. MCFALL. 1991. Platelet-derived growth factor and transforming growth factor-beta regulate plasminogen activator inhibitor-1 synthesis in vascular smooth muscle cells. J. Biol. Chem. **266:** 9419–9427.
3. BUSSO, N., E. NICODEME, et al. 1994. Urokinase and type I plasminogen activator inhibitor production by normal human hepatocytes: modulation by inflammatory agents. Hepatology **20:** 186–190.
4. LE-MAGUERESSE, B.B., G. PERNOD, et al. 1997. Tumor necrosis factor-alpha regulates plasminogen activator inhibitor-1 in rat testicular peritubular cells. Endocrinology **138:** 1097–1105.
5. EDDY, A.A. 2002. Plasminogen activator inhibitor-1 and the kidney. Am. J. Physiol. Renal Physiol. **283:** F209–F220.
6. O'BRIEN, R.M. & D.K. GRANNER. 1996. Regulation of gene expression by insulin. Physiol. Rev. **76:** 1109–1161.
7. ALEXANDER-BRIDGES, M., N.K. MUKHOPADHYAY, et al. 1992. Growth factor-activated kinases phosphorylate IRE-ABP. Biochem. Soc. Trans. **20:** 691–693.
8. O'BRIEN, R.M. & D.K. GRANNER. 1991. Regulation of gene expression by insulin. Biochem. J. **278:** 609–619.
9. O'BRIEN, R.M., M.T. BONOVICH, et al. 1991. Signal transduction convergence: phorbol esters and insulin inhibit phosphoenolpyruvate carboxykinase gene

transcription through the same 10-base-pair sequence. Proc. Natl. Acad. Sci. USA **88**: 6580–6584.
10. O'BRIEN, R.M., R.S. STREEPER, et al. 2001. Insulin-regulated gene expression. Biochem. Soc. Trans. **29**: 552–558.
11. KLEMM, D.J., W.J. ROESLER, et al. 1998. Insulin stimulates cAMP-response element binding protein activity in HepG2 and 3T3-L1 cell lines. J. Biol. Chem. **273**: 917–923.
12. DURHAM, S.K., A. SUWANICHKUL, et al. 1999. FKHR binds the insulin response element in the insulin-like growth factor binding protein-1 promoter. Endocrinology **140**: 3140–3146.
13. VULIN, A.I. & F.M. STANLEY. 2002. A Forkhead/winged helix-related transcription factor mediates insulin-increased plasminogen activator inhibitor-1 gene transcription. J. Biol. Chem. **277**: 20169–20176.
14. BRUNET, A., A. BONNI, et al. 1999. Akt promotes cell survival by phosphorylating and inhibiting a Forkhead transcription factor. Cell **96**: 857–868.
15. KOPS, G.J., N.D. DE RUITER, et al. 1999. Direct control of the Forkhead transcription factor AFX by protein kinase B. Nature **398**: 630–634.
16. OUYANG, L., K.K. JACOB & F.M. STANLEY. 1996. GABP mediates insulin-increased prolactin gene transcription. J. Biol. Chem. **271**: 10425–10428.
17. CHAPMAN, S.C., J.E. AYALA, et al. 1999. Multiple promoter elements are required for the stimulatory effect of insulin on human collagenase-1 gene transcription. Selective effects on activator protein-1 expression may explain the quantitative difference in insulin and phorbol ester action. J. Biol. Chem. **274**: 18625–18634.
18. STREEPER, R.S., S.C. CHAPMAN, et al. 1998. A phorbol ester-insensitive AP-1 motif mediates the stimulatory effect of insulin on rat malic enzyme gene transcription. Mol. Endocrinol. **12**: 1778–1791.
19. BARROSO, I. & P. SANTISTEBAN. 1999. Insulin-induced early growth response gene (Egr-1) mediates a short term repression of rat malic enzyme gene transcription. J. Biol. Chem. **274**: 17997–18004.
20. KIETZMANN, T., A. SAMOYLENKO, et al. 2003. Hypoxia-inducible factor-1 and hypoxia response elements mediate the induction of plasminogen activator inhibitor-1 gene expression by insulin in primary rat hepatocytes. Blood **101**: 907–914.
21. ORTIZ, L., P. AZA-BLANC, et al. 1999. The interaction between the forkhead thyroid transcription factor TTF-2 and the constitutive factor CTF/NF-1 is required for efficient hormonal regulation of the thyroperoxidase gene transcription. J. Biol. Chem. **274**: 15213–15221.
22. THOMPSON, M.J., M.W. ROE, et al. 1994. Insulin and other growth factors induce binding of the ternary complex and a novel protein complex to the c-fos serum response element. J. Biol. Chem. **269**: 21127–21135.
23. WANG, D. & H.S. SUL. 1995. Upstream stimulatory factors bind to insulin response sequence of the fatty acid synthase promoter. USF1 is regulated. J. Biol. Chem. **270**: 28716–28722.
24. WANG, D. & H.S. SUL. 1997. Upstream stimulatory factor binding to the E-box at -65 is required for insulin regulation of the fatty acid synthase promoter. J. Biol. Chem. **272**: 26367–26374.
25. SCHNEIDER, D.J. & B.E. SOBEL. 1991. Augmentation of synthesis of plasminogen activator inhibitor type 1 by insulin and insulin-like growth factor type I: implications for vascular disease in hyperinsulinemic states. Proc. Natl. Acad. Sci. USA **88**: 9959–9963.

26. LI, X.N., H.E. GRENETT, et al. 1997. Genotype-specific transcriptional regulation of PAI-1 expression by hypertriglyceridemic VLDL and Lp(a) in cultured human endothelial cells. Arterioscler. Thromb. Vasc. Biol. **17:** 3215–3223.
27. GRENETT, H.E., R.L. BENZA, et al. 1999. Expression of plasminogen activator inhibitor type I in genotyped human endothelial cell cultures: genotype-specific regulation by insulin. Thromb. Haemost. **82:** 1504–1509.
28. BANFI, C., P. ERIKSSON, et al. 2001. Transcriptional regulation of plasminogen activator inhibitor type 1 gene by insulin: insights into the signaling pathway. Diabetes **50:** 1522–1530.
29. ZELZER, E., Y. LEVY, et al. 1998. Insulin induces transcription of target genes through the hypoxia-inducible factor HIF-1alpha/ARNT. EMBO J. **17:** 5085–5094.
30. JIANG, B.H., G. JIANG, et al. 2001. Phosphatidylinositol 3-kinase signaling controls levels of hypoxia-inducible factor 1. Cell Growth Differ. **12:** 363–369.
31. STIEHL, D.P., W. JELKMANN, et al. 2002. Normoxic induction of the hypoxia-inducible factor 1alpha by insulin and interleukin-1beta involves the phosphatidylinositol 3-kinase pathway. FEBS Lett. **512:** 157–162.
32. TREINS, C., S. GIORGETTI-PERALDI, et al. 2002. Insulin stimulates hypoxia-inducible factor 1 through a phosphatidylinositol 3-kinase/target of rapamycin-dependent signaling pathway. J. Biol. Chem. **277:** 27975–27981.
33. RICHARD, D.E., E. BERRA & J. POUYSSEGUR. 2000. Nonhypoxic pathway mediates the induction of hypoxia-inducible factor 1alpha in vascular smooth muscle cells. J. Biol. Chem. **275:** 26765–26771.
34. GORLACH, A., I. DIEBOLD, et al. 2001. Thrombin activates the hypoxia-inducible factor-1 signaling pathway in vascular smooth muscle cells: role of the p22(phox)-containing NADPH oxidase. Circ. Res. **89:** 47–54.
35. FINK, T., A. KAZLAUSKAS, et al. 2002. Identification of a tightly regulated hypoxia-response element in the promoter of human plasminogen activator inhibitor-1. Blood **99:** 2077–2083.
36. DIMOVA, E.Y., U. MOLLER, et al. 2005. Transcriptional regulation of plasminogen activator inhibitor-1 expression by insulin-like growth factor-1 via MAP kinases and hypoxia-inducible factor-1 in HepG2 cells. Thromb. Haemost. **93:** 1176–1184.
37. DIMOVA, E.Y. & T. KIETZMANN. 2006. Cell-type-dependent regulation of the hypoxia-responsive plasminogen activator inhibitor-1 gene by upstream stimulatory factor-2. J. Biol. Chem. **281:** 2999–3005.
38. IMMENSCHUH, S., V. HINKE, et al. 1998. Transcriptional activation of the haem oxygenase-1 gene by cGMP via a cAMP response element/activator protein-1 element in primary cultures of rat hepatocytes. Biochem. J. **334:** 141–146.
39. KIETZMANN, T., U. ROTH, et al. 1997. Arterial oxygen partial pressures reduce the insulin-dependent induction of the perivenously located glucokinase in rat hepatocyte cultures: mimicry of arterial oxygen pressures by H2O2. Biochem. J. **321:** 17–20.
40. SAMOYLENKO, A., U. ROTH, et al. 2001. The upstream stimulatory factor-2a inhibits plasminogen activator inhibitor-1 gene expression by binding to a promoter element adjacent to the hypoxia-inducible factor-1 binding site. Blood **97:** 2657–2666.
41. KIETZMANN, T., A. SAMOYLENKO. et al. 2003. Hypoxia-inducible factor-1 and hypoxia response elements mediate the induction of plasminogen activator inhibitor-1 gene expression by insulin in primary rat hepatocytes. Blood **101:** 907–914.

42. KIETZMANN, T., U. ROTH & K. JUNGERMANN. 1999. Induction of the plasminogen activator inhibitor-1 gene expression by mild hypoxia via a hypoxia response element binding the hypoxia inducible factor-1 in rat hepatocytes. Blood **94:** 4177–4185.
43. SUWANICHKUL, A., M.L. CUBBAGE & D.R. POWELL. 1990. The promoter of the human gene for insulin-like growth factor binding protein-1. Basal promoter activity in HEP G2 cells depends upon liver factor B1. J. Biol. Chem. **265:** 21185–21193.
44. SUWANICHKUL, A., L.A. DEPAOLIS, *et al.* 1993. Identification of a promoter element which participates in cAMP-stimulated expression of human insulin-like growth factor-binding protein-1. J. Biol. Chem. **268:** 9730–9736.
45. NAKAE, J., B.C. PARK & D. ACCILI. 1999. Insulin stimulates phosphorylation of the forkhead transcription factor FKHR on serine 253 through a Wortmannin-sensitive pathway. J. Biol. Chem. **274:** 15982–15985.
46. GUO, S., G. RENA, *et al.* 1999. Phosphorylation of serine 256 by protein kinase B disrupts transactivation by FKHR and mediates effects of insulin on insulin-like growth factor-binding protein-1 promoter activity through a conserved insulin response sequence. J. Biol. Chem. **274:** 17184–17192.
47. TOMIZAWA, M., A. KUMAR, *et al.* 2000. Insulin inhibits the activation of transcription by a C-terminal fragment of the forkhead transcription factor FKHR. A mechanism for insulin inhibition of insulin-like growth factor-binding protein-1 transcription. J. Biol. Chem. **275:** 7289–7295.
48. RENA, G., S. GUO, *et al.* 1999. Phosphorylation of the transcription factor forkhead family member FKHR by protein kinase B. J. Biol. Chem. **274:** 17179–17183.
49. BIGGS, W.H., 3RD, J. MEISENHELDER, *et al.* 1999. Protein kinase B/Akt-mediated phosphorylation promotes nuclear exclusion of the winged helix transcription factor FKHR1. Proc. Natl. Acad. Sci. USA **96:** 7421–7426.
50. TAKAISHI, H., H. KONISHI, *et al.* 1999. Regulation of nuclear translocation of forkhead transcription factor AFX by protein kinase B. Proc. Natl. Acad. Sci. USA **96:** 11836–11841.
51. DU, K., S. HERZIG, *et al.* 2003. TRB3: a tribbles homolog that inhibits Akt/PKB activation by insulin in liver. Science **300:** 1574–1577.

Role of Mitogen-Activated Protein Kinases, NF-κB, and AP-1 on Cerulein-Induced IL-8 Expression in Pancreatic Acinar Cells

KYUNG DON JU,[a] JI HOON YU,[a] HYEYOUNG KIM,[b] AND KYUNG HWAN KIM[a]

[a]*Department of Pharmacology and Institute of Gastroenterology, Brain Korea 21 Project for Medical Science, Yonsei University College of Medicine, Seoul 120-752, Korea*

[b]*Department of Food and Nutrition, Brain Korea 21 Project, Yonsei University College of Human Ecology, Seoul 120-749, Korea*

> ABSTRACT: The cholecystokine (CCK) analogue cerulein causes pathophysiological, morphological, and biochemical events similar to various aspects of human pancreatitis. Doses of CCK or cerulein beyond those that cause the maximum pancreatic secretion of amylase and lipase result in pancreatitis, which is characterized by a dysregulation of the digestive enzyme production and cytoplasmic vacuolization and the death of acinar cells, edema formation, and an infiltration of inflammatory cells into the pancreas. This study aims to investigate whether cerulein induces IL-8 expression in pancreatic acinar cells, and whether cerulein-induced IL-8 expression is inhibited in the cells transfected with mutant genes for c-jun (TAM-67), or IκBα (MAD-3) or treated inhibitors of mitogen-activated protein kinases (MAPKs). As a result, cerulein induced IL-expression, which was inhibited in the cells transfected with TAM-67 or MAD-3 or treated inhibitors of MAPK. In conclusion, activation of MAPK, nuclear factor-κB (NF-κB), and activator protein-1 (AP-1) may be the upstream signaling for cerulein-induced IL-8 expression in pancreatic acinar cells.
>
> KEYWORDS: cerulein; IL-8; AR42J cells; mitogen-activated protein kinases; NF-κB; AP-1

INTRODUCTION

Acute pancreatitis involves a complex cascade of events starting from inflammatory response in pancreatic acinar cells. Cell injury is most likely to be

Address for correspondence: Kyung Hwan Kim, Department of Pharmacology, Yonsei University College of Medicine, Seoul 120-752, Korea. Voice: 82-2-2228-1732; fax : 82-2-313-1894.
 e-mail: hwan444@yumc.yonsei.ac.kr

one of the earliest events in the pathogenesis of acute pancreatitis.[1,2] However, the exact mechanisms of the development of acute pancreatitis are still a subject of debate. Among a number of animal models of experimental pancreatitis that exhibit a biochemical, morphological, and pathophysiological similarity to various aspects of human pancreatitis, cerulein-included pancreatitis was shown to be one of the best-characterized and widely used experimental models.[3] Cholecystokinin (CCK) is the major hormonal regulator of pancreatic enzyme secretion, acting via mobilization of intracellular calcium. In pancreatitis patients, high plasma CCK levels were observed.[4,5]

Reactive oxygen species (ROS) could act as a molecular trigger of various inflammatory responses. They can attack the biologic membranes directly, triggering the accumulation of neutrophils and their adherence to the capillary wall.[2] Therefore, it is possible that ROS play a critical role in perpetuating pancreatic inflammation and the development of extrapancreatic complication. ROS has been considered to be an important regulator in the development and pathogenesis of pancreatitis and an activator of the nuclear transcription factor-κB (NF-κB) regulating inflammatory cytokine gene expression.[6] Oxidant-sensitive transcription factors, NF-κB and activator protein-1 (AP-1) have been considered as the regulators of inducible genes, such as chemokines. Since oxygen radicals are considered as an important regulator in the pathogenesis of cerulean-induced pancreatitis, chemokines such as interleukin-8 (IL-8) may be regulated by NF-κB and/or AP-1.

Chemokines have a broad range of effects on the recruitment and function of specific populations of leukocytes at the site of inflammation, and also play an important role in the initiation and maintenance of the host inflammatory response. Chemokines are subdivided on a structural basis into the C-X-C subfamily, in which the first two or four conserved cysteine residues are separated by another amino acid, and the C-C subfamily, in which the first two cysteine residues are adjacent.[7,8] The C-X-C chemokines act as a potential chemoattractors of neutrophils (e.g., IL-8), whereas the C-X-C chemokines function mainly as chemoattractants for monocytes, eosinophils, T cells, and basophils (e.g., monocyte chemoattractant protein-1). The chemokines secreted locally in the pancreas may mediate the rapid influx and accumulation of neutrophils, monocytes, and other inflammatory cells that play a significant role in tissue destruction and repair in acute pancreatitis. A chemokine such as IL-8 may be regulated by transcription factor NF-κB and/or AP-1. Recently, various studies reported chemokine expression in the pancreatic acinar cells in a rat experimental pancreatitis model. However, chemokine secretion and its regulatory mechanisms remain unclear. NF-κB is a member of the Rel family including p50 (NF-κB1), p52 (NF-κB2), Rel A (p65), c-Rel, Rel B, and *Drosophila* morphogen dorsal gene product. Its activity is controlled by its cytoplasmic inhibitory protein IκBα. Previously, we found that cerulein induced the NF-κB activation in the pancreatic acinar cells.[9] MAD-3 is a mutant gene for IκBα, which substitutes two serine residues to alanine. Transfection with MAD-3 into the cells inhibits NF-κB activation. Activator protein-1 (AP-1) is a redox-sensitive transcription factor. TAM-67 is a dominant-negative mutant lacking the transactivation domain of c-jun. Transfection with TAM-67 inhibits the function of wild-type c-Jun or c-Fos through either a quenching or blocking mechanism. Mitogen-activated protein kinase (MAPK) pathways transduce a large variety of external signals, leading to a wide range of cellular responses, including growth, differentiation, inflammation, and apoptosis.

Our previous study showed that cerulein induces large amounts of ROS, activates oxidant-sensitive transcription factor NF-κB, and induces cytokine expression.[6,9] Therefore, it is hypothesized that NF-κB, AP-1, and MAPK may be key regulators in chemokine expression. This study aims to investigate the signal transduction for IL-8 expression in cerulein-stimulated pancreatic acinar AR42J cells by determining mRNA expression of IL-8. Wild-type cells with or without treatment of MAPK inhibitors and the cells transfected with mutant genes for c-jun (TAM-67), or IκBα (MAD-3) were used.

MATERIALS AND METHODS

Cell Lines and Culture Condition

The rat pancreatic acinar AR42J cells (pancreatoma, ATCC CRL 1492) were obtained from the American Type Culture Collection (Manassas, VA, USA) and cultured in Dulbecco's modified Eagle's medium (Sigma, St. Louis, MO, USA) supplemented with 10% fetal bovine serum (GIBCO-BRL, Grand Island, NY, USA) and antibiotics (100 U/mL penicillin and 100 μg/mL streptomycin). Cells were used for experiments after incubation for 20–24 h at 37°C in a humidified atmosphere of 90% air–10% CO2.

Experimental Protocol

To determine whether cerulein induces the expression of IL-8 in AR42J cells, the cells were cultured in the presence of cerulein for various time periods and at various concentrations of cerulein. AR42J cells were treated with MAPK inhibitors or transfected with mutant genes for c-Jun, and IκBα and cultured in the absence and presence of cerulein for 4 h. mRNA expression of IL-8 was determined by RT-PCR analysis. The MAP kinase inhibitors, U0126 (an ERK inhibitor, Catalog No. 9903, Cell Signaling Technology, Inc., Beverly, MA, USA), SB203580 (a p38 inhibitor, Catalog No. 559389, Calbiochem Biochemicals, San Diego, CA, USA), and SP600125 (a JNK inhibitor, Cell Signaling Technology, Inc., Beverly, MA, USA) were purchased and dissolved in dimethylsulfoxide at 50 mM of stock solution. The inhibitors, at 20 μM final concentration, were pretreated in the culture medium 2 h before the treatment with cerulein.

Transfection with Mutant Genes for c-Jun (TAM-67) or IκBα (MAD-3)

A mutated IκBα gene, MAD-3 double-point mutant (substitution of two serine residues at positions 32 and 36 by alanine residues) construct was prepared as described previously. The control vector pcDNA3 (Invitrogen Corp., Carlsbad, CA, USA) was transfected to the cells instead of mutant genes for c-Jun (TAM-67) and IκBα (MAD-3). These cells were considered as a relative control and named pcN-3. Subconfluent AR42J cells, plated in a 12-well plate (10^5 cells per plate), were transfected with each 10 mg of expression construct using DOTAP (N-[1-(2,3-dioleoyloxy) propyl]-N,N,N trimethyl ammonium methylsulfate) (Boehringer-Mannheim, Pentzberg, Germany) for 12 h. After transfection, the cells were trypsinized and plated at 10^5 cells per 12-well plate. The cells were cultured in the presence or absence of cerulein, and the level of mRNA of IL-8 was determined at 4-h culture in the cells transfected with control vector (pcDNA-3) or mutant genes (TAM-67, MAD-3). AR42J cells without transfection were cultured in the presence of cerulein (control) or absence of cerulein (none).

Reverse Transcription-Polymerase Chain Reaction (RT-PCR)

Gene expressions of specific mRNAs were assessed using RT-PCR standardized by coamplifying with GAPDH, which served as an internal control. Total RNAs isolated from the cells were reverse-transcribed into cDNAs and used for PCR with rat-specific primers for IL-8 and GAPDH. Sequences of primers used in PCR are as follows: IL-8 forward ACGCTGGCTTCTGACAA-CACTAGT, reverse; CTTCTCTGTCCTGAGACGAGAAGG GAPDH forward ACCACAGTCCATGCCAT CAC, reverse; TCCACCACCCTGTTGCT-GTA. The predicted sizes of PCR products were 499 bp in IL-8 and 460 bp in GAPDH. After coamplifying by 32–35 cycles, PCR products were separated on 1.5% agarose gels and visualized by UV transillumination.

RESULTS

Dose-dependent mRNA expression of IL-8 (FIG. 1A) and time-dependent mRNA expression of IL-8 (FIG. 1B) were shown in cerulein-treated AR42J cells. RT-PCR analysis shows that maximum IL-8 mRNA expression was observed at the concentration of 10^{-8}M at 4-h culture. Cerulien-induced IL-8

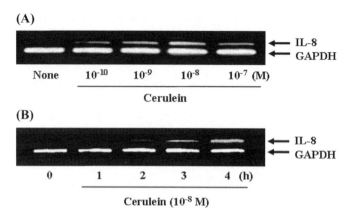

FIGURE 1. Dose- and time-dependent induction of IL-8 mRNA in cerulein-stimulated AR42J cells. For concentration-response of the cells for IL-8 mRNA expression, the cells were cultured for 4 h with various concentrations of cerulein, from 10^{-10} M to 10^{-7} M (**A**). For time-response of the cells for IL-8 mRNA expression, the cells were cultured for 4 h with cerulein (10^{-8} M) (**B**). mRNA expression of IL-8 was determined by RT-PCR analysis. The internal control (GAPDH) was coamplified with IL-8.

mRNA expression was inhibited in the cells transfected with IκB mutant gene (MAD-3) and c-jun dominant-negative gene (TAM-67) as compared with the cells transfected with control vector (pcN-3) (FIG. 2A). Treatment of MAPK inhibitors (U01226 as an ERK inhibitor, SP600125 as a JNK inhibitor, and SB203580 as a p39 inhibitor) significantly suppressed cerulein-induced IL-8 mRNA expression (FIG. 2B).

DISCUSSION

Despite considerable progress in understanding the pathophysiology of pancreatitis, the mechanism of pancreatitis remains unclear. Some evidence shows that the influx and accumulation of inflammatory cells and chemokine expression may be the important event in the course of pancreatitis.[10–13] Cerulein-induced chemokine expression may be regulated by NF-κB, AP-1, and MAPK in pancreatic acinar cells. It is well known that activation of chemokines, cytokines, and pancreatic enzymes characterize the cause of the disease. Although the importance of proinflammatory chemokines in the pathogenesis of acute pancreatitis has been suggested, the exact mechanism has not been fully understood. Many inflammatory genes are induced by NF-κB, AP-1, and MAPK activation in inflammatory process. In patients with acute pancreatitis, IL-8 level was increased and used for the prediction for severity of the disease. It remains unclear whether increased systemic levels are a direct consequence of an outpouring of chemokine from the pancreas, a reflection

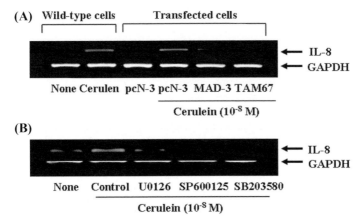

FIGURE 2. Cerulein-induced IL-8 expression in AR42J cells transfected MAD-3 or TAM 67 and treated with MAPK inhibitors. The cells were seeded in 12-wells culture plated at 10^5 cells per well and cultured to 80% confluency. The cells transfected with IκBmutant gene (MAD-3) and c-jun dominant-negative gene (TAM-67) or control vector (pcN-3) (**A**) or treated with MAPK inhibitors (U01226 as an ERK inhibitor, SP600125 as a JNK inhibitor, and SB203580 as a p39 inhibitor) (**B**) and stimulated with cerulein. mRNA expression of IL-8 was determined by RT-PCR analysis. The internal control (GAPDH) was coamplified with IL-8. None, the wild-type cells without cerulein treatment; Cerulein (**A**) or Control (**B**), the wild-type cells with cerulein treatment; pcN-3, the cells transfected with control vector with cerulein treatment; MAD-3, the cells transfected with MAD-3 with cerulein treatment; TAM-67, the cells transfected with TAM-67 with cerulein treatment; U01226, SP600125, and SB203580, the wild-type cells treated with each MAPK inhibitors and cerulein stimulation.

of systemic leukocyte activation with increased chemokine synthesis, or a combination of both. Regardless, increase in systemic chemokoine levels may play an important role in distant organ injury and may, in part, be responsible for distant organ damage secondary to acute pancreatitis. Therefore, the identification of signaling mediators for IL-8 expression may be important to clarify the pathomechanism of pancreatitis. The present results show that cerulein activates MAPK, NF-κB, and AP-1 and thus induces L-8 expression in pancreatic acinar cells. Inhibition of NF-κB, AP-1, or MAPK might alleviate the inflammatory response in pancreatic acinar cells by suppressing IL-8 expression.

ACKNOWLEDGMENTS

This study was supported by a grant (2005-041-E00092) from the Korea Research Foundation made in the program year of 2005 (to K.H.K.). This study was supported by the Brain Korea 21 Project, Yonsei University.

REFERENCES

1. SCHOENBERG, M.H., M. BRUCHLER, M. GASPAR, et al. 1990. Oxygen free radicals in acute pancreatitis of the rat. Gut **31:** 1138–1143.
2. PETER, W.F., D.K. ENGLISH & K. WONG 1993. Free radicals and inflammation: superoxide dependent activation of a neutrophil chemotactic factor in plasma. Proc. Natl. Acad. Sci. USA **77:** 750–754.
3. WISNER, J., D. GREEN, L. FERRELL & I. RENNER. 1988. Evidence for a role of oxygen derived free radicals in the pathogenesis of cerulein induced acute pancreatic in rats. Gut **29:** 1516–1523.
4. HOFBAUER, B., A. SALUJA, M.M. LERCH, et al. 1998. Intra-acinar cell activation of trypsinogen during cerulein-induced pancreatitis in rats. Am. J. Physiol. **275:** G352–G362.
5. LERCH, M.M. & G. ADLER 1994. Experimental animal model of acute pancreatitis. Int. J. Pancreatol. **15:** 159–170.
6. YU, J.H., J.W. LIM, W. NAMKUUNG, et al. 2002. Suppression of cerulein-induced cytokine expression by antioxidants in pancreatic acinar cells. Lab. Invest. **82:** 1359–1368.
7. BAGGIOLIMI, M., B. DEWALD & B. MOSER. 1994. Interleukin-8 and related chemotactic cytokines-CXC and CC chemokines. Adv. Immunol. **55:** 97–179.
8. BETTINA, R., B. KATIJA, M.K. COLIN, et al. 2003. CC-chemokine activation in acute pancreatitis: enhanced release of monocyte chemoattractant protein-1 in patient with local and systemic complication. Intens. Care Med. **29:** 622–629.
9. YU, J.H., J.W. LIM, H. KIM & K.H. KIM. 2005. NADPH oxidase mediates interukin-6 expression in cerulein-stimulated pancreatic acinar cells. Int. J. Biochem. Cell Biol. **37:** 1458–1469.
10. GROSS, V., H.G. LESER, A. HEINISCH & J. SCHOLMERICH. 1993. Inflammatory mediator and cytokines—new aspects of the pathophysiology and assessment of severity of acute pancreatitis. Hepatogastroenterology **40:** 522–530.
11. STEINLE, A.U., H. WEIDENBACH, M. WAGNER, et al. 1999. NF-kB/Rel activation in cerulein pancreatitis. Gastroenterology **116:** 420–430.
12. YU, J.H., J.W. LIM, H. KIM & K.H. KIM. 2005. NADPH oxidase and apoptosis in cerulein-stimulated pancreatic acinar AR2J cells. Free Rad. Biol. Med. **39:** 590–602.
13. BLINMAN, T.A., I. GUKOVSKI, M. MOURIA, et al. 2000. Activation of pancreatic acinar cells in isolation from tissue: cytokine upregulation via p38 MAP kinase. Am. J. Physiol. Cell Physiol. **279:** C1993–C2003.

Upregulation of VEGF by 15-Deoxy-$\Delta^{12,14}$-Prostaglandin J_2 via Heme Oxygenase-1 and ERK1/2 Signaling in MCF-7 Cells

EUN-HEE KIM, HYE-KYUNG NA, AND YOUNG-JOON SURH

Research Institute of Pharmaceutical Sciences, College of Pharmacy, Seoul National University, Seoul 151-742, Korea

ABSTRACT: The vascular endothelial growth factor (VEGF) induces angiogenesis in ischemic or inflamed tissues during tumor growth. 15-Deoxy-$\Delta^{12,14}$-prostaglandin J_2 (15d-PGJ_2), an endogenous ligand of peroxisome proliferator-activated receptor (PPAR) γ, has been reported to upregulate VEGF synthesis through the induction of heme oxygenase (HO)-1. In this work, we found that treatment of human breast cancer (MCF-7) cells with 15d-PGJ_2 led to time-dependent increases in the expression of HO-1. The PPAR γ antagonist GW9662 and N-acetylcysteine failed to block induction of HO-1 by 15d-PGJ_2. Elevated expression or activity of HO-1 has been reported to stimulate proliferation and to accelerate angiogenesis in several tumor cells. The induction of HO-1 expression preceded the upregulation of VEGF in MCF-7 cells stimulated with 15d-PGJ_2. In another experiment, 15d-PGJ_2 induced phosphorylation of extracellular signal-regulated kinase (ERK1/2) in 12 h. Treatment of MCF-7 cells with U0126 or transient transfection with dominant negative ERK (DN-ERK) abrogated 15d-PGJ_2-induced VEGF expression. To determine whether the induction of HO-1 is responsible for ERK1/2 activation, the HO-1 inhibitor, zinc protoporphyrin (ZnPP) was used. The phosphorylation of ERK1/2 by 15d-PGJ_2 was abolished by ZnPP. These results suggest that 15d-PGJ_2 upregulates VEGF expression via induction of HO-1 and ERK-1 and -2 phosphorylation, which may contribute to increased angiogenesis of the tumor cells.

KEYWORDS: 15-deoxy-$\Delta^{12,14}$-prostaglandin J_2; heme oxygenase-1; VEGF; ERK; MCF-7 cells

Address for correspondence: Prof. Young-Joon Surh, College of Pharmacy, Seoul National University, Shinlim-dong, Kwanak-ku, Seoul 151-742, Korea. Voice: +82-2-880-7845; fax: +82-2-874-9775.
e-mail: surh@plaza.snu.ac.kr

INTRODUCTION

Angiogenesis is implicated in the pathogenesis of a variety of disorders, such as proliferative retinopathies, age-related macular degeneration, rheumatoid arthritis, and malignant transformation.[1] Vascular endothelial growth factor (VEGF) has been identified as the central mediator of angiogenesis.[2,3] VEGF expression is elevated in many malignancies including colorectal breast, lung, and other tumors.[4,5] VEGF production is upregulated by several factors, including growth factors,[2] hormones,[6] and inflammatory cytokines.[7]

15-Deoxy-$\Delta^{12,14}$-prostaglandin J_2 (15d-PGJ_2), a cyclopentenone prostaglandin of J_2 series, has been reported to have a broad spectrum of biological effects, including apoptosis, cell proliferation, inflammation, and induction of antioxidant enzymes.[8] Besides the intrinsic role in mediating the proliferation and apoptosis, 15d-PGJ_2 may contribute to the pathophysiology of carcinogenesis.[9,10] 15d-PGJ_2 was found to stimulate the expression of VEGF in endothelial cells, human androgen-independent prostate cancer (PC3) cells and the 5637 urinary bladder carcinoma cell line.[11,12] The upregulation of VEGF production by 15d-PGJ_2 was reduced by the heme oxygenase (HO)-1 inhibitor or the carbon monoxide (CO) scavenger hemoglobin.[13] Therefore, these results suggest that upregulation of VEGF can be controlled through the induction of HO-1 and increased synthesis of CO formed as a byproduct.[14] However, the molecular event that links HO-1 induction and subsequent VEGF synthesis has not been identified. We report herein that 15d-PGJ_2 upregulates VEGF expression through HO-1-driven phosphorylation of extracellular signal-regulated kinase (ERK1/2) in human breast cancer calls.

MATERIALS AND METHODS

Chemical and Biochemical Reagents

15d-PGJ_2 and GW9662 were purchased from Cayman Chemical Co. (Ann Arbor, MI). RPMI 1640 medium and fetal bovine serum were obtained from Gibco BRL (Grand Island, NY). Antibodies against ERK, pERK, HO-1, VEGF, and actin were obtained from Santa Cruz Biotechnology, Inc. (Santa Cruz, CA). Zinc protoporphyrin (ZnPP) was provided from OXIS International Inc. (Portland, OR). N-Acetyl-$_L$-cysteine (NAC) was obtained from Sigma Chemical Co. (St. Louis, MO). Primers for HO-1, VEGF, and glyceraldehyde-3-phosphate dehydrogenase (GAPDH) were purchased from Bionics (Seoul, Korea).

Cell Culture

MCF-7 cells were maintained routinely in RPMI 1640 medium supplemented with 10% fetal bovine serum and 100 ng/mL penicillin/streptomycin/fungizone mixture at 37°C in a humidified atmosphere of 5%

CO_2/95% air. The cells were plated at an appropriate density according to each experimental scale.

Western Blot Analysis

MCF-7 cells (2×10^5 cells/mL) were plated in a 60-mm dish and were treated with 15d-PGJ$_2$ under specified conditions. After rinse with phosphate-buffered saline (PBS), the cells were exposed to the lysis buffer (20 mM Tris-HCl (pH 7.5), 150 mM NaCl, 1 mM Na$_2$EDTA, 1 mM EGTA, 1% Triton, 2.5 mM sodium pyrophosphate, 1 mM β-glycerophosphate, 1 mM Na$_3$VO$_4$, 1 μg/mL leupeptin, and protease inhibitors) (Cell Signaling Technology, Beverly, MA) in the ice for 15 min. After centrifugation at 12,000 g for 15 min, the supernatant was separated and stored at −70°C until use. The protein concentration was determined by using the BCA Protein Assay Kit (Pierce, Rockford, IL). Protein samples were electrophoresed on a 12% SDS-polyacrylamide gel and transferred to polyvinylidene difluoride (PVDF) membrane at 300 mA for 3 h. Blots were incubated in fresh blocking buffer (0.1% Tween-20 in PBS containing 5% nonfat dry milk, pH 7.4) for 1 h followed by incubation with the indicated antibodies in PBS with 3% nonfat dry milk. After washing with PBST three times, blots were incubated with horseradish peroxidase-conjugated secondary antibody in PBS with 3% nonfat dry milk for 1 h at room temperature. Blots were washed again three times in PBST buffer, and transferred proteins were detected with an Enhanced Chemiluminescence Detection Kit (Amersham Pharmacia Biotech, Buckinghamshire, UK).

Reverse Transcription-Polymerase Chain Reaction (RT-PCR)

Total RNA was isolated from MCF-7 cells using TRIzol® reagent (Invitrogen, Carlsbad, CA) according to the manufacturer's protocol. One microgram of total RNA was reverse transcribed with murine leukemia virus reverse transcriptase (Promega, Madison, WI) at 42°C for 50 min and at 72°C for 15 min. One microliter of cDNA was amplified in sequential reactions: 95°C for 1 min, 60°C for 1 min, and 72°C for 1 min, for 25 cycles of HO-1; 95°C for 1 min, 65°C for 1 min, and 72°C for 1 min, for 25 cycles of VEGF; 94°C for 1 min, 56°C for 2 min, and 72°C for 2 min, for 26 cycles of the housekeeping gene, GAPDH, followed by a final extension at 72°C for 10 min. The primers used for the RT-PCR reactions are as follows (forward and reverse, respectively): *HO-1*, 5'-CAG GCA GAG AAT GCT GAG TTC-3' and 5'-GAT GTT GAG CAG GAA CGC AGT-3'; *VEGF*, 5'-GAG AAT TCG GCC TCC GAA ACC ATG AAC TTT CTG T-3' and 5'-GAG CAT GCC CTC CTG CCC GGC TCA CCG C-3'; *GAPDH*, 5'-AAG GTC GGA GTC AAC GGA TTT-3' and 5'-GCA GTG AGG GTC TCT CT-3'.

Amplification products were analyzed on 1.0% agarose gel electrophoresis, stained with ethidium bromide, and photographed under ultraviolet light.

Transient Transfection of Dominant Negative ERK (DN-ERK)

MCF-7 cells were plated at a confluence of 60% in 6-well plate and grown in RPMI supplemented with 10% heat-inactivated fetal bovine serum at 37°C in a humidified atmosphere of 5% CO_2/95% air. Transient transfections were performed using the N-[1-(2,3-Dioleolloxy)propyl]-N,N,N-trimethylammonium methylsulfate (DOTAP) liposomal transfection reagents according to the instructions supplied by the manufacturer (Roche, Diagnostic GmbH, Mannheim, Germany). After 8- to 12-h transfection, cells were treated with 15d-PGJ_2 for additional 24 h, and the cell lysates were analyzed by Western blotting according to the methods described previously.

RESULTS AND DISCUSSION

HO-1 is known to display potent antioxidant and anti-inflammatory functions, which are mainly mediated by the heme degradation products including biliverdin/bilirubin and CO.[15] In this study, treatment of MCF-7 cells with 15d-PGJ_2 resulted in the induction of HO-1 in a time-dependent manner (FIG. 1 A). The induction of HO-1 protein expression correlated well with the increased levels of *HO-1* mRNA (FIG. 1 B). 15d-PGJ_2-dependent induction of HO-1 was not affected by the treatment of GW9662 (the PPAR γ antagonist) and the antioxidant NAC (FIGS. 1 C and D). These results suggest that induction of HO-1 by 15d-PGJ_2 in MCF-7 cells is independent of PPAR γ and reactive oxygen species (ROS).

Angiogenesis is the process of new vessel formation from preexisting capillaries.[16] Physiological angiogenesis in adults is necessary during the female reproductive cycle,[17] wound healing,[18] hair growth,[19] and bone formation.[20] However, dysregulated angiogenesis can cause many abnormal disorders, such as cancer, obesity, arthritis, blindness, and so on (reviewed in Ref. 21). Angiogenesis has been reported to be regulated by numerous angiogenic factors or mediators. As a prime mediator of angiogenesis, VEGF induces angiogenesis in ischemic or inflamed tissues, wound healing, rheumatoid arthritis, or diabetic retinopathy as well as during carcinogenesis (reviewed in Ref. 2). Recently, it has been reported that induction of HO-1 expression is associated with VEGF expression (reviewed in Ref. 22). HO-1 is a critical player of the cellular response of blood vessels to stress stimuli.[23–25] In this work, we noted that 15d-PGJ_2 induced expression of VEGF at concentrations and time capable of inducing HO-1 expression (FIG. 2 A and B).

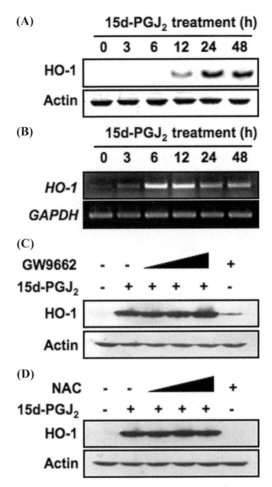

FIGURE 1. 15d-PGJ$_2$-induced expression of HO-1 in MCF-7 cells. (A) Western blot analysis of total cell extract was performed with anti-HO-1 antibody. MCF-7 cells were treated with 30 μM of 15d-PGJ$_2$ for the indicated time periods. (B) The mRNA expression of HO-1 was assessed by RT-PCR under the conditions as described in the section on "Materials and Methods." MCF-7 cells were treated with 30 μM of 15d-PGJ$_2$ for 0, 3, 6, 12, 24, and 48 h, and total RNA was extracted with Trizol. GAPDH was used for normalization. MCF-7 cells were treated with 30 μM of 15d-PGJ$_2$ in the absence or presence of GW9662 (1, 10, or 25 μM) (C) or NAC (1, 5, or 10 mM) (D) for 24 h at 37°C. GW9662 or NAC was added to the media 30 min before the 15d-PGJ$_2$ treatment.

The transcriptional activators of the *VEGF* gene include hypoxia-inducible factor (HIF)-1, Sp-1, and activator protein-1 (AP-1).[26] Fukuda *et al.* demonstrated for the first time that signaling from constitutively active forms of a G protein–coupled receptor, Raf-1, or Ras to ERK1/2 can stimulate HIF-1α and VEGF expression.[27] In addition, it has been reported that phosphorylation of

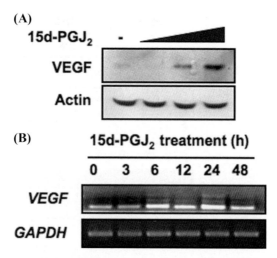

FIGURE 2. Upregulation of VEGF in 15d-PGJ$_2$-stimulated MCF-7 cells. (**A**) Western blot analysis of total cell extract was performed with anti-VEGF antibody. MCF-7 cells were treated with 0, 3, 10, or 30 μM of 15d-PGJ$_2$ for 24 h. (**B**) MCF-7 cells exposed to 30 μM 15d-PGJ$_2$ were harvested at the indicated time intervals, and the total RNA samples were analyzed by RT-PCR for VEGF mRNA transcript as described previously. GAPDH was used as an equal loading control.

SP-1 by ERK1/2 is a crucial event for the upregulation of VEGF.[28,29] In an attempt to identify the signaling pathways responsible for VEGF induction by 15d-PGJ$_2$, we examined its effect on the activation of the ERK1/2 pathway. As shown in FIGURE 3 A, 15d-PGJ$_2$ activated ERK1/2 via phosphorylation. In a subsequent experiment, cells were preincubated for 30 min with U0126 (MEK inhibitor) followed by treatment with 30 μM 15d-PGJ$_2$ for 24 h. The pharmacologic inhibition of ERK1/2 with U0126 abrogated 15d-PGJ$_2$-induced VEGF expression (FIG. 3 B). To verify the role of ERK1/2 in the induction of VEGF expression by 15d-PGJ$_2$, MCF-7 cells were transfected transiently with DN-ERK. DN-ERK transfection markedly diminished the expression of the VEGF protein (FIG. 3 C). These findings indicate that ERK is an essential upstream enzyme responsible for inducing the expression of VEGF by 15d-PGJ$_2$. However, the pharmacologic inhibition of the ERK1/2 barely affected HO-1 induction by 15d-PGJ$_2$ (data not shown), suggesting that the ERK1/2 pathway is unlikely to be involved in 15d-PGJ$_2$-induced HO-1 upregulation in MCF-7 cells. To delineate the significance of ERK activation and HO-1 induction by 15d-PGJ$_2$, we investigated the effects of ZnPP, the HO-1 inhibitor, on ERK phosphorylation. Treatment of ZnPP reduced 15d-PGJ$_2$-induced phosphorylation of ERK in a concentration-dependent manner (FIG. 3 D). These findings indicate that phosphorylation of ERK1/2 by 15d-PGJ$_2$ is downstream of HO-1 expression.

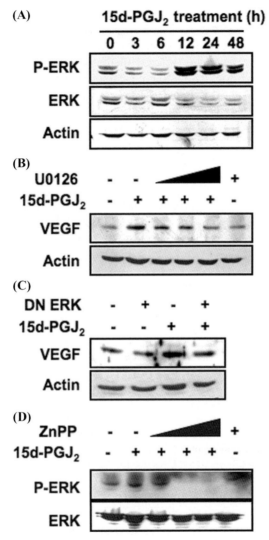

FIGURE 3. The involvement of HO-1-induced ERK activation in VEGF upregulation by 15d-PGJ$_2$. (**A**) The time-related activation of ERK was assessed by measuring the respective phosphorylated form. MCF-7 cells were stimulated with 30 μM 15d-PGJ$_2$ for the indicated times and then immunoblotted with specific antibodies that recognize phospho-ERK (P-ERK). (**B**) The effect of the MEK inhibitor on the induction of VEGF was assessed by Western blot analysis in MCF-7 cells exposed to 30 μM 15d-PGJ$_2$ for 24 h in the presence of U0126 (0, 0.1, 1, or 5 μM). (**C**) MCF-7 cells were transfected transiently with DN-ERK or control plasmid (CEP). The expression of VEGF protein was monitored by Western blot analysis. (**D**) The effect of the HO-1 inhibitor on the phosphorylation of ERK was assessed by Western blot analysis exposed to 30 μM 15d-PGJ$_2$ in the presence of ZnPP (0, 1, 5, or 10 μM).

CO, a diatomic gas, is the one of the byproducts of heme degradation catalyzed by HO-1. CO at low concentrations is known to exert the anti-inflammatory and antiapoptotic effects.[30] The vasodilatory and antiapoptotic effects of CO have been shown to be mediated via production of cGMP.[31] Li et al. reported the cytoprotective function of CO in peroxynitrite-treated PC12 cells.[32] Moreover, CO plays a key role in the induction of VEGF by HO-1 activity.[13] Therefore, we suggest that CO released from HO-1-catayzed reaction mediates phosphorylation of ERK1/2 in 15d-PGJ$_2$-stimulated MCF-7 cells which, in turn, induces VEGF expression. Additional elucidation of detailed molecular mechanisms and the precise molecular link associated with angiogenic activities of CO is the subject of ongoing investigation.

In conclusion, 15d-PGJ$_2$ activates ERK1/2 signaling and subsequently induces VEGF expression via upregulation of HO-1 in human breast cancer calls.

ACKNOWLEDGMENT

This study was supported by grant RO2-2004-000-10197-0 from the Basic Research Program of the Korea Science Engineering Foundation.

REFERENCES

1. FOLKMAN, J. 1995. Angiogenesis in cancer, vascular, rheumatoid and other disease. Nat. Med. **1:** 27–31.
2. FERRARA, N., H.P. GERBER & J. LECOUTER. 2003. The biology of VEGF and its receptors. Nat. Med. **9:** 669–676.
3. CARMELIET, P. 2005. VEGF as a key mediator of angiogenesis in cancer. Oncology **69**(Suppl 3): 4–10.
4. LEE, J.C. et al. 2000. Prognostic value of vascular endothelial growth factor expression in colorectal cancer patients. Eur. J. Cancer **36:** 748–753.
5. POON, R.T., S.T. FAN & J. WONG. 2001. Clinical implications of circulating angiogenic factors in cancer patients. J. Clin. Oncol. **19:** 1207–1225.
6. SHIBUYA, M. 2001. Structure and function of VEGF/VEGF-receptor system involved in angiogenesis. Cell. Struct. Funct. **26:** 25–35.
7. NEUFELD, G. et al. 1999. Vascular endothelial growth factor (VEGF) and its receptors. FASEB J. **13:** 9–22.
8. STRAUS, D.S. & C.K. GLASS. 2001. Cyclopentenone prostaglandins: new insights on biological activities and cellular targets. Med. Res. Rev. **21:** 185–210.
9. BENNETT, A. 1986. The production of prostanoids in human cancers, and their implications for tumor progression. Prog. Lipid Res. **25:** 539–542.
10. MARNETT, L.J. 1992. Aspirin and the potential role of prostaglandins in colon cancer. Cancer Res. **52:** 5575–5589.
11. HASLMAYER, P. et al. 2002. The peroxisome proliferator-activated receptor γ ligand 15-deoxy-Delta^{12,14}-prostaglandin J$_2$ induces vascular endothelial growth factor

in the hormone-independent prostate cancer cell line PC 3 and the urinary bladder carcinoma cell line 5637. Int. J. Oncol. **21:** 915–920.

12. YAMAKAWA, K. *et al.* 2000. Peroxisome proliferator-activated receptor-γ agonists increase vascular endothelial growth factor expression in human vascular smooth muscle cells. Biochem. Biophys. Res. Commun. **271:** 571–574.

13. JOZKOWICZ, A. *et al.* 2003. Heme oxygenase and angiogenic activity of endothelial cells: stimulation by carbon monoxide and inhibition by tin protoporphyrin-IX. Antioxid. Redox. Signal. **5:** 155–162.

14. JOZKOWICZ, A. *et al.* 2000. Ligands of peroxisome proliferator-activated receptor-γ increase the generation of vascular endothelial growth factor in vascular smooth muscle cells and in macrophages. Acta Biochim. Pol. **47:** 1147–1157.

15. MAINES, M.D. 1988. Heme oxygenase: function, multiplicity, regulatory mechanisms, and clinical applications. FASEB J. **2:** 2557–2568.

16. DULAK, J. *et al.* 2005. Heme oxygenase-1 and angiogenesis: therapeutic implications. *In* Heme Oxygenase. L.E. Otterbein & B.S. Zuckerbraun, Eds.: 245–270. Nova Science Publisher. New York.

17. CARMELIET, P. 2003. Angiogenesis in health and disease. Nat. Med. **9:** 653–660.

18. SCHWENTKER, A., *et al.* 2002. Nitric oxide and wound repair: role of cytokines? Nitric Oxide **7:** 1–10.

19. YANO, K., L.F. BROWN & M. DETMAR. 2001. Control of hair growth and follicle size by VEGF-mediated angiogenesis. J. Clin. Invest. **107:** 409–417.

20. GERBER, H.P., *et al.* 1999. VEGF couples hypertrophic cartilage remodeling, ossification and angiogenesis during endochondral bone formation. Nat. Med. **5:** 623–628.

21. FERRARA, N. & R.S. KERBEL. 2005. Angiogenesis as a therapeutic target. Nature **438:** 967–974.

22. DULAK, J. *et al.* 2004. Complex role of heme oxygenase-1 in angiogenesis. Antioxid. Redox. Signal. **6:** 858–866.

23. MAINES, M.D. 1997. The heme oxygenase system: a regulator of second messenger gases. Annu. Rev. Pharmacol. Toxicol. **37:** 517–554.

24. OTTERBEIN, L.E. *et al.* 2003. Heme oxygenase-1: unleashing the protective properties of heme. Trends Immunol. **24:** 449–455.

25. SIKORSKI, E.M. *et al.* 2004. The story so far: molecular regulation of the heme oxygenase-1 gene in renal injury. Am. J. Physiol. Renal Physiol. **286:** F425–F441.

26. KIMURA, H. *et al.* 2000. Hypoxia response element of the human vascular endothelial growth factor gene mediates transcriptional regulation by nitric oxide: control of hypoxia-inducible factor-1 activity by nitric oxide. Blood **95:** 189–197.

27. FUKUDA, R. *et al.* 2002. Insulin-like growth factor 1 induces hypoxia-inducible factor 1-mediated vascular endothelial growth factor expression, which is dependent on MAP kinase and phosphatidylinositol 3-kinase signaling in colon cancer cells. J. Biol. Chem. **277:** 38205–38211.

28. BERRA, E. *et al.* 2000. Signaling angiogenesis via p42/p44 MAP kinase and hypoxia. Biochem. Pharmacol. **60:** 1171–1178.

29. MILANINI-MONGIAT, J., J. POUYSSEGUR & G. PAGES. 2002. Identification of two Sp1 phosphorylation sites for p42/p44 mitogen-activated protein kinases: their implication in vascular endothelial growth factor gene transcription. J. Biol. Chem. **277:** 20631–20639.

30. RYTER, S.W., D. MORSE & A.M. CHOI. 2004. Carbon monoxide: to boldly go where NO has gone before. Sci. STKE. **230:** RE6.

31. FORESTI, R. *et al.* 2004. Vasoactive properties of CORM-3, a novel water-soluble carbon monoxide-releasing molecule. Br. J. Pharmacol. **142:** 453–460.
32. LI, M.H., Y.N. CHA & Y.-J. SURH. 2006. Carbon monoxide protects PC12 cells from peroxynitrite-induced apoptotic cell death by preventing the depolarization of mitochondrial transmembrane potential. Biochem. Biophys. Res. Commun. **342:** 984–990.

SDF-1 Controls Pituitary Cell Proliferation through the Activation of ERK1/2 and the Ca^{2+}-Dependent, Cytosolic Tyrosine Kinase Pyk2

ALESSANDRO MASSA,[a] SILVIA CASAGRANDE,[b] ADRIANA BAJETTO,[a] CAROLA PORCILE,[a] FEDERICA BARBIERI,[a] STEFANO THELLUNG,[a] SARA ARENA,[a] ALESSANDRA PATTAROZZI,[a] MONICA GATTI,[a] ALESSANDRO CORSARO,[a] MAURO ROBELLO,[b] GENNARO SCHETTINI,[a] AND TULLIO FLORIO[a]

[a]*Section of Pharmacology, Department Oncology Biology and Genetics, University of Genova, 16132 Genova, Italy*

[b]*Unit INFM, Department of Physics, University of Genova, 16132 Genova, Italy*

ABSTRACT: Stromal cell-derived factor-1 (SDF-1) is a chemokine of the CXC subfamily that exerts its effects via CXCR4, a G-protein-coupled receptor. CXCR4 is often expressed by tumor cells, and its activation causes tumor cell proliferation. Using GH4C1 cells, here we show that SDF-1 induced cell proliferation in a dose-dependent manner. Thus, we evaluated the intracellular signaling involved in this effect. SDF-1 increased cytosolic $[Ca^{2+}]$ and activated Pyk2, ERK1/2, and BK_{Ca} channels. To correlate these intracellular effectors with the proliferative activity of SDF-1, we inhibited their activity using BAPTA-AM (Ca^{2+} chelator), PD98059 (MEK inhibitor), salicylate (Pyk2 inhibitor), and TEA (K^+ channel blocker). All these compounds reverted SDF-1-induced proliferation, suggesting the involvement of multiple intracellular pathways. To identify a possible crosstalk and a molecular ordering among these pathways, we tested these antagonists on SDF-1-dependent activation of ERK1/2, Pyk2, and BK_{Ca} channels. We report that the inhibition of $[Ca^{2+}]_i$ increase or the blockade of BK_{Ca} channel activity did not affect ERK1/2 activation by SDF-1; Pyk2 activation was purely Ca^{2+}-dependent, not involving ERK1/2 or BK_{Ca} channels; and BK_{Ca} channel activity was antagonized by Pyk2 but not by ERK1/2 inhibitors. These data suggest that SDF-1-dependent increase of $[Ca^{2+}]_i$ activates Pyk2, which, in turn, regulates BK_{Ca} channel activity. Conversely, ERK1/2 activation is an independent phenomenon. In conclusion, we demonstrate that SDF-1 induces proliferation of GH4C1 cells, suggesting that the

Address for correspondence: Tullio Florio, M.D., Ph.D, Section of Pharmacology, Department of Oncology, Biology, and Genetics, University of Genova, Viale Benedetto XV, 2, 16132 Genova, Italy. Voice and fax: +39-010-3538806.
 e-mail: tullio.florio@unige.it

Ann. N.Y. Acad. Sci. 1090: 385–398 (2006). © 2006 New York Academy of Sciences.
doi: 10.1196/annals.1378.042

activation of CXCR4 may represent a novel regulatory mechanism for pituitary cell proliferation which may contribute to pituitary adenoma development.

KEYWORDS: SDF-1; cell proliferation; pituitary cells; ERK1/2; Pyk2

INTRODUCTION

Chemokines are small peptides that regulate adhesion and transendothelial migration of leukocytes, and immune and inflammatory reactions. Chemokines were reported to play an important role in cancer progression, being involved in cell proliferation, modulation of the angiogenic/angiostatic process, and control cell migration and metastasis.[1,2]

Different tumor types including breast, prostate, pancreatic, and ovarian carcinomas as well as neuroblastoma, glioblastoma, and some leukemias express chemokines and chemokine receptors.[2–6]

Stromal cell–derived factor-1 (SDF-1) was characterized as a pre-B-cell stimulatory factor cloned from bone marrow cell supernatants.[7] SDF-1 activity is mediated by interacting with the CXCR4 receptor, a G-protein-coupled receptor (GPCR) that was reported to be the chemokine receptor most frequently expressed by different tumors. For example, CXCR4 receptors are the almost constantly expressed in glioblastomas and breast and ovarian carcinomas; and recent data involved its activation in tumor cell proliferation,[6,8,9] migration and invasion,[9] metastasization[5,10] and in tumoral neo-angiogenesis.[11]

Human pituitary adenomas are benign neoplasms representing the most common tumors of the sellar region, accounting approximately for 15% of all primary intracranial tumors. Pituitary adenomas are mainly classified according to the characteristic clinical syndromes that accompany tumor hormone production. The tumor-dependent increase in growth hormone secretion, resulting in acromegaly or gigantism, represents, after alterations of prolactin release, one of the most common hormone-secreting pituitary adenomas. About 25–30% of the pituitary adenomas are clinically classified as nonsecreting, although the majority of these adenomas do synthesize and secrete an α subunit or the entire gonadotrophins. Therefore, they are defined as clinically "nonfunctioning pituitary adenomas."[12]

Despite the extensive studies performed to elucidate the molecular alterations occurring in pituitary adenomas, their pathogenesis is still obscure.[13] However, besides modifications in the expression of genes controlling the cell cycle, it is well known that tumorigenesis and tumor progression can be triggered by the uncontrolled mitogenic activity of the growth factor or cytokines that may favor the clonal expansion of mutated cells and thus tumor progression (promotion).[14] To date, no data were reported on the possible role of SDF1/CXCR4 in anterior pituitary function and, possibly, in the genesis of pituitary adenomas, although the expression of CXCR4 was reported in the rat pituitary adenoma.[15]

In this article, we report evidence of the role of SDF-1 in pituitary cell proliferation using the rat pituitary adenoma-derived cell line GH4C1, and we characterize the signal transduction mechanisms activated after the binding of SDF-1 to its receptor CXCR4, correlating the intracellular signaling identified with the biological effects of the peptide.

MATERIALS AND METHODS

Reagents and Materials

Anti-phospho-ERK1/2 and anti-ERK1/2 antibodies were purchased from New England BioLabs (Beverly, MA, USA), anti-phospho-Pyk2 from Biosource Europe S.A. (Nivelles, Belgium), anti-Pyk2 from Transduction Labs (Lexington KY, USA); PD98059 and BAPTA-AM from Calbiochem (San Diego, CA, USA), and SDF-1 from PeproTech EC Ltd. (London, UK). All other chemicals were purchased from Sigma Chemical Co. (Milano, Italy), unless otherwise specified.

Cell Cultures

GH4C1 cells were cultured in Ham's F10 medium supplemented with 10% fetal calf serum, as reported.[16] When indicated, pertussis toxin was added to the cell culture 24 h before the SDF-1 stimulation at a concentration of 180 ng/mL.[17]

[^{3}H]-Thymidine Incorporation Assay

DNA synthesis activity was measured by means of the [^{3}H]-thymidine uptake assay.[18] Cells were plated at 5 × 10^4/well in 24-well plates, serum-starved for 48 h before being treated with SDF-1 for 16 h; in the last 4 h cells were pulsed with 1 μCi/mL of [^{3}H]-thymidine (Amersham Pharmacia Biotech, Milano, Italy). At the end of the incubation, cells were trypsinized (15 min at 37°C), extracted in 10% trichloroacetic acid, and filtered under vacuum through fiber glass filters (GF/A; Whatman International Ltd., Maidstone, UK). The filters were then washed sequentially under vacuum with 10% and 5% trichloroacetic acid and 95% ethanol. A trichloroacetic acid–insoluble fraction was then counted in a scintillation counter.

Electrophysiological Studies

Solutions

The standard external solution consisted of (mM): 135 NaCl, 5.4 KCl, 1.8 $CaCl_2$, 1 $MgCl_2$, 5 HEPES, 10 glucose. The pH was adjusted to 7.4 using NaOH. The pipettes were filled with a corresponding solution, whereas the K^+ concentration was increased to 140 mM by equimolar replacement of NaCl with KCl. Tested drugs were added to the bath to obtain the final concentrations indicated.

Electrophysiology

We used the patch clamp technique in the cell-attached configuration. The patch electrodes were connected to an EPC-7 (List-Medical) amplifier. Patch pipettes were manufactured from borosilicate glass capillaries (TW150-3, World Precision Instruments, Inc., Sarasota, FL, USA) with a programmable Sachs and Flaming puller (model PC-84), and the tips were fired-polished with a microforge (MF-83; Narishige). The holding potential was set to 0 mV in all the experiments reported.

Ion currents were recorded with Labmaster D/A, A/D converter driven by p-Clamp 7 software (Axon Instruments, Burlingame, CA, USA). Capacitance-transient neutralization and series resistance compensation were optimized. Single-channel currents were amplified and filtered with a low pass filter (ITHACO, 4382 Dual 24 dB/octave filter) at a cut-off frequency of 1 kHz with a sampling rate of 13.3 kHz. For each cell the recording time was 2 min before and after SDF-1α treatment.

Analysis was performed using the SIGMA PLOT (Jandel Scientific, Erkrath, Germany) software and p-Clamp 6 (Axon Instruments), measuring channel current amplitudes and the open probability of single channels.

Measurement of Intracellular Ca^{2+} Concentration ($[Ca^{2+}]_i$), at Single-Cell Level

Cells were plated on 25-mm clean glass coverslips, previously coated with poly-L-lysine (10 μg/mL) and transferred to 35-mm petri dishes, and after 24 h cells were serum-starved for a further 24 h. On the day of the experiment the cells were washed for 10 min with a balanced salt solution containing (mM) HEPES 10, pH 7.4; NaCl 150; KCl 5.5; $CaCl_2$ 1.5; $MgSO_4$ 1.2; glucose 10. Then cells were loaded with Fura-2 penta-acetoxymethyl ester (4 μM) (Calbiochem, Laufentigen, Switzerland) for 60 min. Fluorescence measurements were performed as previously reported.[19] For the calibration of fluorescence signals, we used cells loaded with Fura-2; Rmax and Rmin are ratios at saturating and zero $[Ca^{2+}]_i$, respectively, and were obtained by perfusing the cells with a salt solution containing $CaCl_2$ (10 mM), digitonin (2.5 μM), and ionomycin (2 μM) and subsequently with a Ca^{2+}-free salt solution containing EGTA (10 mM). The values of obtained Rmax and Rmin, expressed as gray level mean, were used to calculate the $[Ca^{2+}]_i$, using the Quanticell software, according to the equation of Grynkiewicz.[20]

Reverse Transcriptase (RT)-PCR

Total RNA was isolated from GH4C1 cells using the acid phenol extraction. Before cDNA synthesis, the RNA was treated with 40 U of RNAse-free DNAse-I (Roche, Basel, Switzerland) for 45 min at 37°C in 25 mM Tris–HCl (pH 7.2), 20 mM $MgCl_2$, 0.1 mM EDTA. Total RNA (5 μg) was reverse-transcribed in a 20-μL reaction volume containing 50 mM Tris–HCl (pH 8.3), 8 mM $MgCl_2$, 50 mM NaCl, 1 mM dithiothreitol, 1 mM dNTPs, 22U RNAse inhibitor, 2.5 μM oligo dT (16-mer) and 10 U RT (Amersham, Milano, Italy) for 40 min at 42°C. To control whether contaminating genomic DNA was present, RNA samples not subjected to RT were included in the PCR amplification. The gene-specific primers used for CXCR4 and SDF-1 amplification are: sense 5′-ggccctcaagaccacagtca-3′ and antisense 5′-ttagctggagtgaaaacttgaag-3′ for CXCR4 and sense 5′-atgaacgccaaggtcgtggtc-3′ and antisense 5′-ggtctgttgtgcttacttgttt-3′ for SDF-1.[21] PCR amplification was performed in a reaction mixture containing 10 mM Tris–HCl (pH 8.8), 50 mM KCl, 0.1% Triton X-100, 3 mM $MgCl_2$, 0.2 mM dNTPs, 1 μM of each primer, 1 U Taq DNA polymerase (Roche Basel, Switzerland). The PCR program was as follows: one cycle (5 min at 94°C) followed by 30 cycles (94°C 30 sec; 60°C 30 sec; 72°C 30 sec). Amplification of β-actin was used as a positive control for the PCR reaction using the following primers: sense 5′-tccggagacggggtca-3′ and antisense 5′-cctgcttgctgatcca-3′.

Statistical Analysis

Experiments were performed in quadruplicate and repeated at least three times. Data are expressed as mean ± SE values and statistical significance was assessed by ANOVA for independent groups. A p value of ≤ 0.05 was considered statistically significant.

RESULTS

SDF-1 Induces Proliferation of GH4C1 Pituitary Adenoma Cells in a G-Protein-Dependent Manner

To test whether SDF-1 may represent a mitogenic factor for pituitary cells, we studied the ability of this chemokine to induce proliferation of the rat pituitary

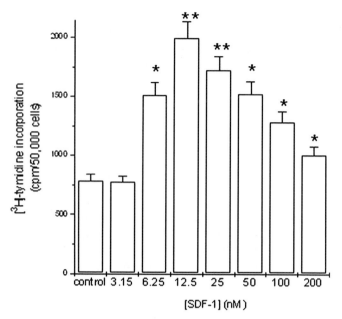

FIGURE 1. SDF-1 regulation of GH4C1 cell proliferation. Dose response (3.15–200 nM) effect of SDF-1 on the proliferation of GH4C1 cells. GH4C1 cell proliferation was evaluated by means of [^{3}H]-thymidine uptake assay. SDF-1 induced a dose-dependent increase in DNA synthesis with a maximal effect at a concentration of 12.5 nM. Data are expressed as cpm/50,000 cells *$P < 0.05$, and **$P < 0.01$ versus basal value.

adenoma cell line GH4C1 by means of [^{3}H]-thymidine incorporation assay. First, we demonstrated by RT-PCR analysis that GH4C1 cells express CXCR4 but not its ligand, SDF-1 (data not shown). GH4C1 cells were serum-starved for 48 h and then treated for 16 h with SDF-1. SDF-1 increased DNA synthesis in a dose-dependent manner with a maximum effect at the concentration of 12.5 nM (FIG. 1), while at higher concentrations (up to 200 nM) a reduction in the proliferative response occurred, likely due to CXCR4 downregulation, as observed in other tumor cell lines (glioblastoma, ovarian carcinoma).[4,6] The pretreatment with pertussis toxin completely prevented the proliferative effects of SDF-1, demonstrating that these effects were mediated by a pertussis toxin–sensitive G protein activated via CXCR4 (data not shown).

SDF-1 Treatment Increases GH4C1 Intracellular Ca^{2+} Concentration ($[Ca^{2+}]_i$) Inducing Ca^{2+} Release from the Intracellular Stores

Thus, we studied the intracellular pathways implicated in GH4C1 cell proliferation, analyzing the effects of SDF-1 on the intracellular Ca^{2+} homeostasis, since the modulation of the Ca^{2+} levels is known to represent a major transduction mechanism activated by CXCR4.[22] In microfluorimetric experiments, we show that the treatment of GH4C1 cells with SDF-1 (50 nM) induced a significant increase in $[Ca^{2+}]_i$ (basal = 132 ± 22 nM; SDF-1 = 375 ±25 nM, $n = 5$). This effect was mainly dependent on the release from the intracellular stores since in experimental conditions in which Ca^{2+} was removed from the external medium, although starting from a lower basal level, the treatment with SDF-1 was still able to increase the $[Ca^{2+}]_i$ (data not shown).

Ca^{2+}-Mediated Intracellular Signaling Induced by SDF-1 Treatment of GH4C1 Cells

The increase in $[Ca^{2+}]_i$ induced by SDF-1 may affect the activity of a number of intracellular second messengers involved in the control of cell proliferation. Thus, we analyzed the effects of SDF-1 on the activation of the Ca^{2+}-dependent cytosolic tyrosine kinase, Pyk2, known to be activated by G-protein-coupled receptors including CXCR4[23] and ERK1/2 activation, which was previously reported to be involved in the proliferative effects of SDF-1 in glioma cells.[6]

In Western blot experiments we demonstrated that SDF-1 dose-dependently (maximum effect at the concentration of 12.5 nM) activated both Pyk2 (FIG. 2A) and ERK1/2 (FIG. 2B). Pyk2 was activated/phosphorylated after

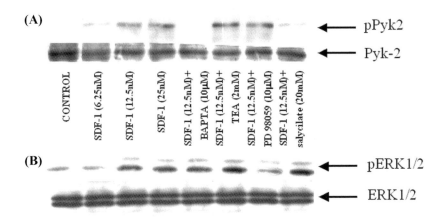

FIGURE 2. SDF-1 regulation of PkK2 and ERK1/2 activity in GH4C1 cells and crosstalk between these intracellular pathways. Shown are dose–response of SDF-1 (6.25–25 nM) on Pyk2 (**A**) and ERK1/2 (**B**) phosphorylation/activation and the effect of the pretreatment with Pyk2 inhibitor salicylate (20 mM), the Ca^{2+} chelator BAPTA-AM (10 μM), the K^+ channel inhibitor TEA (2 mM), and the MEK inhibitor PD98059 (10 μM) on SDF-1 (12.5 nM) effects. Pyk2 and ERK1/2 activation is evaluated by Western blot, using phosphospecific antibodies. All of the antagonists were added to the cell culture medium 20 min before SDF-1. Total Pyk2 and ERK1/2 expression (*lower lanes* in both **A** and **B** *panels*) was evaluated to demonstrate equal protein loading in the gels.

10 min of SDF-1 treatment at the concentration of 12.5 nM and reached a maximal stimulation after 30 min of treatment (data not shown). SDF-1 activation of ERK1/2 was already detectable after 5 min of treatment, lasted up to 10 min, and then declined after 20 min (data not shown), thus being much more rapid than the activation of Pyk2. The evaluation of the total expression of Pyk2 and ERK1/2 confirmed the equal loading of proteins in the different lanes (FIG. 2A, B).

Liu et al.[24] reported that $[Ca^{2+}]_i$ increase, induced by CXCR4 stimulation, may lead to the activation of the large-conductance Ca^{2+}-activated K^+ channels (BK_{Ca}) known to be involved in the proliferation of different cell types, including the GH4C1-related cell line GH3.[25] The activity of BK_{Ca} channels was tested after SDF-1 treatment in electrophysiology experiments, using the cell-attached patch clamp configuration. We observed at the resting membrane potential spontaneously active BK_{Ca} channels in more than 80% of the cell-attached patches examined, but no patches with active channel were identified when 2 mM tetraethyl ammonium (TEA) was added to the electrode-filled solution (data not shown). Basal conductance of single channels was 148 ± 14 pS ($n = 10$).

BK_{Ca} activity increased transiently after SDF-1 (25 nM) external application for the whole period of recording: the frequency of the channel openings was increased with an initial period of high channel activity followed by sporadic openings (FIG. 3). The single channel current slightly increased after SDF-1 treatment without reaching a statistically significant difference, while the open probability increased more than fourfold during SDF-1 stimulation.

Thus, in GH4C1 cells, SDF-1 treatment activated Pyk2, ERK1/2, and BK_{Ca} channels.

Effect of the Ca^{2+}-Mediated Intracellular Signaling on the Proliferative Activity of SDF-1

To correlate the effects of SDF-1 on cell proliferation with the second messengers activated by the chemokine, we measured the SDF-1-increase in [^{3}H]-thymidine incorporation in the presence or absence of drugs able to interfere with the increase in $[Ca^{2+}]_i$ or the activation of ERK1/2, Pyk2, and BK_{Ca} channels. GH4C1 proliferation induced by SDF-1 was completely inhibited by BAPTA/AM (10 μM), a cell-permeable Ca^{2+} chelator, salicylate (20 mM), a rather selective inhibitor of Pyk2,[26] PD98059 (10 μM), a MAP kinase kinase (MEK) inhibitor, and TEA (2 mM), a K^+ channel inhibitor, all added 20 min before SDF-1 treatment (TABLE 1). These results show that all these pathways are involved in the proliferative effects of SDF-1.

Thus, we tried to establish a molecular ordering in the intracellular signaling of SDF-1. As far as Pyk2 activation is concerned, we found that SDF-1

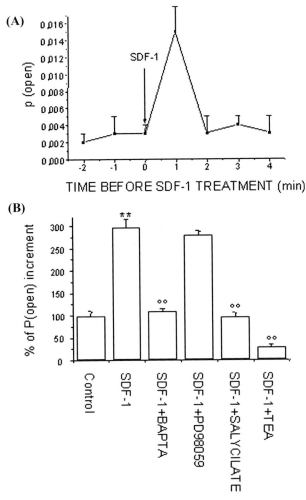

FIGURE 3. SDF-1 regulation BK_{Ca} channel activity in GH4C1 cells. (**A**) Time course of the mean open channel probability before and after the addition of SDF-1 (100 nM, *arrow*). Each point is expressed as the average ± SE ($n = 5$). (**B**) Summary of data showing the effect of SDF-1 (25 nM) on BK_{Ca} channel activity and the effects of BAPTA-AM (10 μM), PD98059 (10 μM), salicylate (20 mM), and TEA (2 mM) on SDF-1 (25 nM)-induced channel activity. Each point is expressed as a percentage ± SE compared with the controls (100%) ($n = 5$). BAPTA-AM, salicylate, and TEA completely prevented SDF-1 activation of BK_{Ca} channel, whereas PD98059 was ineffective. $^{**}P < 0.01$ versus control value; $^{\circ\circ}P < 0.01$ versus SDF-125 nM value.

effects were completely reversed by BAPTA/AM and salicylate pretreatment, while PD98095 and TEA were ineffective (FIG. 2A). Conversely, SDF-1-dependent activation of ERK1/2 was completely abolished by PD98059 but not by BAPTA/AM, salicylate, or TEA (FIG. 2B).

TABLE 1. Intracellular mechanism involved in SDF-1-induced GH4C1 proliferation

Basal	100 ± 5
SDF-1 (12.5 nM)	$134 \pm 2^{\circ\circ}$
SDF-1 (12.5 nM) + BAPTA-AM (10 μM)	$103 \pm 3^{**}$
SDF-1 (12.5 nM) + PD98059 (10 μM)	$98 \pm 11^{*}$
SDF-1 (12.5 nM) + tetraethyl ammonium (2 mM)	$105 \pm 10^{*}$
SDF-1 (12.5 nM) + salicylate (20 mM)	$109 \pm 4^{*}$

Different signal transduction inhibitors were used to dissect the intracellular pathways activated by SDF-1 (12.5 nM) to induce GH4C1 cell proliferation.

The compounds used were: BAPTA-AM (10 μM), an intracellular Ca^{2+} chelator, PD98059 (10 μM), a MEK inhibitor, TEA (2 mM), a K^+ inhibitor and salicylate (20 mM), a Pyk2 inhibitor.

All of the antagonists were added to the cell culture medium 20 min before SDF-1.

$^{\circ\circ}P < 0.01$ versus basal value; $^{*}P < 0.05$ and $^{**}P < 0.01$ versus SDF-1 stimulation.

FIGURE 3B shows the mean percentage of opening probability increment of BK_{Ca} channels recorded in the presence of 25 nM SDF-1 and the intracellular pathway inhibitors in the external solution. The pretreatment of GH4C1 cells with TEA, BAPTA-AM, and salicylate, but not with PD98059, completely blocked the activating effects of SDF-1 on BK_{Ca} channels (FIG. 3B).

From these results we demonstrate that SDF-1-stimulated GH4C1 proliferation independently involves ERK1/2 and Pyk2/BK_{Ca} channel pathways. We proposed that ERK1/2 activation by SDF-1 is an autonomous phenomenon from the signaling by which the chemokine controls BK_{Ca} channel activity, which, conversely, requires the sequential increase in $[Ca^{2+}]_i$ and the activation of Pyk2, as demonstrated by the inhibitory effects of BAPTA-AM and salicylate.

DISCUSSION

Much evidence shows that the genesis of pituitary adenomas involves a multistep process including both initiating and promoting events.[14] There is a general consensus that the majority of pituitary adenomas derives from a single transformed cell in which a gain of function, as far as proliferative potential, allows its clonal expansion in the presence of promoting factors.[14] Among the promoting factors, hypothalamic and locally produced growth factors play a pivotal role.[14] Here we identify SDF-1 as a possible novel growth factor for pituitary cells. Previous studies identified the presence of binding sites for SDF-1 in rat pituitary adenomas, by means of autoradiographic studies, and hypothalamus, using immunohistochemistry techniques.[15,27] There is increasing evidence of the role of SDF1/CXCR4 signaling in the growth of different tumoral histotypes.[4,6,9] To study the effects of CXCR4 activation in pituitary cells at cellular and molecular level, we analyzed the effects of SDF-1 on the proliferation of the rat pituitary adenoma cell line GH4C1. We show that these cells express CXCR4 mRNA but do not secrete SDF-1, thus representing a good model for assessing the biological activity of this chemokine, avoiding possible constitutive receptor desensitization or downregulation observed *in vitro* for other cell types.[6,22] Our experiments demonstrate that SDF-1 treatment induces proliferation in GH4C1 cells, suggesting a role for this chemokine in the regulation of pituitary functioning. We show that low concentrations (maximal effect 12.5 nM) of SDF-1 are able to induce a significant proliferation of GH4C1 cells, through the activation of a pertussis toxin–sensitive G protein, coupled to CXCR4. Moreover, we identified multiple intracellular pathways activated by SDF-1 that participate in the chemokine-dependent pituitary cell proliferation. In fact, on one hand, inhibitors of ERK1/2, Pyk2, and K^+ channels, as well as the Ca^{2+} chelator BAPTA, all reversed the DNA synthesis induced by the chemokine, while, on the other, the treatment with SDF-1 induced a significant activation of ERK1/2, an increase of $[Ca^{2+}]_i$, a Ca^{2+}-dependent phosphorylation/activation of the cytosolic tyrosine kinase, Pyk2, as well as increased the activity of BK_{Ca} channels. All these intracellular pathways were previously reported to mediate cell proliferation in many cell types and, more importantly, were reported to represent CXCR4 intracellular effectors.[23,24,28,29] Then we tried to correlate the effects of these second messengers with the proliferative activity of SDF-1. We show that GH4C1 cell proliferation induced by SDF-1 is dependent on multiple intracellular effectors, since in the presence of blockers of ERK1/2 (PD98059), Pyk2 (salicylate), or K^+ channels (TEA) or the Ca^{2+} chelator BAPTA-AM, the chemokine was unable to elicit GH4C1, cell proliferation.

As far as ERK1/2 activation is concerned, in our experimental model no cross-modulation was identified by the other SDF-1-dependent effectors: on one hand the MEK inhibitor PD98059 affected neither Pyk2 nor BK_{Ca} activation (being both Ca^{2+}-dependent processes) and, on the other hand, ERK1/2 activity was not inhibited by preventing the $[Ca^{2+}]_i$ rise using the intracellular Ca^{2+} chelator BAPTA-AM, the Pyk2 inhibitor salicylate, or the K^+ channel blocker TEA.

Conversely, the activation of BK_{Ca} channels by SDF-1 was inhibited by salicylate, demonstrating, as already described for other K^+ channels,[30] that it required the activation of Pyk2. Finally, Pyk2 activation was a completely Ca^{2+}-dependent process (being inhibited by BAPTA) and was not affected by the inhibition of K^+ channels, locating the activity of this kinase upstream of BK_{Ca} channel activation.

Thus, in GH4C1 cells, ERK1/2 and Pyk2/BK_{Ca} channels contribute independently to the SDF-1-dependent cell proliferation, but their coordinated activities are necessary for the final biological effect.

In conclusion, in this work we propose that SDF-1, either coming from the systemic circulation or directly from the hypothalamus, may represent a powerful proliferative agent for pituitary cells, contributing to the regulation of

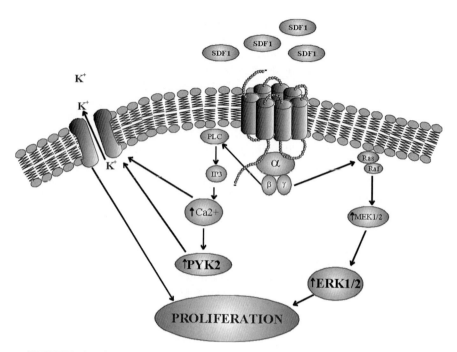

FIGURE 4. Diagrammatic representation of SDF-1 intracellular signaling involved in GH4C1 cell proliferation. SDF-1 activation of CXCR4, via a pertussis toxin–sensitive G protein, leads to the activation of two independent pathways that converge only on the final biological effect (activation of cell proliferation). The first pathway results in the classical activation of the MAP kinase cascade, while the second one is Ca^{2+}-dependent, in which the increase in $[Ca^{2+}]_i$ causes the activation of PyK2, which, in turn, controls the activity of the BK_{Ca} channels. Although there is no cross-activity between these two intracellular pathways the activation of both of them is necessary for SDF-1 to induce GH4C1 cell proliferation.

pituitary function and possibly participating in the genesis of pituitary adenomas. Moreover, we identified two main intracellular pathways involved in the CXCR4-dependent cell proliferation, the first involving the activation of ERK1/2 and the second being Ca^{2+}-dependent and resulting in the sequential activation of the tyrosine kinase Pyk2 and the Ca^{2+}-regulated K^+ channels, BK_{Ca} (FIG. 4).

ACKNOWLEDGMENTS

This work was supported by an Italian Association for Cancer Research (AIRC 2004) grant to T.F.

REFERENCES

1. MULLER, A., B. HOMEY, H. SOTO, et al. 2001. Involvement of chemokine receptors in breast cancer metastasis. Nature **410:** 50–56.
2. BALKWILL, F. 2004. The significance of cancer cell expression of the chemokine receptor CXCR4. Semin. Cancer Biol. **14:** 171–179.
3. PORCILE, C., A. BAJETTO, S. BARBERO, et al. 2004. CXCR4 activation induces epidermal growth factor receptor transactivation in an ovarian cancer cell line. Ann. N.Y. Acad. Sci. **1030:** 162–169.
4. PORCILE, C., A. BAJETTO, F. BARBIERI, et al. 2005. Stromal cell-derived factor-1α (SDF-1α /CXCL12) stimulates ovarian cancer cell growth through the EGF receptor transactivation. Exp. Cell Res. **24:** 241–253.
5. GEMINDER, H., O. SAGI-ASSIF, L. GOLDBERG, et al. 2001. A possible role for CXCR4 and its ligand, the CXC chemokine stromal cell-derived factor-1, in the development of bone marrow metastases in neuroblastoma. J. Immunol. **167:** 4747–4757.
6. BARBERO, S., R. BONAVIA, A. BAJETTO, et al. 2003. Stromal cell-derived factor 1α stimulates human glioblastoma cell growth through the activation of both extracellular signal-regulated kinases 1/2 and Akt. Cancer Res. **63:** 1969–1974.
7. BAJETTO, A., R. BONAVIA, S. BARBERO, et al. 2001. Chemokines and their receptors in the central nervous system. Front. Neuroendocrinol. **22:** 147–184.
8. HALL, J.M. & K.S. KORACH. 2003. Stromal cell-derived factor 1, a novel target of estrogen receptor action, mediates the mitogenic effects of estradiol in ovarian and breast cancer cells. Mol. Endocrinol. **17:** 792–803.
9. SCOTTON, C.J., J.L WILSON, K. SCOTT, et al. 2002. Multiple actions of the chemokine CXCL12 on epithelial tumor cells in human ovarian cancer. Cancer Res. **62:** 5930–5938.
10. HELBIG, G., K.W CHRISTOPHERSON, II, P. BHAT-NAKSHATRI, et al. 2003. NF-κB promotes breast cancer cell migration and metastasis by inducing the expression of the chemokine receptor CXCR4. J. Biol. Chem. **278:** 21631–21638.
11. SALCEDO, R., K. WASSERMAN, H.A YOUNG, et al. 1999. Vascular endothelial growth factor and basic fibroblast growth factor induce expression of CXCR4 on human endothelial cells: *in vivo* neovascularization induced by stromal-derived factor-1alpha. Am. J. Pathol. **154:** 1125–1135.
12. GITTOES, N.J. 1998. Current perspectives on the pathogenesis of clinically nonfunctioning pituitary tumours. J. Endocrinol. **157:** 177–186.
13. ASA, S.L. & S. EZZAT. 1998. The cytogenesis and pathogenesis of pituitary adenomas. Endocr. Rev. **19:** 798–827.
14. FAGLIA, G. & A. SPADA. 2001. Genesis of pituitary adenomas: state of the art. J. Neurooncol. **54:** 95–110.
15. BANISADR, G., D. SKRZYDELSKI, P. KITABGI, et al. 2003. Highly regionalized distribution of stromal cell-derived factor-1/CXCL12 in adult rat brain: constitutive expression in cholinergic, dopaminergic and vasopressinergic neurons. Eur. J. Neurosci. **18:** 1593–1606.
16. FLORIO, T., B.A PERRINO, et al. 1996. Cyclic 3,5 adenosine monophosphate and cyclosporin A inhibit cellular proliferation and serine/threonine protein phosphatase activity in pituitary cells. Endocrinology **137:** 4409–4418.
17. SCHETTINI, G., T. FLORIO, O. MEUCCI, et al. 1989. Somatostatin inhibition of adenylate cyclase activity in different brain areas. Brain Res. **492:** 65–71.

18. FLORIO, T., M.G PAN, B. NEWMAN, et al. 1992. Dopaminergic inhibition of DNA synthesis in pituitary tumor cells is associated with phosphotyrosine phosphatase activity. J. Biol. Chem. **267:** 24169–24172.
19. FLORIO, T., S. THELLUNG, S. ARENA, et al. 1999. Somatostatin and its analog lanreotide inhibit the proliferation of dispersed human non-functioning pituitary adenoma cells *in vitro*. Eur. J. Endocrinol. **141:** 396–408.
20. GRYNKIEWICZ, G., M. POENIE, & R.Y. TSIEN. 1985. A new generation of Ca^{2+} indicators with greatly improved fluorescence properties. J. Biol. Chem. **260:** 3440–3450.
21. BARBERO, S., A. BAJETTO, R. BONAVIA, et al. 2002. Expression of the chemokine receptor CXCR4 and its ligand stromal cell-derived factor 1 in human brain tumors and their involvement in glial proliferation *in vitro*. Ann. N. Y. Acad. Sci. **973:** 60–69.
22. BAJETTO, A., R. BONAVIA, S BARBERO, et al. 1999. Expression of chemokine receptors in the rat brain. Ann. N. Y. Acad. Sci. **876:** 201–209.
23. BAJETTO, A., S. BARBERO, R. BONAVIA, et al. 2001. Stromal cell-derived factor-1α induces astrocyte proliferation through the activation of extracellular signal-regulated kinases 1/2 pathway. J. Neurochem. **77:** 1226–1236.
24. LIU, Q.H., D.A WILLIAMS, C. MCMANUS, et al. 2000. HIV-1 gp120 and chemokines activate ion channels in primary macrophages through CCR5 and CXCR4 stimulation. Proc. Natl. Acad. Sci. USA **97:** 4832–4837.
25. HUANG, M.H., S.N WU, C.P. CHEN, et al. 2002. Inhibition of Ca^{2+}-activated and voltage-dependent K^+ currents by 2-mercaptophenyl-1,4-naphthoquinone in pituitary GH3 cells: contribution to its antiproliferative effect. Life Sci. **70:** 1185–1203.
26. WANG, Z. & P. BRECHER. 2001. Salicylate inhibits phosphorylation of the nonreceptor tyrosine kinases, proline-rich tyrosine kinase 2 and c-Src. Hypertension **37:** 148–153.
27. BANISADR, G., E. DICOU, T. BERBAR, et al. 2000. Characterization and visualization of $[^{125}I]$ stromal cell-derived factor-1α binding to CXCR4 receptors in rat brain and human neuroblastoma cells. J. Neuroimmunol. **110:** 151–160.
28. ROLAND, J., B.J MURPHY, B. AHR, et al. 2003. Role of the intracellular domains of CXCR4 in SDF-1-mediated signaling. Blood **101:** 399–406.
29. SELA, U., S. GANOR, I. HECHT, A. BRILL, et al. 2004. Allicin inhibits SDF-1α-induced T cell interactions with fibronectin and endothelial cells by down-regulating cytoskeleton rearrangement, Pyk-2 phosphorylation and VLA-4 expression. Immunology **111:** 391–399.
30. BYRON, K.L. & P.A. LUCCHESI. 2002. Signal transduction of physiological concentrations of vasopressin in A7r5 vascular smooth muscle cells A role for PYK2 and tyrosine phosphorylation of K^+ channels in the stimulation of Ca^{2+} spiking. J. Biol. Chem. **277:** 7298–7307.

Insulin Primes Human Neutrophils for CCL3-Induced Migration

Crucial Role for JNK 1/2

FABRIZIO MONTECUCCO, GIORDANO BIANCHI, MARIA BERTOLOTTO, GIORGIO VIVIANI, FRANCO DALLEGRI, AND LUCIANO OTTONELLO

Clinic of Internal Medicine I, Department of Internal Medicine, University of Genoa, Medical School, Genoa, Italy

ABSTRACT: The present article shows that a short-term exposure of purified human neutrophils to recombinant insulin conferred on these cells both the ability to migrate and the capacity to mobilize $[Ca^{2+}]i$ in response to CCL3, a chemokine *per se* ineffective with native neutrophils. Furthermore, the effects of recombinant insulin were reproduced by short-term incubation with sera from adult patients with metabolic syndrome, known to be characterized by a hyperinsulinemic state. A strict linear correlation ($P < 0.01$) between sera insulin levels and sera's ability to induce neutrophil locomotion was indeed found. Our data also suggest that (i) insulin primed neutrophils for migration to CCL3 via the selective activation of JNK 1/2, as shown by the use of inhibitors and kinase activation assay; (ii) the activation of Src kinases was necessary but not sufficient for CCL3-induced locomotory activity; (iii) PI3K-Akt, ERK 1/2, and p38 MAPK were not involved in insulin-induced migratory competence. In summary, we provided evidence that the exposition of neutrophils to insulin, as it occurs in hyperinsulinemic conditions, confers the competence of the cells to migrate in response to CCL3, known to be generated near atherosclerotic plaques. As neutrophils have been recently suggested to be involved in breaking unstable atherosclerotic plaques, the present findings contribute to the understanding of the pathophysiology of plaque instability. Finally, biochemical analysis herein carried out raises the hypothesis of JNK 1/2 as an attractive therapeutic target.

KEYWORDS: neutrophil; insulin; migration; CCL3; MAPK; Src kinases; metabolic syndrome; atherosclerosis

Address for correspondence: Luciano Ottonello, M.D., Dipartimento di Medicina Interna e Specialità Mediche, viale Benedetto XV n. 6, I-16132 Genova, Italy. Voice: +39-010-3538686; fax: +39-010-3538686.
e-mail: otto@unige.it

INTRODUCTION

Recent studies have shown that neutrophils also play an important role in atherosclerosis, by infiltrating and breaking unstable atherosclerotic plaques.[1] Nevertheless, the mechanisms underlying neutrophil recruitment at sites of atheroma are largely undefined.[2] On the other hand, the metabolic syndrome, among various conditions associated with increasing rate of atherosclerotic lesions, is characterized by high level of circulating insulin.[3,4] Because insulin is also known to be capable of favoring actin reorganization in neutrophils, this hormone might be a factor involved in the control of migration of neutrophils to arterial walls.[5] Consequently, we studied the capacity of insulin to modulate neutrophil migration, using as chemotactic factor CCL3, a chemokine produced in atherosclerotic plaques and *per se* incapable of promoting neutrophil locomotion.[6,7]

PATIENTS AND METHODS

Neutrophils, obtained from healthy volunteers after informed consent, were isolated by dextran sedimentation and subsequent centrifugation on a Ficoll-Hypaque density gradient. Contaminating erythrocytes were removed by hypotonic lysis. Serum was obtained from 11 adult (age 25–65 years) patients with metabolic syndrome (diagnosed by AT III and WHO criteria) and from four healthy controls after informed consent.[4] Insulin serum concentrations were determined by radioimmune assay with insulin Bridge Kit from Adaltis (Montreal, Quebec, Canada). Neutrophil locomotion was studied in duplicate according to the leading front method using blind well chambers (NeuroProbe, Cabin John, MD) with a 3-μm pore-size cellulose ester filter (Millipore, Milan, Italy) separating the cells from the chemoattractant. After incubation in the absence or presence of appropriate reagents, the distance (μm) traveled by the leading front of cells was measured at ×400 magnification.[7] Intracellular $[Ca^{++}]_i$ was determined in fura-2 AM-loaded neutrophils by monitoring fluorescence changes before and after addition of CCL3, in the absence or in presence of insulin.[8] In selective experiments, neutrophils were preincubated with PD098059 (MEK, a kinase directly activating ERK, inhibitor, from Biomol Research Laboratories, Plymouth Meeting, PA), SB203580 (p38 MAPK inhibitor, Biomol Research Laboratories, Inc.), LY294002 (PI3K inhibitor, Sigma Chimica, Milano, Italy), 4-amino-1-teert-butyl-3-(1′-naphtyl)pyrazolo [3,4-α]pyrimidine (PP1 analogue, Src kinase inhibitor, Calbiochem, San Diego, CA), 1L-6-hydroxymethyl-*chiro*-inositol 2 [(R)-2-O-methyl-3O-octadecylcaromate (Akt inhibitor, Calbiochem), and SP600125 (JNK inhibitor, Calbiochem). The activation of intracellular kinases was studied by Western blot and *in vitro* kinase assays with the appropriate specific antibodies (the antiphosphorylated ERK AF1018 and antiphosphorylated p38 MAPK AF869 Abs were from R&D

System, Minneapolis, MN; the antiphosphorylated Akt 1/2/3 sc-16646-R, the anti-Hck N-30 and the antiphosphorylated JNK 1/2 sc-6254 mAbs were from Santa Cruz Biotechnology, Santa Cruz, CA).[7] Insulin was from Lilly France S.A.S. TAK-779 was kindly donated by Takeda Chemical Industries, Osaka, Japan. CCL3 was from PeproTech Inc., Rocky Hill, NJ. Differences between groups were analyzed by one-way ANOVA with Bonferroni's post test. Correlations were calculated by the Spearman test. Statistical analyses were performed using GraphPad InStat version 3.05 for Windows XP, GraphPad Software (San Diego, CA). Differences were accepted as significant when $P < 0.05$.

RESULTS

Insulin Primes Human Neutrophils for Migration to CCL3

Although freshly purified neutrophils from healthy volunteers did not migrate toward CCL3, the cells became capable of migrating to the chemokine (30 nM) when preincubated for 15 min with different doses of insulin in a dose-dependent manner (FIG. 1A). Consistent with these findings, neutrophils treated with 300 μU/mL of insulin and exposed to different doses of CCL3 displayed the bell-shaped dose–response curve characteristic for chemoattractants (data not shown).

Insulin Primes Neutrophils for Migration to CCL3 via CCR5

CCL3 is a specific ligand for two different CCRs (CCR1 and CCR5). We have previously shown that freshly purified human neutrophils express CCR5, but not CCR1.[7] In agreement with these findings, FIGURE 1B shows that the selective antagonist of CCR5, TAK-779 (10 nM), significantly inhibited the migration of insulin-treated neutrophils to CCL3. Furthermore, after insulin preincubation, neutrophils acquired the capacity to migrate also to CCL4, the selective CCR5 agonist devoid of CCR1 affinity, in a TAK-779-inhibitable manner (data not shown).

Insulin Primes Human Neutrophils for CCL3-Induced Intracellular Calcium Mobilization

FIGURE 1C shows that CCL3 stimulation did not induce any increase of $[Ca^{2+}]i$ in neutrophils (upper line). Similarly, no $[Ca^{2+}]i$ mobilization was observed in response to insulin (lower line). On the contrary, after exposure to 300 μU/mL of insulin, neutrophils mounted a rapid increase of $[Ca^{++}]_i$ in response to CCL3 (lower line). Insulin-treated neutrophils also mounted a rapid increase of $[Ca^{++}]_i$ in response to CCL4 (data not shown). In accord with these data, TAK-779 inhibited $[Ca^{++}]_i$ induced by CCL3 in neutrophils exposed to insulin (data not shown).

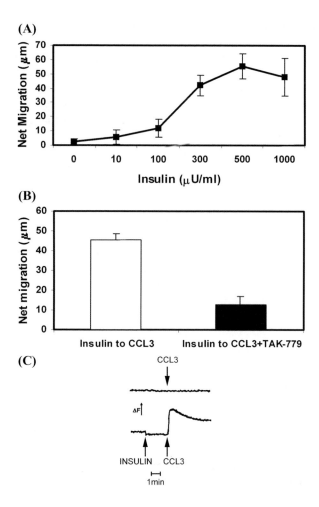

FIGURE 1. (**A**) Effects of different doses of insulin on neutrophils, locomotory response to 30 nM CCL3. Neutrophils were incubated with insulin for 15 min, washed once, resuspended in medium and exposed to 30 nM CCL3. Results are expressed as net migration obtained by subtracting spontaneous locomotion from CCL3-induced locomotion (mean ± SEM, $n = 3$). (**B**) Effects of the selective CCR5 antagonist TAK-779 on insulin-induced neutrophil migration toward CCL3. Neutrophils were incubated with 300 μU/mL insulin for 15 min, washed once, resuspended in medium and exposed 30 nM CCL3 in absence (*open bar*) and presence (*black bar*) of 10 nM TAK-779. Results are expressed as net migration obtained by subtracting spontaneous locomotion from CCL3-induced locomotion. Insulin-treated neutrophil migration to CCL3 versus insulin-treated neutrophil migration to CCL3 plus TAK-779: $P < 0.05$ (mean ± SEM, $n = 3$). (**C**) Effects of insulin and CCL3 on neutrophil intracellular calcium levels. Neutrophils (2.5×10^6) were loaded with fura-2 AM and fluorescence changes (ΔF) were monitored before and after exposure to 30 nM CCL3 (*upper line*) and to 300 μU/mL insulin, followed by 30 nM CCL3 (*lower line*). Arrows indicates the stimulus addition. Results represent one of three experiments that yielded similar results.

Insulin Primes Neutrophils for CCL3 Migration via JNK 1/2 Activation

FIGURE 2A shows that the selective inhibitor of JNK 1/2 SP600125 at the concentration of 20 μM efficiently inhibited CCL3-induced migration of neutrophils exposed to insulin. Furthermore, FIGURE 2B shows that insulin (lane 3), but not CCL3 (lane 2), was capable of inducing JNK 1/2 phosphorylation. These data suggest that insulin priming for CCL3-mediated neutrophil migration is related to insulin's ability to activate JNK 1/2.

Src Kinase Activation Is Involved in Insulin-Mediated Neutrophil Migration to CCL3

FIGURE 2A shows that the Src kinase inhibitor PP1 analogue (20 μM) inhibited insulin-mediated neutrophil migration to CCL3. On the other hand, when the activation of Hck (a member of Src kinases) was examined by the kinase assay (FIG. 2B), two well-detectable bands of autophosphorylation and enolase phosphorylation were found after CCL3 (lane 2) and insulin (lane 3) triggering.

PI3K-Akt, p38 MAPK, and ERK 1/2 Activation Are Not Involved in Insulin-Mediated Neutrophil Migration to CCL3

We next examined the PI3K-Akt system, a major pathway involved in several functional responses of neutrophils.[9] As shown in FIGURE 2A, 10 μM LY294002 (PI3K inhibitor), 40 μM AktI (Akt inhibitor), 1 μM SB203580 (p38 MAPK inhibitor), and 25 μM PD098059 (a selective inhibitor of MEK, a kinase directly activating ERK1/2) did not inhibit migration of insulin-treated neutrophils to CCL3. Furthermore, as shown in FIGURE 2B, both CCL3 (lane 2) and insulin (lane 3) induced Akt activation. Finally, insulin was incapable of activating p38 MAPK and ERK1/2 (lane 3), while CCL3 (lane 2) activated p38 and, partially, ERK 1/2.

Sera from Metabolic Syndrome Patients Induce Neutrophil Migration to CCL3

Neutrophils exposed to sera from healthy controls were incapable of migrating to 30 nM CCL3 (net migration 8.9 ± 3.5 μm/45 min, $x \pm$ SEM, $n = 4$). On the other hand, after treatment with sera from patients with the metabolic syndrome, neutrophil migration to 30 nM CCL3 was 30.73 ± 4.09 μm/45 min ($x \pm$ SEM, $n = 11$, $P < 0.05$). Insulin concentration in sera from metabolic syndrome patients was 34.18 ± 4.81 μU/mL, $x \pm$ SEM, $I = 11$, range 13.8–67.5 μU/mL. A direct correlation was found between net migration to CCL3 of neutrophils treated with metabolic syndrome sera and insulin concentrations in the same serum samples (Spearman's $r = 0.7974$; $P = 0.0033$).

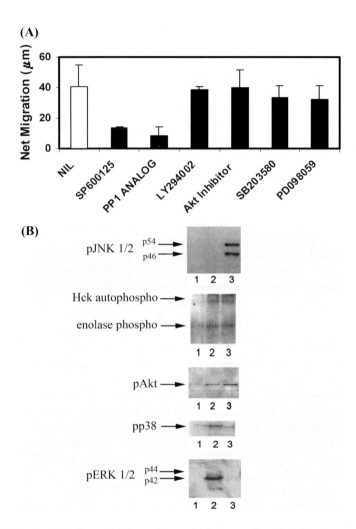

FIGURE 2. (**A**) Effects of different kinase inhibitors on insulin-primed neutrophil migration to CCL3. Neutrophils were preincubated with medium (*open bars*) or in presence (*black bars*) of 20 μM SP600125 (JNK inhibitor), 20 μM PP1 analogue (Src kinase inhibitor), 10 μM LY294002 (PI3K inhibitor), 40 μM Akt inhibitor, 1 μM SB203580 (p38 MAPK inhibitor), 25 μM PD098059 (MEK, a kinase activating ERK inhibitor) for 30 min. Then 300 μU/mL insulin was added for 15 min. After washing once, neutrophils were resuspended in medium and exposed to 30 nM CCL3. Results are expressed as net migration obtained by subtracting spontaneous locomotion from CCL3-induced locomotion. Migration in absence versus migration in presence of SP600125 or PP1analogue: $P < 0.05$. Migration in absence versus migration in presence of LY294002, Akt inhibitor, SB20358, or PD098059: n. s. (mean ± SEM, $n = 3$). (**B**) *In vitro* kinase assay and Western blot analysis of kinase activation in neutrophils. Neutrophils were treated with control medium (*lane 1*), CCL3 (30 nM, 5 min, *lane 2*) or insulin (300 μU/mL, 3 min, *lane 3*). Results represent one of three experiments that yielded similar results.

DISCUSSION

In the present article, we show that a short-term incubation of human neutrophils with insulin renders the cells capable of migrating in response to CCL3. This is in agreement with previous observations showing that insulin induces F-actin polymerization in human neutrophils.[9] In other words, a short-term exposure of neutrophils to insulin switches on the locomotory responsiveness of the cells to CCL3, known to be generated in arterial walls near atherosclerotic plaques.[6,10] CCL3 is the specific ligand for CCR1 and CCR5.[11] CCL3-mediated locomotory response herein observed depends on CCR5 solely.[7] This is suggested by two different sets of data: (i) the selective CCR5 antagonist TAK-779 blocked the migration of insulin-treated neutrophils toward CCL3; (ii) insulin predisposed neutrophils to migrate also to CCL4, that is, the CCR5 specific chemokine devoid of activity toward CCR1.

It is well-known that classical neutrophil chemoattractants induce $[Ca^{++}]_i$ mobilization.[12] Our data show that, although neither insulin nor CCL3 induces calcium fluxes in neutrophils, a short-term exposure of the cells to insulin resulted in a subsequent and well-detectable spike of $[Ca^{++}]_i$ mobilization by CCL3. As for the locomotory response, $[Ca^{++}]_i$ mobilization in insulin-treated neutrophils is CCR5-dependent, as suggested by the capacity of TAK-779 to prevent CCL3 as well as CCL4-mediated $[Ca^{++}]_i$ flux.

To understand the intracellular molecular processes responsible for the ability of insulin to capacitate CCR5 responsiveness to its ligands, we focused on various kinases known to be involved in neutrophil locomotory response.[13,14] Our data show that insulin activity is JNK 1/2 dependent. This is suggested by two independent findings: (i) insulin strongly activated neutrophil JNK 1/2, as detected by Western blot analysis; (ii) the JNK inhibitor SP600125 prevented the ability of insulin to promote neutrophil response to CCL3. These findings are in agreement with the general concept of JNKs as a link between inflammatory responses and metabolic diseases.[15]

Although Src kinases are actually involved in insulin activity, their activation is insufficient to give the locomotory competence to neutrophils without the intervention of JNK. In fact, Src kinases were found to be directly activated by CCL3, *per se* incapable of promoting neutrophil migration in the absence of insulin priming, that is, in the absence of JNK activation. On the other hand, as stated above, insulin action is independent of PI3K-Akt, p38 MAPK, and ERK 1/2 activation. In fact, the selective inhibitors of these kinases were found to be incapable of blocking insulin action on neutrophils. Consequently, although insulin was found to induce Akt phosphorylation, the activation of this kinase has no direct role as far as the locomotory response is concerned. This is also consistent with recent studies showing that PI3K-Akt is not activated by insulin in neutrophils.[9]

It is of note that the herein observed activity of insulin can be reproduced by using sera from patients suffering from metabolic syndrome, which is characterized by high insulinemic levels.[4] In fact, sera from these patients were found to promote the locomotory competence in normal neutrophils to CCL3. However, as insulin levels in these sera were found to be lower than the concentrations of recombinant insulin used in our *in vitro* assays, it is likely that the observed effects of sera may reflect a cooperative action of insulin together with other presently unidentified agents. On the other hand, as far as neutrophils are concerned, recombinant insulin might have also an activity lower than that of the natural hormone.

In summary, as neutrophils have been recently suggested to be involved in breaking unstable atherosclerotic plaques, the present findings contribute to the understanding of the pathophysiology of the plaque instability.[1,16] In particular, our data provide an explanation for the mechanisms whereby circulating neutrophils can be recruited in arterial walls. Moreover, biochemical analysis herein carried out calls attention to JNK 1/2 as an attractive and possible therapeutic target.

REFERENCES

1. BUFFON, A. *et al.* 2002. Widespread coronary inflammation in unstable angina. N. Engl. J. Med. **347:** 5–12.
2. CHARO, F.I. & M.B. TAUBMAN. 2004. Chemokines in the pathogenesis of vascular disease. Circ. Res. **95:** 858–866.
3. DANDONA, P. *et al.* 2005. Metabolic syndrome: a comprehensive perspective based on interactions between obesity, diabetes, and inflammation. Circulation **111:** 1448–1454.
4. ECKEL, R.H., S.M. GRUNDY & P.Z. ZIMMET. 2005. The metabolic syndrome. Lancet **365:** 1415–1428.
5. KUTSUNA, H. *et al.* 2004. Actin reorganization and morphological changes in human neutrophils stimulated by TNF, GM-CSF and G-CSF: role of mitogen-activated protein kinases. Am. J. Physiol. Cell. Physiol. **286:** 55–64.
6. HAYES, I.M. *et al.* 1998. Human vascular smooth muscle cells express receptors for CC chemokines. Arterioscler. Thromb. Vasc. Biol. **18:** 397–403.
7. OTTONELLO, L. *et al.* 2005. CCL3 (Mip-1α) induces in vitro migration of GM-CSF-primed human neutrophils via CCR5-dependent activation of ERK 1/2. Cell Signal. **17:** 355–363.
8. OTTONELLO, L. *et al.* 2004. Leptin as a uremic toxin interferes with neutrophil chemotaxis. J. Am. Soc. Nephrol. **15:** 2366–2372.
9. CHODNIEWICZ, D. & D. V. ZHELEV. 2003. Novel pathways of F-actin polymerization in the human neutrophil. Blood **102:** 2251–2258.
10. HAYES, I.M. *et al.* 1999. Human vascular smooth muscle cells express receptors for CC chemokines. Arterioscler. Thromb. Vasc. Biol. **18:** 397–403.
11. MURPHY, P.M. *et al.* 2000. International Union of Pharmacology. XXII. Nomenclature for chemokine receptors. Pharmacol. Rev. **52:** 145–176.
12. ROSE, J.J. *et al.* 2004. On the mechanism and significance of ligand-induced internalisation of human neutrophil chemokine receptors CXCR1 and CXCR2. J. Biol. Chem. **279:** 24372–24386.

13. ZU, Y.L. et al. 1998. P-38 mitogen-activated protein kinase activation is required for Human neutrophil function triggered by TNF-α or FMLP stimulation. J. Immunol. **160:** 1982–1989.
14. XU, J. et al. 2005. Neutrophil microtubules suppress polarity and enhance directional migration. Proc. Natl. Acad. Sci. USA **102:** 6884–6889.e
15. LIU, G. & C.M. RONDINONE. 2005. JNK: bridging the insulin signaling and inflammatory pathway. Curr. Opin. Investig. Drugs **6:** 979–987.
16. NARUKO, T. et al. 2002. Neutrophil infiltration of culprit lesions in acute coronary syndromes. Circulation **106:** 2894–2900.

Doxorubicin-Induced MAPK Activation in Hepatocyte Cultures Is Independent of Oxidant Damage

ROSAURA NAVARRO, ROSA MARTÍNEZ, IDOIA BUSNADIEGO, M. BEGOÑA RUIZ-LARREA, AND JOSÉ IGNACIO RUIZ-SANZ

Department of Physiology, Faculty of Medicine and Dentistry, University of the Basque Country, 48080-Bilbao, Spain

ABSTRACT: Doxorubicin (DOX) is a potent anticancer drug, whose clinical use is limited on account of its toxicity. DOX cytotoxic effects have been associated with reactive oxygen species (ROS) generated during drug metabolism. ROS induce signaling cascades leading to changes in the phosphorylation status of target proteins, which are keys for cell survival or apoptosis. The mitogen-activated protein kinase (MAPK) cascades are routes activated in response to oxidative stress. In this work, the effects of DOX on cytotoxicity, indicators of oxidative stress (malondialdehyde -MDA- and GSH), and the phosphorylation status of extracellular signal-regulated kinases (ERKs), c-Jun N-terminal kinases (JNKs), and p38 kinases were analyzed in primary cultures of rat hepatocytes. DOX (1–50 μM) did not modify lactate dehydrogenase (LDH) release into the medium, the levels of MDA (determined by high-performance liquid chromatography [HPLC]) or the intracellular GSH during the incubation time up to 6 h. GSH levels from mitochondria extracted by Percoll gradient from cultured hepatocytes were not modified by DOX, thus excluding its depletion or any impaired mitochondrial uptake. Characterization of proteins by Western blot analysis revealed that DOX increased phosphorylation of p38 kinases and JNK1 and JNK2 in a dose- and time-dependent manner. DOX also increased ERK2 phosphorylation at latter time points. In conclusion, DOX triggers activation of ERK, JNK, and p38 kinases in primary cultures of rat hepatocytes independently of oxidant damage.

KEYWORDS: doxorubicin; mitogen-activated kinase; Western blot; rat hepatocyte culture

Address for correspondence: M. Begoña Ruiz-Larrea, Department of Physiology, Medicine and Dentistry School, University of the Basque Country, 48080-Bilbao, Spain. Voice: +34-946012829; fax: +34-946015662.
 e-mail: mbego.ruizlarrea@ehu.es

Ann. N.Y. Acad. Sci. 1090: 408–418 (2006). © 2006 New York Academy of Sciences.
doi: 10.1196/annals.1378.044

INTRODUCTION

Reactive oxygen species (ROS) are continuously generated in humans either as a consequence of the cellular intrinsic metabolism or derived from pro-oxidant exposure. Transient fluctuations in ROS elicit a wide variety of responses, such as the upregulation of adhesion molecules, inflammatory responses, and cell survival, through regulation of cell signaling.[1] ROS activate signaling cascades, leading to changes in the phosphorylation status of target proteins, which are keys for cell survival or apoptosis. However, when present at high and/or sustained levels, ROS can cause severe damage to lipids, proteins, and DNA. Cells contain antioxidant defenses that minimize ROS fluctuations, but often ROS generation exceeds the antioxidant capacity, which originates a state termed *oxidative stress*. Cell survival depends on the capacity for adapting or resisting the stress, and, thus, numerous responsive mechanisms have evolved that are rapidly activated against these aggressions. Several routes are associated preferentially with a higher survival, while others are associated with cell death. Among the routes that are activated in response to oxidative stress, a signal transduction pathway is the mitogen-activated protein kinase (MAPK) cascade. MAPKs constitute a large number of serine/threonine kinases involving three major subfamilies, which can be classified as: extracellular signal-regulated kinases (ERKs), c-Jun N-terminal kinases (JNKs), and the p38 kinases. ERKs, JNKs and p38 are activated via independent, although sometimes overlapping, signaling cascades involving protein phosphorylations catalyzed by other kinases upstream. The ERK pathway is most commonly linked to the regulation of cell proliferation, while the JNK and p38 kinase pathways are more strongly associated with stress.[2] For this reason, JNK and p38 kinases are often grouped together and referred to as stress-activated protein kinases (SAPKs). ERKs, JNKs, and p38 kinases have been shown to be activated in response to oxidant injury in several cell types.

Doxorubicin (DOX) is a potent anticancer drug, with limited clinical use, because of the toxicity it produces in several cells and tissues. This toxicity has been ascribed to the reactive species derived from the metabolic bioactivation of the anthracycline. DOX is predominantly metabolized in hepatocytes, where it undergoes reversible oxido-reduction reactions with formation of a semiquinone intermediate and other ROS.[3] In this work, we have studied in primary cultures of rat hepatocytes the effects of DOX on (*a*) cytotoxicity and indicators of oxidative stress (malondialdehyde and GSH), and (*b*) the activation of mitogen-activated kinase cascades, including ERK, p38 kinases, and JNK.

MATERIALS AND METHODS

Reagents

Tissue culture media were from SEROMED-Biochrom (LKB Biochrom Ltd., Germany). Gentamycin was from Gibco. Butylated hydroxytoluene

(BHT), 1,1,3,3-tetraethoxypropane, thiobarbituric acid (TBA), RPMI 1640, amphotericin B, penicillin, streptomycin, dexamethasone, and insulin were from Sigma (St. Louis, MO, USA). Standard analytical grade laboratory reagents were obtained from Merck (Darmstadt, Germany). Collagenase was from Boheringer (Mannheim, Germany). Doxorubicin was from TEDEC-MEIJI FARMA, S.A. (Madrid, Spain). The antibodies used in this study, anti-p38 MAPK and anti-p44/42 MAPK, and the secondary antibodies were purchased from Cell Signaling Technology, Inc. Anti-SAPK/JNK antibody was from Promega (Madison, WI, USA). High-performance liquid chromatography (HPLC) grade acetonitrile was purchased from Scharlau (Spain).

Preparation of Hepatocyte Primary Cultures and DOX Treatment

Hepatocytes were isolated from male Sprague–Dawley rats (180–200 g) by collagenase perfusion, as described previously.[4] The hepatocyte viability was determined by means of the trypan blue exclusion test and was typically greater than 90%. Purified hepatocytes were plated on 100×20 mm culture dishes (Falcon) at a density of 10×10^6 cells/plate in RPMI 1640 adhesion buffer, pH = 7.4, containing fetal bovine serum (10%), glucose (2 mg/mL), pyruvate (0.11 mg/mL), dexamethasone (1 μM), gentamycin (5 mg/L), penicillin:streptomycin (5 U/mL: 5 mg/mL), amphotericin B (0.5 μg/mL), and insulin (0.7 μM). After 5–6-h incubation at 37°C in a 5% CO_2, 95% air atmosphere, the medium was replaced with fresh medium without amphotericin and supplemented with 0.2% fatty acid free bovine serum albumin and 10 nM insulin. Cultures were equilibrated for 12 h at 37°C in a 5% CO_2, 95% air atmosphere. Reactions started by the addition of DOX (10–20–50 μM), dissolved in the incubation medium, to hepatocyte cultures. At time intervals cells were harvested, washed in PBS, and collected by centrifugation.

Mitochondrial Isolation

After incubation of cells, hepatocytes were resuspended in 1 mL of sonication buffer (10 mM Tris/HCl, pH 7.4, plus 250 mM sucrose and 2 mM EGTA). Samples were sonicated at 6–7 μm for 3 sec during 3 cycles with intervals of 3 sec, then spinned, and the supernatants were used for isolating mitochondria in Percoll gradient, as described by Vargas.[5] The final pellet was resuspended in 200 μL of sonication buffer. One aliquot was taken for protein determination,[6] and the rest was used for determining GSH.[7]

Cell Viability

The cell viability was measured in terms of lactate dehydrogenase (LDH) activity leakage into the medium.[8] Viability was expressed as the percentage of cellular LDH activity respecting total (cells + medium) LDH activity.

Total and Mitochondrial GSH Determinations

After incubations, GSH was determined in supernatants of perchlorized hepatocytes as described previously.[7] For the mitochondrial GSH analysis, perchloric acid (20%) plus 2 mM EDTA was added to the mitochondrial suspensions. The tubes were vortexed and centrifuged at 12,500 × g for 5 min at 4°C. Supernatants were used for determining GSH. Mitochondrial GSH was expressed as nmol/mg protein.

Malodialdehyde (MDA) Analysis

MDA was measured by an isocratic HPLC method using UV-visible detection. Analyses were performed on a Kontron Instrument HPLC system. Detection was carried out using a UV-visible HPLC-Detector 430 (Kontron). The system was controlled, and the data were collected and integrated, using KromaSystem 2000 software.

For preparation of the samples, an aliquot of 50 μL (2.5 × 10^6 cells/mL) was mixed with 500 μL of 0.2% TBA in 2 M sodium acetate buffer pH 3.5, in the presence of 0.05% BHT and 1 mM EDTA. The sample mixture was heated in a water bath at 95°C for 60 min. After cooling, the samples were centrifuged at 16,000 g at 4°C for 10 min, and the supernatant used for HPLC analysis. Samples were filtered (0.45 μm) and 100 μL were injected into HPLC system using an autosampler. The MDA-(TBA)$_2$ adduct was separated using a C18 column (Water Spherisorb ODS2, 5-μm particle size, 150 × 4.6 mm) at ambient temperature. The column was protected by a guard column with the same packing. The mobile phase consisted in 50 mM KH_2PO_4 pH 6.8/acetonitrile (83/17, v/v) and was filtered (0.45 μm) and degassed before use. The flow rate was 1 mL/min and the MDA-(TBA)$_2$ adduct was detected at 532 nm. At the end of each HPLC run, the column was washed with acetonitrile/milliQ water (70/30, v/v) to eliminate carryover and to extend the column life. After the daily work, the column was washed with acetonitrile/milliQ water (17/83, v/v). The concentration of MDA was calculated from the peak area, based on a calibration curve prepared by using 1,1,3,3-tetra-ethoxypropane as MDA standard. Calibration curves were run daily. Values were expressed as nmol of MDA/million cells.

Characterization of Proteins by Western Blotting

Hepatocytes were lysed for 30 min at 4°C in 20 mM Hepes pH 7.5, containing 10 mM EGTA, 40 mM β-glycerophosphate, Nonidet P-40 (1%), 1 mM NaF, 2.5 mM $MgCl_2$, 2 mM $NaVO_4$, 1 mM DTT, protease inhibitor cocktail (Sigma, St. Louis, MO, USA). Samples were centrifuged at 14,000 g for 10 min at 4°C, and cell proteins collected in the supernatant. Protein content was assayed by the method of Bradford.[6]

Twenty-five micrograms of protein were loaded per lane on a 12% SDS-polyacrylamide gel. Proteins were separated electrophoretically and transferred to PVDF membranes (Immobilon-P, Millipore) using the semidry transfer system from BioRad. These membranes were blocked with 5% BSA in TBS (10 mM Tris-HCl, 150 mM NaCl), pH 7.5, 0.05% Tween 20 for 1 h. Primary antibodies against phospho-p44/42 MAPK (Thr202/Tyr204), phospho-p38 MAPK (Thr180/Tyr182), and phospho-JNK (dually phosphorylated in the peptide containing the Thr183/Pro184/Tyr185 motif of JNK) were applied overnight at dilutions of 1:1,000, 1:2,000 and 1:5,000, respectively. After washing, appropriate secondary antibodies (anti-IgG-peroxidase linked) were applied at 1:2,000 dilution for 1 h. Blots were washed, incubated in commercially enhanced chemiluminescence reagents (ECL, Amersham Pharmacia Biotech AB, Uppsala, Sweden), and exposed to autoradiographic film, and the bands were quantified by bioimage analysis. Membranes were then washed and incubated in stripping buffer (62.5 mM Tris-HCl, pH 6.7, 2% SDS, 100 mM 2-mercapthoethanol) at 50°C for 30 min. After washing, membranes were blocked and treated with primary antibodies against p38 MAPK, JNK, and p44/42 MAPK at dilutions of 1:1,000 and the secondary antibodies as described before. MAPK activation was expressed as the ratio of the phosphorylated form of the kinase and the total fraction.

Statistical Analysis

Data were expressed as mean ± SEM from three to nine experiments. Statistical significance was determined by the Student's *t*-test for paired data. A *P*-value less than 0.05 was considered to be statistically significant.

RESULTS

Cell Death, Total GSH, and Lipid Peroxidation

To determine the effects of DOX on hepatocyte indicators of oxidative stress, cell cultures were exposed to the drug (0–50 μM) for different periods up to 6 h and cell viability, total GSH, and lipid peroxidation determined. DOX did not modify either the LDH leakage into the medium (FIG. 1A) or endogenous GSH (FIG. 1B). Lipid peroxidation was estimated by measuring hepatocyte MDA by HPLC. As can be seen in FIGURE 1C, DOX had no effect on MDA.

Mitochondrial GSH

GSH is synthesized exclusively in cytosol and part of it is sequestered in mitochondria by a transport system.[9,10] Mitochondrial GSH plays an important

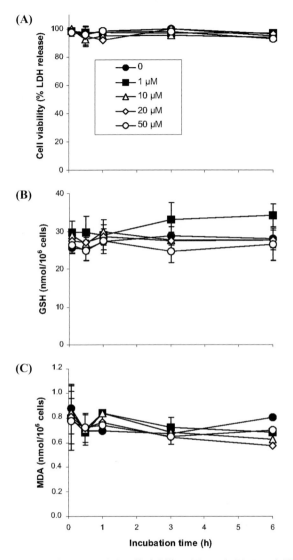

FIGURE 1. Effects of DOX on (**A**) cell viability, (**B**) total GSH, and (**C**) lipid peroxidation of primary cultures of rat hepatocytes. Hepatocyte cultures were treated with DOX at the indicated concentrations. At each time, cells were processed and LDH, GSH, and malondialdehyde were determined as described in MATERIALS AND METHODS. Data are mean ± SEM for two to five experiments.

role in antioxidative protection and the maintenance of vital cell functions,[11] its selective depletion having been demonstrated to be associated with the progression of liver damage.[12] We determined mitochondrial GSH levels from cultured hepatocytes exposed to DOX. The anthracyclin did not modify

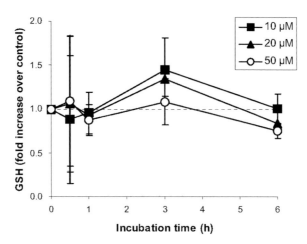

FIGURE 2. Mitochondrial GSH from cultured hepatocytes incubated with DOX. Cell cultures were treated with DOX at the indicated concentrations. At each time, mitochondrial fractions were separated from hepatocytes and GSH was determined as described in MATERIALS AND METHODS. Data are expressed as fold increase over controls. Data are mean ± SEM for three experiments.

mitochondrial GSH levels of cultured hepatocytes during 6 h (FIG. 2), thus excluding its depletion or any impaired mitochondrial uptake. Results indicated that DOX at the concentrations used did not induce cell toxicity or oxidant damage expressed in terms of total and mitochondrial GSH decreases and increased lipid peroxidation.

MAPKs Phosphorylation

To determine the effects of DOX on MAPKs in cultured hepatocytes, Western blot analysis was performed using the phospho- or the total forms of antibodies against the MAPKs, ERK2, p38, and JNK1 and JNK2. Cell lysates were prepared after 30 min, 1 h, 3 h, and 6 h of treatment with DOX (10, 20, and 50 μM). The anthracyclin stimulated phosphorylation of p38 MAPK and JNK in a dose- and time-dependent manner (FIG. 3). JNK1 and p38 activations were statistically different from controls after 30 min of treatment with the drug, while JNK2 significantly increased from the first time point. ERK2 phosphorylation showed a different profile. Thus, DOX increased ERK2 activity at latter time points (3 and 6 h), with maximal effect at 3 h.

DISCUSSION

The quinone DOX exerts toxic effects by multiple mechanisms. One of the hypotheses explaining the toxic effects involves the free radical-based

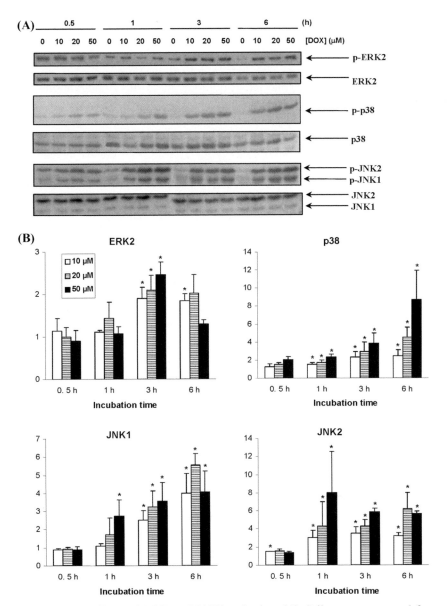

FIGURE 3. Effects of DOX on MAPK activation. (**A**) Cell extracts prepared from hepatocyte cultures treated with DOX (10, 20, and 50 μM) for the indicated times were subjected to inmmunoblot analysis using antibodies against phospho- or the total form of the MAPKs (ERK2, p38 MAPK, and JNK1 and JNK2). (**B**) Quantification of Western blot data. Levels of the active forms of MAPKs were normalized with respect to the total forms of kinases and expressed as relative fold increase in comparison to control samples, which were arbitrarily set to 1.0. Values correspond to the mean ± SEM of three to nine independent experiments.

mechanisms. Accordingly, the anthracycline exerts toxic actions by oxygen-free radicals generated in reactions catalyzed by reductases.[3] The single-electron transfer reaction, principally catalyzed by microsomal cytochrome P450 reductases, renders a semiquinone radical, which can be reoxidized by molecular oxygen, giving the parent quinone and a superoxide anion (O_2^-) (*redox cycle*). Superoxide anions are converted by superoxide dismutase (SOD) to hydrogen peroxide (H_2O_2), which, in the presence of iron generates the even more reactive and damaging species, the hydroxyl radical. Cells contain detoxifying enzymes that reduce O_2^- and H_2O_2 to water (SOD, catalase, and glutathione peroxidase). The cytotoxicity of DOX has been reported mainly in cardiomyocytes. Consistent with the free radical-based mechanisms, cardiomyocytes are not equipped with sufficient antioxidant enzymes and thus are subjected to DOX toxicity. The two-electron reduction of DOX, catalyzed by cytosolic NADPH-dependent reductases, renders doxorubicinol, the C13-ol-derived hydroquinone. Although doxorubicinol is considered to be redox-inactive toward oxygen, it exerts inhibitory actions on mitochondrial ATPase and other ionic transport systems, these actions having been involved in doxorubicinol cardiotoxicity.[13–15] Few reports have described toxic effects of DOX in other nontransformed cells. We showed that DOX did not induce cell toxicity or oxidative stress expressed in terms of lipid peroxidation or GSH depletion in cultured hepatocytes (FIGS.1 and 2). In the liver, the principal detoxifying organ, xenobiotics are converted into more polar and water-soluble products to be excreted in the urine. Hepatocytes are well equipped with antioxidant defenses; thus free radical-mediated processes would be avoided. DOX was shown to reduce the activities of P450 isoenzymes and also to increase specific cytosolic reductase activities in cultures of hepatocytes.[16] Such alterations in enzyme activities would favor the production of doxorubicinol. In contrast to cardiomyocytes, acute DOX treatment only slightly affected ATPase activity in hepatocytes and did not alter inner membrane activities to dramatic extents.[17] To our knowledge, no alterations by DOX or doxorubicinol have been described on ionic gradients in hepatocytes. As a toxin defense system, liver cells also enhance P-glycoprotein expression, a DOX-transport system, permitting the efflux of the drug out of the cell,[18,19] which alternatively could explain the lack of toxic effects on hepatocytes.

Despite the lack of effect on lipid peroxidation and intracellular GSH, DOX significantly stimulated MAPK cascades (ERK, p38 MAPK, JNK) (FIG. 3), which are involved in regulating cell survival. ERK2 phosphorylation was increased at the latter time point of 3 h. Although DOX did not induce oxidative stress, moderate production of ROS is not excluded. Mild concentrations of free radicals can also act as messengers capable of inducing signal transduction pathways. Recently, O_2^- has been reported to have a regulatory role in ERK signaling in a mouse myoblast cell line.[20] In these transformed cells, an O_2^- donor was able to induce ERK phosphorylation, while the SOD mimetic MnTBAP prevented its activation. A similar mechanism of MAPK activation elicited by O_2^- could also take place in our conditions. The crosstalk between DOX and the MAPK cascades in hepatocytes awaits further investigation.

ACKNOWLEDGMENTS

This work was supported by the Basque Government (Research Project and Predoctoral Training Grant to R.N.) and the University of the Basque Country (UPV00081.327-E-15294/2003). We are grateful to Montserrat Busto for her skillful technical assistance.

REFERENCES

1. MATSUZAWA, A. & H. ICHIJO. 2005. Stress-responsive protein kinases in redox-regulated apoptosis signaling. Antioxid. Redox Signal. **7:** 472–481.
2. MARTINDALE, J.L. & N.J. HOLBROOK. 2002. Cellular response to oxidative stress: signaling for suicide and survival. J. Cell. Physiol. **192:** 1–15.
3. MINOTTI, G., P. MENNA, E. SALVATORELLI, et al. 2004. Anthracyclines: molecular advances and pharmacologic developments in antitumor activity and cardiotoxicity. Pharmacol. Rev. **56:** 185–229.
4. RUIZ-LARREA, M.B., M.J. GARRIDO & M. LACORT. 1993. Estradiol-induced effects on glutathione metabolism in rat hepatocytes. J. Biochem. **113:** 563–567.
5. VARGAS, A.M. 1982. Rapid preparation of metabolically active mitochondria from control and hormone-treated rat liver cells. J. Biochem. Biophys. Methods **7:** 1–6.
6. BRADFORD, M.M. 1976. A rapid and sensitive method for the quantitation of microgram quantities of protein utilizing the principle of protein dye binding. Anal. Biochem. **72:** 248–254.
7. BRIGELIUS, R., C. MUCKEL, T.P. AKERBOOM & H. SIES. 1983. Identification and quantitation of glutathione in hepatic protein mixed disulfides and its relationship to glutathione disulfide. Biochem. Pharmacol. **32:** 2529–2534.
8. FARBER, J.L. & E.E. YOUNG. 1981. Accelerated phospholipid degradation in anoxic rat hepatocytes. Arch. Biochem. Biophys. **211:** 312–330.
9. MARTENSSON, J., J.C. LAI & A. MEISTER. 1990. High-affinity transport of glutathione is part of a multicomponent system essential for mitochondrial function. Proc. Natl. Acad. Sci. USA **87:** 7185–7189.
10. FERNANDEZ-CHECA, J.C., C. GARCIA-RUIZ, M. OOKHTENS & N. KAPLOWITZ. 1991. Impaired uptake of glutathione by hepatic mitochondria from chronic ethanol-fed rats. Tracer kinetic studies *in vitro* and *in vivo* and susceptibility to oxidant stress. J. Clin. Invest. **87:** 397–405.
11. NAGAI, H., K. MATSUMARU, G. FENG & N. KAPLOWITZ. 2002. Reduced glutathione depletion causes necrosis and sensitization to tumor necrosis factor-alpha-induced apoptosis in cultured mouse hepatocytes. Hepatology **36:** 55–64.

12. HIRANO, T., N. KAPLOWITZ, H. TSUKAMOTO, et al. 1992. Hepatic mitochondrial glutathione depletion and progression of experimental alcoholic liver disease in rats. Hepatology **16:** 1423–1427.
13. BOUCEK, R.J.J., R.D. OLSON, D.E. BRENNER, et al. 1987. The major metabolite of doxorubicin is a potent inhibitor of membrane-associated ion pumps. A correlative study of cardiac muscle with isolated membrane fractions. J. Biol. Chem. **262:** 15851–15856.
14. OLSON, R.D., P.S. MUSHLIN, D.E. BRENNER, et al. 1988. Doxorubicin cardiotoxicity may be caused by its metabolite, doxorubicinol. Proc. Natl. Acad. Sci. USA **85:** 3585–3589.
15. DODD, D.A., J.B. ATKINSON, R.D. OLSON, et al. 1993. Doxorubicin cardiomyopathy is associated with a decrease in calcium release channel of the sarcoplasmic reticulum in a chronic rabbit model. J. Clin. Invest. **91:** 1697–1705.
16. SCHROTEROVA, L., H. KAISEROVA, V. BALIHAROVA, et al. 2004. The effect of new lipophilic chelators on the activities of cytosolic reductases and P450 cytochromes involved in the metabolism of anthracycline antibiotics: studies in vitro. Physiol. Res. **53:** 683–691.
17. BAROGI, S., A. BARACCA, M. CAVAZZONI, et al. 2000. Effect of the oxidative stress induced by adriamycin on rat hepatocyte bioenergetics during ageing. Mech. Ageing Dev. **113:** 1–21.
18. LE BOT, M.A., D. KERNALEGUEN, J. ROBERT, et al. 1994. Modulation of anthracycline accumulation and metabolism in rat hepatocytes in culture by three revertants of multidrug resistance. Cancer Chemother. Pharmacol. **35:** 53–58.
19. ALBERTUS, J.A. & R.O. LAINE. 2001. Enhanced xenobiotic transporter expression in normal teleost hepatocytes: response to environmental and chemotherapeutic toxins. J. Exp. Biol. **204:** 217–227.
20. HUANG, W.C., C.C. CHIO, K.H. CHI, et al. 2002. Superoxide anion-dependent Raf/MEK/ERK activation by peroxisome proliferator activated receptor gamma agonists 15-deoxy-delta(12,14)-prostaglandin J(2), ciglitazone, and GW1929. Exp. Cell. Res. **277:** 192–200.

Superoxide Anions Are Involved in Doxorubicin-Induced ERK Activation in Hepatocyte Cultures

ROSAURA NAVARRO, IDOIA BUSNADIEGO, M. BEGOÑA RUIZ-LARREA, AND JOSÉ IGNACIO RUIZ-SANZ

Department of Physiology, Medicine and Dentistry School, University of the Basque Country, 48080-Bilbao, Spain

ABSTRACT: Doxorubicin (DOX), an antineoplastic agent widely used for the treatment of cancer, belongs to the anthracycline family of antitumor antibiotics. DOX may undergo one-electron reduction to the corresponding semiquinone free radical by flavin-containing reductases. Under aerobic conditions, the semiquinone radical reacts rapidly with oxygen to generate superoxide anion, undergoing redox cycling. At moderate concentrations, reactive oxygen species (ROS) play an important role as regulatory mediators in signaling processes. We have shown that DOX increased phosphorylation of enzymes comprising mitogen-activated protein (MAP) kinase cascades in primary hepatocyte cultures, and that this action was independent of oxidant damage. In particular, extracellular signal-regulated kinase (ERK) was phosphorylated by the drug treatment. In this work, we have determined the possible involvement of particular free radicals in DOX-induced ERK phosphorylation in hepatocyte cultures by using specific free radical scavengers. The levels of ERK phosphorylation were measured by Western blot analysis with an anti-Thr^{202}/Tyr^{204}-phosphorylated p44/p42 MAPK antibody. Deferoxamine (DFO; iron chelator), catalase (hydrogen peroxide-removing enzyme), or α-tocopherol (peroxyl-radical scavenger) did not affect DOX-increased ERK phosphorylation levels. However, the cell-permeable superoxide dismutase mimetic MnTBAP and the flavin-containing enzyme inhibitor diphenyleneiodonium reverted DOX-induced effects. These results suggest that superoxide anions, probably generated by DOX metabolism, are involved in the effects of the anthracycline on the MAP kinase cascade activation.

KEYWORDS: doxorubicin; MAP kinase; oxygen radicals; hepatocyte

Address for correspondence: José Ignacio Ruiz-Sanz, Department of Physiology, Medicine and Dentistry School, University of the Basque Country, 48080-Bilbao, Spain. Voice: +34-946015795; fax: +34-946015662.
e-mail: joseignacio.ruizs@ehu.es

Ann. N.Y. Acad. Sci. 1090: 419–428 (2006). © 2006 New York Academy of Sciences.
doi: 10.1196/annals.1378.045

INTRODUCTION

Doxorubicin (DOX) is the most commonly used anthracycline antibiotic, originally isolated from fermentation products of *Streptomyces peucetus*. It is the drug primarily used in the treatment of patients with lymphomas, breast cancer, and sarcomas. Its ability to inhibit topoisomerase II appears to be the primary mechanism for tumor cytotoxicity. DOX, because of its quinone moiety, undergoes one- and two-electron reductions producing reactive compounds, which damage macromolecules and lipid membranes.[1] Reduced metals (such as iron) are critical components in the formation of these reactive intermediates.[2] DOX and other anthracyclines are cardiotoxic, this cardiotoxicity being associated with the generation of free-radical intermediates. DOX is metabolized predominantly by the liver to the major metabolite, doxorubicinol, and several cytotoxic aglycone derivatives.[3] Hepatocytes, in contrast to myocardial cells, are well provided with free-radical scavenging defenses. Reactive oxygen species (ROS)-dependent redox cycling is considered to be critical for the regulation of protein interactions, which modulate the activity of important proteins in signal transduction and carcinogenesis, including the activity of protein kinase C and mitogen-activated protein kinases (MAPKs), especially extracellular signal-regulated kinase (ERK).[4,5] Three major subfamilies of MAPKs have been described: p42/p44 ERK1/2,[6] c-Jun N-terminal kinases (JNK),[7] and p38 kinases.[8] The activated MAPKs affect the phosphorylation of Ser-Pro and Thr-Pro motifs in the substrate proteins and are regulated by a variety of extracellular stimuli, including growth factors, mitogens, cytokines, and environmental stresses.[9] The MAPKs have been shown to be activated in response to oxidant-induced alterations in the cellular redox state.[10]

In this study we examined the effects of DOX on ERK activity in primary cultures of rat hepatocytes and the putative implication of ROS in this effect.

MATERIALS AND METHODS

Reagents and Antibodies

Tissue culture media were from SEROMED-Biochrom (KG, Berlin, Germany). Gentamycin was from Gibco Life Technologies (Grand Island, NY, USA). Penicillin, streptomycin, dexamethasone, insulin, protease inhibitor cocktail, deferoxamine (DFO), and diphenyleneiodonium (DPI) were from Sigma (St. Louis, MO, USA). Collagenase was from Boehringer (Mannheim, Germany). DOX was from TEDEC-MEIJI FARMA, S.A. (Madrid, Spain). ECL reagents were purchased from Pierce (Rockford, IL, USA). Primary antibodies against p44/p42 MAPK and phospho-p44/p42 MAPK (Thr^{202}/Tyr^{204}) were acquired from Cell Signaling (Beverly, MA, USA), and horseradish peroxidase-conjugated anti-rabbit and anti-mouse antibodies were purchased from Amersham (Arlington Heights, IL, USA). Mn(III) tetrakis (4-benzoic acid) porphyrin (MnTBAP) was from Alexis; bisindolylmaleimide I hydrochloride and PD98059 were from Calbiochem (San Diego, CA, USA). Standard analytical-grade laboratory reagents were obtained from Merck (Darmstadt, Germany).

Hepatocyte Cultures

Hepatocytes were isolated from male Sprague-Dawley rats (180–200 g) by collagenase perfusion, as described previously.[11] The hepatocyte viability was determined by means of the trypan blue exclusion test and was typically greater than 90%. Purified hepatocytes were plated on 100 × 20-mm culture dishes (Falcon) at a density of 10×10^6 cells/plate in RPMI 1640 adhesion medium, supplemented with 10% fetal bovine serum, glucose (2 mg/mL), pyruvate (0.11 mg/mL), dexamethasone (1 μM), gentamycin (5 mg/L), penicillin:streptomycin (5 U/mL: 5 mg/mL), amphotericin B (0.5 μg/mL), and insulin (0.7 μM). After 5–6 h incubation at 37°C in a 5% CO_2, 95% air atmosphere, the medium was replaced with fresh medium without amphotericin and supplemented with 0.2% fatty acid free bovine serum albumin and 10 nM insulin. Cultures were equilibrated overnight at 37°C in a 5% CO_2, 95% air atmosphere. Treatments were started by the addition of 10-μM DOX, dissolved in the incubation medium, to hepatocyte cultures. At time intervals, cells were harvested, washed in phosphate-buffered saline (PBS), and collected by centrifugation.

Western Blot Analysis

Cells on 100-mm dishes were washed three times in ice-cold PBS, scraped from the dishes, and then collected in extraction buffer (20 mM Hepes pH 7.5 containing 10 mM EGTA, 40 mM β-glycerophosphate, 1% Nonidet P-40, 1 mM NaF, 2.5 mM $MgCl_2$, 2 mM $NaVO_4$, 1 mM DTT, and protease inhibitor cocktail). After incubation on ice for 30 min, samples were centrifuged at $14{,}000 \times g$ for 10 min at 4°C, and the amount of proteins in the cleared lysates was quantitated by the technique of Bradford.[12] An equal amount of proteins (25 μg) was then separated on 12% SDS-PAGE gel, and then transferred electrophoretically onto polyvinylidene fluoride (PVDF) membranes (0.4 mm Immobilon-P, Millipore [Eschborn, Germany]) using the semidry transfer system from BioRad (Richmond, CA, USA). These membranes were blocked with 5% bovine serum albumin (BSA) in TBS (10 mM Tris-HCl, 150 mM NaCl), pH 7.5, 0.05% Tween 20 for 1 h and subsequently probed overnight at 4°C with antibody against phospho-ERK in Tris-buffered saline at 1:1,000 dilution. After washing, the antibody–antigen complexes were detected using goat anti-mouse IgG peroxidase conjugates. The results were visualized by chemiluminescence using enhanced chemiluminescence (ECL) reagents according to the manufacturer's instructions. Relative protein levels were determined by scanning densitometry analysis.

Membranes were then washed and incubated in stripping buffer (62.5 mM Tris-HCl, pH 6.7, 2% SDS, 100 mM 2-mercapthoethanol) at 50°C for 30 min. After washing, membranes were blocked and reprobed with primary antibodies against ERK at 1:1,000 dilution and the secondary antibodies as described before. MAPK activation was expressed as the ratio of the phosphorylated form of the kinase relative to the total fraction. The given values represent the mean (in relative units), normalized to control values, of at least three independent experiments from which similar results were obtained.

Statistical Analysis

Statistical significance was analyzed by two-tailed Student's t-test. Data were expressed as the mean ± SEM. Differences were considered statistically significant at $P < 0.05$.

RESULTS

DOX Induction of ERK Phosphorylation in Hepatocytes

To test the time course of DOX-induced ERK phosphorylation, hepatocytes were treated with 10 μM of the anthracycline during 1, 3, and 6 h. As shown in FIGURE 1, DOX induced phosphorylation of ERK1/2 (Thr^{202}/Tyr^{204}), which was evident at a time point of 3 h. The augmentation of phosphorylation was maintained at least for a further 18 h (data not shown).

Effects of Antioxidants on the ERK Phosphorylation Induced by DOX in Hepatocytes

DOX is associated with the generation of ROS. We therefore analyzed whether ROS were involved in DOX-induced ERK phosphorylation in hepatocytes. We treated the cells with the peroxyl radical-scavenging α-tocopherol, the hydrogen peroxide-scavenging enzyme catalase, and the iron-chelating agent DFO. These agents did not abrogate the DOX-induced ERK phosphorylation (FIG. 2).

Taking into account the anthraquinone feature of DOX, its metabolism could render superoxide anions (O_2^-) as primary ROS through redox cycling. We treated the hepatocytes with the membrane-permeable superoxide

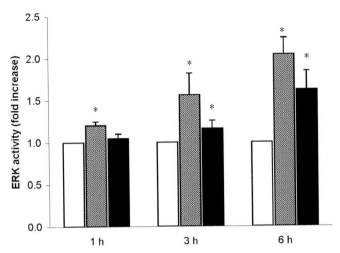

FIGURE 1. Doxorubicin induction of ERK phosphorylation in hepatocytes. Primary rat hepatocyte cultures were incubated with 10 μM doxorubicin for 1, 3, and 6 h. Cell lysates (25 μg) were prepared and subjected to Western blot analysis with phospho-p44/p42 or p44/p42 antibodies as indicated. Densitometric data are shown as mean ± SEM bars. All experiments were performed at least three times. *$P < 0.05$. Control (□); phospho-p44 (▨); phospho-p42 (■).

dismutase mimetic Mn(III) tetrakis (4-benzoic acid) porphyrin (MnTBAP). We also used the NAD(P)H oxidase inhibitor DPI on DOX-induced phosphorylation of p42/p44. The anthracycline-induced activation of ERK was blocked by MnTBAP, and pretreatment with the NAD(P)H oxidase inhibitor DPI completely inhibited DOX-induced ERK phosphorylation (FIG. 3). Taken together, our results suggest that O_2^- production is required for ERK activation. This superoxide production is probably mediated by NAD(P)H oxidase.

DOX-Induced ERK Phosphorylation Is Mediated by MEK and PKC

We tested whether DOX-induced ERK phosphorylation occurs through the classical MAPK pathway by using the specific pharmacological inhibitors PD98059, an inhibitor of MEK1/2 (upstream activator of ERK1/2), and bisindolylmaleimide, a protein kinase C (PKC) inhibitor. As shown in FIGURE 4 both compounds totally blocked ERK activation.

DISCUSSION

Oxidative stress refers to a disrupted redox equilibrium between the production of free radicals and the ability of cells to protect against damage caused by these species. Oxidative stress has been implicated as a major cause of

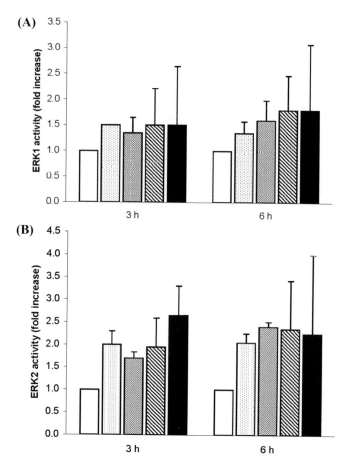

FIGURE 2. Effects of antioxidants on doxorubicin-induced ERK phosphorylation. Hepatocytes were pretreated with α-tocopherol (0.25 mM), catalase (1000 U/mL), or deferoxamine (100 μM) ■ for 30 min, followed by 10-μM doxorubicin exposure for 3 and 6 h. Western blot analysis was then performed. Control (no additions) □, doxorubicin (10 μM). Densitometric data are shown as mean ± SEM bars. All experiments were performed at least three times.

cellular injuries in a variety of human diseases. We were unable to detect any modification in oxidative stress markers (increased malondialdehyde or decreased glutathione) or cytotoxic effects in hepatocytes treated with DOX.[13] However, we hypothesize that DOX may cause mild increases in ROS production, triggering the signaling pathways that increase the ERK phosphorylation status. Hence, a moderate ROS production may suffice for the activation of ERK, but may not be sufficient to induce oxidative stress that would cause cytotoxicity. The redox cycle of DOX should be considered as the source of ROS, since DOX has both the para-quinone and para-hydroquinone residues.

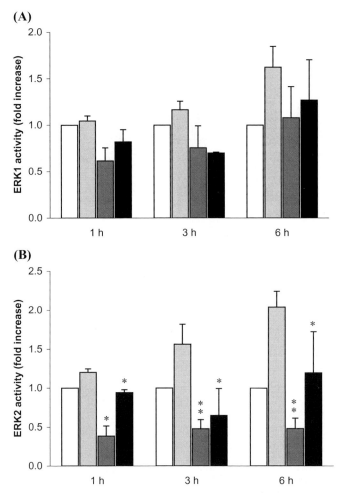

FIGURE 3. Effects of MnTBAP and DPI on DOX-induced ERK phosphorylation. Hepatocytes were pretreated for 30 min with the NAD(P)H oxidase inhibitor diphenyleneiodonium (10 μM) ▨ or the membrane-permeable superoxide dismutase mimetic MnTBAP (500 μM) ■, followed by 10 μM doxorubicin for up to 6 h. Then Western blots were probed with phospho-specific antibodies against ERK1/2 (Thr^{202}/Tyr^{204}). An anti-ERK1/2 antibody served as a protein-loading control. Control (no additions) □, doxorubicin (10 μM) ▦. Densitometric data are shown as mean ± SEM bars for three experiments.

DOX reacts at the quinone residue with cytochrome P450 reductase in the presence of NADPH to form the semiquinone radical via a one-electron reduction, resulting in the generation of O_2^-.[14,15] Superoxide anions are also generated by the oxidative activation of DOX, that is, a one-electron oxidation of para-hydroxy residue.[16] MnTBAP is a stable and cell-permeable superoxide dismutase mimetic which permeates biological membranes and scavenges

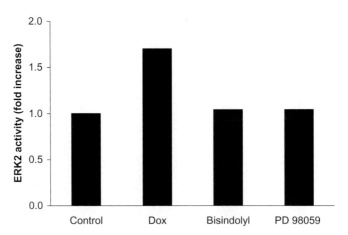

FIGURE 4. Doxorubicin-induced ERK phosphorylation dependence on protein kinase C. Hepatocytes were pretreated for 30 min with the PKC inhibitor, bisindolylmaleimide (25 μM) or the MEK1/2, inhibitor PD 98059 (25 μM), followed by 10 μM doxorubicin for 6 h. Then Western blots were probed with phospho-specific antibodies against ERK1/2 (Thr^{202}/Tyr^{204}). An anti-ERK1/2 antibody served as a protein loading control.

superoxide anions and peroxynitrite *in vitro*.[17] Thus, MnTBAP scavenges intracellular superoxide anions and prevents the formation of hydroxyl radicals. MnTBAP exerts potent protective effects in systems where the injury is likely to be mediated by superoxide alone.[18] The role of ROS, particularly the role of O_2^- and H_2O_2, have been demonstrated in various models of DOX-induced cell death. We have seen that the superoxide dismutase mimetic, MnTBAP, significantly reduced DOX-induced phosphorylation of ERK1/2, thus indicating the involvement of O_2^- in the signaling cascade leading to ERK1/2 activation.

Another source of ROS is the activation of NAD(P)H oxidase. It has been reported that DNA damage induced by DOX hyperactivates poly(ADP-ribose) polymerase.[19] The subsequent depletion of both NAD^+ and $NADP^+$ in cells treated with the anthracycline may result in the activation of NAD(P)H oxidase to maintain the cellular redox balance by converting NAD(P)H to $NAD(P)^+$.[20] In our system, we have demonstrated that DPI totally prevents the DOX-induced activation of ERK. Our results suggest that ROS, particularly superoxide, is required for ERK activation. This superoxide production is probably mediated by NADPH oxidase. However, although DPI is usually considered as an NAD(P)H oxidase inhibitor, it actually inhibits all flavin-based enzymes, among them NADPH cytochrome P450 reductases.[21,22]

Several papers have reported that the MAPK signaling pathway may play an important role in the activities and effects of chemotherapeutic drugs. MAPK signaling pathways are clearly involved in a host of cellular functions, including cell growth, differentiation, development, and apoptosis. ERK activation is generally assumed to inhibit apoptosis, but recent studies have demonstrated that sustained ERK activation is also involved in the apoptotic process.[23] Endogenous ROS generation has been reported to support sustained MAPK activation via the inhibition of CD45 and other tyrosine phosphatases, and to contribute to the induction of distinct MAPK activation profiles via differential signaling pathways.[5,24] In our study, DOX-induced ERK phosphorylation was inhibited by PD98059 and bisindolylmaleimide. These results clearly show the PKC and MEK requirement for the DOX-induced upregulation of ERK signaling.

In conclusion, we show that DOX elicits ERK phosphorylation and that superoxide anions initiate the MEK-ERK1/2 signaling pathway triggered by DOX.

ACKNOWLEDGMENTS

This work was supported by the Basque Government (Research Project and Predoctoral Training Grant to R.N.) and the University of the Basque Country (UPV00081.327-E-15294/2003). We are grateful to Montserrat Busto for her skillful technical assistance.

REFERENCES

1. MYERS, C.E., W.P. MCGUIRE, R.H. LISS, et al. 1997. Adriamycin: the role of lipid peroxidation in cardiac toxicity and tumor response. Science **197:** 165–167.
2. MINOTTI, G. 1989. Adriamycin-dependent release of iron from microsomal membranes. Arch. Biochem. Biophys. **268:** 398–403.
3. DODION, P., A.L. BERNSTEIN, B.M. FOX & N.R. BACHUR. 1987. Loss of fluorescence by anthracycline antibiotics: effects of xanthine oxidase and identification of the nonfluorescent metabolites. Cancer Res. **47:** 1036–1039.
4. WARD, N.E., K.R. GRAVITT & C.A. O'BRIAN. 1995. Irreversible inactivation of protein kinase C by a peptide-substrate analog. J. Biol. Chem. **270:** 8056–8060.
5. ZHOU, B., Z.X. WANG, Y. ZHAO, et al. 2002. The specificity of extracellular signal-regulated kinase 2 dephosphorylation by protein phosphatases. J. Biol. Chem. **277:** 31818–31825.
6. BOULTON, T.G., S.H. NYE, D.J. ROBBINS, et al. 1991. ERKs: a family of protein-serine/threonine kinases that are activated and tyrosine phosphorylated in response to insulin and NGF. Cell **65:** 663–675.
7. DAVIS, R.J. 1999. Signal transduction by the c-Jun N-terminal kinase. Biochem. Soc. Symp. **64:** 1–12.
8. NEBREDA, A.R. & A. PORRAS. 2000. p38 MAP kinases: beyond the stress response. Trends Biochem. Sci. **25:** 257–260.
9. WHITMARSH, A.J. & R.J. DAVIS. 1996. Transcription factor AP-1 regulation by mitogen-activated protein kinase signal transduction pathways. J. Mol. Med. **74:** 589–607.

10. FIALKOW, L., C.K. CHAN, D. ROTIN, et al. 1994. Activation of the mitogen-activated protein kinase signaling pathway in neutrophils. Role of oxidants. J. Biol. Chem. **269:** 31234–31242.
11. RUIZ-LARREA, M.B., M.J. GARRIDO & M. LACORT. 1993. Estradiol-induced effects on glutathione metabolism in rat hepatocytes. J. Biochem. **113:** 563–567.
12. BRADFORD, M.M. 1976. A rapid and sensitive method for the quantitation of microgram quantities of protein utilizing the principle of protein dye binding. Anal. Biochem. **72:** 248–254.
13. NAVARRO, R., R. MARTÍNEZ, I. BUSNADIEGO, et al. 2006. Doxorubicin-induced MAPK activation in hepatocyte cultures is independent of oxidant damage. Proceeding from the Signal Transduction Pathways as Therapeutic Targets meeting, Luxembourg, January 25–28, p. 209.
14. GEWIRTZ, D.A. 1999. A critical evaluation of the mechanisms of action proposed for the antitumor effects of the anthracycline antibiotics adriamycin and daunorubicin. Biochem. Pharmacol. **57:** 727–741.
15. JUNG, K. & R. RESZKA. 2001. Mitochondria as subcellular targets for clinically useful anthracyclines. Adv. Drug. Deliv. Rev. **49:** 87–105.
16. RESZKA, K.J., M.L. MCCORMICK & B.E. BRITIGAN. 2001. Peroxidase- and nitrite-dependent metabolism of the anthracycline anticancer agents daunorubicin and doxorubicin. Biochemistry **40:** 15349–15361.
17. CUZZOCREA, S., G. COSTANTINO, E. MAZZON, et al. 1999. Beneficial effects of Mn(III)tetrakis (4-benzoic acid) porphyrin (MnTBAP), a superoxide dismutase mimetic, in zymosan-induced shock. Br. J. Pharmacol. **128:** 1241–1251.
18. ZINGARELLI, B., B.J. DAY, J. CRAPO, et al. 1997. The potential involvement of peroxynitrite in the pathogenesis of endotoxic shock. Br. J. Pharmacol. **120:** 259–264.
19. PACHER, P., L. LIAUDET, P. BAI, et al. 2002. Activation of poly(ADP-ribose) polymerase contributes to development of doxorubicin-induced heart failure. J. Pharmacol. Exp. Therap. **300:** 862–867.
20. MIZUTANI, H., S. TADA-OIKAWA, Y. HIRAKU, et al. 2005. Mechanism of apoptosis induced by doxorubicin through the generation of hydrogen peroxide. Life Sci. **76:** 1439–1453.
21. BRANDES, R.P. 2003. A radical adventure: the quest for specific functions and inhibitors of vascular NAPDH oxidases. Circ. Res. **92:** 583–585.
22. RAMJI, S., C. LEE, T. INABA, et al. 2003. Human NADPH-cytochrome P450 reductase overexpression does not enhance the aerobic cytotoxicity of doxorubicin in human breast cancer cell lines. Cancer Res. **63:** 6914–6919.
23. YANG, Y.L. & X.M. LI. 2000. The IAP family: endogenous caspase inhibitors with multiple biological activities. Cell. Res. **10:** 169–177.
24. LEE, K. & W.J. ESSELMAN. 2002. Inhibition of PTPs by H_2O_2 regulates the activation of distinct MAPK pathways. Free Radic. Biol. Med. **33:** 1121–1132.

MAPKinase Gene Expression, as Determined by Microarray Analysis, Distinguishes Uncomplicated from Complicated Reconstitution after Major Surgical Trauma

E. MARION SCHNEIDER,[a] MANFRED WEISS,[b] WEIDONG DU,[a] GERHARD LEDER,[c] KLAUS BUTTENSCHÖN,[c] ULRICH C. LIENER,[c] AND UWE B. BRÜCKNER[c]

[a]*Section of Experimental Anesthesiology, University Clinic Ulm, Ulm 89075, Germany*

[b]*Department of Clinical Anesthesiology, University Clinic Ulm, Ulm 89075, Germany*

[c]*Department of Visceral Surgery, University Clinic Ulm, Ulm 89075, Germany*

ABSTRACT: Microarray expression analysis was performed in patients with major surgical trauma to identify signaling pathways which may be indicative for complicated versus uneventful reconstitution post trauma. In addition to a generalized upregulation of nonspecific stress response genes in all patients, a remarkable number of differences in gene expression patterns were found in individual patients. Some of the differing genes were associated with uncomplicated convalescence such as upregulation of both the ERK5 pathway (MAPK7 [mitogen-activated protein kinase-7]) and transcription factors which stimulate hematopoiesis and tissue reconstitution (MEF2, BMP-2, TNFRSF11A [RANK], and RUNX-1). Chemokine genes active in stem cell recruitment from the bone marrow as well as dendritic cell and natural killer (NK) cell maturation (SCYA14 [HCC-1]), and activators of the lymphoid compartment (TNFRSF7 [CD27], CD3zeta and perforin [PRF1]) were increased. In contrast, all these transcripts were downregulated in complicated reconstitution and later development of septic shock. Moreover, p38 kinase (MAPK14), S100 molecules, and members of the lipoxygenase pathway were associated with a more eventful outcome. Microarray expression studies are a promising tool for screening and then selecting differentially regulated genes in favorable as compared to complicated reconstitution post trauma.

Address for correspondence: E. Marion Schneider, Ph.D., Department of Experimental Anesthesiology, University Clinic Ulm, Steinhoevelstrasse 9, 89075 Ulm, Germany. Voice: +49-731-500-60080; fax: +49-731-500-60080.
 e-mail: marion.schneider@uni-ulm.de

Ann. N.Y. Acad. Sci. 1090: 429–444 (2006). © 2006 New York Academy of Sciences.
doi: 10.1196/annals.1378.046

KEYWORDS: trauma-induced gene expression; microarray; sepsis, MAP-kinase; tissue regeneration; cytokines; transcription factors; reconstitution; hematopoiesis; outcome

INTRODUCTION

In trauma, immune activation signals originating from damaged tissue lead to inflammation by cytokines and oxidative and osmotic stress.[1–3] The modulatory effects of anesthetic drugs on stress responses have been recently discussed with regard to basic knowledge on ion channel effects,[4] but their effect on trauma-conditioned sepsis and septic shock has not been studied so far. Either anesthesia or surgical trauma (or both) play an important role in post trauma–associated immunosuppression[5,6] and increased incidence of nosocomial infections. Microarray-assisted analysis ideally requires large numbers of patients with a high proportion of similar characteristics to be tested in multiple gene expression studies to identify markers as has been recently shown.[7] However, high costs and ethical restrictions may hamper extended studies of this kind. Recent reports demonstrate that valuable conclusions can also be obtained from small-sized studies.[8] The current report contributes to this issue by studying patients with clinically well-defined major trauma who markedly differed in their postoperative reconstitution. Transcription profiling may lead to the identification of transcripts associated with a favorable reconstitution phase and others with a complicated course and various episodes of sepsis and septic shock.

Following approval of a larger patient population, association markers could support a better stratification of patients, as it would allow existing therapies to be more rationally applied and future therapies to be more appropriately tested.[8–10] Most modern analyses of patients with severe trauma and sepsis emphasize the involvement of immune markers.[10,11] Previous expression profiling in rats and mice demonstrated tissue-specific expression signatures.[12] Blood samples, in particular, can be readily obtained for diagnosis and have the advantage that the normal variation in gene expression is comparably low.[13]

The present study has been performed to find out whether differences in the activation profile of regenerative versus degenerative processes can be found in patients after trauma. Expression microarrays have been proven to be robust and realistic as a tool. A necessary proof is the translation into protein-determined physiology followed by clinical validation.[14,15]

MATERIAL AND METHODS

Patients

Upon approval by the Independent Ethics Committee of the University of Ulm and with signed written informed consent, three patients with different surgical trauma, age, anesthesia, and outcome were included in this prospective pilot study. Patient 1 (Pat. 1) had general anesthesia with the volatile anesthetic desflurane combined with thoracic epidural anesthesia. Patient 2 (Pat. 2) received total intravenous anesthesia (TIVA) with propofol, and in addition, prednisolone, theophylline, beta-blocker (metoprolol, 5 mg), and anticholinergic drugs given intraoperatively. Patient 3 (Pat. 3) underwent general anesthesia with the volatile anesthetic desflurane, and additionally received transfusion of two units of erythrocyte concentrates. Blood samples (25 mL) were drawn immediately before and exactly 24 h after surgery, independently of the time of surgical intervention.

Methods

RNA was prepared from peripheral whole blood drawn into PAXgene™ tubes (PreAnalytiX GmbH, purchased via Qiagen, Hilden, Germany) of three patients before and after trauma as well as from Ficoll™-separated peripheral blood mononuclear cells of four pooled buffy coat preparations of healthy donors using RNAeasy columns (Qiagen, Hilden, Germany). RNA yields ranged between 20 and 80 μg per patient's sample and 400 μg of the controls. The cDNA were prepared by Superscript II (Stratagene, La Jolla, CA) reverse transcription with incorporation of Amino-allyl-dUTP nucleotides during reverse transcription. The resulting cDNA from patients and control samples were labeled separately with Cy3 and Cy5, respectively, to generate differentially fluorescent-labeled cDNA. The cDNA samples were then purified using a polymerase chain reaction (PCR) clean-up kit (Qiagen, Hilden, Germany) and free dye was removed. After denaturation and hybridization, the hybridized microarray slides were washed, dried, and scanned using a Genepix 4000B, ScanPro 20 microarray scanner (Fuji Medical Systems, Stamford, CT). Red (Cy5, patient sample) and green (Cy3, control samples) spots indicate relative amounts of cDNA. To control variation in dye incorporation, dye-swap approaches were performed so that the control cDNA was labeled with Cy5 and patients' cDNA with Cy3. The expression microarray applied contained about 500 cDNA probes derived from human genes encoding cytokines, chemokines, receptors, and enzymes involved in inflammatory responses, proliferation and apoptosis, cell cycle genes, transcription factors, and negative control cDNAs from *Arabidopsis thaliana*. All cDNA spots were tested in triplicate. Mean values of the three spots were calculated, and data were further processed if the standard deviation was less than 5%.[8] The comparison of transcript levels between samples from pre- versus postoperative patients' isolates was performed by comparing the transcriptional activity measured for each cDNA isolate with the controls. This set of data has been processed using AIDA™ software on scanned microarrays. Stimulation indices (SI) of genes are defined: SI = ratio of the relative transcriptional expression levels described as the quotient of postoperative values (expression of patient isolate [Cy5 fluorescence] divided by expression of the control [Cy3fluorescence]) divided by the preoperative levels {expression of patient isolate [Cy5 fluorescence]} divided by expression of the control [Cy3fluorescence]). Values higher than 1.0 indicate upregulation, and values less than 1.0 refer to relative downregulation of the respective transcript.

RESULTS

Clinical Course of the Patients

Pat. 1 was a 59-year-old woman (BMI 38.1) who underwent gastrectomy and creation of a Rodino pouch with the diagnosis of gastric adenocarcinoma (pT2N0M0); time of surgery was 440 min. As premedication she received chlorazepate (abenzodiazepine). Sufentanil and fentanyl were used for epidural anesthesia. Generalized anesthesia and analgesia were induced by desflurane, remifentanyl, and metamizol. Additional drugs were clonidine, urapidil, cafedrine, furosemide, and antibiotics. On day 7, the patient needed relaparotomy on account of anastomotic leakage and subsequent development of septic shock. To counter massive infection with *Enterobacter* species, treatment with piperacillin plus combactam and gentamycin was initiated. Subsequently, because of persistent acute respiratory failure and detection of massive infection with *Enterococci* species and candidiasis, tracheotomy was performed and antibiosis switched to cefotaxim. On day 12, a further laparotomy on account of anastomotic leakage was necessary. Candidiasis was treated with diflucane and amphotericin B. On day 30, the patient was released from the ICU.

Pat. 2 (BMI 77.3) was a 38-year-old woman who suffered from hypertension and excessive obesity. Gastric bypass and cholecystectomy were performed. As premedication, she also received chlorazepate. Propofol and fentanyl were used to induce anesthesia. For anesthesia and analgesia she received propofol, remifentanil, and metamizol. Additional drugs were applied intraoperatively and included prednisolone, theophylline, salbutamol, metoprolol, clonidine, urapidil, cafedrine, and antibiotics. Intraoperatively, she also received 250 mg of prednisone, and 1,500 mL of plasma expander (HES). Postoperatively, she received metoprolol and the calcium antagonist nifedipin. Recovery phase was uncomplicated and no intensive care management was necessary.

Pat. 3 (BMI 22.3) was an 81-year-old female who underwent total endoprosthetic replacement of the right hip on account of underlying osteoarthritis. Premedication was with rofecoxib and the ACE blocker cilazapril. Propofol and fentanyl were used to induce anesthesia. Anesthesia and analgesia were obtained by desflurane and remifentanyl. She also received the local anesthetic ropivacaine for 3-in-1 blockade. Intraoperatively, only cafedrine (aphosphodiesterase inhibitor) was given and 1,000 mL of HES were infused.

Two units of erythrocytes were transfused. Recovery phase was uncomplicated, requiring no intensive care management.

MAPKinase Elements and Transcription Factors

Both the range of SI levels as well as their pre- and postoperative differences are similar in Pats. 1 and 3, but differ greatly in Pat. 2. Major results are summarized in TABLE 1. Except for PTK (Polo-like kinase)(MAP3K11), which was slightly upregulated in all patients, other transcripts contributing to MAPKinase pathways behaved in an individual pattern. However, patients with similarly severe trauma share a differential regulation of TGF-β-activated kinase (TAK1), an element feeding the c-Jun N-terminal kinases (JNK) and p38 pathway, and MAP3K4 (ERK3-like), feeding JNK as well as MAP3K1. When comparing patients with respect to different anesthetic regimens but similar trauma (Pats. 1 and 2), we found that MAP3K13 and MAP3K10 were downregulated in association with volatile anesthesia but not with TIVA. With TIVA, *c-fos* and *c-jun* were decreased. A pathway related to tissue remodeling by ERK5 and the transcription factors MEF2A, MEF2B, CREBL2, and CREB3 was activated in the TIVA patient. Along the same line, BMP2 and RANK were found to be increased. In summary, transcripts of different stages in the ERK5 pathway were revealed to be simultaneously increased in association with TIVA regimen and positive outcome (TABLE 1).

Cyclic adenosine monophosphate (AMP)-regulated transcription factors were included to monitor possible effects of perioperative medication. Specifically, neither NFkB-inducing kinase (NIK) nor the p105 subunit of NFkB were upregulated in any of the patients.

Infection- and Inflammation-Related Transcripts

PTX3, S100, IL-8 receptor, G-CSF receptor, and enzymes involved in arachidonic acid metabolism (ALOX5, ALOX5AP) were differentially regulated in Pats. 1 and 2.

Genes affecting hematopoietic stem cells belong to the small inducible cytokine family (SCY), such as HCC1, recruiting CD34, and CD164, regulating homing. Their activation differed in Pats. 1 and 2. Other genes linking inflammation to recruitment and activation of lymphocytes, such as CD27, CD3, IL-2R, perforin, and growth factors, such as G-CSF, were also distinct. Finally, various cytokine receptors were almost exclusively stimulated in Pat. 2 but not in Pat. 1.

Transcription factor activation analysis demonstrated a remarkable upregulation of Jun-B and TCF8 in Pat. 1. CREBL2 and RUNX-1 were the only factors markedly higher in the two patients with uncomplicated recovery.

TABLE 1. Stimulation indices (SI) of transcripts (post/pretrauma levels)

	Name	P1 unfavorable outcome [SI]	P2 favorable outcome [SI]	P3 minor trauma [SI]
p38 pathway				
PLK/Polo-like kinase	MAP3K11	1.07	1.17	2.79
MEKK1	MAP3K8	1.29	0.47	1.00
TAK1	MAP3K7	0.90	0.46	
p38 kinase	MAPK14	1.83		0.62
TWEAK/TNF ligand superfamily 12	TNFSF12	1.13	0.73	3.84
ERK pathway				
ERK2	MAPK1	1.06	1.73	0.70
ERK1	MAPK3	1.12	0.98	0.36
ERK3-like	MAPK4	0.87	0.81	
ERK3	MAPK6	1.05	0.12	
ERK5	MAPK7	0.76	4.23	2.51
NIK/NFκB pathway				
NIK/NFκB inducible kinase	MAP3K14	0.85	0.86	0.37
Transcription factor	NFKB1	0.81	0.58	
Tissue regeneration				
ERK5	MAPK7	0.76	4.23	2.51
TNF-R superfamily 11A/RANK	TNFRSF11A	0.79	2.47	1.92
TGF-β binding protein 3	LTBP3	0.63	3.48	
Bone morphogenic protein 2	BMP2	0.65	4.55	
Myocyte-specific enhancer factor 2	MEF2A	0.80	2.15	0.29
Effect on stem- & hematopoietic cells				
HCC-1 small inducible cytokine 14	SCYA14	0.72	2.16	
Sialomucin	CD164	1.91	0.34	0.29
CD27	TNFRSF7	0.57	1.23	1.75
CD3zeta	CD3Z	0.75	3.72	
Perforin1	PRF1	1.00	2.17	1.75
NO synthase 2	NOS2A	0.69	1.51	
Early activation antigen p69	CD69	1.47	0.20	
Infection associated				
Pentaxin-related protein 3	PTX3	0.62	1.45	0.22
Calgranulin A	S100A12	7.73	0.57	3.50
Calgranulin C	S100A8	9.89	0.34	2.67
IL8R	IL8RA	1.47	0.36	0.90
G-CSF receptor	CSF3R	2.08	0.72	1.29
Arachidonate 5-lipoxygenase	ALOX5	1.66	0.92	1.10
Activating protein of ALOX5	ALOX5AP	2.15	0.79	1.73
Pro- and anti-inflammatory cytokines				
TGF-ß	TGFB1	1.18	1.22	1.79
Small inducible protein 16	SCYA16	0.85	2.66	
Eotaxin	SCYA11	0.64	2.25	
Granulocyte colony-stimulating factor (G-CSF)	CSF3	0.67	2.21	

Continued

TABLE 1. Continued

	Name	P1 unfavorable outcome [SI]^{a,b}	P2 favorable outcome [SI]	P3 minor trauma [SI]
IL-2 receptor alpha chain	IL2RA	0.75	1.90	
IL-5 receptor alpha	IL5RA	0.84	3.59	
IL12-R β-chain	IL12RB2	0.76	2.16	
IL-17 receptor	IL17R	0.65	2.47	
IL-18 receptor type1	IL18R1	0.72	1.65	
Cytokine receptor gamma-chain, CD132	IL2RG	0.44	1.36	1.00
IL10R alpha chain	IL10RA	0.96	0.54	1.19
IL10R ß-chain	IL10RB	0.97	0.67	2.42
IL1 receptor type 1	IL1R1	0.72	0.33	
IL-6	IL6	0.93	0.74	
IL-6receptor alpha, p80	IL6RA	1.00		0.49
Transcription factors				
Transcription factor 8	TCF8	2.69	0.52	
cAMP-dependent transcription factor	ATF2	1.04	1.16	1.20
	ATF3	1.04	0.43	
	ATF4	0.99	0.64	5.56
	ATF5	0.88		1.62
	ATF6	1.06	0.64	
	ATF7	0.77	0.48	
Cyclin-dependent kinase inhibitor 1B	CDKN1B	1.37	0.17	
Basic leucine zipper protein, LZIP	CREB3	0.68	1.35	1.82
cAMP-responsive element binding protein-like 2	CREBL2	0.67	2.85	3.78
Proto-oncogene c-fos	FOS	1.16	0.49	0.19
Proto-oncogene c-jun	JUN	0.98	0.51	
Jun-B	JUNB	2.91		0.68
Ets domain protein1 ELK-1	ELK1	0.77	0.55	
Pancreatic tumor-related protein	FKSG12	0.72		1.50
Myocyte-specific enhancer factor 2A	MEF2A	0.80	2.15	0.29
Myocyte-specific enhancer factor 2B	MEF2B	1.04	1.29	
Runt-related transcription factor, 1, oncogene AML-1	RUNX1	0.73	5.33	2.56
Nuclear factor of activated T cells, cytoplasmic 3	NFATC3	1.17	0.49	5.05
Nuclear receptor 3 corticoid 1	NR3C1	0.71	0.54	

NOTE: Values higher than 1.0 indicate upregulation, and values less than 1.0 refer to relative downregulation of the respective transcript.

^{a}Stimulation indices (SI) of genes regulated before and after surgical trauma in Pats. 1, 2, and 3, respectively.

^{b}SI = ratio of the relative transcriptional expression levels described as the quotient of postoperative values divided by the preoperative levels.

ATF4 appeared to be the most prominent transcription factor activated in Pat. 3, but ATF2, ATF5, and CREB3 were also upregulated. Steroid receptor NR3C1 was downmodulated in both of the major trauma patients (TABLE 1).

DISCUSSION

Numerous microarray-based assays have been developed to classify and diagnose diseases, most frequently cancer subtypes.[15,16] Their success depends on the robustness of the gene expression signature.[17] Gene expression profiling, especially when performed with automated microsystems, can be applied in reproducible ways (i) to investigate drug actions nonclinically and clinically, (ii) to stratify and predict patient response, and (iii) to further understand and diagnose diseases.[18]

The modulation of immune functions induced by trauma, surgical interventions, and anesthesia is thought to play a crucial role in the development of postoperative inflammatory and immunodeficiency syndromes. We then studied the individual stress response in two patients with similar surgical trauma but different anesthetic regimens. Another patient with a minor surgical trauma was included to selectively address transcriptional regulation by volatile anesthesia but less trauma.

MAPKinase Enzyme Activation Pathways and Subsequent Transcription Factors

Pathways of MAPKinase enzymes and subsequent transcription factors activated and involved in patients after surgical trauma are outlined in a simplified cartoon (FIG. 1). Note that pathways activated were distinct in individual patients.

Transcripts Associated with Distinct Signal Transduction Pathways

JNK Pathway

The JNK pathway, inducible by stress factors, cytokines as well as ceramide, does not appear to be markedly activated in any of these patients (MAP3K8, MAP3K4, c-Jun, Elk-1, and ATF2) (TABLE 1; FIG. 1)

p38 Pathway

This pathway as determined by MAP3K11 (PLK, Polo-like kinase), MAP3K7 (TAK1), MAPK14 (p38), and human tumor necrosis factor (TNF)-like weak inducer of apoptosis (TWEAK) (synonym for TNFS12) is apparently

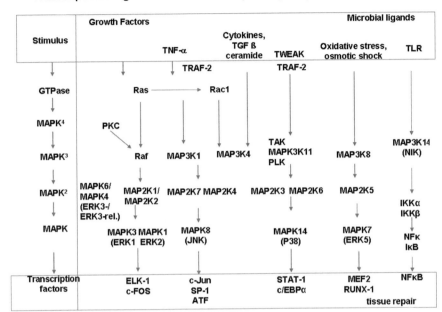

FIGURE 1. Transcriptional regulation of MAPKinase enzymes in postoperative patients. The different activation of these enzymes via various stimuli in postoperative patients may result in an activation of distinct transcription factors.

not activated in Pat. 2, who experienced a major trauma and received TIVA. In Pat. 3 who had a minor trauma and volatile anesthesia, we found activation of TWEAK (SI = 3.84) and MAP3K11 (synonym for PLK) (SI = 2.79), which may be involved in the p38 MAPKinase pathway associated with the stimulation of proliferative and inflammatory events.[19,20] p38 (MAPK14) was upregulated in Pat. 1, who suffered from various episodes of septic shock after surgery (TABLE 1). The p38 enzyme is a major player in the JNK-SAPK (stress-activated protein kinase) pathway related to apoptosis and including the *c-jun* and ATF2 transcription factors,[21] as well as inflammation in general.[22,23] Although p38 has also been described as acting in tissue protection and antiapoptosis,[24] our results clearly point to negative effects induced by p38.

ERK1/2 Pathway

Enzymes playing a role in the ERK1/2 pathway, which is linked to activation of Ras via growth factors, include MAPK1 (ERK2) and MAPK3 (ERK1). These kinases were found to be upregulated in Pat. 2 (ERK 2; SI = 1.7; TABLE 1), suggesting the involvement of a growth factor-regulated pathway.

None of the other two patients displayed a higher transcriptional activity of ERK2 and ERK1 post trauma (TABLE 1).

NIK/NFkB Pathway

NFkB activation is most relevant for trauma-associated stimulation pathways.[25] Our transcriptional analysis concentrated on MAP3K4 (NIK, NFκB inducible kinase) and NFκB (p105 Rel), which may be linked to the stimulation via toll-like receptors (TLRs), but also to tumor necrosis factor (TNF)-α, and TNF receptor-associated factor (TRAF)–regulated pathways.[26] There was no specific upregulation of NIK and TRAF-associated complexes in any of our patients (TABLE 1).

Transcripts Associated with Tissue Regeneration

ERK 5 (MAPK7) was expressed in Pat. 2 and was upregulated after trauma (SI = 4.23). We cannot exclude that this transcriptional activation was not related to metabolism and the high BMI of 77.3 in this patient. However, in Pat. 3, there was also an upregulation after trauma (SI = 2.51). ERK 5 is a member of a signaling cascade that may be induced by a number of stress factors including TRAF2, indicating induction by oxidative stress and osmotic shock (FIG. 1). Data generated in the ERK5 knockout mouse have revealed that this transcription factor is important in embryonic development,[27] and may be relevant to defined events in tissue reconstitution, especially muscle.[28] In Pat. 2, other genes upregulated with fast and uncomplicated reconstitution are involved in tissue remodeling and reconstitution, such as TNFRSF11A (RANK)[29] (SI = 2.47). In Pat. 3, we found an SI of 1.92 for RANK, whereas Pat. 1 showed reduced transcriptional activity for RANK post trauma (SI = 0.79).

Because Pat. 3 and Pat. 1 behaved differently, the activation or repression of this remodeling factor is apparently not due to the anesthetic regimen. The transforming growth factor (TGF)-β binding protein 3 (LTBP3)[30] and BMP2 were found to be upregulated in Pat. 2 (SI = 3.48 and 4.55, respectively). Pat. 1, however, again revealed reduced transcriptional activity (SI = 0.63 and 0.65, respectively). BMP2 plays a role in bone remodeling.[31] Along these lines, the myocyte-specific enhancer factor2 (MEF2A) was upregulated in Pat. 2 (SI = 2.15), but not in the two patients with volatile anesthesia (SI = 0.29 in Pat. 3, and 0.80 in Pat. 1, respectively). MEF2A is, as mentioned earlier for ERK5, involved in muscle reconstitution,[32] but may also stimulate neuronal cell survival.[33,34] In summary, the upregulation of MEF2A and MAPK7 (ERK5) highlights an important pathway deeply involved in tissue repair and protective mechanisms.[35]

Effect on Stem Cells and Other Hematopoietic Cells

The small inducible cytokine SCYA14 (HCC-1), a chemokine that stimulates hematopoietic stem cell proliferation,[36] was stimulated in Pat. 2 (SI = 2.16), but downregulated in Pat. 1 (SI = 0.72) (TABLE 1). In contrast, CD164, the sialomucin that regulates the homing of CD34 and myeloid progenitors in the bone marrow, was reduced in Pats. 2 and 3, but upregulated in Pat. 1 (TABLE 1). Reduced CD164 may be relevant to explain facilitated release from the bone marrow compartment as well as from endothelial lining cells.

Stimulation of NK-, and T cells was documented in Pat. 2 by upregulation of TNFRSF7 (CD27),[37] which plays a role in T and NK activation, and of the CD3zeta chain, as well as by a remarkable upregulation of perforin (SI = 2.17). There is compelling evidence that NK cell activation,[38] perforin, and also T cell activation (CD3) may not only be associated with increased cellular cytotoxicity, but may also control sepsis-associated immunodeficiency. In Pat. 1, all these transcripts were either unaltered (perforin) or downregulated. Nevertheless, the activation antigen CD69 (lymphocyte-activation antigen) was increased in Pat. 1, suggesting that the lack of CD3, CD27, and perforin upregulation is not due to less lymphocyte activation as such (TABLE 1).

Transcripts Associated with Infections

PTX3 represents a transcript associated with an improved immune function in trauma and sepsis,[39] and was found to be increased in Pat. 2 and downregulated in Pats. 1 and 3. PTX3 is a TNF-inducible protein expressed on macrophages and dendritic cells. Its upregulation may indicate positive immune function. The importance of dendritic cells in various aspects of tissue damage and sepsis has been emphasized.[40]

In Pat. 1, S100 molecules were upregulated; these are calgranulin A and C, which act as calcium-dependent antimicrobial proteins and indicate immune activation due to infections, and S100 is related to a higher risk of complicated reconstitution post trauma.[41] The spectrum of activated genes in Pat. 1 includes enzymes of the arachidonic acid pathway such as ALOX5 (arachidonate 5-lipoxygenase) and its activating protein ALOX5AP. By contrast, all these transcripts were specifically downregulated in Pat. 2, in whom no infection occurred.

Pro- and Anti-inflammatory Cytokines

Pat. 2 had higher transcriptional levels of the cytokine families and receptors, such as small inducible cytokines (SCYA16), SCYA11 (eotaxin), G-CSF, and interleukin receptors 2, 5, 12, 17, and 18, respectively. The common cytokine gamma chain was increased as well. Anesthetic regimen may be responsible for modulatory effects on stress hormone levels and cytokine transcription via IL-10.[42,43] Pat. 2 had TIVA (propofol, prednisolone, theophylline, and anticholinergic drugs given intraoperatively), whereas Pat. 1 had general anesthesia with the volatile anesthetic desflurane combined with thoracic epidural anesthesia. The third patient, Pat. 3, also underwent general anesthesia with the volatile anesthetic desflurane, but also received ropivacaine for regional anesthesia.

Transcription Factors

Pat. 2 showed downregulation of mRNA of transcription factor 8 (TCF8), which represses IL-2 transcription. Moreover, the cyclic adenosine 3′,5′-phsophate (cAMP)-dependent transcription factor 3 (ATF3) and the cyclin-dependent kinase inhibitor 1B (CDKN1B) were decreased in Pat. 2. Cyclic AMP-modulating drugs, such as catecholamines and β-adrenoceptor blockers (metoprolol for example, Pat. 2), may affect cAMP responses. Indeed, cAMP-responsive transcription factors (ATFs) were homogeneously downregulated in Pat. 2 (receiving metoprolol) but not in Pat. 1. Despite the markedly shorter surgical intervention in Pat. 3, ATF2, 4 (SI = 5.56) and ATF5 were upregulated in this patient. ATF2 has been reported to be involved not only in trauma and reperfusion injury, but also in brain development and, recently, ERK1 has been demonstrated to be identical to a p46 c-Jun/ATF-2 kinase.[24,44] ATF7 regulates the turnover of epithelial tissues as well as their differentiation.[45] Of the cAMP/calcium response elements binding protein family (pCREB), CREB2, and CREB3 were selectively activated in Pats. 2 and 3 to a marginal extent, but these factors were downregulated in Pat. 1. CREB2-binding may inhibit TNF-α and IL-8 production through the generation of cAMP and activation of protein kinase A.[46]

Stress factors are responsible for the extremely rapid induction of *c-fos* and *c-jun*, whose expression has been shown to be amplified and prolonged by anesthesia in a rat model.[47] We performed our analyses 24 h after operation and found no evidence for major activation of stress-associated transcription factors in any of the patients: *C-fos* and *c-jun* were less in Pats. 2 and 3 but not in Pat. 1. The well-known anti-inflammatory effects of opioids may explain why trauma-driven inflammatory responses have been attenuated in our patients, all of whom received remifentanyl. In addition, one patient had sufentanil for epidural anesthesia (Pat. 1), and in two patients, fentanyl was used to induce anesthesia (Pats. 2 and 3). One differential effect was found regarding NOS2, which was increased in Pat. 2 but reduced in Pat. 1 postoperatively. In *ex vivo* studies, morphine inhibited lipopolysaccharide-induced NFkB nuclear translocation via nitric oxide (NO) in human blood neutrophils and monocytes in a time-, concentration-, and naloxone-sensitive-dependent manner.[48]

It is possible that the anti-inflammatory effects of intravenous anesthetics may operate synergistically with opioids because no inflammation was found in Pat. 2, who received TIVA.

Evidence for opioid-specific effects on NFAT were also found, because NFATC3 was upregulated in Pats. 1 and 3, but downregulated in Pat. 2.

Three prototype stress hormones—dexamethasone, norepinephrine, and epinephrine—may further act as anti-inflammatory drugs as demonstrated in *ex vivo* stimulation assays.[49] Pat. 2 received 250 mg of prednisone intraoperatively, resulting in a reduced level of glucocorticoid receptor (NR3C1) gene expression as well as of *c-fos* and *c-jun* (TABLE 1). *In vitro* studies showed that glucocorticoids suppressed the LPS-stimulated secretion of TNF-α from human monocytes, largely through antagonizing transactivation by c-Jun/ATF-2 and NFκB complexes at binding sites on the TNF-α promoter.[50]

CONCLUSIONS

Novel working hypotheses can be efficiently generated by defined array-based screening experiments. In our investigation, differences in ERK3 and ERK5 pathways have been identified that have not been reported previously. ERK5 transcriptional regulation together with MEF2 and BMP-2 correlated with an uncomplicated outcome, whereas p38, S100, prostaglandin synthesis, and cAMP-responsive transcription factors were found in an unfavorable outcome. A major influence by anesthetic regimen as well as analgesia is very likely. In addition to monitoring the signatures generated by transcription factors, understanding the fine regulation of a multiplicity of cytokine family members on the protein levels would be a remarkable step in future.[10] To maintain adequate excellence of the protein analysis, array techniques have to be improved for their sensitivity. A variety of smart experimental approaches to this end will become realistic in the near future.[51]

ACKNOWLEDGMENTS

We thank Pete Symmons and Julien Landré (SIRS Lab Gmbh, Jena, Germany) for critically reading the manuscript and Eva Moeller (SIRS Lab Gmbh, Jena, Germany) for her array analysis.

REFERENCES

1. PERL, M., F. GEBHARD, M.W. KNOFERL, *et al.* 2003. The pattern of preformed cytokines in tissues frequently affected by blunt trauma. Shock **19**: 299–304.
2. KEHLET, H. 1997. Multimodal approach to control postoperative pathophysiology and rehabilitation. Br. J. Anaesth. **78**: 606–617.

3. MATZINGER, P. 2002. An innate sense of danger. Ann. N. Y. Acad. Sci. **961:** 341–342.
4. FRIEDERICH, P. 2003. Basic concepts of ion channel physiology and anaesthetic drug effects. Eur. J. Anaesthesiol. **20:** 343–353.
5. BRAND, J.M., C. FROHN, J. LUHM, et al. 2003. Early alterations in the number of circulating lymphocyte subpopulations and enhanced proinflammatory immune response during opioid-based general anesthesia. Shock **20:** 213–217.
6. LIENER, U.C., U.B. BRUCKNER, M.W. KNOFERL, et al. 2002. Chemokine activation within 24 hours after blunt accident trauma. Shock **17:** 169–172.
7. PATHAN, N., C.A. HEMINGWAY, A.A. ALIZADEH, et al. 2004. Role of interleukin 6 in myocardial dysfunction of meningococcal septic shock. Lancet **363:** 203–209.
8. PRUCHA, M., A. RURYK, H. BORISS, et al. 2004. Expression profiling: toward an application in sepsis diagnostics. Shock **22:** 29–33.
9. TABRIZI, A.R., B.A. ZEHNBAUER, B.D. FREEMAN, et al. 2001. Genetic markers in sepsis. J. Am. Coll. Surg. **192:** 106–117; quiz 145–146.
10. KNIGHT, P.R., A. SREEKUMAR, J. SIDDIQUI, et al. 2004. Development of a sensitive microarray immunoassay and comparison with standard enzyme-linked immunoassay for cytokine analysis. Shock **21:** 26–30.
11. WARD, P.A. 2004. The dark side of C5a in sepsis. Nat. Rev. Immunol. **4:** 133–142.
12. CHINNAIYAN, A.M., M. HUBER-LANG, C. KUMAR-SINHA, et al. 2001. Molecular signatures of sepsis: multiorgan gene expression profiles of systemic inflammation. Am. J. Pathol. **159:** 1199–1209.
13. WHITNEY, A.R., M. DIEHN, S.J. POPPER, et al. 2003. Individuality and variation in gene expression patterns in human blood. Proc. Natl. Acad. Sci. USA **100:** 1896–1901.
14. ROSENWALD, A., G. WRIGHT, W.C. CHAN, et al. 2002. The use of molecular profiling to predict survival after chemotherapy for diffuse large-B-cell lymphoma. N. Engl. J. Med. **346:** 1937–1947.
15. BULLINGER, L., K. DOHNER, E. BAIR, et al. 2004. Use of gene-expression profiling to identify prognostic subclasses in adult acute myeloid leukemia. N. Engl. J. Med. **350:** 1605–1616.
16. GOLUB, T.R. 2003. Mining the genome for combination therapies. Nat. Med. **9:** 510–511.
17. COBB, J.P. & T.G. BUCHMAN. 2001. MicroArRAY of hope. Shock **16:** 264–265.
18. PETRICOIN, E.F. 3RD, J.L. HACKETT, L.J. LESKO, et al. 2002. Medical applications of microarray technologies: a regulatory science perspective. Nat. Genet. **32**(Suppl.): 474–479.
19. MCGILVRAY, I.D., V. TSAI, J.C. MARSHALL, et al. 2002. Monocyte adhesion and transmigration induce tissue factor expression: role of the mitogen-activated protein kinases. Shock **18:** 51–57.
20. GEORGESCU, S.P., I. KOMURO, Y. HIROI, et al. 1997. Downregulation of polo-like kinase correlates with loss of proliferative ability of cardiac myocytes. J. Mol. Cell Cardiol. **29:** 929–937.
21. JOHNSON, M.E., J.C. SILL, C.B. UHL, et al. 1996. Effect of volatile anesthetics on hydrogen peroxide-induced injury in aortic and pulmonary arterial endothelial cells. Anesthesiology **84:** 103–116.
22. KHADAROO, R.G., J. PARODO, K.A. POWERS, et al. 2003. Oxidant-induced priming of the macrophage involves activation of p38 mitogen-activated protein kinase through an Src-dependent pathway. Surgery **134:** 242–246.

23. KHADAROO, R.G., R. HE, J. PARODO, et al. 2004. The role of the Src family of tyrosine kinases after oxidant-induced lung injury in vivo. Surgery **136:** 483–488.
24. ZHANG, D., J. LI, Z. JIANG, et al. 2003. The relationship between tumor necrosis factor-alpha gene polymorphisms and acute severe pancreatitis. Chin. Med. J. (Engl.) **116:** 1779–1781.
25. O'SUILLEABHAIN, C.B., S. KIM, M.R. RODRICK, et al. 2001. Injury induces alterations in T-cell NFkappaB and AP-1 activation. Shock **15:** 432–437.
26. CHEN, D., L.G. XU, L. CHEN, et al. 2003. NIK is a component of the EGF/heregulin receptor signaling complexes. Oncogene **22:** 4348–4355.
27. REGAN, C.P., W. LI, D.M. BOUCHER, et al. 2002. Erk5 null mice display multiple extraembryonic vascular and embryonic cardiovascular defects. Proc. Natl. Acad. Sci. USA **99:** 9248–9253.
28. DINEV, D., B.W. JORDAN, B. NEUFELD, et al. 2001. Extracellular signal regulated kinase 5 (ERK5) is required for the differentiation of muscle cells. EMBO Rep. **2:** 829–834.
29. LEE, K.S., S.H. HONG & S.C. BAE. 2002. Both the Smad and p38 MAPK pathways play a crucial role in Runx2 expression following induction by transforming growth factor-beta and bone morphogenetic protein. Oncogene **21:** 7156–7163.
30. ZHANG, X., Z. CHEN, H. HUANG, et al. 2002. DNA microarray analysis of the gene expression profiles of naive versus activated tumor-specific T cells. Life Sci. **71:** 3005–3017.
31. TATSUYAMA, K., Y. MAEZAWA, H. BABA, et al. 2000. Expression of various growth factors for cell proliferation and cytodifferentiation during fracture repair of bone. Eur. J. Histochem. **44:** 269–278.
32. CALVO, S., J. STAUFFER, M. NAKAYAMA, et al. 1996. Transcriptional control of muscle plasticity: differential regulation of troponin I genes by electrical activity. Dev. Genet. **19:** 169–181.
33. LIU, L., J.E. CAVANAUGH, Y. WANG, et al. 2003. ERK5 activation of MEF2-mediated gene expression plays a critical role in BDNF-promoted survival of developing but not mature cortical neurons. Proc. Natl. Acad. Sci. USA **100:** 8532–8537.
34. GARCIA-LORA, A., M. MARTINEZ, S. PEDRINACI, et al. 2003. Different regulation of PKC isoenzymes and MAPK by PSK and IL-2 in the proliferative and cytotoxic activities of the NKL human natural killer cell line. Cancer Immunol. Immunother. **52:** 59–64.
35. LIU, L.X., Z.H. LIU, H.C. JIANG, et al. 2003. Gene expression profiles of hepatoma cell line HLE. World J. Gastroenterol. **9:** 683–687.
36. NOMIYAMA, H., S. FUKUDA, M. IIO, et al. 1999. Organization of the chemokine gene cluster on human chromosome 17q11.2 containing the genes for CC chemokine MPIF-1, HCC-2, HCC-1, LEC, and RANTES. J. Interferon. Cytokine. Res. **19:** 227–234.
37. OBATA-ONAI, A., S. HASHIMOTO, N. ONAI, et al. 2002. Comprehensive gene expression analysis of human NK cells and CD8(+) T lymphocytes. Int. Immunol. **14:** 1085–1098.
38. GODSHALL, C.J., M.J. SCOTT, P.T. BURCH, et al. 2003. Natural killer cells participate in bacterial clearance during septic peritonitis through interactions with macrophages. Shock **19:** 144–149.

39. DIAS, A.A., A.R. GOODMAN, J.L. DOS SANTOS, *et al.* 2001. TSG-14 transgenic mice have improved survival to endotoxemia and to CLP-induced sepsis. J. Leukoc. Biol. **69:** 928–936.
40. EFRON, P. & L.L. MOLDAWER. 2003. Sepsis and the dendritic cell. Shock **20:** 386–401.
41. FRIES, M., D. KUNZ, A.M. GRESSNER, *et al.* 2003. Procalcitonin serum levels after out-of-hospital cardiac arrest. Resuscitation **59:** 105–109.
42. SCHNEEMILCH, C.E. & U. BANK. 2001. Release of pro- and anti-inflammatory cytokines during different anesthesia procedures. Anaesthesiol. Reanim. **26:** 4–10.
43. GILLILAND, H.E., M.A. ARMSTRONG, U. CARABINE, *et al.* 1997. The choice of anesthetic maintenance technique influences the antiinflammatory cytokine response to abdominal surgery. Anesth. Analg. **85:** 1394–1398.
44. ANGELASTRO, J.M., T.N. IGNATOVA, V.G. KUKEKOV, *et al.* 2003. Regulated expression of ATF5 is required for the progression of neural progenitor cells to neurons. J. Neurosci. **23:** 4590–4600.
45. PETERS, C.S., X. LIANG, S. LI, *et al.* 2001. ATF-7, a novel bZIP protein, interacts with the PRL-1 protein-tyrosine phosphatase. J. Biol. Chem. **276:** 13718–13726.
46. FARMER, P. & J. PUGIN. 2000. Beta-adrenergic agonists exert their "antiinflammatory" effects in monocytic cells through the IkappaB/NF-kappaB pathway. Am. J. Physiol. Lung. Cell Mol. Physiol. **279:** L675–L682.
47. HAMAYA, Y., T. TAKEDA, S. DOHI, *et al.* 2000. The effects of pentobarbital, isoflurane, and propofol on immediate-early gene expression in the vital organs of the rat. Anesth. Analg. **90:** 1177–1183.
48. WELTERS, I.D., A. MENZEBACH, Y. GOUMON, *et al.* 2000. Morphine inhibits NF-kappaB nuclear binding in human neutrophils and monocytes by a nitric oxide-dependent mechanism. Anesthesiology **92:** 1677–1684.
49. ELENKOV, I.J., D.A. PAPANICOLAOU, R.L. WILDER, *et al.* 1996. Modulatory effects of glucocorticoids and catecholamines on human interleukin-12 and interleukin-10 production: clinical implications. Proc. Assoc. Am. Physicians **108:** 374–381.
50. STEER, J.H., K.M. KROEGER, L.J. ABRAHAM, *et al.* 2000. Glucocorticoids suppress tumor necrosis factor-alpha expression by human monocytic THP-1 cells by suppressing transactivation through adjacent NF-kappa B and c-Jun-activating transcription factor-2 binding sites in the promoter. J. Biol. Chem. **275:** 18432–18440.
51. PAVLICKOVA, P., E.M. SCHNEIDER & H. HUG. 2004. Advances in recombinant antibody microarrays. Clin. Chim. Acta **343:** 17–35.

Effects of Chemical Ischemia on Cerebral Cortex Slices

Focus on Mitogen-Activated Protein Kinase Cascade

ANNA SINISCALCHI,[a] SABRINA CAVALLINI,[a] SILVIA MARINO,[a] SOFIA FALZARANO,[b] LARA FRANCESCHETTI,[b] AND RITA SELVATICI[b]

[a]*Department of Clinical and Experimental Medicine, Section of Pharmacology, University of Ferrara, 44100 Ferrara, Italy*

[b]*Department of Experimental and Diagnostic Medicine, Section of Medical Genetics, University of Ferrara, 44100 Ferrara, Italy*

ABSTRACT: A variety of harmful stimuli, among them energy depletion occurring during transient brain ischemia, are thought to unbalance protein kinase cascades, ultimately leading to neuronal damage. In superfused, electrically stimulated rat cerebral cortex slices, chemical ischemia (CI) was induced by a 5-min treatment with the mitochondrial toxin, sodium azide (10 mM), combined with the glycolysis blocker, 2-deoxyglucose (2 mM). Thereafter, 1 h reperfusion (REP) with normal medium followed. Western blot analysis of $p21^{Ras}$, extracellular signal-regulated protein kinases (ERK)1/2 (p44/42), phospho-ERK1/2, mitogen-activated protein kinase (MAPK)-p38, phospho-p38, stress-activated protein kinases/c-Jun NH_2-terminal protein kinases (SAPK/JNK), phospho-SAPK/JNK was carried out. The level of $p21^{Ras}$ was increased by 40% immediately after CI, and did not return to control values following REP. Both ERK1 and ERK2 levels were reduced by CI and recovered to control values following REP; no significant change in their phosphorylation degree (phosphorylated to total level ratio, about 50% in the controls) was observed. Neither p38 levels, nor phosphorylation degree were changed following CI/REP. The activation of SAPK/JNK was significantly reduced under CI, and did not recover following REP. All CI/REP-induced effects were prevented by the NMDA receptor antagonist MK-801, 10 μM, suggesting the involvement of glutamate. The present findings show that although CI stimulates the $p21^{Ras}$ protein, MAPK levels and/or phosphorylation are reduced, possibly because of acute energy depletion. Because the activation of SAPK/JNK has been related to both apoptosis and neuroprotection, the decrease observed

Address for correspondence: Anna Siniscalchi, Department of Clinical and Experimental Medicine, Section of Pharmacology, University of Ferrara, Via Fossato di Mortara 17, 44100 Ferrara, Italy. Voice: +39-0532-291208; fax: +39-0532-291205.
   e-mail: snn@unife.it

under CI/REP conditions may instead be related to nonapoptotic neuronal death. These results could be of interest in developing preventive treatments for ischemia/REP-induced brain damage.

KEYWORDS: brain ischemia; cerebral cortex slices; chemical ischemia; sodium azide; 2-deoxyglucose; energy failure; reperfusion; $p21^{Ras}$; mitogen-activated protein kinase; ERK1/2 (p44/42); SAPK/JNK; p38; phosphorylation degree; neuronal damage; cell death; glutamate; NMDA receptor; MK-801; neurotoxicity; neuroprotection

INTRODUCTION

The mitogen-activated protein kinases (MAPKs) are a family of serine/threonine kinases involved in intracellular signaling related to cell proliferation, differentiation, inflammatory responses, and cell survival. Eukaryotic cells possess multiple MAPK pathways, which can be grouped into three categories: extracellular signal-regulated protein kinases (ERK); p38-MAPK (or p38); and stress-activated protein kinases/c-Jun NH_2-terminal protein kinases (SAPK/JNK). Each pathway can be activated by diverse stimuli, including growth factors, neurotransmitters, hormones, and cellular stresses such as those associated with cytokines, endotoxin, and oxidants. A correct balance between activation and inhibition of such signaling pathways is needed to ensure an appropriate signaling output.[1–3] The activation of the small G protein $p21^{Ras}$ has been reported to play a role upstream to protein kinases/protein phosphatase cascades, and to be in turn activated by depolarization and by increase in intracellular calcium concentration. Moreover, there are reports that exposure of neuronal cultures to either ischemic conditions or to N-methyl-D-asparate (NMDA) glutamate receptor stimulation results in $p21^{Ras}$ activation.[4–6]

Cultured neuronal cells undoubtedly represent a simple, useful model to study the changes induced by ischemia-like conditions in neuronal functions.[7–9] However, brain slices, in which the complexity of both neuronal circuitry and glial functions is at least partially preserved, are regarded as a more convenient *in vitro* model to study the multifaceted mechanisms of neurotoxicity/neuroprotection.[9–11] The energy failure that occurs during transient brain ischemia *in vivo* can be mimicked, *in vitro*, by treatment with mitochondrial toxins, as an alternative either to oxygen and glucose deprivation, or to high glutamate treatment. In our laboratory, the metabolic poison, sodium azide, combined with the glycolysis blocker, 2-deoxyglucose, has been used to induce chemical ischemia (CI) in rat cerebral cortex slices: a reversible failure in the neurosecretory function, assessed as acetylcholine release, and the involvement of glutamate overflow, calcium overload, and nitric oxide efflux in the mechanisms of CI have been shown.[12] Afterward, the downstream events following CI have been addressed by studying the roles played by the different protein kinase C isozymes.[13] In the present work, we extend the latter studies by investigating CI-induced changes on the $p21^{Ras}$/MAPK pathway. Because glutamate-induced excitotoxicity, mainly via NMDA receptors, is well-known to occur during brain ischemia/reperfusion,[14–17] we felt of interest to test the effectiveness of MK-801, a NMDA receptor antagonist, in preventing the changes triggered by CI.

MATERIALS AND METHODS

Cerebral Cortex Slices: Superfusion and Treatments

The experiments were carried out on male Sprague–Dawley rats (200–300 g), kept under regular lighting conditions (12-h light:12-h darkness cycle), with food and water available *ad libitum*, in accordance with protocols approved by the Ethics Committee of the University of Ferrara. Following decapitation under light ether anaesthesia, 400-μm-thick cerebral cortex slices were prepared and superfused (0.25 mL/min) with Krebs-Ringer buffer (KRB, mM composition: NaCl 118.5, KCl 4.7, $CaCl_2$ 1.25, $MgSO_4$ 1.2, KH_2PO_4 1.2, $NaHCO_3$ 25, glucose 10, pH 7.4), bubbled with 95% O_2, 5% CO_2 at 37°C. A continuous electrical stimulation (5 Hz, 1 msec, 20 mA/cm^2) was applied throughout the experiment.[18] In CI experiments, at the 30th min of superfusion, the medium was substituted for 5 min with a glucose-free, 2-deoxyglucose (2 mM)-containing KRB, to which 10 mM NaN_3 was added. The reperfusion (REP) conditions were mimicked by superfusing the cortical slices with normal, oxygenated KRB for 1 h. The corresponding control slices were superfused with normal, oxygenated KRB throughout the entire experiment. To test the effects of MK-801, the drug was added to the medium from the beginning of the superfusion.

MAP Kinase Analysis by Western Blotting

At the end of the superfusion experiment, the slices were rapidly placed in 1 mL ice-cold lysis buffer containing 5 mM HEPES, 320 mM sucrose, 5 mM glycerolphosphate, 5 mM KF, 2 mM 2-mercaptoethanol (2-ME), 3 mM EGTA, 0.5 mM $MgSO_4$, 2 mM phenyl-methyl-sulphonylfluoride (PMSF), 0.01% leupeptin, 10 μg/mL aprotinin, and then centrifuged at 14,000 rpm for 5 min to pellet nuclei and unbroken cells. The supernatant was recovered in a separate tube and sonicated six times with 10-sec bursts. Protein content was determined by Bradford's method.[19] Samples containing equal amounts of proteins (20 μg) were diluted with one-third of the loading buffer (187.5 mM Tris-HCl, pH 6.8, 15% 2-ME, 0.1% sodium dodecylsulfate, 30% glycerol, 0.003% bromophenol blue), subjected to gel electrophoresis on a 10% gel and then transferred to PVDF membranes (0.2 μm; Bio-Rad, Milan, Italy) by electroblotting. Membranes were immersed overnight in a Tris buffered saline solution (TBS: 20 mM

Tris, 137 mM NaCl) pH 7.6 containing 5% blotting grade blotter (Bio-Rad) at 4°C to block nonspecific binding, washed three times with TBS plus 0.1% Tween 20 (TBS-T) and incubated for 2 h at room temperature with rabbit polyclonal antibodies against p38, pospho-p38, p44/42 (ERK1/2), phospho-p44/42, JNK, phospho-JNK (Cell Signaling) used 1:1000 in TBS-T buffer. Following washes with TBS-T buffer, a 1:6000 dilution of horseradish peroxidase labelled anti-rabbit IgG was added at room temperature for 1 h. ECL^{TM} Western blotting detection reagents (Amersham, Milan, Italy) were used to visualize specific hybridization signals. Densitometric analysis of autoradiographic bands was performed with a Bio-Rad densitometer GS700 and expressed as optical density (OD) units/mm.[2]

Statistics

Data are given as means ± SEM. The significance of differences between treated and control samples, run in parallel, has been assessed with the Student's t-test for nonpaired data.

Materials

Leupeptin, 2-mercapto-ethanol, phenyl-methyl-sulphonylfluoride, phosphatidyl-serine, and sodium azide were purchased from Sigma Chemical Company (St. Louis, MO); 2-deoxyglucose and Nonidet P-40 from ICN Pharmaceutical (Milan, Italy); MK-801 (dizocilpine maleate) from Tocris (Northpoint, UK); all other chemicals were from standard commercial sources and were of the highest purity available.

RESULTS

As shown in FIGURE 1, p21^{Ras} levels were significantly increased immediately after 5-min CI and remained elevated following 1-h REP. The NMDA receptor antagonist MK-801, 10 μM, prevented CI/REP effects, demonstrating the involvement of glutamate.

TABLE 1 shows the changes observed in MAPKs total protein levels. Both ERK1 and ERK2 (p44/42) total levels were significantly reduced by CI conditions; REP restored total protein levels to values which did not significantly differ from the controls. Once again, 10 μM MK-801 prevented CI-induced changes. Conversely, p38-MAPK levels were not significantly affected by any experimental condition. Similarly, immunoblot analysis of total SAPK/JNK did not show any significant difference between control, CI, and REP groups, although a trend to reduction by CI was apparent (TABLE 1).

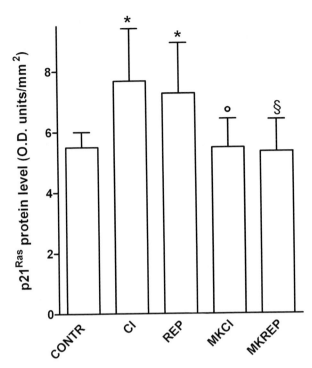

FIGURE 1. $p21^{Ras}$ protein levels in rat cerebral cortex slices. Effects of 5-min CI (10 mM sodium azide + 2 mM 2-deoxyglucose), of 1-h REP, and of the pretreatment with 10 μM MK-801 (NMDA receptor antagonist). Histograms represent OD units/mm^2 (mean ± SEM) in arbitrary scale. *$P < 0.05$ versus control values; °$P < 0.05$ versus CI; §$P < 0.05$ versus REP, Student's t-test.

To determine whether the activation of ERK1/2, p38-MAPK, and SAPK/JNK was altered by CI, cerebral cortex slice homogenates were subjected to Western blot analysis using antibodies against their active, phosphorylated forms. TABLE 2 reports the phosphorylation degrees (i.e., the ratios between phosphorylated forms to total protein levels × 100) of ERK1, ERK2, and p38-MAPK. Note that under the control conditions ERK1/2 were partially activated (phosphorylation degree about 50%), while p38-MAPK was almost completely in the nonphosphorylated form. The data reported in TABLE 2 demonstrate unchanged phosphorylation degrees for ERK1/2 and p38-MAPK under CI/REP conditions.

Interestingly, a reduction in the phosphorylation degree immediately after CI, persistent after REP, was displayed by SAPK/JNK (FIG. 2). MK-801 fully prevented CI/REP-induced effects. It should be noted that phospho-SAPK/JNK showed higher densitometric readings than total SAPK/JNK; this is probably a function of differences in the reactivities of the antibodies raised against the phosphorylated compound to the total proteins.[20]

TABLE 1. MAP Kinases total protein levels in rat cerebral cortex slices

MAP Kinase	Control	CI	REP	CI + MK-801
ERK1	10.52 ± 3.7	5.11 ± 0.85*	7.70 ± 0.73°	7.49 ± 0.76°
ERK2	17.57 ± 2.5	7.65 ± 1.6*	12.09 ± 0.22°°	12.48 ± 0.66°
p38	10.81 ± 1.2	10.96 ± 1.1	11.40 ± 1.6	11.72 ± 2.3
SAPK/JNK	2.95 ± 0.88	2.28 ± 0.46	3.12 ± 0.99	2.43 ± 1.04

Effects of 5-min CI (10 mM sodium azide + 2 mM 2-deoxyglucose), of 1-h REP, and of the pretreatment with 10 μM MK-801 (NMDA receptor antagonist).
Values are means ± SEM of OD units/mm².
*$P < 0.01$ vs. control values, °$P < 0.05$, °°$P < 0.01$ vs. CI, Student's t-test.

DISCUSSION

In the present work an *in vitro* model has been used to investigate the role played by the p21^{Ras}-MAPK cascade in the events that follow brain ischemia-REP. Sodium azide, whose mechanism of action mainly relies on cytochrome c oxidase—respiratory chain complex IV—inhibition,[21] has already been used either alone or combined with the glycolysis blocker, 2-deoxyglucose[22] to induce CI both *in vitro* [12,23–26] and *in vivo*. [27,28]

The principal results obtained in the rat cerebral cortex slices, submitted to 5-min CI, followed by 1-h REP, were: (i) persistent activation of p21^{Ras}; (ii) transient reduction of ERK1/2 total levels; (iii) decrease in SAPK/JNK phosphorylation degree. All these changes were prevented by the NMDA glutamate receptor antagonist, MK-801, in agreement with the central role recognized for glutamate-induced excitotoxicity in ischemic neuronal damage.[14–17]

The increase observed under CI/REP conditions in the level of the small G-protein p21^{Ras}, which operates upstream to the MAPK cascade,[4–6] was expected. In fact, we have previously found enhanced glutamate efflux during CI *in vitro*.[12] Moreover, it has been shown that sodium azide induces cell depolarization and increases intracellular calcium concentration in neuronal cells in culture, at least in part via NMDA receptor stimulation.[23,25] All these have been shown to induce activation of p21^{Ras}.[4–6]

TABLE 2. MAP kinase phosphorylation degrees in rat cerebral cortex slices

MAP kinase	Control	CI	REP
ERK1	50.0 ± 14	54.9 ± 9	39.0 ± 6
ERK2	55.1 ± 8	65.1 ± 13	53.5 ± 18
p38	17.4 ± 2	15.8 ± 2	16.9 ± 2

Effects of 5-min CI (10 mM sodium azide + 2 mM 2-deoxyglucose), of 1-h REP, and of the pretreatment with 10 μM MK-801 (NMDA receptor antagonist).
Values represent the percentage ratios between phosphorylated and nonphosphorylated forms (mean ± SEM).

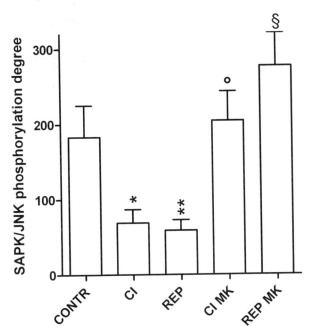

FIGURE 2. SAPK/JNK phosphorylation degree in rat cerebral cortex slices. Effects of 5-min CI (10 mM sodium azide + 2 mM 2-deoxyglucose), of 1-h REP, and of the pretreatment with 10 μM MK-801 (NMDA receptor antagonist). Histograms represent the percentage ratios between phosphorylated and nonphosphorylated forms (mean ± SEM). $P < 0.05$; **$P < 0.01$ versus control values; °$P < 0.01$ versus CI; §$P < 0.01$ versus REP, Student's t-test.

However, under the present experimental conditions, p21^{Ras} activation does not appear to be linked to the activation of MAPKs.

In fact, the total protein levels of ERK1/2 were reduced by CI and were restored by REP, suggesting a connection with acute energy depletion: it should be noted that in our experimental model the cortical slices were continuously electrically stimulated, hence they were metabolically active and highly dependent on energy supply.[18] Moreover, it is interesting to note that inhibition in protein synthesis during CI has been reported.[29]

On the other hand, the phosphorylation degrees of ERK1/2 as well as of p38 did not change under CI/REP conditions, ruling out a pivotal role for these kinases and for their substrates both in the effects triggered by CI and in the neurotoxic/neuroprotective mechanisms taking place during REP. These findings are at variance with the important roles attributed to ERK1/2 and to p38 activation in the mechanisms of cell damage/neuroprotection in different experimental models of brain ischemia and neurodegeneration.[30–32] In a previous study, we observed that even a brief (5 min) period of CI was able to induce a profound derangement in protein kinase C, indicative of irreversible cell damage, while the changes induced by oxygen-glucose deprivation were interpreted as a trend to react to the ischemic insult.[13] Thus, the discrepancies observed in the present study between data in the existing literature could be explained by the features of the experimental model employed, which mimics an acute, severe insult, instead of a more prolonged, milder one. Further experiments, with a longer treatment of cortical slices and a lower sodium azide concentration, should answer this question.

Activation of SAPK/JNKs has been related to both proapoptotic and neuroprotective actions. Although SAPK/JNKs exist in different isoforms, possibly mediating even opposite actions,[33,34] the net role they play in ischemic/REP conditions is likely to be neuroprotective. However, the ultimate consequence may vary depending on the environmental conditions and the co-ordination of other signaling molecules.[20] A downregulation, possibly resulting from the activation of the protein phosphatase 2A, was found by Liu et al.[10] in anoxic slices, and JNK1 has been reported to be required for maintenance of cytoskeleton integrity in neuronal cells.[35] Accordingly, the reduction in SAPK/JNK phosphorylation displayed under CI/REP conditions could be explained by nonapoptotic death, to which neurons submitted to brief severe CI undergo (unpublished observations).

In conclusion, ischemic injury to the brain induces a complex response, resulting in activating/silencing both death and survival signaling pathways[36]: dysfunction in protein kinase cascades, and consequent inbalance in protein kinases and protein phosphatases,[10,20] represents one of the factors involved in such a response. A full knowledge of the substrates phosphorylated by the single kinases and of the roles they play in the development of neuronal damage is still required in view of future treatment strategies for brain ischemia targeting kinase pathways.

The simple experimental protocol utilized in this study models the mitochondrial dysfunction taking place during severe ischemia. Although the combined treatment with sodium azide and 2-deoxyglucose possibly triggers irreversible neuronal damage, the effects displayed on different parameters examined in this and in previous studies (neurosecretory function, PKCs, MAPKs) could be prevented by pharmacological treatment. Thus, the protocol of CI/REP in brain slices can be considered suitable for the screening of potential neuroprotective drugs.

ACKNOWLEDGMENTS

This work was supported by University of Ferrara grants to Anna Siniscalchi and by the E. and E. Rulfo for Medical Genetics, Parma, Italy and Fondazione Cassa di Risparmio of Ferrara, Italy to Rita Selvatici.

REFERENCES

1. SEGER, R. & E.G. KREBS. 1995. The MAPK signaling cascade. FASEB J. **9:** 726–735.
2. WIDMANN, C., S. GIBSON, et al. 1999. Mitogen activated protein kinase: conservation of a three-kinase module from yeast to human. Physiol. Rev. **79:** 143–180.
3. KYRIAKIS, J.M. & J. AVRUCH. 2001. Mammalian mitogen-activated protein kinase signal transduction pathways activated by stress and inflammation. Physiol. Rev. **81:** 807–869.
4. AGELL, N., O. BACHS, et al. 2002. Modulation of the Ras/Raf/MEK/ERK pathway by Ca^{2+} and calmodulin. Cell Signal. **14:** 649–654.
5. SWEATT, J.D. 2001. The neuronal MAP kinase cascade: a biochemical signal integration system subserving synaptic plasticity and memory. J. Neurochem. **76:** 1–10.
6. YUN, H.-Y., M. GONZALEZ-ZULUETA, et al. 1998. Nitric oxide mediates N-methyl-D-aspartate receptor-induced activation of p21. Proc. Natl. Acad. Sci. **95:** 5773–5778.
7. CHIHAB, R., C. BOSSENMEYER, et al. 1998. Lack of correlation between the effects of transient exposure to glutamate and those of hypoxia/reoxygenation in immature neurons *in vitro*. J. Neurochem. **71:** 1177–1186.
8. BICKLER, P.E., C.S. FAHLMAN, et al. 2004. Hypoxia increases calcium flux through cortical neuron glutamate receptors via protein kinase C. J. Neurochem. **88:** 878–884.
9. SELVATICI, R., S. MARINO, et al. 2003. Protein kinase C activity, translocation, and selective isoform subcellular redistribution in the rat cerebral cortex after *in vitro* ischemia. J. Neurosci. Res. **71:** 64–71.
10. LIU, R., J.-J. PEI, et al. 2005. Acute anoxia induces tau dephosphorylation in rat brain slices and its possible underlying mechanisms. J. Neurochem. **94:** 1125–1234.
11. SBRENNA, S., G. CALO, et al. 1998. Experimental protocol for studying delayed effects of in vitro ischemia on neurotransmitter release from brain slices. Brain Res. Protocols **3:** 61–67.
12. CAVALLINI, S., M. MARTI, et al. 2005. Effects of chemical ischemia in cerebral cortex slices. Focus on nitric oxide. Neurochem. Int. **47:** 482–490.
13. SELVATICI, R., S. FALZARONO, et al. 2006. Differential activation of protein kinase C isoforms following chemical ischemia in rat cerebral cortex slices. Neurochem. Int. In press.
14. CHOI, D.W. 1990. The role of glutamate neurotoxicity in hypoxic-ischemic neuronal death. Annu. Rev. Neurosci. **13:** 171–182.
15. HARDINGHAM, G.E. & H. BADING. 2003. The Yin and Yang of NMDA receptor signalling. Trends Neurosci. **26:** 81–89.
16. LO, E.H., T. DALKARA, et al. 2003. Mechanism, challenges and opportunities in stroke. Nat. Rev. Neurosci. **4:** 399–415.
17. BONDE, J., J. NORABERG, et al. 2005. Ionotropic glutamate receptors and glutamate transporters are involved in necrotic neuronal cell death induced by oxygen-glucose deprivation of hippocampal slice cultures. Neuroscience **136:** 779–794.
18. BADINI, I., L. Beani, et al. 1997. Post-ischemic recovery of acetylcholine release *in vitro*: influence of different excitatory amino acid receptor subtype antagonists. Neurochem. Int. **31:** 817–824.

19. BRADFORD, M.M. 1976. A rapid and sensitive method for the quantitation of microgram quantities of protein utilizing the principle of protein-dye binding. Anal. Biochem. **72:** 248–254.
20. LEE H.-P., Y.-C. JUN, *et al*. 2005. Dysregulation of MAPK pathways in prion disease J. Neurochem. **95:** 584–593.
21. DURANTEAU, J., N.S. CHANDEL, *et al*. 1998. Intracellular signaling by reactive oxygen species during hypoxia in cardiomyocytes. J. Biol. Chem. **273:** 11619–11624.
22. CHI, M.M., M.E. PUSATERI, *et al*. 1987. Enzymatic assays for 2-deoxyglucose and 2-deoxyglucose 6-phosphate. Anal. Biochem. **161:** 508–513.
23. ANDO, H. 1990. Novel effect of azide on sodium channel of *Xenopus* oocytes. Biochem. Biophy. Res. Commun. **172:** 300–305.
24. VARMING, T., J. DREJER, *et al*. 1996. Characterization of a chemical anoxia model in cerebellar granule neurons using sodium azide: protection by nifedipine and MK-801. J. Neurosci. Res. **44:** 40–46.
25. JORGENSEN, N.K., S.F. PETERSEN, *et al*. 1999. Increases in [Ca2+]i and changes in intracellular pH during chemical anoxia in mouse neocortical neurons in primary culture. J. Neurosci. Res. **56:** 358–370.
26. GRAMMATOPOULOS, T.N., V. JOHNSON, *et al*. 2004. Angiotensin type 2 receptor neuroprotection against chemical hypoxia is dependent on the delayed rectifier K^+ channel, Na^+/Ca^{2+} exchanger and Na^+/K^+ ATPase in primary cortical cultures. Neurosci. Res. **50:** 299–306.
27. BENNETT, M.C., G.W. MLADY, *et al*. 1996. Chronic in vivo sodium azide infusion induces selective and stable inhibition of cytochrome c oxidase. J. Neurochem. **66:** 2606–2611.
28. VECSEI, L., J. TAJTI, *et al*. 2001. Sodium azide treatment decreases striatal and cortical concentrations of alpha-tocopherol in rats. J. Neural Transm. **108:** 273–278.
29. MAUS, M., Y. TORRENS, *et al*. 2005. 2-deoxyglucose and NMDA inhibit protein synthesis in neurons and regulate phosphorylation of elongation factor-2 by distinct mechanisms. J. Neurochem. doi 10.1111/J.1471-4159.2005.03601.x
30. HETZMAN, M. & A. GOZDZ. 2004. Role of extracellular signal regulated kinases 1 and 2 in neuronal survival. Eur. J. Biochem. **271:** 2050–2055.
31. CHU, C.T., D.J. LEVINTHAL, *et al*. 2004. Oxidative neuronal injury. The dark side of ERK1/2. Eur. J. Biochem. **271:** 2060–2066.
32. ZARUBIN, T. & J. HAN. 2005. Activation and signaling of the p38 MAP kinase pathway. Cell Res. **15:** 11–18.
33. BRECHT, S., R. KIRCHHOF, *et al*. 2005. Specific pathophysiological functions of JNK isoforms in the brain. Eur. J. Neurosci. **21:** 363–377.
34. WAETZIG, V. & T. HERDENGEN. 2005. Context-specific inhibition of JNKs: overcoming the dilemma of protection and damage. Trends Pharmacol. Sci. **26:** 455-461.
35. CHANG, L., Y. JONES, *et al*. 2003. JNK1 is required for maintenance of neuronal microtubules and controls phosphorylation of microtubule-associated proteins. Dev. Cell **4:** 521–533.
36. BUTLER, T.L., C.A. KASSED, *et al*. 2003. Signal transduction and neurosurvival in experimental models of brain injury. Brain Res. Bull. **5:** 339–351.

Amyloid Precursor Protein Modulates ERK-1 and -2 Signaling

VALENTINA VENEZIA,[a] MARIO NIZZARI,[a] EMANUELA REPETTO,[a] ELISABETTA VIOLANI,[a] ALESSANDRO CORSARO,[a] STEFANO THELLUNG,[a] VALENTINA VILLA,[a] PIA CARLO,[a] GENNARO SCHETTINI,[a] TULLIO FLORIO,[a] AND CLAUDIO RUSSO[b]

[a]*Department of Oncology, Biology and Genetics, University of Genova, L.go R. Benzi 10, 16132 Genova, Italy*

[b]*Department of Health Sciences, University of Molise, Campobasso, Italy*

ABSTRACT: The amyloid precursor protein (APP) is a transmembrane protein with a short cytoplasmic tail whose physiological function is unclear, although it is well documented that the proteolytic processing of APP could influence the development of Alzheimer's disease (AD) through the formation of membrane-bound C-terminal fragments (CTFs) and of β-amyloid peptides (Aβ). We have recently shown that tyrosine-phosphorylated APP and CTFs may interact with Grb2 and ShcA adaptor proteins and that this coupling occurs at a higher extent in AD subjects only. To study the interaction between APP or CTFs and ShcA/Grb2 and to investigate their molecular target we have used as experimental model two different cell lines: H4 human neuroglioma cells and APP/APLP null mouse embryonic fibroblast cells (MEFs). Here we show that in H4 cells APP interacts with Grb2; conversely in APP/APLP-null MEF cells this interaction is possible only after the reintroduction of human APP by transfection. We have also shown that in MEF cells the transfection of a plasmid encoding for human APP wild-type enhances the phosphorylation of ERK-1 and -2 as revealed by Western blotting and immunofluorescence experiments. Finally, also in H4 cells the overexpression of APP upregulates the levels of phospho-ERK-1 and -2. In summary our data suggest that APP may influence phospho-ERK-1 and -2 signaling through its binding with Grb2 and ShcA adaptors. The meaning of this event is not clear, but APP interaction with these adaptors could be relevant to regulate mitogenic pathway.

KEYWORDS: amyloid precursor protein; Alzheimer's disease; adaptor proteins; phospho-ERK-1 and -2

Address for correspondence: Claudio Russo, Ph.D., Department of Health Sciences, University of Molise, Via De Sanctis, 86100 Campobasso, Italy. Voice: +39-010-35338253; fax: +39-010-3538806.
e-mail: claudio.russo@unimol.it

INTRODUCTION

Alzheimer's disease (AD) is a heterogeneous neurodegenerative disorder with insidious onset and irreversible progression. The prognosis of the disease is an inexorable decline of cognitive functions and to date there are no effective therapeutics.[1,2] The amyloid precursor protein (APP) is a type I transmembrane protein that undergoes enzymatic processing to form many membrane-bound carboxy-terminal fragments (CTFs), which can induce a direct apoptotic neurodegeneration when overexpressed *in vitro* and *in vivo*,[3] and highly insoluble peptides named β-amyloid peptides (Aβ), which constitute the core of plaques detected in the brain parenchyma of AD patients.[2,4]

The cytodomain of APP is the center of a complex network of interactions with several proteins, in particular, the last 20 amino acids of the carboxy-terminal region of APP constitute a docking site for X11,[5] Fe65,[6] mDab,[7] c-Abl,[8] JIP-1,[9] and Numb.[10] The phosphorylation of the two tyrosines of the YENPTY motif generates two different docking sites, which can be potentially recognized by the Src homology-2 domain (SH2) or by the phosphotyrosine-binding domain (PTB) of several intracellular adaptors. We have recently identified two new binding partners for APP: ShcA[11] and Grb2,[12] whose interaction with APP requires the specific tyrosine phosphorylation of Tyr 682 of APP.[13] The fact that the amounts of these complexes[14] are significantly increased in AD brain suggests a pathogenic correlation. Since the family of Shc and Grb2 adaptors do usually connect growth factor receptors to specific signaling pathways (typically Ras/mitogen-activated protein kinase [MAPK]), we investigated the effect of the interaction between APP and ShcA/Grb2 on the activation of the downstream target of this pathway: ERK-1 and -2. Here we provide evidence that the overexpression of APP695 by transient transfection significantly enhanced the basal activation of ERK-1 and -2 MAPK Thus we can hypothesize that the selective phosphorylation of APP, or of its CTFs, might couple them via ShcA and/or Grb2 to these cellular pathways, the understanding of which may have implications both for the normal biological function of APP and for its pathological role in the genesis of AD.

MATERIALS AND METHODS

Cell Culture

APP/APLP-null immortalized mice embryonic fibroblasts cells (MEF 1.1, provided by Dr. U. Mueller) and human neuroglioma cells (H4) were cultured in Dulbecco's-modified Eagle's medium (DMEM) (EuroClone, Paignton-Devon, UK) supplemented with 10% fetal bovine serum (Gibco/Life Technologies, Rockville, MD, USA), 0.5 mM L-glutamine, 0.5 mM penicillin, and 0.5 mM streptomycin (EuroClone, Paignton-Devon, UK).

SDS-PAGE, Immunoblotting, and Coprecipitation

Cells were lysed in a buffer containing 100 mM NaCl, 10 mM Tris, 10 mM EDTA, 0.5% NP40, 0.5% cholic acid, pH 7.6. After a 10-min centrifugation at 200 g, cell lysates were either cold-methanol-precipitated, and the resulting pellets analyzed by Western blotting after protein counting (Bio-Rad Protein Assay, Bio-Rad Laboratories, Hercules, CA, USA) or immunoprecipitated with different antibodies as specified below. The antigen–antibody complexes were collected by protein A-agarose beads, which were then electrophoresed by Tris-tricine SDS-PAGE and electroblotted on PVDF membrane, and proteins were probed with specific antibodies, hybridized with a secondary antibody HRP-conjugated, and detected by chemiluminescence (ECL, Amersham Pharmacia, Uppsala, Sweden).

Antibodies and Drugs

A specific polyclonal antibody for the C terminus of APP, the 51-2700 (Zymed Laboratories, South San Francisco, CA, USA), was used 1:1,000 in immunoblotting and 1:100 in immunoprecipitation. Anti-ShcA antibody (Upstate Biotechnology, Lake Placid, NY, USA) was used 1:200 for immunoprecipitation. Monoclonal antibody specific for Grb2 (Transduction Laboratories, Lexington, KY, USA) was used 1:100 for immunoprecipitation. The phospho-ERK-1 and -2 and unphosphorylated ERK-1 and -2 specific antibodies (New England Biolabs, Beverly, MA, USA) were used both 1:1,000 for immunodetection and 1:100 for immunofluorescence. The monoclonal antibody specific for γ-tubulin (Sigma, St. Louis, MO, USA) was used 1:200 in immunofluorescence. All the chemicals were from Sigma unless otherwise specified.

DNA Transfection

APP/APLP1-null immortalized mice embryonic fibroblasts (MEF 1.1 cells provided by Dr. U. Mueller), were grown in DMEM and transfected with a plasmid encoding APP695 under CMV promoter (Invitrogen S.r.l., Milan, Italy) as previously described.[8] The construct was sequenced by dideoxy terminator method from TIB Molbiol S.r.l. (Genova, Italy). The plasmid DNA was propagated by standard procedure on XL10-Gold competent cells and purified on Qiagen midi-column (Qiagen, Hilden, Germany).

Cells were seeded at 2×10^6 cells/100-mm cell culture dish and transfected with 5 µg DNA vector encoding for human APP695 mixed with 6 µL jet-PEI™ (Polyplus-Transfection, Illkirch, France) for 24 h and then were lysed as previously described.

Immunofluorescence and Confocal Analysis

The cells were seeded at 10^5 cells/well on glass coverslips in 12-well plates and transfected with 1 μg DNA vector encoding for APP695 mixed with 1.2 μL jetPEI™ (Polyplus-Transfection) for 24 h. Cells were cold-methanol-fixed for 3 min and then were incubated for 1 h with an antibody specific for the phosphorylated isoform of ERK-1 and -2 (New England Biolabs) and with an antibody specific for γ-tubulin (Sigma). Finally, the cells were incubated for 45 min with the specific secondary antibodies Alexa Fluor® 568 goat anti-rabbit and Alexa Fluor® 488 goat anti-mouse (Molecular Probes, Invitrogen Detection Technology, Invitrogen S.r.l., Milan, Italy). Samples were analyzed on BioRad-MRC 1024 ES confocal microscope (60× objective, argon laser sequential excitation at 468 and 580 nm), 0.2-μm step in z-plane acquisition. Co-localization analysis was performed with the Laserpix software (Bio-Rad Laboratories) in all the z-plane sections analyzed in a series of triplicate experiments.

RESULTS

To investigate the specificity of the previously described interactions between APP and ShcA/Grb2, we performed experiments in the human neuroglioma cell line (H4) and in APP/APLP–null mouse fibroblasts (MEF 1.1). The studies on MEF 1.1 cells were performed upon transient transfection with a vector encoding for human APP695 wild-type or with an empty vector as control of our experimental settings. After being transfected, cells were then lysed and immunoprecipitated with an antibody specific for the C-terminal region of APP (51-2700 antibody from Zymed Laboratories). Only in transfected cells did Zymed antibody coprecipitate different APP and CTF bands (FIG. 1). It is interesting to note that APP695 is fully processed in MEF 1.1 cells since both the mature and the immature isoforms are detectable.

In a recent study we have shown that only Aβ-containing CTFs are tyrosine-phosphorylated in human brain.[15] We have also shown that these species are increasingly coupled to the ShcA-Grb2 transduction machinery in AD subjects only.[11] To verify whether in this model APP could also interact with ShcA, we immunoprecipitated cell lysates with an antibody specific for ShcA. Immunoblotting with the 51-2700 antibody specific for the C-terminal region of APP revealed that two bands migrating at about 100 kDa were coprecipitated by ShcA antibody only in transfected cells, while no detectable APP bands were observed in the empty vector-transfected cells, thus confirming that the binding between these proteins is specific (FIG. 2). The first band interacting with ShcA likely corresponds to the immature form of APP and the second slower migrating band, which is present in a significantly lower amount, likely corresponds to a glycosylated or to posttranslationally modified APP

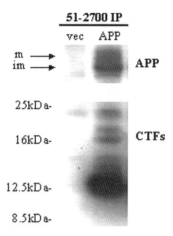

FIGURE 1. MEF1.1 cell lysates of empty vector (*vec*) and human APP695 (*APP*)-transfected cells were immunoprecipitated with 51-2700 antibody specific for the C terminus of APP. The immunostaining with the same antibody reveals that APP and CTF bands are coprecipitated only in APP695-transfected cells, while no APP and CTF bands are detectable in empty vector-transfected cells. This experiment confirms that MEF1.1 cells are actually APP/APLP–null and suggests that the exogenous APP is fully processed in these cells since both mature (*m*) and immature (*im*) APP isoforms are detectable.

isoforms (FIG. 2). We then verified the presence of complexes between ShcA and CTFs. Immunoprecipitation of cell lysates with ShcA antibody followed by immunoblotting with 51-2700 antibody specific for the C-terminal region of APP revealed one CTF band corresponding to the C99 fragment, only upon APP695 transfection (FIG. 2).

To verify whether Grb2 could also directly interact with APP or its CTFs, we immunoprecipitated cell lysates from empty vector- and APP695-transfected cells with an antibody specific for Grb2. We can observe that Grb2 recognized and coprecipitated either the mature or the immature APP isoforms only in transfected cells. This experiment showed the possibility for Grb2 to directly interact with a subset of APP holoprotein and with its C99 fragment. The direct interaction between Grb2 and APP is also confirmed by our FRET experiments, in which Grb2 directly interacted with APP (data not shown).

To get insight into the underneath signal and into the significance of such interaction for the AD, we used human H4 neuroglioma cells, wild-type or transfected with a vector encoding for human APP695 to mimic a model in which APP level is increased.

The immunoprecipitation with the 51-2700 antibody specific for APP and for both β- and α-CTFs, followed by immunostaining of electrophoresed proteins with the same antibody, confirms the overexpression of APP and CTFs in transfected cells (data not shown). To verify whether in this *in vitro* model APP overexpression could affect APP or CTF interaction with ShcA/Grb2

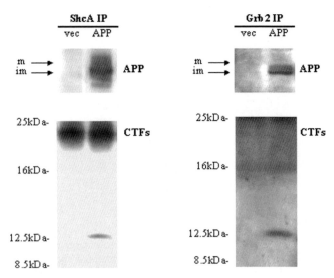

FIGURE 2. MEF1.1 cell lysates of empty vector- (*vec*) and human APP695 (*APP*)-transfected cells were immunoprecipitated with an anti-ShcA antibody. The immunodetection performed with 51-2700 antibody specific for APP and its CTF reveals that no APP and CTF bands are coprecipitated by ShcA in empty vector-transfected cells, while ShcA coprecipitates two different APP bands likely corresponding to the mature (*m*) and the immature (*im*) isoforms of APP and one CTF band only in APP695-transfected cells likely corresponding to C99 fragment. The same results were carried out from the immunoprecipitation experiment performed on cell lysates with an antibody specific for Grb2. Also in this case, as previously shown for ShcA, this adaptor interacts and coprecipitates APP and CTF bands only in APP-transfected cells.

adaptor proteins and their signaling, we carried out an immunoprecipitation experiment with an anti-ShcA antibody, looking at the coprecipitation of APP or CTFs. Cell lysates from H4 wild-type or transfected with APP695 were coimmunoprecipitated with antibodies specific for ShcA and Grb2 and then were analyzed by Western blotting with antibodies specific for human APP. Our data show that APP interacted with ShcA and Grb2 either in H4 wild-type and in H4 transfected with APP, but these interactions were enhanced in transfected cells (data not shown).

Since the MAPK pathway is the major target of the ShcA/Grb2 signaling upon activation of TRK receptors, we have analyzed the ERK-1 and -2 status in MEF1.1 and in H4 cells in our experimental settings. Our hypothesis is that ShcA and Grb2 may transduce for a signal pathway targeted to ERK-1 and -2 involving APP and CTFs. The significance of these data for AD is supported by the fact that ShcA and phospho-ERK-1 and -2 are both increased in the AD brain.[11] To evaluate the effect of the interactions between APP and ShcA/Grb2 we have analyzed by Western blotting cell lysates from MEF1.1 and H4 cells either transfected with empty vector or APP695 examining ERK-1 and -2 status.

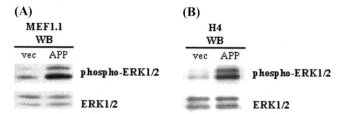

FIGURE 3. (**A**) Western blotting analysis with antibodies specific for unphosphorylated (*lower panel*) and phosphorylated ERK-1 and -2 isoforms (*upper panel*) shows that human APP695 transfection of MEF1.1 cells (APP) significantly upregulates phospho-ERK-1 and -2 levels (*upper panel*) in comparison to empty vector-transfected cells (vec). Unphosphorylated ERK-1 and -2 (*lower panel*) is shown as loading control for phospho-ERK-1 and -2 activation. (**B**) Also in H4 cells the transfection with human APP695 (APP) significantly upregulates phospho-ERK-1 and -2 levels (*upper panel*) in comparison to empty vector-transfected cells (vec). Unphosphorylated ERK-1 and -2 (*lower panel*) is shown as loading control for phospho-ERK-1 and -2 activation.

The immunoblotting with an antibody specific for the phosphorylated isoform of the MAPK ERK-1 and -2 revealed that the transfection with APP695 induced the activation of ERK-1 and -2, which was increasingly located at nuclear levels in these conditions either in MEF1.1 (FIG. 3A) and in H4 (FIG. 3B).

To deeply investigate the role of APP on the activation of this intracellular pathway we have analyzed by confocal microscopy cold methanol-fixed MEF1.1 cells transiently transfected with a plasmid encoding for APP695 and then we have immunodecorated these cells with antibodies specific for the phosphorylated isoforms of ERK-1 and -2 and for γ-tubulin, which is a centrosomal marker, considering that phospho-ERK-1 and -2 translocates to the nucleus and also to the centrosome upon activation in mitotic cells.[16] In accord with Western blotting results (FIG. 3) we also found out by immunofluorescence that the phosphorylation of ERK-1 and -2 was enhanced in APP695-transfected cells in comparison to empty vector-transfected cells and that phospho-ERK-1 and -2 translocates to the centrosome in mitotic cells (FIG. 4).

DISCUSSION

The cytoplasmic domain of APP is involved in a complex protein network, the specific modulation of which may have a functional significance for the development of AD. Among all these interacting proteins, only ShcA and Grb2 require the specific tyrosine phosphorylation of Tyr 682 of APP[13,14] and the amount of ShcA and CTFs/ShcA/Grb2 complexes is significantly increased in AD brain as compared to control,[11] thus suggesting a pathogenic correlation.

The goal of our study was to see in our model whether APP695 could bind to ShcA and Grb2 and to investigate the final target of this intracellular

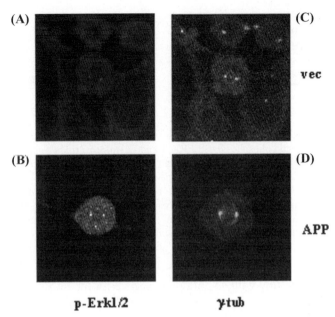

FIGURE 4. MEF1.1 cells upon transfection with empty vector (vec) and APP695 (APP) were immunofluorescently labeled with antibodies specific for the phosphorylated isoform of ERK-1 and -2 (conjugated to Alexa Fluor® 568) and for γ-tubulin (conjugated to Alexa Fluor® 488) and were observed under a confocal microscope. This experiment reveals that transfection with APP695 induces the activation of phospho-ERK-1 and -2 (**B**) in comparison to empty vector-transfected cells (**A**). It is important to note that the activation of ERK-1 and -2 is more evident in mitotic cells than in cells in other phases of the cell cycle as shown by the staining performed with an antibody specific for γ-tubulin (**C** and **D**), which is a marker for centrosome.

pathway. Here we show that APP and its CTFs interacted with these two adaptors only in APP-transfected cells (FIG. 2). Therefore we focused our study on the investigation of the possible target of this mechanism activated by these interactions.

The Grb2 involvement in the activation of the MAPK pathways cascade is well known.[17–24] It is worth noting that MAPK activation is increased in the AD brain[11] and that activated MAPKs can participate in the abnormal hyperphosphorylation of tau in AD.[25,26] Besides its involvement in signal transduction pathways mediated by tyrosine–kinase receptors, Grb2 may also anchor to a number of proteins involved in cell signaling and vesicular trafficking, such as dynamin and synapsin[27,28] or to proteins regulating endocytic trafficking, cytoskeletal dynamics, cell cycle, metastatic proliferation, and even the morphology of dendrites and axons.[18–20,29–34]

In this article, we report that APP695 signaling activity triggered by its binding with ShcA and Grb2 adaptor proteins in MEF1.1 and H4 cells is targeted to ERK-1 and -2 phosphorylation and to its consequent activation (FIG. 3). At this point in the study we can only hypothesize that this interaction (as others in which Grb2 is involved) might be linked to cell proliferation[28] and further studies are needed to decipher the significance of Grb2–APP interaction for the AD pathogenesis.

In our system it is the tyrosine at residue 682 of APP that, upon phosphorylation, regulates the interaction with Grb2 (unpublished observation), as previously shown for ShcA,[14] possibly through an interaction with its SH2 domain. In fact, Grb2 possesses a SH2 domain that may directly bind the pYENP motif in a manner similar to that shown for the SH2 domain of c-Abl, which binds APP cytodomain upon tyr-682 phosphorylation.[8] Further studies are needed to clarify this aspect and the implication that the activation of this pathway may modulate both APP's degradation and its yet unclear signaling activity.

The comprehension of the molecular mechanisms that regulate APP processing and signaling might be crucial for understanding not only AD pathogenesis, but also the regulation of cell properties critical for adhesion and survival of highly proliferating tumor cells, as shown in this work.

ACKNOWLEDGMENTS

This work was supported by grants from the Alzheimer Association (IIRG-02-3976, E.C. contract No. LSHM-CT-2003-503330/APOPIS, MIUR FIRB RBNE01FEJ7_006).

REFERENCES

1. GOLDE, T.E. 2003. Alzheimer disease therapy: can the amyloid cascade be halted? J. Clin. Invest. **111:** 11–18.
2. SELKOE, D.J. 2002. Deciphering the genesis and fate of amyloid beta-protein yields novel therapies for Alzheimer disease. J. Clin. Invest **110:** 1375–1381.
3. MCPHIE, D.L., T. GOLDE, C.B. ECKMAN, et al. 2001. Beta-secretase cleavage of the amyloid precursor protein mediates neuronal apoptosis caused by familial Alzheimer's disease mutations. Brain Res. Mol. Brain Res. **97:** 103–113.
4. OSTER-GRANITE, M.L., D.L. MCPHIE, J. GREENAN & R.L. NEVE. 1996. Age-dependent neuronal and synaptic degeneration in mice transgenic for the C terminus of the amyloid precursor protein. J. Neurosci. **16:** 6732–6741.
5. BORG, J.P., J. OOI, E. LEVY & B. MARGOLIS. 1996. The phosphotyrosine interaction domains of X11 and FE65 bind to distinct sites on the YENPTY motif of amyloid precursor protein. Mol. Cell Biol. **16:** 6229–6241.
6. ZAMBRANO, N., J.D. BUXBAUM, G. MINOPOLI, et al. 1997. Interaction of the phosphotyrosine interaction/phosphotyrosine binding-related domains of Fe65 with wild-type and mutant Alzheimer's beta-amyloid precursor proteins. J. Biol. Chem. **272:** 6399–6405.

7. HOWELL, B.W., L.M. LANIER, R. FRANK, et al. 1999. The disabled 1 phosphotyrosine-binding domain binds to the internalization signals of transmembrane glycoproteins and to phospholipids. Mol. Cell Biol. **19:** 5179–5188.
8. ZAMBRANO, N., P. BRUNI, G. MINOPOLI, et al. 2001. The beta-amyloid precursor protein APP is tyrosine-phosphorylated in cells expressing a constitutively active form of the Abl protoncogene. J. Biol. Chem. **276:** 19787–19792.
9. SCHEINFELD, M.H., R. RONCARATI, P. VITO, et al. 2002. Jun NH2-terminal kinase (JNK) interacting protein 1 (JIP1) binds the cytoplasmic domain of the Alzheimer's beta-amyloid precursor protein (APP). J. Biol. Chem. **277:** 3767–3775.
10. RONCARATI, R., N. SESTAN, M.H. SCHEINFELD, et al. 2002. The gamma-secretase-generated intracellular domain of beta-amyloid precursor protein binds Numb and inhibits Notch signaling. Proc. Natl. Acad. Sci. USA **99:** 7102–7107.
11. RUSSO, C., V. DOLCINI, S. SALIS, et al. 2002. Signal transduction through tyrosine-phosphorylated C-terminal fragments of amyloid precursor protein via an enhanced interaction with Shc/Grb2 adaptor proteins in reactive astrocytes of Alzheimer's disease brain. J. Biol. Chem. **277:** 35282–35288.
12. VENEZIA, V., C. RUSSO, E. REPETTO, et al. 2004. Apoptotic cell death influences the signaling activity of the amyloid precursor protein through ShcA and Grb2 adaptor proteins in neuroblastoma SH-SY5Y cells. J. Neurochem. **90:** 1359–1370.
13. TARR, P.E., C. CONTURSI, R. RONCARATI, et al. 2002. Evidence for a role of the nerve growth factor receptor TrkA in tyrosine phosphorylation and processing of beta-APP. Biochem. Biophys. Res. Commun. **295:** 324–329.
14. TARR, P.E., R. RONCARATI, G. PELICCI, et al. 2002. Tyrosine phosphorylation of the beta-amyloid precursor protein cytoplasmic tail promotes interaction with Shc. J. Biol. Chem. **277:** 16798–16804.
15. RUSSO, C., S. SALIS, V. DOLCINI, et al. 2001. Amino-terminal modification and tyrosine phosphorylation of [corrected] carboxy-terminal fragments of the amyloid precursor protein in Alzheimer's disease and Down's syndrome brain. Neurobiol. Dis. **8:** 173–180.
16. WILLARD, F.S. & M.F. CROUCH. 2001. MEK, ERK, and p90RSK are present on mitotic tubulin in Swiss 3T3 cells: a role for the MAP kinase pathway in regulating mitotic exit. Cell Signal. **13:** 653–664.
17. CATTANEO, E. & P.G. PELICCI. 1998. Emerging roles for SH2/PTB-containing Shc adaptor proteins in the developing mammalian brain. Trends Neurosci. **21:** 476–481.
18. DANKORT, D., B. MASLIKOWSKI, N. WARNER, et al. 2001. Grb2 and Shc adapter proteins play distinct roles in Neu (ErbB-2)-induced mammary tumorigenesis: implications for human breast cancer. Mol. Cell Biol. **21:** 1540–1551.
19. JIANG, X., F. HUANG, A. MARUSYK & A. SORKIN. 2003. Grb2 regulates internalization of EGF receptors through clathrin-coated pits. Mol. Biol. Cell **14:** 858–870.
20. PUTO, L.A., K. PESTONJAMASP, C.C. KING & G.M. BOKOCH. 2003. p21-activated kinase 1 (PAK1) interacts with the Grb2 adapter protein to couple to growth factor signaling. J. Biol. Chem. **278:** 9388–9393.
21. YANG, N., Y. HUANG, J. JIANG & S.J. FRANK. 2004. Caveolar and lipid raft localization of the growth hormone receptor and its signaling elements: impact on growth hormone signaling. J. Biol. Chem. **279:** 20898–20905.

22. VAN DER, G.P., S. WILEY, G.D. GISH, et al. 1996. Identification of residues that control specific binding of the Shc phosphotyrosine-binding domain to phosphotyrosine sites. Proc. Natl. Acad. Sci. USA **93:** 963–968.
23. YOKOTE, K., S. MORI, K. HANSEN, et al. 1994. Direct interaction between Shc and the platelet-derived growth factor beta-receptor. J. Biol. Chem. **269:** 15337–15343.
24. ROZAKIS-ADCOCK, M., R. FERNLEY, J. WADE, et al. 1993. The SH2 and SH3 domains of mammalian Grb2 couple the EGF receptor to the Ras activator mSos1. Nature **363:** 83–85.
25. GUISE, S., D. BRAGUER, G. CARLES, et al. 2001. Hyperphosphorylation of tau is mediated by ERK activation during anticancer drug-induced apoptosis in neuroblastoma cells. J. Neurosci. Res. **63:** 257–267.
26. HARDY, J. & D.J. SELKOE. 2002. The amyloid hypothesis of Alzheimer's disease: progress and problems on the road to therapeutics. Science **297:** 353–356.
27. MCPHERSON, P.S., A.J. CZERNIK, T.J. CHILCOTE, et al. 1994. Interaction of Grb2 via its Src homology 3 domains with synaptic proteins including synapsin I. Proc. Natl. Acad. Sci. USA **91:** 6486–6490.
28. YOON, S.Y., M.J. JEONG, J. YOO, et al. 2001. Grb2 dominantly associates with dynamin II in human hepatocellular carcinoma HepG2 cells. J. Cell Biochem. **84:** 150–155.
29. BENESCH, S., S. LOMMEL, A. STEFFEN, et al. 2002. Phosphatidylinositol 4,5-biphosphate (PIP2)-induced vesicle movement depends on N-WASP and involves Nck, WIP, and Grb2. J. Biol. Chem. **277:** 37771–37776.
30. CHENG, A.M., T.M. SAXTON, R. SAKAI, et al. 1998. Mammalian Grb2 regulates multiple steps in embryonic development and malignant transformation. Cell **95:** 793–803.
31. LAMPRECHT, R., C.R. FARB & J.E. LEDOUX. 2002. Fear memory formation involves p190 RhoGAP and ROCK proteins through a GRB2-mediated complex. Neuron **36:** 727–738.
32. MARTINU, L., A. SANTIAGO-WALKER, H. QI & M.M. CHOU. 2002. Endocytosis of epidermal growth factor receptor regulated by Grb2-mediated recruitment of the Rab5 GTPase-activating protein RN-tre. J. Biol. Chem. **277:** 50996–51002.
33. SAUCIER, C., V. PAPAVASILIOU, A. PALAZZO, et al. 2002. Use of signal specific receptor tyrosine kinase oncoproteins reveals that pathways downstream from Grb2 or Shc are sufficient for cell transformation and metastasis. Oncogene **21:** 1800–1811.
34. SHEN, T.L. & J.L. GUAN. 2001. Differential regulation of cell migration and cell cycle progression by FAK complexes with Src, PI3K, Grb7 and Grb2 in focal contacts. FEBS Lett. **499:** 176–181.

Index of Contributors

Accorsi, A., 59–68, 217–225, 226–233
Aceto, A., 276–291
Akiyama, H., 311–317
Alber, D., 113–119
Albertini, M.C., 59–68, 217–225, 226–233
Alexia, C., 1–17
Amelio, D., 305–310
Ammendola, S., 50–58
Andrikovics, H., 344–354
Angyal, A., 326–331
Ardizzone, N., 305–310
Arena, S., 385–398
Arend, A., 253–264
Aunapuu, M., 253–264
Auricchio, F., 194–200

Bajetto, A., 332–343, 385–398
Balestrieri, E., 130–137
Balietti, M., 79–88
Baltaziak, M., 18–25
Barbieri, F., 332–343, 385–398
Baumann, T., 138–146
Behne, D., 113–119
Bergamaschi, A., 50–58, 59–68, 69–78, 217–225, 226–233
Bertolotto, M., 399–407
Bertoni-Freddari, C., 26–34, 79–88
Besançon, F., 203–208
Bianchi, G., 399–407
Bors, A., 344–354
Brózik, A., 344–354
Brückner, U.B., 429–444
Bras, M., 1–17
Busnadiego, I., 408–418, 419–428
Buttenschön, K., 429–444

Campanella, C., 305–310
Cappello, F., 305–310
Carlo, P., 455–465
Casagrande, S., 385–398
Casey, N.P., 344–354
Casoli, T., 26–34, 79–88
Castedo, M., 35–49
Cavallini, S., 445–454

Cerella, C., 50–58, 59–68, 69–78, 217–225, 226–233
Chevanne, M., 108–112
Chiovitti, K., 276–291
Cho, S.O., 298–304
Coppola, S., 69–78
Coquelle, A., 35–49
Cordisco, S., 217–225, 59–68
Corsaro, A., 276–291, 385–398, 455–465
Corti, A., 305–310

D'Alessio, M., 50–58, 59–68, 69–78, 217–225, 226–233
D'Ettorre, G., 130–137
Dallegri, F., 399–407
Daniel, F., 1–17
De Angelis, C., 26–34
De Luca, A., 305–310
De Nicola, M., 50–58, 59–68, 69–78, 217–225, 226–233
Deheuninck, J., 188–202
Di Felice, V., 305–310
Di Stefano, G., 26–34, 79–88
Diederich, M., xxvii–xxviii
Dimanche-Boitrel, M.-T., 108–112, 209–216
Dimova, E.Y., 355–367
Dorcaratto, A., 332–343
Du, W., 245–252, 429–444
Duraj, E., 18–25

Fafeur, V., 188–202
Fagerholm, S.C., 318–325
Fallot, G., 1–17
Falzarano, S., 445–454
Fattoretti, P., 26–34, 79–88
Favalli, C., 130–137
Florio, T., 276–291, 332–343, 385–398, 455–465
Floros, K.V., 89–97
Florou, D., 89–97
Fontaine, A., 209–216
Foveau, B., 188–202
Franceschetti, L., 445–454

Gaboriau, F., 108–112
Gahmberg, C.G., 318–325
Gatti, M., 385–398
Geiszt, M., 344–354
Ghazi-Khansari, M., 98–107
Ghibelli, L., 50–58, 59–68, 69–78, 217–225, 226–233
Giorgetti, B., 79–88
Glander, H.-J., 138–146
Gorria, M., 108–112
Grbavac, I., 113–119
Grdiša, M., 120–129
Grelli, S., 130–137
Grossi, Y., 79–88
Grunewald, S., 138–146
Gueorguieva, M., 234–244

Hegedűs, C., 344–354
Hirano, M., 311–317
Hosseini, M.-J., 98–107
Hsu, H.-Y., 168–176
Huc, L., 108–112
Hwan Kim, K., 368–374

Ji, Z., 188–202
Joos, T., 168–176
Ju, K.D., 368–374

Kühbacher, M., 113–119
Kanczuga-Koda, L., 18–25, 265–275
Katano, H., 311–317
Kerckaert, J.-P., 188–202
Kietzmann, T., 355–367
Kim, E.-H., 375–384
Kim, H., 292–297, 298–304, 368–374
Kim, J.-H., 147–160
Kim, K.H., 292–297, 298–304
Kim, M.H., 182–187
Kim, S.-H., 147–160
Kiss, E., 161–167
Koda, M., 18–25, 265–275
Kokk, K., 253–264
Kozma, A., 344–354
Kriebel, F., 168–176
Kroemer, G., 35–49
Ku, J.-S., 147–160
Kumei, Y., 311–317
Kyriakopoulos, A., 113–119

Lagadic-Gossmann, D., 108–112, 209–216
Lamy, T., 209–216
Lapshina, M.A., 177–181
Lasfer, M., 1–17
Lauria, F., 130–137
Le Moigne, G., 209–216
Leder, G., 429–444
Lee, E.-O., 147–160
Lee, E.Y., 182–187
Lee, H.-J., 147–160
Lee, M.-H., 147–160
Leroy, C., 188–202
Lesniewicz, T., 18–25
Liener, U.C., 429–444
Liigant, A., 253–264
Lunardi, G., 332–343

Ma, X., 245–252
Macchi, B., 130–137
Magócsi, M., 344–354
Magrini, A., 50–58, 59–68, 69–78, 217–225, 226–233
Marcheselli, F., 79–88
Marino, S., 445–454
Martínez, R., 408–418
Massa, A., 332–343, 385–398
Mastino, A., 130–137
Mathieu, J., 203–208
Matkó, J., 161–167
Matteucci, C., 130–137
Mearelli, C., 50–58
Medgyesi, D., 326–331
Meurette, O., 209–216
Mikecin, A., 120–129
Minutolo, A., 130–137
Mohammadi-Bardbori, A., 98–107
Montalbano, A., 305–310
Montecucco, F., 399–407
Montella, F., 130–137
Morita, S., 311–317
Mouhamad, S., 35–49
Musiatowicz, B., 18–25
Musiatowicz, M., 18–25

Német, K., 344–354
Na, H.-K., 375–384
Navarro, R., 408–418, 419–428

INDEX OF CONTRIBUTORS

Nizzari, M., 276–291, 455–465
Nuccitelli, S., 217–225, 226–233
Nurmi, S.M., 318–325

Ottonello, L., 399–407
Oyha, K., 311–317

Paasch, U., 138–146
Paludi, D., 276–291
Parkhomenko, I.I., 177–181
Paternoster, L., 226–233
Pattarozzi, A., 332–343, 385–398
Perna, E., 26–34
Poldoja, E., 253–264
Porcile, C., 332–343, 385–398
Poznic, M., 120–129

Quief, S., 188–202

Radogna, F., 217–225, 226–233
Ranftler, C., 234–244
Ravetti, J.L., 332–343
Rebillard, A., 209–216
Recchioni, R., 79–88
Reveneau, S., 188–202
Rissel, M., 108–112
Robello, M., 385–398
Ruiz-Larrea, M.B., 408–418, 419–428
Ruiz-Sanz, J.I., 408–418, 419–428
Russo, C., 276–291, 455–465

Sármay, G., 161–167
Sármay, G., 161–167
Sarmay, G., 326–331
Schettini, G., 276–291, 332–343, 385–398, 455–465
Schneider, E.M., 168–176, 245–252, 429–444
Scorilas, A., 89–97
Selstam, G., 253–264
Selvatici, R., 445–454
Sergent, O., 108–112
Shimokawa, H., 311–317
Simovart, H.E., 253–264
Siniscalchi, A., 445–454
Solazzi, M., 26–34, 79–88

Spaziante, R., 332–343
Sulkowska, M., 265–275
Sulkowski, S., 265–275
Surh, Y.-J., 375–384
Szymanska, M., 18–25

Talieri, M., 89–97
Tamouza, H., 1–17
Tapfer, H., 253–264
Tekpli, X., 108–112
Terentiev, A.A., 177–181
Thellung, S., 276–291, 385–398, 455–465
Thomadaki, H., 89–97
Tota, B., 305–310
Tulasne, D., 188–202

Vadrot, N., 1–17
Vella, S., 130–137
Venezia, V., 276–291, 455–465
Viaud, S., 35–49
Villa, V., 276–291, 455–465
Villenet, C., 188–202
Violani, E., 455–465
Vitale, I., 35–49
Vivet, S., 35–49
Viviani, G., 399–407
Vorlaender, K., 245–252
Vullo, V., 130–137

Weiss, M., 168–176, 245–252, 429–444
Wenda, N., 113–119
Wincewicz, A., 18–25, 265–275
Wittemann, S., 168–176
Wolf, C., 113–119
Węsierska-Gądek, J., 234–244

Yang, D.-C., 147–160
Yoon, J.-H., 298–304
Yu, J.H., 292–297, 368–374

Zitvogel, L., 35–49
Zona, G., 332–343
Zummo, G., 305–310

Erratum

In [1], the following error was published on page 352.

MOTOHIKO OGAWA, KOTARO MATSUMOTO, PAROLA PHILIPPE, DIDIER RAOULT, AND PHILIPPE BROUQUI

The text was incorrect and should have read:

MOTOHIKO OGAWA, KOTARO MATSUMOTO, P. PAROLA, DIDIER RAOULT, AND PHILIPPE BROUQUI

We apologize for this error.

REFERENCE

1. OGAWA, M. *et al.* 2006. Expression of rOmpA and rOmpB Protein in Rickettsia massiliae during the Rhipicephalus turanicus Life Cycle Ann. N.Y. Acad. Sci. **1078:** 352–356.

Erratum

In [1], the following error was published on page 191.

J. RAUD, J. THORLACIUS, X. XIE, L. LINDBOM, AND P. HEDQVIST

The text was incorrect and should have read:

J. RAUD, H. THORLACIUS, X. XIE, L. LINDBOM, AND P. HEDQVIST

We apologize for this error.

REFERENCE

1. RAUD, J. *et al.* 1994. Interactions between histamine and leukotrienes in the microcirculation: aspects of relevance to acute allergic inflammation. Ann. N.Y. Acad. Sci. **744:** 191–198.